Yeast Hybrid Technologies

Other BioTechniques® Books

Gene Cloning and Analysis by RT-PCR
P.D. Siebert and J.W. Larrick (Eds.)

Apoptosis Detection and Assay Methods
L. Zhu and J. Chun (Eds.)

Protein Staining and Identification Techniques
R.C. Allen and B. Budowle

Immunological Reagents and Solutions: A Laboratory Handbook
B.B. Damaj

Affinity and Immunoaffinity Purification Techniques
T.M. Phillips and B.F. Dickens

Ribozyme Biochemistry and Biotechnology
G. Krupp and R. Gaur (Eds.)

Antigen Retrieval Techniques: Immunohistochemistry and Molecular Morphology
S.-R. Shi, J. Gu, and C.R. Taylor (Eds.)

Bioinformatics: A Biologist's Guide to Biocomputing and the Internet
S. Brown

Gene Transfer Methods: Introducing DNA into Living Cells and Organisms
P.A. Norton and L.F. Steel (Eds.)

SNP and Microsatellite Genotyping: Markers for Genetic Analysis
A. Hajeer, J. Worthington, and S. John (Eds.)

Viral Vectors: Basic Science and Gene Therapy
A. Cid-Arregui and A. García-Carrancá

Yeast Hybrid Technologies

Edited by

Li Zhu
Genetastix Corporation
San Jose, CA, USA

Gregory J. Hannon
Cold Spring Harbor Laboratory
Cold Spring Harbor, NY, USA

A BioTechniques® Books Publication
Eaton Publishing

Li Zhu, PhD
Genetastix Corporation
Suite 303
780 Montague Expressway
San Jose, CA, USA 95131

Gregory J. Hannon, PhD
Cold Spring Harbor Laboratory
1 Bungtown Road
Cold Spring Harbor, NY, USA 11724

Library of Congress Cataloging-in-Publication Data
Yeast hybrid technologies / edited by Li Zhu, Gregory J. Hannon.
     p. cm.
   Includes bibliographical references and index.
   ISBN 1-881299-15-5
     1. Protein binding--Laboratory manuals. 2. Yeast fungi--Biotechnology--Laboratory
   manuals. 3. Transcription factors--Laboratory manuals. I. Zhu, Li, 1949- II. Hannon,
   Gregory J., 1964-

   QP551 .Y397 2000
   572'.6--dc21
                                                                                       00-047610

ISBN 1-881299-15-5

Printed in the United States of America

9 8 7 6 5 4 3 2 1

Eaton Publishing
BioTechniques Books Division
154 E. Central Street
Natick, MA 01760
www.BioTechniques.com

Francis W. Eaton: *Publisher and President*
Stephen Weaver: *Director and Editor-in-Chief*
Christine McAndrews: *Managing Editor*
Chanc E VanWinkle: *Project Editor*
Sandy Lamont: *Production Manager*
Ken Strom: *Cover Designer*

iv

Contributors

Eva Chan
Rigel, Inc.
San Francisco, CA, USA

Rene Chan
Department of Biology
University of California-San Diego
La Jolla, CA, USA

Hwai Wen Chang
Digital Gene Technologies
La Jolla, CA, USA

Chawnshang Chang
Department of Pathology
University of Rochester
Rochester, NY, USA

Robert Cormack
Max-Planck-Institut für
  Züchtungsforschung
Abeteilung Biochemie
Köln, Germany

Toshihiko Ezashi
Department of Animal Sciences
University of Missouri
Columbia, MO, USA

Abdallah Fanidi
CLONTECH Laboratories
Palo Alto, CA, USA

Andrew Farmer
CLONTECH Laboratories
Palo Alto, CA, USA

Naohiro Fujimoto
University of Rochester
Rochester, NY, USA

Mark Hannink
Biochemistry Department
University of Missouri-Columbia
Columbia, MO, USA

Gregory J. Hannon
Cold Spring Harbor Laboratory
Cold Spring Harbor, NY, USA

Ann E. Holtz
CLONTECH Laboratories
Palo Alto, CA, USA

Yi-Ren Hong
Graduate Institute of Biochemistry
Kaohsiung Medical University
Kaohsiung, Taiwan, China

Shao-bing Hua
Genetastix Corporation
San Jose, CA, USA

Jianing Huang
CLONTECH Laboratories
Palo Alto, CA, USA

Philip James
Department of Biomolecular
  Chemistry
University of Wisconsin
Madison, Wisconsin, USA

Shinji Kamada
Department of Medical Genetics
Biomedical Research Center
Osaka University Graduate School
  of Medicine
Osaka, Japan

Hong-Yo Kang
University of Rochester
Rochester, NY, USA

Gouzel Karimova
Unité de Biochimie Cellulaire,
Institut Pasteur
Paris, France

Eungseok Kim
University of Rochester
Rochester, NY, USA

Nancianne Knipfer
CLONTECH Laboratories
Palo Alto, CA, USA

Joachim Kremerskothen
Zentrum für Molekulare
  Neurobiologie Hamburg
University of Hamburg
Hamburg, Germany

Dietmar Kuhl
Zentrum für Molekulare
  Neurobiologie Hamburg
University of Hamburg
Hamburg, Germany

Danial Ladant
Unité de Biochimie Cellulaire
Institut Pasteur
Paris, France

Candice A. Leanna
Bio-Technical Resources
Manitowoc, WI, USA

Jian Liao
CLONTECH Laboratories
Palo Alto, CA, USA

Ulrike Löhr
Department of Biochemistry and
  Molecular Biology
Mount Sinai School of Medicine
New York, NY, USA

Olivier Louvet
Quantum-Appligene
Illkirch, France

Ying Luo
Rigel, Inc.
San Francisco, CA, USA

Jun Ma
Division of Developmental Biology
Children's Hospital Research
  Foundation
University of Cincinnati College
  of Medicine
Cincinnati, OH, USA

Marie-Claude Marsolier
Service de Biochimie et de
  Génétique Moléculaire
CEA-Saclay
Gif-sur-Yvette, France

Kristen Mayo
CCS Associates, Inc.
Mountain View, CA, USA

Kazuo Nishimura
Department of Pathology
University of Rochester
Rochester, NY, USA

Leslie Pick
The Brookdale Center
Department of Biochemistry
  and Molecular Biology
Mount Sinai School of Medicine
New York, NY, USA

Ulrich Putz
Zentrum für Molekulare
  Neurobiologie Hamburg,
University of Hamburg
Hamburg, Germany

Meng Sheng Qiu
Department of Anatomical
  Sciences and Neurobiology
Health Sciences Center
University of Louisville
Louisville, KY, USA

R. Michael Roberts
Department of Animal Sciences
University of Missouri
Columbia, MO, USA

Erik R. Sampson
University of Rochester
Rochester, NY, USA

Paul Skehel
Division of Neurophysiology
National Institute for Medical
  Research
London, England, UK

Igor Stagljar
Institute of Veterinary
  Biochemistry
University of Zurich-Irchel
Zurich, Switzerland

Stephan te Heesen
Origin(Schweiz)AG
Zurich, Switzerland

Yoshihide Tsujimoto
CREST of Japan Science and
  Technology Corp. (JST)
Osaka, Japan

Hiroshi Uemura
University of Rochester
Rochester, NY, USA

Chihuei Wang
University of Rochester
Rochester, NY, USA

Xin Wang
University of Rochester
Rochester, NY, USA

Shian-Jang Yan
University of Rochester
Rochester, NY, USA

Shuyuan Yeh
University of Rochester
Rochester, NY, USA

Yan Yu
OriGene Technologies, Inc.
Rockville, MD, USA

Jie Zhang
Guilford Pharmaceuticals
Baltimore, MD, USA

Helen Zhou
Aviron, Inc.
Mountain View, CA, USA

Li Zhu
Genetastix Corporation
San Jose, CA, USA

Contents

Preface

During the past seven years, both of us (L.Z. and G.H.) were actively involved in the development and application of the yeast two-hybrid system. This book is a result of efforts that started in early 1997, a time when yeast two-hybrid system technology became popular.

During the mid 1980's, the biomedical research community that was working on transcription regulation of eukaryotic organisms made the discovery that transcription regulators possess two distinct functions: DNA-binding and transcription activation (or repression). Furthermore, these two different functions are performed by two distinct domain structures. Interestingly, the two domains are physically separable and functionally independent. These findings prepared for the distillation of the two-hybrid concept that was intelligently adapted to detect protein–protein interactions. This concept was first established by Stanley Fields in 1987 and was experimentally proved by his group in 1989.

In Stanley Fields' model, he uses a well-studied yeast transcription factor, Gal4. This protein contains two domain structures. One is a DNA-binding domain and the other an activation domain. These two domains make a fully functional activator protein if they are covalently or non-covalently associated. Neither of these two domains by themselves can activate the reporter(s) that are already built into the yeast reporter strain. To assay for protein interactions, they are individually fused with two proteins of interest. One protein of interest is linked to the DNA-binding domain of Gal4, and the other is linked to the activation domain of Gal4. Thus, two artificial fusion genes are formed, and fusion proteins are expressed in the yeast reporter strain. Assume that when they are introduced into the yeast separately, none of the two-hybrid proteins can activate the reporter gene. But simultaneously introduced two-hybrid proteins will activate the reporter gene if the two proteins under study can physically associate. The reason is very simple: the two proteins of interest bring the two Gal4 functional domains into such a close proximity that they form a functional activator. Stanley Fields further demonstrated in 1991 that the system is also capable of screening for novel protein interaction partners if a library screening process is incorporated into the system. In this application, a cDNA library is built into the system by fusion with the GAL4 activation domain. So the other protein of interest, fused with the GAL4 DNA-binding domain, becomes the "bait." Rare clone(s) in the cDNA library may encode a protein that binds the bait and can be "fished out" from the large pool of the fusion library. This development added much greater power to the original two-hybrid system model.

In 1993 the two of us (L.Z. and G.H.) had met each other after being introduced by Stanley Fields. G.H. by then was a post-doctorate researcher at Cold Spring Harbor Laboratory and had made several popular two-hybrid cDNA libraries. L.Z. was working at CLONTECH Laboratories, which made the first commercial introduction of two-hybrid reagents. The alignment of academic and industrial laboratories accelerated the flow of this technology from a few pioneering labs to a wide research community. A member of a pioneering lab, Dr. Jun Ma, who himself has made significant contributions to the underlying principles of the two-hybrid systems, wrote an excellent introduction in Chapter 1 of this book.

In the next seven years we saw a flourish of two-hybrid system applications. With the two-hybrid system, researchers were armed with a novel genetic system to study protein–protein interactions. For the first time, they no longer had to express proteins as a common requirement for in vitro protein binding assays. Also, they no longer had to rely on the availability of antibodies. Now, they can study the protein interaction in vivo instead of in vitro only.

During the last seven years, the two-hybrid system experienced a series of evolutions. The system was modified extensively by many groups including ours. Section II of this book contains five chapters that describe in detail several of the most updated protocols of the two-hybrid system. Variations of the two-hybrid system have been developed to study membrane proteins or proteins that have activation function, DNA–protein interaction, RNA–RNA interaction, as well as systems that can search for molecules that can interrupt the interaction between a pair of proteins. Sections III and IV describe several modified yeast two-hybrid systems and specialty hybrid systems. In the past several years, the two-hybrid system has been made more sensitive, less prone to false-positive results, and more adaptive to large-scale use. This book is a collective effort made by many researchers who have worked in this field. We hope it will provide an expert guide to those that may start to use the two-hybrid system or its variants.

The two-hybrid system was originally developed in yeast. Most variants of the two-hybrid system (one- and three-hybrid systems) were also first developed in yeast. This is why this book has the title "Yeast Hybrid Technologies." However, the principle of the two-hybrid system has been successfully adapted to bacterial and mammalian cells. This book contains several chapters in Section V that work with model organisms other than yeast. Nevertheless, yeast is still the most matured and the most widely used host system.

Many molecular biologists, including us, were not trained for yeast manipulations. One of us (L.Z.) was trained originally as an immunologist. The first job he did involved a YAC library construction in 1990. The experience gained from this library construction prepared L.Z. for the emergence of the yeast two-hybrid system. The lesson we both have learned from our own past is that success depends on two elements: the opportunity you have which is pure luck to anyone, and how well you are prepared for that opportunity, which is not pure luck. Only those that are well prepared will be able to take advantage of opportunities. To L.Z., although the time he spent on the YAC library construction might not have been sufficiently rewarding, the experience obtained from this project became a critical factor underlying the competitive success of the MATCHMAKER™ system.

This book gives a detailed guide for those who wish to practice the two-hybrid system but still lack yeast training. In our experience, yeast is a very simple and pleasant organism to work with. The organism has been studied for a long period of time. The genetic markers employed in the two-hybrid systems are very stable. It is largely true that the wide acceptance of two-hybrid technologies has made yeast a very popular laboratory workhorse in the 1990's.

The yeast two-hybrid system and other associated systems described in this book have contributed to many important discoveries in the last decade. It is perhaps safe to say that the yeast two-hybrid system technology is one of the most important technical developments in the 1990's in the field of biological research. The last section of this book (Section VI) contains four chapters that describe novel applications of the yeast two-hybrid systems. We can foresee that this technology will bring up more exciting discoveries, especially in the new era of functional genomics and proteomics.

Although the current prevailing two-hybrid system is much more advanced than the original system invented by Stanley Fields and his colleagues, we still should recognize his distinguished contribution. We admire his historical contribution to the technology. We want to thank Phil James for writing the Foreword. Phil's lab developed a most sensitive yeast nutritional reporter (Ade) that significantly improved the performance of library screening with the two-hybrid system.

One of us (L.Z.) would like to thank the following staff at CLONTECH: Kenneth Fong for his critical support; Anne Scholz for her involvement in the early stage of marketing development of MATCHMAKER systems; Ann Holtz for her long time dedication to the MATCHMAKER system renovation; Kristen Mayo for editing the Appendices; Eddy Garcia for the design of the cover page; and Ingrid Chen for taking the yeast image used on the cover page. We also wish to thank Mr. Steve Weaver of BioTechniques Books for his support and encouragement. Without his support, this book would not have been possible. Besides many other BioTechniques Books staff members who made significant contributions to this book, we want to thank particularly Chanc VanWinkle for her excellent proofreading and other editorial work.

Li Zhu and Greg Hannon
September 2000

Foreword

In its early years, research in the field of molecular biology was defined by the paradigm "one gene, one protein, one graduate student", and studying the regulation of biological processes generally meant investigating the control of nucleic acid function. During the 1990's the focus of regulatory studies shifted to the functional interactions of proteins with one another and now it is uncommon for a project to be limited to a single gene or protein. This is the result of advances in our understanding of the complexity of biological systems, but also it is the result in no small part of the introduction of the two hybrid system by Stan Fields and his colleagues in 1989. The two hybrid system has been one of the major success stories of the past decade, inspiring numerous modifications, improvements, and new applications, and generating over 2,700 publications at a rate that continues to accelerate.

Prior to the introduction of the two hybrid system there were two methods by which one could identify protein–protein interactions. The first, if your favorite organism was amenable, was to carry out extensive genetic screens to identify genes whose products might interact, and then to develop biochemical assays with the hopes of showing the interaction was direct. The second approach was to isolate candidates via co-immuno-precipitation or affinity chromatography. The desired result was an unknown protein whose only easily determined feature was molecular weight. In the best cases both approaches could be used and might even provide complementary results, but often this was a long and involved project.

The two hybrid system changed not only the efficiency but also the sensitivity and scope of these investigations. As an in vivo assay that identifies only direct interactions, it combines the best of both prior methods. Yet it is not limited to genetically tractable systems, it can be used with protein fragments or peptide aptamers, and it provides immediate access to the gene corresponding to each interacting partner. These features allow for rapid re-screening of newly identified proteins and straightforward mapping of interaction domains. Best of all, it is fast and sensitive—interactions that previously took years to identify now are found in months or weeks, along with many others that previously went undetected.

It is not surprising then that the two hybrid system has become one of the most widely used experimental methods. Its popularity has led to a wide range of improvements to the basic method as well as modifications that allow screening for DNA and RNA interactions, assays in the cytosol rather than the nucleus, and screening in bacterial or mammalian hosts. The potential of the two hybrid system for drug discovery is evidenced by its rapid inclusion in the strategies of many biotech and pharmaceutical companies. As we move ahead, the two hybrid system will complement the advances taking place in genome sequencing. In fact, projects are already underway to create genome-scale maps of protein–protein interactions.

Despite the popularity of the two hybrid system and its straightforward theoretical basis, two hybrid experiments are not simple to carry out successfully, and there are many "tricks of the trade". This volume provides both an extensive overview of the two hybrid system and a practical guide to its applications. It is organized into six sections encompassing 21 chapters. Sections one and two provide a comprehensive background on the original, transcription-based yeast two-hybrid assay and state-of-the-art protocols for its implementation. Sections three, four, and five describe many of the important modifications that add functionality to the system, including non-nuclear and non-yeast assays and detection of protein–nucleic acid interactions. Finally, section six reviews a number of new applications that exemplify the potential of the two hybrid system. Overall these chapters provide a comprehensive background in the methodology of two hybrid screening, and I am certain that both beginners and experienced practitioners of the two hybrid system will find this book a valuable resource.

Philip James
Department of Biomolecular Chemistry
University of Wisconsin
Madison, WI, USA

Section I

Introduction:
History and Principle

1

Yeast Transcriptional Activation and the Two-Hybrid System

Jun Ma

Division of Developmental Biology, Children's Hospital Research Foundation, University of Cincinnati College of Medicine, Cincinnati, OH, USA

1. INTRODUCTION

When yeast cells are grown in media containing galactose as the only carbon source, a subset of genes is specifically turned on (21). These genes are required for utilizing galactose and are thus called the *GAL* genes; *GAL1*, for example, encodes the enzyme galactokinase. The induction of the *GAL* genes requires a transcriptional activator protein called GAL4 that binds specifically to the upstream activation sequences (UASs) located near the promoters. A repressor protein called GAL80 inhibits the action of GAL4, thus silencing these genes when cells are grown in media lacking galactose. The molecular analysis of eukaryotic transcriptional activators, in particular GAL4, during the mid- to late 1980s facilitated the development of the yeast two-hybrid system, a powerful genetic tool for detecting protein–protein interactions. This chapter briefly reviews the major events leading to the proposal of the yeast two-hybrid system, provides a list of the basic components, and addresses a few selected issues.

2. THE DEVELOPMENT OF THE YEAST TWO-HYBRID SYSTEM

One of the major advances in the field of eukaryotic transcription during the early 1980s was the identification of DNA fragments that can increase the expression of their linked genes. As we know now, these DNA fragments—called UASs in yeast and enhancers in higher eukaryotes—contain DNA sequences that are recognized specifically by transcriptional activators.

2.1 Separable DNA-Binding and Activation Domains

Despite the identification of enhancers and UASs, it was then largely unclear how transcriptional activators, bound at these DNA sequences, might activate transcription in eukaryotic cells. Fortunately, studies of activators in bacteria had provided

Yeast Hybrid Technologies
Edited by L. Zhu and G.J. Hannon
© 2000 Eaton Publishing, Natick, MA

important lessons. Most notably, it had been proposed that a bacterial activator (lambda repressor, so called because this protein can also repress transcription) used two distinct surfaces of the protein to recognize DNA and to activate transcription. This idea was based mainly on the finding that specific mutations of the protein (called positive control mutations) could abolish its activation function without affecting its DNA-binding ability (18,40). The location of these mutations in lambda repressor further suggested that the activation surface may interact directly with the RNA polymerase.

In 1985, Brent and Ptashne described an experiment that significantly advanced the field of eukaryotic transcriptional activation and laid a foundation stone for the yeast two-hybrid system (4). It was known that the yeast activator GAL4 specifically recognized DNA sites within the *GAL* UASs (15, 27). However, it was not known whether or not the mechanisms of transcriptional activation in eukaryotes and bacteria were fundamentally similar. The authors tested this idea by constructing a hybrid protein, LexA-GAL4, that contained the DNA-binding domain (DBD) of the bacterial protein LexA (residues 1–87) fused to the carboxy terminal portion of GAL4 (residues 147–881). They demonstrated that this hybrid protein activated transcription of a *GAL1* reporter gene in yeast cells if, and only if, LexA-binding sites were placed upstream of the *GAL1* minimal promoter (Figure 1, A and B). This experiment illustrated that the carboxy terminal portion of GAL4 contained an activation function that was capable of stimulating gene expression when attached to a heterologous DBD. It also suggested that the two essential functions of GAL4, DNA-binding and transcriptional activation, were carried on physically separable domains. Experiments by Keegan et al. (23) further supported this latter idea: the DBD of GAL4 (residues 1–147) could bind its DNA target sites specifically but failed to activate transcription in yeast cells. Together, these experiments suggested that transcriptional activation in eukaryotes, like that in bacteria, may involve an interaction between the transcriptional activation domain (TAD) of an activator and the transcription machinery, as opposed to DNA structural perturbations induced by the activator's DBD. Subsequent studies better defined two relatively small TADs of GAL4 (residues 148–196 for TAD1 and residues 768–881 for TAD2), and isolated short acidic TADs encoded by bacterial genomic DNA (32,33), which benefited the construction of future hybrid proteins.

2.2 An Artificial Composite Activator

A major issue concerning GAL4 function was how its activity was repressed by GAL80 when yeast cells were grown in media lacking galactose. A systematic deletion analysis of GAL4, designed to define its TADs, also revealed a 30 amino acid region (residues 851–881) that was responsible for GAL80 regulation (30). GAL4 derivatives lacking this region escape the repressive effect of GAL80, thus activating transcription even in the absence of galactose. Both genetic and biochemical experiments had previously suggested that GAL80 and GAL4 interacted with each other (28,35). Based on the observation that the region recognized by GAL80 is an essential part of GAL4's TAD2, it was proposed that GAL80 inhibited GAL4 activity by physically masking its activation function so that GAL4, though bound to the UAS (15,27), was unable to activate transcription (30). According to this model, GAL80 dissociates from GAL4 when cells are grown in the presence of galactose, thus unmasking its activation domain(s). A similar model was concurrently proposed based on the finding that genetically isolated GAL4 mutations that escaped GAL80 repression were mapped to the GAL4's TAD2 (22). Subsequent studies suggested that GAL80 may interact with GAL4 differently, rather than simply dissociate, when yeast cells are grown in the presence of galactose (24,38).

The interaction between GAL4 and GAL80 represented an opportunity to further elucidate the molecular properties of eukaryotic activators, leading to an experiment that laid another foundation stone for the yeast two-hybrid system. As discussed above, it was known that the TAD and DBD were two essential but physically separable domains of a eukaryotic activator. There were also examples of natural activators (e.g., the herpes viral activator VP16) that did not seem to recognize DNA directly but rather interacted with other DNA-bound proteins (39,47), thus suggesting that a TAD might not even need to be covalently linked to a DBD to activate transcription. However, it had not been determined whether a composite activator, with its DBD and TAD residing on two separate molecules, could be constructed experimentally. In 1988, Ma and Ptashne reported an experiment designed to test this idea (31). They constructed a hybrid protein, GAL80-B42, that contained the bacterially-derived TAD B42 (33) attached to GAL80. They demonstrated that GAL80-B42, a protein incapable of recognizing DNA (28), activated transcription of a *GAL1* reporter gene if, and only if, the yeast cells also contained a GAL4 derivative capable of interacting with both DNA and GAL80 (Figure 1C). This study established for the first time that it was experimentally feasible to artificially bring together, through protein–protein interaction, an activator's two essential domains that reside on separate molecules.

2.3 The Yeast Two-Hybrid System

In 1989, Fields and Song made the first reported proposal that the concept of bringing together the TAD and DBD, residing on separate molecules, could be used as a general tool to study protein-protein interactions (12). Such a proposal officially marked the birth of the yeast two-hybrid system. To test their idea, they fused SNF1 and SNF4, two proteins known to interact, to GAL4's DBD and TAD2, respectively. They demonstrated that, as expected, while either hybrid protein alone failed to activate transcription, both proteins together activated transcription of a *GAL1* reporter gene in yeast cells. In the proposed yeast two-hybrid system, SNF1 and SNF4 could be replaced by any pair of interacting proteins (or protein domains) of interest to bring together the TAD and DBD for gene activation in yeast (Figure 1D).

Since the proposal of the yeast two-hybrid system, many laboratories have successfully utilized it to detect and dissect interactions between previously known proteins. More importantly, several laboratories pioneered the construction of cDNA libraries that expressed TAD hybrid proteins and also worked on the development of the yeast two-hybrid system as a gene cloning method to identify interacting proteins (6,7,9,10,16,17). Numerous novel genes have been isolated using this strategy, and proteins previously unsuspected to interact have been functionally connected. In addition, many two-hybrid-related methods have been suggested for an even wider spectrum of applications (see other chapters of this book for further details). The yeast two-hybrid system has now become a powerful and versatile genetic tool for studying protein–protein interactions.

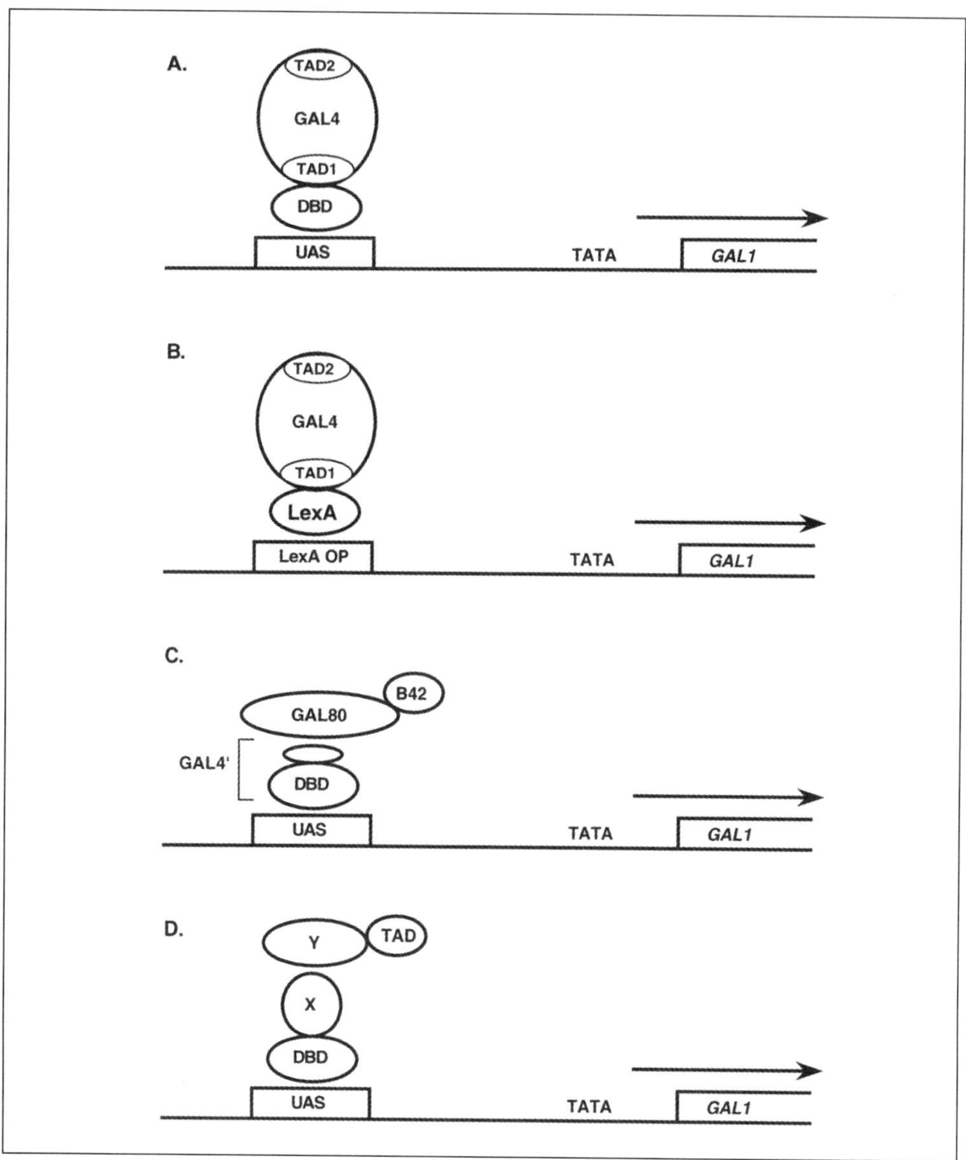

Figure 1. Functional domains of GAL4 and the yeast two-hybrid system. (A) *Transcriptional activation by GAL4.* The yeast activator protein GAL4 stimulates transcription by binding to the upstream activation sequence (UAS) located 5′ of the yeast *GAL1* gene. DBD represents the DNA-binding domain of GAL4, while TAD1 and TAD2 represent its two transcriptional activation domains. **(B)** *Separable domains of GAL4.* The LexA-GAL4 hybrid protein activates transcription by binding to the LexA operator (LexA OP) placed upstream of a *GAL1* reporter gene. This experiment demonstrates that the two essential domains of GAL4, DBD and TAD, are functionally separable and that the TAD can be brought to the DNA when attached to a heterologous DBD. **(C)** *An artificial composite activator.* The bacterially-derived TAD B42, which is fused to the repressor protein GAL80, is brought to the DNA by a GAL4 derivative (GAL4′) that contains two domains: one that interacts with the DNA and the other with GAL80. This experiment demonstrates that an activator's two essential domains can reside on separate molecules and be artificially brought together by protein–protein interaction. **(D)** *The yeast two-hybrid system.* The DBD and TAD of an activator are fused respectively to X and Y, any two interacting proteins (or protein domains) of interest. While either hybrid protein alone fails to activate transcription, both proteins together reconstitute a composite activator capable of activating transcription in yeast cells. X is often referred to as the "bait" and Y the "prey". There are currently two major designs of the yeast two-hybrid system categorized according to the DBD used: GAL4 DBD-based or LexA DBD-based. Other variations are based mainly on the choices of different yeast host strains, reporter genes, TADs, and plasmids expressing the hybrid proteins.

3. BASIC COMPONENTS FOR THE YEAST TWO-HYBRID SYSTEM

This section briefly discusses the four basic components for the yeast two-hybrid system: a proper yeast host strain, a reporter gene, a plasmid expressing the DBD hybrid protein, and a plasmid expressing the TAD hybrid protein. Detailed descriptions of these components, as well as experimental procedures, are found in other chapters of this book.

3.1 Yeast Strain

The yeast strains used in the two-hybrid system should have proper nutritional markers for plasmid selection as well as for reporter gene readout selection. In addition, they should be compatible with specific designs; for example, if the two-hybrid system is based on GAL4 DBD, to avoid any background interference problem, the yeast strain should not have the endogenous *GAL4* gene.

3.2 Reporter

A reporter gene provides a readout mechanism to detect the interaction between the two hybrid proteins expressed in the yeast cells. Although there are many different designs of the two-hybrid system, they can be grouped into two broad classes depending on the DBD used: GAL4 DBD-based or LexA DBD-based. The reporter gene must have proper DNA sites located upstream. For a system that is based on LexA DBD, the reporter gene must have LexA-binding sites located upstream; alternatively, a GAL4 DBD-based design requires the reporter gene to have the *GAL* UAS or synthetic GAL4-binding sites (called "the 17-mers") located upstream. One of the most commonly used reporter genes is the bacterial *lacZ* gene because its encoded enzyme, β-galactosidase, can be detected easily both in liquid and on plates. A second selectable reporter can often be used in conjunction with the *lacZ* reporter, mainly to facilitate library screening. For example, a yeast strain may contain both *lacZ* and *LEU2* reporter genes, each containing appropriate DNA sites upstream of their promoters. In such a design, positive cells, in which the two hybrid proteins interact, not only turn blue on X-gal plates but also gain the ability to grow on plates lacking leucine. These reporters can be either carried on replicating plasmids or integrated into the yeast genome. Unlike reporter constructs carried on replicating plasmids whose copy numbers may fluctuate in cells, integrated reporter constructs maintain a constant copy number and, therefore, often give more consistent readout signals.

3.3 DBD Plasmid

This plasmid expresses the DBD hybrid protein in the yeast two-hybrid system. The protein attached to the DBD is often called the "bait". Ideally, the DBD-bait hybrid protein should not activate reporter gene transcription strongly by itself. Often, this hybrid protein is designed to contain only an individual functional domain (or fragment) of the bait protein of interest; this may help circumvent several problems, including protein instability, toxicity, and strong activation by the DBD-bait protein alone. The availability of two different designs of the two-hybrid system (one based on GAL4 DBD and the other on LexA DBD) increases its usefulness because one may explore the alternative design if the first design fails to work properly due to problems associated with, for example, protein instability and non-functional fusion conformations.

3.4 TAD Plasmid

Several different TADs have been used in the yeast two-hybrid system, including the herpes viral activator VP16 (44), TAD2 of GAL4 (32), and the bacterially derived TAD B42 (33). These TADs have different activating strengths (VP16 being the strongest and B42 being the weakest), which may allow fine tuning of the sensitivity of the system. A nuclear localization sequence is often attached to these fusion proteins to ensure their proper localization to the nucleus. When the yeast two-hybrid system is used for gene cloning purposes, the cDNA fusion libraries are often carried on the TAD plasmid. Such a design allows one to test, and modify if necessary, the DBD-bait hybrid protein prior to screening. In addition, this design avoids a major false positive problem inherent to the alternative design because many protein sequences can function as TADs when attached to a DBD (33). In designs based on LexA DBD (but not GAL4 DBD), the inducible *GAL1* promoter can be used to permit controlled expression of the TAD hybrid proteins.

4. TRANSCRIPTIONAL ACTIVATION AND FLEXIBILITY OF THE YEAST TWO-HYBRID SYSTEM

Once an activator binds to its DNA sites near a promoter, exactly how does it activate transcription? At the transcription initiation step, activators have been proposed to stimulate transcription by recruiting one or more general transcription factors to the promoter, thus facilitating the formation of the pre-initiation complexes (42,49). Several general transcription factors have been shown to interact directly with the TADs of activators, including the TATA box binding protein (TBP), TBP-associated factors (TAFs), and transcription factor IIB (TFIIB). In addition, biochemical experiments have suggested that activators can induce conformational changes of the pre-initiation complex formed at the promoter (37). Such conformational changes may help convert an inactive, "closed" complex to an active, "open" complex capable of initiating transcription efficiently. Moreover, it has been been suggested that some activators work by recruiting protein complexes that alter chromatin structure around gene promoters for better access by the general transcriptional factors. Other steps of transcription, e.g., elongation, can also be positively affected by activators (37). These models share a common mechanistic feature: an activator's TAD must interact specifically with other proteins near a promoter, either general transcription factors or factors involved in chromatin remodeling.

Our current understanding of how activators may stimulate transcription also helps explain why the yeast two-hybrid system is highly flexible and versatile; it can detect the interaction between proteins virtually regardless of their organismic origins, native cellular functions, and the hybrid proteins' configurations. It has been shown that some TADs (e.g., VP16) can interact with multiple target proteins in the transcription machinery and work on multiple steps in the transcriptional activation process (37). The versatile nature of the TADs contributes strongly to the flexibility of the yeast two-hybrid system. Physical properties of the molecular components in the two-hybrid system, particularly the DNA molecule's ability to twist and loop, also contribute to its overall flexibility.

5. SENSITIVITY OF THE YEAST TWO-HYBRID SYSTEM

The yeast two-hybrid system is also known for its sensitivity. It has been shown that weak interactions, with dissociation constants in the micromolar range, can be detected by this method (11,48). In addition, the yeast two-hybrid system can reveal interactions undetectable by other biochemical methods (13). A consideration of the context in which the protein–protein interaction events take place may help explain the high sensitivity of the yeast two-hybrid system. These interactions occur on DNA with both ends anchored to DNA—one through the DBD at the upstream DNA sites and the other through the general transcription factors or chromatin remodeling factors at or near the promoter. In such a "self-locking" system, even very weak interactions may be stabilized by the DNA. This idea may also help explain why the interactions between TADs and the general transcription factors are often revealed to be relatively weak.

An understanding of the sources of the system's sensitivity also provides possibilities for its further fine tuning. The following parameters may be modified to modulate the sensitivity of the system; some, but not all, of these modifications have been experimentally tested (11): *(i) Number of DNA binding sites.* The number of DNA binding sites for the DBD hybrid protein determines how many TADs are brought to the DNA through protein–protein interaction. Therefore, more DNA binding sites make the system more sensitive, thus permitting the detection of weaker protein–protein interactions. *(ii) Reporter gene promoter strength.* A stronger core promoter with consensus TATA box and optimal flanking sequences should contribute more free energy to the transcription system and, therefore, increase the sensitivity. *(iii) TAD strength.* A stronger TAD would interact with the general transcription (or chromatin remodeling) machinery more efficiently, thus enhancing the system's sensitivity. *(iv) Distance.* The distance between the core promoter and the DNA sites for the DBD hybrid protein can affect activation levels, especially for weak activators in yeast cells (43). Thus, a shortened distance should increase the sensitivity to allow for the detection of weak interactions. *(v) Reporter gene status.* Reporter genes carried on plasmids are thought to be under less repressive effects from chromatin structure than integrated reporter genes and, additionally, may exist in multiple copies in cells. Therefore, a reporter gene carried on a replicating plasmid can be more sensitive (11). *(vi) Choice of DBD.* The DNA-binding property of an activator may influence how well it activates transcription (34). Both GAL4 and LexA bind DNA as dimers and, therefore, their dimerization domains contribute positively to DNA-binding and the system's sensitivity. [LexA(1–202) and GAL4(1–147) or GAL4(1–98) contain dimerization domains, whereas LexA(1–87) and GAL4(1–74) do not (4,5,26)]. *(vii) Protein levels.* High protein levels can facilitate their interactions in cells and, therefore, increase the system's sensitivity (1). In addition, it is preferable to have the TAD hybrid protein in excess over the DBD hybrid protein for strong activation (13).

It should be noted that background problems are generally associated with high sensitivity. Increased sensitivity raises the transcription levels activated by the DBD-bait hybrid protein alone, and possibly increases the chance of isolating false positives in gene cloning experiments. In addition, strong activation domains like VP16 and TAD2 of GAL4 may cause undesirable toxic effects (14); therefore, in some systems only moderate TADs like B42 are used. Another problem associated with a system that is overly sensitive is that it may be difficult to differentiate between strong and weak interactions (11). Nevertheless, a systematic consideration of the sensitivity of the two-hybrid system may facilitate tailored designs to best fit any specific needs. However, it should be noted that, since the two-hybrid system is a tool based on experimentation, one cannot predict the outcome of a given test a priori.

6. MOLECULAR PROXIMITY—A GENERAL PRINCIPLE

I would like to conclude this chapter by further highlighting one important lesson that we have learned from transcriptional activation and the yeast two-hybrid system. The essential event in the two-hybrid system is to bring the TADs to DNA near gene promoters. Such an action effectively increases the local concentration of the TADs, placing them in proximity to, and facilitating the interaction with, their physiological targets in the general transcription (or chromatin remodeling) machinery. Molecular proximity, or increased local concentration, is a strategy widely used by many regulatory processes in biology (8,41). The concept of molecular proximity has been used in the design of in vivo interaction methods that are based on readout mechanisms other than transcriptional activation in yeast, e.g., membrane receptor activation (46), activation of Ras signaling by recruiting Sos (2,19), and bacterial gene repression (20,29). These and other methods, including the yeast two-hybrid and its related systems, permit the detection of interactions among a wide range of molecular entities, including proteins, small ligands, and RNA molecules (3,25,36,45). The principle of molecular proximity provides a basis for the future design of other useful methods to analyze molecular interactions and to dissect biological processes.

ACKNOWLEDGMENTS

I thank I. Cartwright, Z. Liu, J. Molkentin, L. Zhu, and members of this laboratory for helpful discussions and comments on the manuscript. The work in this laboratory is supported by NIH grants R01-GM56215, R01-GM052467, and P30-ES06096.

REFERENCES

1. **Allen, J.B., M.W. Walberg, M.C. Edwards, and S.J. Elledge.** 1995. Finding prospective partners in the library: the two-hybrid system and phage display find a match. TIBS *20*:511-516.
2. **Aronheim, A., E. Zandi, H. Hennemann, S.J. Elledge, and M. Karin.** 1997. Isolation of an AP-1 repressor by a novel method for detecting protein-protein interactions. Mol. Cell. Biol *17*:3094-3102.
3. **Belshaw, P.J., S.N. Ho, G.R. Crabtree, and S.L. Schreiber.** 1996. Controlling protein association and subcellular localization with a synthetic ligand that induces heterodimerization of proteins. Proc. Natl. Acad. Sci. USA *93*:4604-4607.
4. **Brent, R. and M. Ptashne.** 1985. A eukaryotic transcriptional activator bearing the DNA specificity of a prokaryotic repressor. Cell *43*:729-736.
5. **Carey, M., H. Kakidani, J. Leatherwood, F. Motashari, and M. Ptashne.** 1989. An amino-terminal fragment of GAL4 binds DNA as a dimer. J. Mol. Biol. *209*:423-432.
6. **Chevray, P.M. and D. Nathans.** 1992. Protein interaction cloning in yeast: Identification of mammalian proteins that react with the leucine zipper of Jun. Proc. Natl. Acad. Sci. USA *89*:5789-5793.
7. **Chien, C.-T., P.L. Bartel, R. Sternglanz, and S. Fields.** 1991. The two-hybrid system: A method to identify and clone genes for proteins that interact with a protein of interest. Proc. Natl. Acad. Sci. USA *88*:9578-9582.
8. **Crabtree, G.R. and S.L. Schreiber.** 1996. Three-part inventions: intracellular signaling and induced proximity. TIBS *21*:418-422.
9. **Dalton, S. and R. Treisman.** 1992. Characterization of SAP-1, a protein recruited by serum response factor to the *c-fos* serum response element. Cell *68*:597-612.

10. **Durfee, T., K. Becherer, P.-L. Chen, S.-W. Yeh, Y. Yang, A.E. Kilburn, W.-H. Lee, and S.J. Elledge.** 1993. The retinoblastoma protein associates with the protein phosphate type 1 catalytic subunit. Genes Dev. *7*:555-569.

11. **Estojak, J., R. Brent, and E. Golemis.** 1995. Correlation of two-hybrid affinity data with in vitro measurements. Mol. Cell. Biol. *15*:5820-5829.

12. **Fields, S. and O. Song.** 1989. A novel genetic system to detect protein-protein interactions. Nature *340*:245-246.

13. **Fields, S. and R. Sternglanz.** 1994. The two-hybrid system: an assay for protein-protein interactions. Trends Genet. *10*:286-291.

14. **Gill, G. and M. Ptashne.** 1988. Negative effect of the transcriptional activator GAL4. Nature *334*:721-724.

15. **Giniger, E., S.M. Varnum, and M. Ptashne.** 1985. Specific DNA binding of GAL4, a positive regulatory protein of yeast. Cell *40*:767-774.

16. **Gyuris, J., E. Colemis, H. Chertkov, and R. Brent.** 1993. Cdi1, a human G1 and S phase protein phosphatase that associates with Cdk2. Cell *75*:791-803.

17. **Hannon, D.J., D. Demetrick, and D. Beach.** 1993. Isolation of the Rb-related p130 through its interaction with CDK2 and cyclins. Genes Dev. *7*:2378-2391.

18. **Hochschild, A., N. Irwin, and M. Ptashne.** 1983. Repressor structure and the mechanism of positive control. Cell *32*:319-325.

19. **Holsinger, L.J., D.M. Spencer, D.J. Austin, S.L. Schreiber, and G.R. Crabtree.** 1995. Signal transduction in T lymphocytes using a conditional allele of Sos. Proc. Natl. Acad. Sci. USA *92*:9810-9814.

20. **Hu, J., E. O'Shea, P. Kim, and R. Sauer.** 1990. Sequence requirements for coiled-coils: analysis with λ repressor-GCN4 leucine zipper fusions. Science *250*:1400-1403.

21. **Johnson, M.** 1987. A model fungal gene regulatory mechanism: The *GAL* genes of *Saccharomyces cerevisiae*. Microbiol. Rev. *51*:458-476.

22. **Johnston, S.A., J.M. Salmeron, and S.S. Dincher.** 1987. Interaction of positive and negative regulatory proteins in the galactose regulation of yeast. Cell *50*:143-146.

23. **Keegan, L., G. Gill, and M. Ptashne.** 1986. Separation of DNA binding from the transcriptional-activating function of a eukaryotic regulatory protein. Science *231*:699-704.

24. **Leuther, K.K. and S.A. Johnston.** 1992. Nondissociation of GAL4 and GAL80 in vivo after galactose induction. Science *256*:1333-1335.

25. **Licitra, E.J. and J.O. Liu.** 1996. A three-hybrid system for detecting small ligand-protein receptor interactions. Proc. Natl. Acad. Sci. USA *93*:12817-12821.

26. **Little, J. and D. Mount.** 1982. The Sos regulatory system of *Escherichia coli*. Cell *29*:11-22.

27. **Lohr, D. and J.E. Hopper.** 1985. The relationship of regulatory proteins and DNase I hypersensitive sites in the yeast *GAL*1-10 genes. Nucleic Acids Res. *13*:8409-8423.

28. **Lue, N.F., D.I. Chasman, A.R. Buchman, and R.D. Kornberg.** 1987. Interaction of *GAL4* and *GAL80* gene regulatory proteins in vitro. Mol. Cell. Biol. *7*:3446-3451.

29. **Ma, J.** 1992. Detecting interactions between eukaryotic proteins in bacteria. Gene Expression *2*:139-146.

30. **Ma, J. and M. Ptashne.** 1987. The carboxy-terminal 30 amino acids of *GAL4* are recognized by *GAL80*. Cell *50*:137-142.

31. **Ma, J. and M. Ptashne.** 1988. Converting a eukaryotic transcriptional inhibitor into an activator. Cell *55*:443-446.

32. **Ma, J. and M. Ptashne.** 1987. Deletion analysis of GAL4 defines two transcriptional activating segments. Cell *48*:847-853.

33. **Ma, J. and M. Ptashne.** 1987. A new class of yeast transcriptional activators. Cell *51*:113-119.

34. **Ma, X., D. Yuan, T. Scarborough, and J. Ma.** 1999. Contributions to gene activation by multiple functions of Bicoid. Biochem. J. *338*:447-455.

35. **Nogi, Y., K. Matsumoto, A. Toh-e, and Y. Oshima.** 1977. Interaction of super-repressible and dominant constitutive mutations for the synthesis of galactose pathway enzymes in *Saccharomyces cerevisiae*. Mol. Gen. Genet. *152*:137-144.

36. **Nyanguile, O., M. Uesugi, A.J. Austin, and G.L. Verdine.** 1997. A nonnatural transcription coactivator. Proc. Natl. Acad. Sci. USA *94*:13402-13406.

37. **Orphanides, G., T. Lagrange, and D. Reinberg.** 1996. The general transcription factors of RNA polymerase II. Genes Dev. *10*:2657-2683.

38. **Parthun, M.R. and J.A. Jaehning.** 1992. A transcriptionally active form of GAL4 is phosphorylated and associated with GAL80. J. Mol. Cell. Biol. *12*:4981-4987.

39. **Preston, C.M., M.C. Frame, and M.E.M. Campbell.** 1988. A complex formed between cell components and an HSV structural polypeptide binds to a viral immediate early gene regulatory DNA sequence. Cell *52*:425-434.

40. **Ptashne, M.** 1992. A Genetic Switch. Second ed. Cell and Blackwell Scientific Press, Cambridge.

41. **Ptashne, M. and A. Gann.** 1998. Imposing specificity by localization: mechanism and evolution. Curr. Biol. *8*:R812-R822.

42. **Ptashne, M. and A. Gann.** 1997. Transcriptional activation by recruitment. Nature *386*:569-577.

43. **Ruden, D.M. and M. Ptashne.** 1988. No strict alignment is required between a transcriptional activator binding site and the "TATA box" of a yeast gene. Proc. Natl. Acad. Sci. USA *85*:4262-4266.
44. **Sadowski, I., J. Ma, S. Triezenberg, and M. Ptashne.** 1988. GAL4-VP16 is an unusually potent transcriptional activator. Nature *335*:563-564.
45. **SenGupta, D.J., B. Zhang, B. Kraemer, P. Pochart, S. Fields, and M. Wickens.** 1996. A three-hybrid system to detect RNA protein interactions in vivo. Proc. Natl. Acad. Sci. USA *93*:8496-8501.
46. **Spencer, D.M., T.J. Wandless, S.L. Schreiber, and G.R. Crabtree.** 1993. Controlling signal transduction with synthetic ligands. Science *262*:1019-1024.
47. **Triezenberg, S.J., R.C. Kingsbury, and S.L. McKnight.** 1988. Functional dissection of VP16, the transactivator of herpes simplex virus immediate early gene expression. Genes Dev. *2*:718-729.
48. **Yang, M., Z. Wu, and S. Fields.** 1995. Protein-protein interactions analyzed with the yeast two-hybrid system. Nucleic Acids Res. *23*:1152-1156.
49. **Zawel, L. and D. Reinberg.** 1995. Common themes in assembly and function of eukaryotic transcription complexes. Annu. Rev. Biochem. *64*:533-561.

General Procedures of Two-Hybrid Systems

2

Basic Principles and Procedures of the Yeast Two-Hybrid Assay

Kristen Mayo[1], Ann E. Holtz[1], and Li Zhu[1,2]
[1]CLONTECH Laboratories, Palo Alto, and
[2]Genetastix, San Jose, CA, USA

1. INTRODUCTION

The yeast two hybrid system, first developed by Stan Fields and coworkers (3,4,14,16), provides an in vivo transcriptional assay for specific protein–protein interactions. The original two-hybrid system and its variants provide a sensitive method to detect relatively weak and transient protein–protein interactions. Such interactions may not be biochemically detectable, but may be critical for the proper functioning of complex biological systems (13,22). There are many examples in the literature of the use of two-hybrid technology to test for an interaction between two previously known proteins whose genes have been cloned. Two-hybrid systems have been used successfully to identify protein–protein interactions from a wide variety of biological sources, including yeast, plant, Drosophila, and mammalian. Purified target proteins or antibodies to the protein of interest are not required to obtain a result with the two-hybrid system, although most researchers choose to confirm a two-hybrid result using affinity chromatography or immunoprecipitation (16).

The sensitivity of the two-hybrid assay is primarily attributable to the many-fold amplification of positive signals in vivo (i.e., transcriptional, translational, and enzymatic). In addition, because the two-hybrid assay is performed in vivo, the proteins are more likely to be in their native conformations, which may lead to increased sensitivity and accuracy of detection. The sensitivity of the two-hybrid assay means that it can be used to pinpoint single amino acid residues critical for specific protein–protein interactions and to evaluate protein variants for the relative strength of their interactions (56). The binding data reported by Yang et al. (56) lead them to suggest that protein–protein interactions with dissociation constants (K_d) above approximately 70 µM can be detected using a GAL4-based two-hybrid assay.

This chapter reviews the basic principles of the original two-hybrid system and its variants and provides protocols to perform an interaction assay. See Table 1 for a comparison of the salient features of the GAL4-based system and two LexA-based variants. Chapter 4 provides detailed information on using yeast two-hybrid technol-

Yeast Hybrid Technologies
Edited by L. Zhu and G.J. Hannon
© 2000 Eaton Publishing, Natick, MA

Table 1. Key features of the Classic GAL4 and LexA Yeast Two-Hybrid Systems

	Original GAL4 System	Modified GAL4 System	Original LexA System	Modified LexA System
Developed by:	S. Fields	S. Elledge	R. Brent	Hollenberg
Elements				
BD	GAL4[a]	GAL4[a]	LexA[b]	LexA[b]
AD	GAL4[c]	GAL4[c]	B42AD[d]	VP16AD[e] or GAL4[f]
Binding sequence	*GAL* UAS[g]	*GAL* UAS[g]	LexA ops[h]	LexA ops[h]
Plasmid Selection				
BD	*TRP1*	*TRP1*	*HIS3*	*TRP1*
AD	*LEU2*	*LEU2*	*TRP1*	*LEU2*
Interaction Reporter Genes				
Integrated	*HIS3, lacZ*	*HIS3, lacZ*	*LEU2*	*HIS3, lacZ*
Autonomous	n/a	n/a	*lacZ*	n/a
Recommended Plasmids				
BD	pGBT9	pAS2-1	pLexA	pBTM116
AD	pGAD424	pACT2	pB42AD	pVP16
Reporter (*lacZ*)	n/a	n/a	p8op-lacZ	n/a
Recommended Host Strains				
lacZ only	SFY526	Y187	n/a	n/a
lacZ, HIS3	HF7c[i]	Y190[j] or CG-1945[i,j]	n/a	L40
LEU2	n/a	n/a	EGY48	n/a
lacZ, LEU2	n/a	n/a	EGY48 [p8op-lacZ]	n/a
Additional Features:				
Galactose-Inducible expression of AD fusion?	No	No	Yes	No
CHX counterselection	No	Yes	No	No
Yeast mating protocol	No	Yes	Yes	Yes
Hybrid proteins detectable on Western blots?	No	Yes[k]	Yes	Unknown

[a] GAL4 DNA-binding domain: a.a. 1–147.
[b] The entire prokaryotic LexA protein (202 a.a. residues).
[c] GAL4 activation domain: a.a. 768–881.
[d] An 88-residue acidic *E. coli* peptide (B42) that activates transcription in yeast (39).
[e] Acidic activation domain of the herpes simplex virus VP16 protein (53,54).
[f] GAL4 activation domain: a.a. 768–881 (9).
[g] Either a native GAL UAS or a synthetic UAS G 17-mer consensus sequence.
[h] Two to eight copies of the LexA operator.
[i] Expression of *lacZ* reporter in these strains is somewhat weaker than in SFY526, Y187, and Y190, due to promoter differences.
[j] 3-AT is required in the medium to suppress background growth due to leaky *HIS3* expression.
[k] Using CLONTECH's GAL4 AD and DNA-BD–specific Monoclonal Antibodies (Catalog No.s 5398-1 and 5399-1, respectively).

ogy to screen a library for a novel protein that interacts with a known bait protein. (For additional reviews on yeast two-hybrid systems, see References 1,17,22,31,38, 41,42.) Several of the most recent variations of the two-hybrid system are described in other chapters of this book.

2. PRINCIPLE OF THE TWO-HYBRID ASSAY

The yeast two-hybrid assay is based on the fact that many eukaryotic trans-acting transcription factors are composed of physically separable, functionally independent domains. Such regulators often contain a DNA-binding domain (BD) that binds to a specific enhancer-like sequence, which in yeast is referred to as an upstream activation site (UAS; 26). One or more activation domains (AD) direct the RNA polymerase II complex to transcribe the gene downstream of the UAS (30,35,39). Both the BD and the AD are required to activate a gene and normally, as in the case of the native yeast GAL4 protein, the two domains are part of the same protein. If physically separated by recombinant DNA technology and expressed in the same host cell, the BD and AD peptides do not directly interact with each other and thus cannot activate the responsive genes (7,40). However, if the BD and AD can be brought into close physical proximity in the promoter region, the transcriptional activation function will be restored (Figure 1). In principle, any AD can be paired with any BD to activate transcription, with the BD providing the promoter specificity (7).

In the original two-hybrid system developed by Fields and coworkers, the BD and the AD are both derived from the yeast GAL4 protein (a.a. 1–147 and 768–881, respectively) and a GAL4-responsive element provides the binding site for the GAL BD. In the LexA-based two-hybrid system developed by Roger Brent (19,21,24,42), the BD is provided by the entire prokaryotic LexA protein, which normally functions as a repressor when it binds to LexA operators (12). (With the promoters used in the yeast two-hybrid system, the LexA protein does not act as a repressor.) The AD is an 88-residue acidic *Escherichia coli* peptide (B42) that activates transcription in yeast (39); multiple copies of the LexA operator provide the binding site for the BD. The LexA system of Hollenberg et al. (28,53) is similar to that of Brent, except for the selectable marker on the AD vector (Table 1).

In any of the yeast two-hybrid systems, two different cloning vectors are used to generate separate fusions of the AD and BD to genes encoding proteins that potentially interact with each other. The recombinant hybrid proteins are coexpressed in yeast and are targeted to the yeast nucleus (10,50). An interaction between a bait protein (fused to the BD) and a library-encoded protein (fused to the AD) creates a novel transcriptional activator with binding affinity for the appropriate UAS (Figure 1). This factor then activates reporter genes having the appropriate UAS (or LexA operator) in their promoter and this makes the protein–protein interaction phenotypically detectable. If the two hybrid proteins do not interact with each other, the reporter genes will not be transcibed.

3. REPORTER GENE CONSTRUCTS USED IN TWO-HYBRID SYSTEMS

Most two-hybrid reporter genes are under control of artificial promoter constructs comprised of a TATA and an appropriate UAS (or operator) sequence derived from another gene. For a simple two-hybrid interaction assay, generally one reporter gene is sufficient to identify positives. *lacZ* is a popular choice for this application because β-galactosidase (β-gal) can be detected using a variety of assays that differ in their relative levels of convenience and sensitivity (see Assay for Reporter Gene *[lacZ]* Expression). Furthermore, assaying liquid cultures for β-gal activity yields quantitative data that permits comparison of the relative strength of the two-hybrid interactions observed among several positive clones generated using the same host strain and cloning vectors.

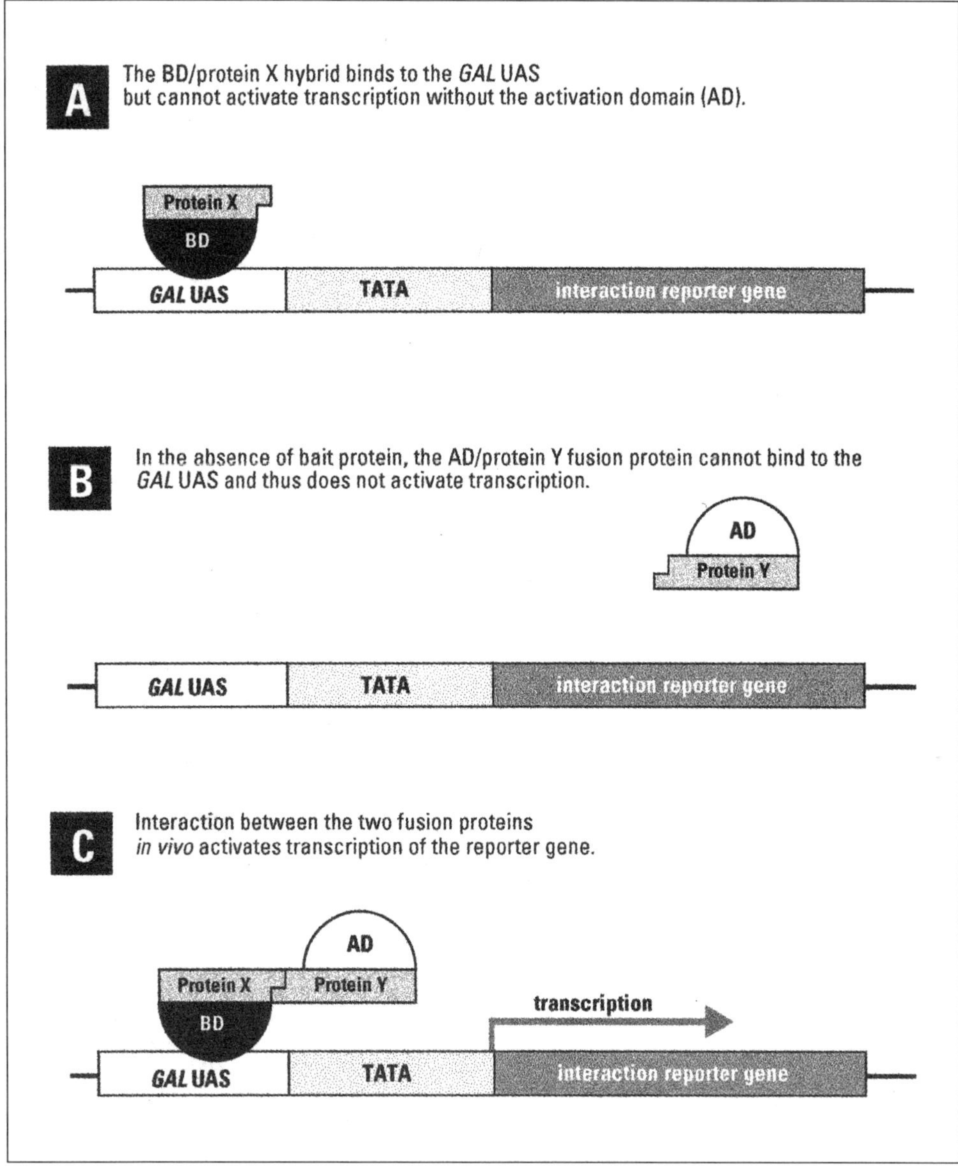

Figure 1. Schematic diagram of GAL4-based yeast two-hybrid systems. This is essentially the same for LexA-based systems, except that the LexA protein and LexA operators are used instead of the GAL4 BD and *GAL* UAS, respectively.

To confirm the specificity of a two-hybrid interaction and to eliminate a class of false positives, most studies use a second reporter gene, whose promoter differs from that of the first in all but the UAS. The use of two reporter genes (with different promoters) is especially critical when screening a library, where 50% or more of the initial positives may turn out to be false upon further testing (see Chapters 3 and 6). Also, when screening a library, it is necessary to use a nutritional reporter gene, such as *HIS3*, *LEU2*, or *ADE2* in the GAL4 systems, or *URA3* in LexA systems, because of the large numbers of clones screened. A nutritional reporter provides an elegant and sensitive growth selection that can identify a single positive transformant out of several million candidate clones. Screening positive clones for expression of the *lacZ* reporter significantly reduces the possibility of a false positive result.

The classic GAL4-based two-hybrid systems utilize two integrated reporter genes (*HIS3* and *lacZ*) under the control of a GAL4-responsive element (either a native *GAL* UAS or a synthetic UAS$_{G\,17\text{-mer}}$ consensus sequence). Most of the GAL4 two-hybrid yeast strains contain the *lacZ* reporter; in addition, some strains (e.g., HF7c, CG-1945, and Y190) contain both the *HIS3* and *lacZ* reporters. The yeast reporter strains used in GAL4-based systems must carry deletions for the *gal4* and *gal80* genes; this prevents interference of native galactose regulatory proteins with the regulatory elements of the two-hybrid system (23,26,34). Thus, nutritional regulation of *GAL* UASs is not a feature of the GAL4-based two-hybrid systems.

The original LexA system uses a plasmid-borne *lacZ* reporter and an integrated *LEU2* reporter, both under the control of multiple LexA operators; their promoters differ in the sequences flanking the LexA operators. The location of the *lacZ* reporter gene on an autonomously replicating, high–copy-number plasmid means that there can be many copies of this reporter in each cell. Thus, in the LexA system, it is possible to assay β-gal activity directly on the culture plate by including X-gal in the medium. However, in some cases such as high background of false positives, the reporter plasmid may be integrated into the host cell (EGY48) genome to create a reporter strain with only one copy of the *lacZ* reporter gene per cell. Host strains used in the LexA system support galactose induction of fusion proteins under control of the *GAL* UAS, as they are wild-type for GAL4 and GAL80 functions. Inducible expression makes the LexA system more suitable for fusion proteins that may be toxic to yeast.

The variability in induced and background levels observed among the various reporter strains is due to differences in the promoters used to drive their reporter genes. For example, the intact *GAL1* UAS is a stronger promoter than the UAS$_{G\,17\text{-mer}}$ consensus sequence and is associated with a higher level of induced expression of the downstream reporter gene. Furthermore, the intact *GAL1* UAS is normally tightly regulated; however, a *GAL1* UAS followed by a TATA box derived from the *HIS3* gene will lead to leaky expression of the downstream reporter gene (e.g., *HIS3* reporter in Y190). For general information on yeast promoters and gene regulation, see References 26 and 51.

In general, the multiple LexA operators used to control the reporter genes in the LexA system provide a higher level of sensitivity and discrimination than can be attained using a GAL4-based system; furthermore, the availability of premade reporter cassettes with various multiples of LexA operators adds flexibility to this system. However, the standard GAL4-based system has been around longer and has a more extensive publication record. At this point, it is difficult to predict which system will give the best results for a particular type of protein–protein interaction. Some interactions may only be detectable in a GAL4 two-hybrid system or a LexA-based system, but not both. Furthermore, there is anecdotal evidence for a small subset of protein–protein interactions that may be detectable using a modified GAL4 system, such as that of S. Elledge and coworkers (2, 25), but not the original GAL4 two-hybrid system of Stan Fields—and vice versa. Thus, if initially you obtain only negative results, you could be surprised by different results if you switch systems. Finally, there are potential limitations inherent to both the GAL4 and LexA systems. For example, test proteins with intrinsic DNA-binding and/or transcriptional activating properties will be difficult or impossible to use in these systems without additional manipulations. This and other potential limitations are discussed further in the Conclusion.

4. PREPARATION FOR A TWO-HYBRID ASSAY

4.1 Selection Media

The selection media for the experimental and control transformations depend on: which system you are using; which plasmids you are using; whether you are selecting for one, two, or three plasmids; whether you are selecting or screening for colonies in which two hybrid proteins are interacting; and, in the case of the LexA system, whether or not you want the AD fusion protein to be expressed. In both types of two-hybrid systems discussed here, the BD and AD cloning plasmids contain different nutritional genes that enable them to grow in synthetically defined medium lacking that specific nutrient. In other two-hybrid systems, the two types of plasmids may be distinguished by different antibiotic resistance genes.

For detailed information on preparing the media, see Appendix F. Alternatively, many types of yeast media are commercially available (CLONTECH, Palo Alto, CA, USA), including YPD (general purpose yeast growth medium), synthetic dropout (SD) base (with or without agar), and many different dropout (DO) supplements. SD medium contains a DO Supplement lacking the appropriate nutrient for plasmid selection and, if appropriate, selection for a two-hybrid interaction. SD media contains 2% glucose as the carbon source for optimal growth; induction medium for the LexA system is SD/Gal/Raff, which contains 2% galactose and 1% raffinose as the carbon source (raffinose is included to improve growth properties of the cells).

4.2 Phenotype Testing

Before you begin working with yeast, we suggest that you test your untransformed host strain(s) for expected phenotype, particularly with respect to nutritional requirements. (For complete genotypes, see Appendix D.) All of the recommended host strains listed in Table 1 are deficient for *TRP* and *LEU* (i.e., they are Trp⁻, Leu⁻) and cannot grow on SD minimal medium lacking those nutrients unless functional *TRP1* and *LEU2* genes are introduced. All of the strains listed are also deficient for *HIS3* (i.e., are His⁻). The GAL4 system host strains HF7c, CG-1945, and Y190 exhibit varying degrees of leaky *HIS3* expression, which must be controlled by including 3-AT in the medium when using the system to screen a library (see Chapter 3). The nontransformed host strain used in the LexA system, EGY48, is deficient for *URA* (i.e., is Ura⁻); however, EGY48 transformed with p8op-lacZ (the *lacZ* reporter plasmid) is Ura⁺ and must be grown on SD medium lacking Ura to maintain the plasmid.

4.3 Construction of Fusion Genes

A brief outline of the protocol is given below (for more detailed information, see Reference 46). The gene for one of the test proteins (X) is fused to the BD in the appropriate BD vector, and the gene for the other protein (Y) is fused to the AD in the appropriate AD vector using standard techniques. These constructs are referred to as BD/X and AD/Y fusion proteins, respectively. The orientation and reading frame of each fusion must be maintained so that hybrid proteins will be expressed. See Figures 3–9 for maps, MCS sequences, and references for the vectors used in the classic GAL4 and LexA systems.

- A fusion gene can easily be generated if compatible restriction sites are present in the test gene and the corresponding vector. If not, the gene fragment can be generated by polymerase chain reaction (PCR) with useful restriction sites incorporated into the primers (47). Often a restriction site at the end of the gene of interest can be changed into a different site or put into a different reading frame by using a PCR primer that incorporates the desired mutation.
- For this application, either vector can be used to make the hybrid with either of the known proteins—unless one has an activation or DNA-binding activity that would interfere with the proper functioning of the two-hybrid system.
    1. Purify the gene fragment, whether generated by PCR or cut out of a plasmid, using any standard method.
    2. Digest the BD or AD vector with the appropriate restriction enzyme(s), treat with phosphatase, and purify.
    3. Ligate the appropriate vector and insert. Transform *E. coli* with the ligation mixtures.
       Note: We recommend the Ligation Express™ Kit (CLONTECH; Catalog No. K1049-1) for rapid and efficient ligation of plasmid vectors to inserts.
    4. Identify insert-containing plasmids by restriction analysis.
    5. Check the orientation and reading frame by sequencing across the junction between the genes encoding the BD and protein X or between the genes encoding the AD and protein Y.

4.4 Verification That the Hybrid Constructs do not Autonomously Activate the Reporter Gene(s)

To obtain meaningful results in any two-hybrid assay, we strongly recommend that you verify that the two hybrid proteins do not activate the reporter gene(s) when the host strain is separately transformed with each plasmid. The following procedure is written for the GAL4 two-hybrid system but can also be used for a LexA system by substituting the appropriate SD selection media (Table 1).

1. Independently transform the BD/X and AD/Y constructs into strain SFY526 (or Y187) using the small-scale yeast transformation protocol (next section). Select for transformants on SD/-Trp and SD/-Leu, respectively. Be sure to include positive and negative controls (e.g., pCL1 and the parental BD and AD cloning vectors; Table 2).
2. Assay the transformants for activation of the *lacZ* reporter gene using the β-gal colony-lift filter assay.

Table 2. Experimental and Control Transformations for a GAL4 Two-Hybrid Interaction Assay

Expt. No.	Plasmid 1 (BD)	Plasmid 2 (AD)	SD Selection Medium	Expected LacZ Phenotype (colony color)	Comment
Experimental:					
1	BD/X	-	-Trp	white	Rules out autonomous activation of reporter gene
2	-	AD/Y	-Leu	white	(same as above)
3	BD/X	AD/Y	-Leu/-Trp	blue[a]	If two hybrids interact
				white	If no interaction
Plasmid Controls:					
4[b]	-	pCL1[c]	-Leu	blue[a]	Control for β-gal assay
5	BD	-	-Trp	white	Control for individual plasmid selection marker and media
6	-	AD	-Leu	white	(same as above)
7[b]	BD	AD	-Leu/-Trp	white	Control for simultaneous selection of both plasmids
Interaction Assay Controls:					
8	BD/p53	-	-Trp	white	Controls to eliminate autonomous activation of the reporter gene
9	-	AD/T-antigen	-Leu	white	(same as above)
10[b]	BD/p53	AD	-Leu/-Trp	white	Controls to eliminate nonspecific interaction with nonhybrid moieties
11[b]	BD	AD/T-antigen	-Leu/-Trp	white	(same as above)
12[b]	BD/p53	AD/T-antigen	-Leu/-Trp	blue[a]	Control for a positive two-hybrid interaction[d]

Table 2. Experimental and Control Transformations for a GAL4 Two-Hybrid Interaction Assay (Continued)

Expt. No.	Plasmid 1 (BD)	Plasmid 2 (AD)	SD Selection Medium	Expected LacZ Phenotype (colony color)	Comment
Controls to Demonstrate Specificity of Interaction:					
13	BD/LamC	-	-Trp	white	Control to eliminate autonomous activation of the reporter gene[e]
14[b]	BD/LamC	AD/T-antigen	-Leu/-Trp	white	Controls to eliminate nonspecific interactions between hybrid proteins
15[b]	BD/X	AD/T-antigen	-Leu/-Trp	white	(same as above)
16[b]	BD/LamC	AD/Y	-Leu/-Trp	white	(same as above)

[a]Blue colonies are also expected to grow on SD medium lacking His due to activation of the *HIS3* reporter (in yeast strains Y190, CG-1945, and Hf7c).

[b]Important controls to include when testing for interaction between two known proteins and when performing transformations to eliminate false positives.

[c]pCL1 encodes the full-length, wild-type GAL4 protein and provides a positive control for the β-gal assay (15). This vector is a derivative of YCp50.

[d]Murine p53 and SV40 large T-antigen are known to interact in a yeast two-hybrid assay (33,36).

[e]Lam C provides a control for a fortuitous interaction between an unrelated protein (i.e., human lamin C) and either the AD/T-antigen control or your AD/Y fusion protein. Lamin C has been reported not to form complexes nor to interact with most other proteins (4,57). This test can be repeated using a plasmid that generates a fusion of a different, presumably noninteracting protein and the BD. In fact, using two or more unrelated, different fusion proteins will improve the reliability of the test.

3. If no autonomous activation is observed for either type of hybrid construct (i.e., the transformant colonies are white), prepare stock plates and liquid cultures for freezing. Use SD/-Trp for the BD hybrid construct and SD/-Leu for the AD hybrid construct. These strains will be used in later experiments.

4. If autonomous activation is observed (i.e., transformant colonies turn blue), it will be difficult to obtain meaningful results. Try switching the test proteins to the other vector (from the BD to the AD plasmid and vice versa). Autonomous activation is more frequently observed with the BD fusion than with the AD fusion because a BD fused with a test protein that has an intrinsic transcriptional activation domain will function as a transcriptional activator specific for this system. Alternatively, it may be possible to remove the activating domain of the bait protein by creating specific deletions within the gene and assaying the deletion constructs for those that no longer have activation function (4).

5. SMALL-SCALE YEAST TRANSFORMATION PROTOCOL

5.1 General Information

Small-scale transformation is used when establishing a single plasmid-containing yeast strain, such as the bait strain, or a specific pair of plasmid-containing yeast strains. Thus, for the application of testing two known proteins for a two-hybrid interaction, a small-scale simultaneous or sequential transformation generally works well. Many such small-scale transformations are needed for pre- or post-library screening procedures and for the pairwise two-hybrid assays that are necessary to eliminate some of the false positives.

The small-scale yeast transformation procedure described here can be used for up to 15 parallel transformations and uses 0.1 μg of each type of plasmid. Depending on the application, the basic yeast transformation method can be scaled up without a decrease in transformation efficiency. If you plan to perform a two-hybrid library screening, you will need a large or library-scale transformation procedure, which will require significantly more plasmid DNA. Please see Chapter 3 for further information on library screening strategies and specific protocols.

There are several methods commonly used to introduce DNA into yeast including the spheroplast method, electroporation, and the lithium acetate (LiAc)-mediated method (reviewed in Reference 23). At CLONTECH, we have found the LiAc method (32), as modified by Schiestl and Gietz (48), Hill et al. (27), and Gietz et al. (18), to be simple and highly reproducible. This method typically results in transformation efficiencies of 10^5 cfu per μg of DNA when using a single type of plasmid. This chapter provides detailed protocols for using the LiAc procedure in a standard plasmid transformation and in a modified transformation to integrate linear DNA into the yeast genome.

In the LiAc transformation method, yeast competent cells are prepared and suspended in a LiAc solution with the plasmid DNA to be transformed, along with excess carrier DNA. Polyethylene glycol (PEG), with the appropriate amount of LiAc, is then added, and the mixture of DNA and yeast is incubated at 30°C. After the incubations, dimethyl sulfoxide (DMSO) is added and the cells are heat shocked, which allows the DNA to enter the cells. The cells are then plated onto the appropriate medium to select for transformants containing the introduced plasmid(s). Because, in yeast, this selection is usually nutritional, an appropriate synthetic dropout (SD) medium is used.

5.2 Simultaneous vs. Sequential Transformations

When the LiAc method is used to simultaneously cotransform yeast competent cells with two plasmids that have different selection markers, the efficiency is typically 10^4 cfu per μg of DNA, as determined by the number of colonies growing on SD medium that selects for both plasmids. This is about an order of magnitude lower than transformation with a single plasmid due to the lower probability that a particular yeast cell will take up both plasmids. (Yeast, unlike bacteria, can support the propagation of more than one plasmid having the same replication origin, i.e., there is no plasmid incompatibility issue in yeast.) In a cotransformation experiment, the efficiency of transforming each type of plasmid should remain at approximately 10^5 cfu per μg of DNA, as determined by the number of colonies growing on SD medium that selects for only one of the plasmids.

Simultaneous cotransformation is generally preferred because it is simpler than sequential transformation and because of the risk that expression of proteins encoded by the first plasmid may be toxic to the cells. If the expressed protein is toxic, clones arising from spontaneous deletions in the first plasmid will have a growth advantage and will accumulate at the expense of clones containing intact plasmids. However, if there is no selective disadvantage to cells expressing the first cloned protein, sequential transformation may be preferred because it uses significantly less plasmid DNA than simultaneous cotransformation. In some cases, such as when one of the two plasmids is the same for several different cotransformations, sequential transformations may be more convenient.

5.3 Integration vs. Nonintegration of Yeast Plasmids

For most yeast transformations performed while using a two-hybrid system, it is not necessary or desirable to have the plasmid integrate into the yeast genome. (In fact, yeast plasmids do not efficiently integrate if they carry a yeast origin of replication and are used uncut.) However, there are two exceptions to this general rule: *(i)* In the one-hybrid system, the researcher must construct his or her own custom reporter plasmid and then integrate it into the yeast host strain before performing the one-hybrid assay (See Chapter 11). *(ii)* In the LexA two-hybrid system, the p8op-lacZ reporter plasmid can be used either as an autonomously replicating plasmid or as an integrated plasmid, depending on the desired level of reporter gene expression. The primary reason to integrate a plasmid in some applications is to generate a stable yeast reporter strain in which only one copy of the reporter gene is present per cell and thereby control the level of background expression. If you have an application that requires integration of a plasmid into the yeast genome, please see the relevant section to follow.

5.4 Transformation Controls

When setting up any type of transformation experiment, we suggest that you include controls for transformation efficiencies. In the case of simultaneous cotransformation, plating an aliquot of the transformation mixture on the appropriate SD media that will select for only one type of plasmid, as well as on medium that will select for both plasmids, will help you determine the transformation efficiencies of each type of plasmid independently as well as that of both plasmids together. That way, if the cotransformation efficiency is low, you may be able to determine whether one of the plasmid types is responsible. Example calculations are shown at the end of the transformation procedure. When screening a library or performing a one- or two-hybrid assay, you will need additional controls, as explained in the relevant sections.

5.5 Reagents and Materials Required

- YPD or the appropriate SD liquid medium (see Appendix F)
- Sterile 1× TE/LiAc (Prepare immediately prior to use from 10× stocks; stock recipes to follow).
- Sterile 1.5-mL microcentrifuge tubes for the transformation
- Appropriate SD agar plates (100-mm diameter)
  Notes:
  - Prepare the selection media and pour the required number of agar plates in advance (Appendix F).
  - Allow SD agar plates to dry (unsleeved) at room temperature for 2–3 days or at 30°C for 3 h prior to plating any transformation mixtures. Excess moisture on the agar surface can lead to inaccurate results due to uneven spreading of cells or localized variations in additive concentrations.
- Appropriate plasmid DNA in solution (check amounts required)
- Appropriate yeast reporter strain for making competent cells (check volume of competent cells required; Steps 1–11 of the small-scale LiAc Yeast Transformation procedure will give you 1.5 mL, enough for 14–15 small-scale transformations)
- Herring testes carrier DNA (10 mg/mL) Sonicated, herring testes carrier DNA in solution can be purchased (CLONTECH; Catalog No. K1606-A) or can be prepared using a standard method (46). Just prior to use, denature the carrier DNA by placing it in a boiling water bath for 20 min and immediately cooling it on ice. Use only high-quality carrier DNA; nicked calf thymus DNA is not recommended.
  - Sterile PEG/LiAc solution (polyethylene glycol/lithium acetate)
  Prepare fresh just prior to use.

	Final Conc.	To prepare 10 mL of solution
PEG 4000	40%	8 mL of 50% PEG*
TE buffer	1×	1 mL of 10× TE
LiAc	1×	1 mL of 10× LiAc

*PEG 3350 (Sigma Chemical, St. Louis, MO, USA; Catalog No. P-3640)

- 10× TE buffer: 0.1 M Tris-HCl, 10 mM EDTA, pH 7.5. Autoclave.
- 10× LiAc: 1.0 M lithium acetate (Sigma Chemical; Catalog No. L-6883) Adjust to pH 7.5 with dilute acetic acid and autoclave.
- 100% DMSO (Sigma Chemical; No. D-8779)
- Sterile 1× TE buffer

- Sterile glass rod, bent Pasteur pipette, or 5-mm glass beads to spread cells on plates.

  Note: CLONTECH's YEASTMAKER™ Yeast Transformation System (Catalog No. K1606-1) contains all of the solutions (except media, H_2O, and DMSO) required for yeast transformation. YEASTMAKER reagents have been optimized for use in the MATCHMAKER™ One- and Two-Hybrid Systems.

5.6 Tips for a Successful Transformation

- Fresh (1–3-week-old) colonies will give best results for liquid culture inoculation. A single colony may be used for the inoculum if it is 2–3 mm in diameter. Scrape the entire colony into the medium. If colonies on the stock plate are smaller than 2 mm, then scrape several colonies into the medium.
- Vigorously vortex-mix liquid cultures to disperse the clumps before using them in the next step.
- The health and growth phase of the cells at the time they are harvested to make competent cells is critical for the success of the transformation. The expansion culture should be in log-phase growth (i.e., OD_{600} between 0.4 and 0.6) at the time the cells are harvested. If they are not, see the Troubleshooting guide.
- When collecting cells by centrifugation, a swinging bucket rotor results in better recovery of the cell pellet.
- For the highest transformation efficiency (as is necessary for library screening), use competent cells within 1 h of their preparation. If necessary, competent cells can be stored at room temperature for several h with a minor reduction in competency.
- To obtain an even growth of colonies on the plates, continue to spread the transformation mixtures over the agar surface until all liquid has been absorbed. Alternatively, use 5-mm sterile glass beads (5–7 beads per 100-mm plate) to promote even spreading of the cells.

5.7 Integrating Plasmids into the Yeast Genome

The small-scale LiAc transformation procedure may be adapted to promote the integration of yeast plasmids as follows:

- Before transformation, linearize 1–4 µg of the reporter vector by digesting it with an appropriate restriction enzyme in a total volume of 40 µL at 37°C for 2 h. Electrophorese a 2-µL sample of the digest on a 1% agarose gel to confirm that the plasmid has been efficiently linearized.

  Notes:
    - If the vector contains a yeast origin of replication (i.e., 2 µ ori), it will be necessary to remove it before you attempt to integrate the vector.
    - The vector should be linearized within the gene encoding the transformation (i.e., nutritional selection) marker. However, if the digestion site is within a region that is deleted in the host strain, the plasmid will not be able to integrate.

- At Step 12, add 1–4 µg of the linearized reporter plasmid + 100 µg of carrier DNA; for each reporter plasmid, also set up a control transformation with undigested plasmid (+ 100 µg carrier DNA).

- At Step 20, resuspend cells in 150 µL of TE buffer.
- Plate the entire transformation mixture onto one plate of the appropriate SD medium to select for colonies with an integrated reporter gene.

5.8 Small-scale LiAc Yeast Transformation Procedure

1. Inoculate 1 mL of YPD or SD with several fresh colonies that are 2–3 mm in diameter.

   Note: For host strains previously transformed with another autonomously replicating plasmid, use the appropriate SD selection medium to maintain the plasmid.

2. Vortex-mix vigorously for 5 min to disperse any clumps.

3. Transfer this into a flask containing 50 mL of YPD or the appropriate SD medium.

4. Incubate at 30°C for 16–18 h with shaking at 250 rpm to stationary phase (OD_{600} >1.5).

5. Transfer 30 mL of overnight culture to a flask containing 300 mL of YPD. Check the OD_{600} of the diluted culture and, if necessary, add more of the overnight culture to bring the OD_{600} up to 0.2–0.3.

6. Incubate at 30°C for 3 h with shaking (230 rpm). At this point, the OD_{600} should be 0.4–0.6. (This is the expansion culture.)

   Note: If the OD_{600} is <0.4, something is wrong with the culture.

7. Place cells in 50-mL tubes and centrifuge at $1000\times g$ for 5 min at room temperature (20°–21°C).

8. Discard the supernatant and add 25–50 mL of sterile TE or distilled H_2O to the tube. Thoroughly resuspend the cell pellets by vortex-mixing.

9. Pool cells in one tube and centrifuge at $1000\times g$ for 5 min at room temperature.

10. Decant the supernatant.

11. Resuspend the cell pellet in 1.5 mL of freshly prepared, sterile 1× TE/LiAc.

12. Add 0.1 µg of plasmid DNA and 0.1 mg of herring testes carrier DNA to a fresh 1.5-mL tube and mix.

    Notes:
    - For simultaneous cotransformation (using two different plasmids), use 0.1 µg of each plasmid (an approximately equal molar ratio) in addition to the 0.1 mg of carrier DNA.
    - For transformations to integrate a reporter vector, use at least 1 µg of linearized plasmid DNA in addition to the carrier DNA.

13. Add 0.1 mL of yeast competent cells to each tube and mix well by vortex-mixing.

14. Add 0.6 mL of sterile PEG/LiAc solution to each tube and vortex-mix at high speed for 10 s.

15. Incubate at 30°C for 30 min with shaking at 200 rpm.

16. Add 70 μL of DMSO. Mix well by gentle inversion. Do not vortex-mix.

17. Heat shock for 15 min in a 42°C water bath.

18. Chill cells on ice for 1–2 min.

19. Centrifuge cells for 5 s at $16\,000\times g$ rpm at room temperature. Remove the supernatant.

20. Resuspend cells in 0.5 mL of sterile 1× TE buffer.

21. Plate 100 μL onto each SD agar plate that will select for the desired transformants. To ensure that you will obtain a plate with well-separated colonies, also spread 100 μL of a 1:1000, 1:100, and 1:10 dilution onto 100-mm SD agar plates. These will also provide controls for cotransformation efficiency.

    Note: If you are performing a cotransformation, plate controls to check transformation efficiency and markers of each plasmid. On separate 100-mm plates, spread 1 μL (diluted in 100 μL H_2O) on medium that will select for a single type of plasmid.

22. Incubate plates, upside down, at 30°C until colonies appear (generally, 2–4 days).

23. To calculate the cotransformation efficiency, count the colonies (cfu) growing on the dilution plate from Step 22 above that has 30–300 cfu.

$$\frac{\text{cfu} \times \text{total suspension vol. (μL)}}{\text{Vol. plated (μL)} \times \text{dilution factor} \times \text{amt. DNA used (μg)*}} = \text{cfu/μg DNA}$$

* In a cotransformation, this is the amount of one of the plasmid types, not the sum of them. If you have used unequal amounts of two plasmids, use the amount of the lesser of the two.

Sample calculation:

- 100 colonies grew on the 1:100 dilution plate (dilution factor = 0.01)
- plating volume: 100 μL
- resuspension volume = 0.5 mL
- amount of limiting plasmid = 0.1 μg

$$\frac{100 \text{ cfu} \times 0.5 \text{ mL} \times 10^3 \text{ μL/mL}}{100 \text{ μL} \times 0.01 \times 0.1 \text{ μg}} = 5 \times 10^5 \text{ cfu/μg DNA}$$

24. Pick the largest colonies and restreak them on the same selection medium for master plates. Seal plates with Parafilm M™ and store at 4°C for 3–4 weeks.

5.9 Troubleshooting Yeast Transformation

The overall transformation efficiency should be at least 10^4 cfu/μg for transformation with a single type of plasmid and 10^3 cfu/μg for simultaneous cotransformation with two types of plasmids. If your cotransformation efficiency is lower than expected, calculate the transformation efficiency of the single plasmids from the number of transformants growing on the appropriate control plates. If the two types of plasmids separately gave transformation efficiencies $>10^5$ cfu/μg, switch to sequential transformation.

If the transformation efficiency for one or both of the separate plasmids is <10^5 cfu/μg, several explanations are possible.

1. Sub-optimal plasmid preparation

   - Repeat the transformation using more (up to 0.5 μg) of the plasmid DNA that had the low transformation efficiency.
   - Check the purity of the DNA and, if necessary, repurify it by ethanol precipitation before using it again.

2. Sub-optimal carrier DNA

   - If you are not already doing so, use YEASTMAKER Carrier DNA, which is available separately (CLONTECH; Catalog No. K1606-A) or as part of the YEASTMAKER Yeast Transformation System (Catalog No. K1606-1), and has been optimized for high transformation efficiencies in two-hybrid systems.
   - If transformation efficiencies are declining in successive experiments, the carrier DNA may be renaturing. Reboil the carrier DNA for 20 min, and then chill it quickly in an ice-water bath.

3. Sub-optimal yeast competent cells

   - Make sure that the expansion culture was in log-phase growth at the time the cells were harvested to make competent cells. If the overnight culture or expansion culture grew slower than expected (or not at all), start over by preparing a fresh overnight culture. Failure to thoroughly disperse the colony used for the inoculum will result in slow growth. If you still have problems obtaining a healthy liquid culture, streak a fresh working stock plate (from the frozen glycerol stock) and inoculate with a fresh colony.
   - Check the liquid medium to make sure it was made correctly. If you suspect that the medium or carbon source stock solutions have been over-autoclaved, remake fresh solutions and either filter sterilize them or adjust the autoclave settings appropriately before autoclaving.
   - The addition of adenine hemisulfate to YPD will enhance the growth of yeast strains that contain the ade2-101 mutation.
   - Check the concentration of the resuspended competent cells using a hemocytometer. If the cell concentration is <1×10^9/mL, spin the cells down again (at 1000× g for 5 min) and resuspend them in a smaller volume of 1× TE/LiAc buffer.
   - Occasionally, there is a contaminant in the water that can affect transformation efficiency and/or cell growth. Prepare all reagents using sterile, deionized, distilled water such as Milli-Q™–filtered. Confirm that your water purification system is functioning properly.

6. INTERACTION ASSAY: EXPERIMENTAL AND CONTROL TRANSFORMATIONS

The following procedure is written for a GAL4 two-hybrid system. It can be used for a LexA system by substituting the appropriate selection media. Omit Ura from the selection medium if you are using EGY48[p8op-lacZ], and use SD/Gal/Raff when you wish to induce expression of the AD fusion plasmid.

1. Cotransform SFY526 (or Y187) with the two types of hybrid plasmids (BD/X and AD/Y) using the small-scale yeast transformation protocol. Plate the cotransformation mixtures on SD/-Leu/-Trp to select for colonies containing both hybrid plasmids. For controls, perform the transformations on lines 4, 7, 10, 11, 12, 14, 15, and 16 in Table 2 using the selection media indicated in that table. For comparison, plate the control transformation mixtures on three types of SD media: *(i)* -Trp (for selection of the BD plasmid); *(ii)* -Leu (for selection of the AD plasmid); and *(iii)* -Leu/-Trp (for selection of both plasmids). Use SD/-His/-Leu/-Trp to test for expression of the *HIS3* reporter (in HF7c, CG-1945, or Y190 transformants).

2. Assay the transformants for activation of the *lacZ* reporter gene using the β-gal colony-lift filter assay or using 5-bromo-4-chloro-3-indolyl-β-D-galactopyranoside (X-gal) in the agar medium (LexA system). For the filter assay, use only fresh (2–4-day-old) transformant colonies growing on the appropriate SD medium. Colonies should be at least 1 mm in diameter before transferring them to the filter. If you wish to obtain quantitative β-gal results, assay liquid cultures using o-nitrophenyl β-D-galactopyranoside (ONPG; Sigma Chemical; Catalog No. N-1127), chlorophenol red-β-D-galactopyranoside (CPRG; Roche Molecular Biochemicals; Catalog No. 884-308), or a chemiluminescent substrate such as Galacton-Star™ as the substrate.

3. Compare the results of your two-hybrid experiment with the positive and negative controls performed in parallel. Refer to Figure 2 and Table 2 for expected results. If the controls do not behave as expected, check the reagents and repeat the assay.

7. ASSAYS FOR REPORTER GENE *(lacZ)* EXPRESSION

7.1 General Information

- X-gal must be used as the β-gal substrate for solid-support assays because of its high degree of sensitivity. (X-gal is approx. 10^6-fold more sensitive than ONPG.) Although more sensitive than X-gal, Galacton-Star™ is not recommended for agar plate and filter assays because it gives troublesome background.

- The filter assay and all of the liquid assays described here use at least one freeze/thaw cycle in liquid nitrogen to lyse the yeast cell walls. Freeze-thaw cycles are a rapid and effective cell lysis method that permits accurate quantification of β-gal activity (49).

- The colony-lift filter assay (6) is primarily used to screen large numbers of cotransformants that survive the *HIS3* growth selection in a GAL4 two-hybrid or one-hybrid library screeening. It can also be used to assay for an interaction between two known proteins in a GAL4 two-hybrid system.

- The in vivo, agar plate assay is primarily used to screen large numbers of cotransformants for the expression of the *lacZ* reporter gene in a LexA two-hybrid library screening when the reporter gene is maintained on an autonomously replicating plasmid. The in vivo assay works for LexA transformants because of the *lacZ* reporter plasmid's high copy number and because of the preamplification step that normally precedes the β-gal assay in this system (see Chapter 3). Because of its relatively low sensitivity, the in vivo, agar plate assay is not suitable for screening transformants in a GAL4-based two-hybrid assay, or in a LexA-based two-hybrid assay when the reporter gene has been integrated into the host genome.

• Liquid cultures may be assayed for β-gal activity to verify and quantify two-hybrid interactions. Because of their quantitative nature, liquid assays can be

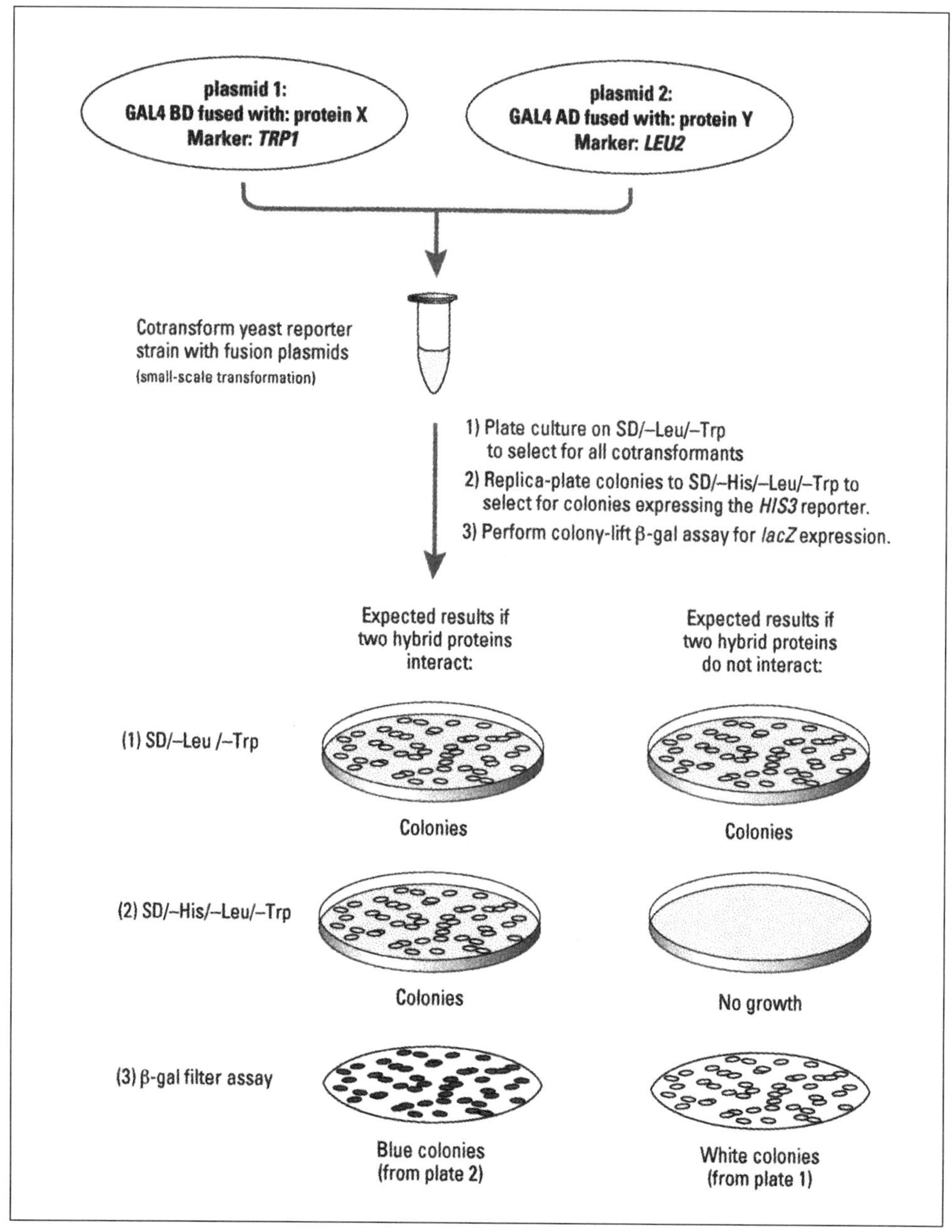

Figure 2. Using the GAL4-based two-hybrid system to test for an interaction between two proteins (X and Y) whose genes have previously been cloned. For LexA-based systems, the procedure is the same, but the selection markers and interaction reporters are different (see Table 1).

used to compare the relative strength of the protein–protein interactions observed in selected transformants. However, there is no direct correlation between β-gal activity and the K_d of an interaction (13). Furthermore, quantitative data cannot be compared between different host strains that have different *lacZ* reporter constructs. In fact, due to promoter strength differences, it may be possible to quantitate the relative strength of interactions in some yeast strains (e.g., Y190, Y187) but not in others (e.g., CG-1945 or HF7c).

- Several substrates are available for liquid β-gal assays, including ONPG, CPRG, and chemiluminescent substrates (e.g., Galacton-Star). The three substrates differ in their relative cost, sensitivity, and reproducibility. Protocols using CPRG are ideal for assaying small numbers of selected transformants. CPRG is approximately 10 times more sensitive than ONPG, although it can be less reproducible than ONPG for strong positive colonies because of CPRG's faster reaction rate. *Note*: β-gal assays using ONPG and CPRG are included in CLONTECH's Yeast Protocols Handbook, which can be accessed through the CLONTECH website (http://www.clontech.com).

- To reduce variability in liquid β-gal assays, we recommend assaying five separate transformant colonies and performing each assay in triplicate.

- The colonies to be assayed for β-gal activity should be growing on the appropriate SD minimal medium. SD medium is used to keep selective pressure on the hybrid plasmids and, in the case of the LexA two-hybrid system, the *lacZ* reporter plasmid up to the time the cells are lysed for the assay. The type of SD medium needed depends on the plasmids and host strains used. Furthermore, when working with a *lacZ* reporter under the control of the inducible *GAL1* promoter (such as in the LexA System), the SD medium must contain galactose (not glucose) as the carbon source. See Table 2 for media recommendations.

7.2 In vivo Plate Assay Using X-Gal in the Medium

Reagents and Materials

- Appropriate SD agar plates containing X-gal (80 mg/L) and 1× BU salts (Appendix F).

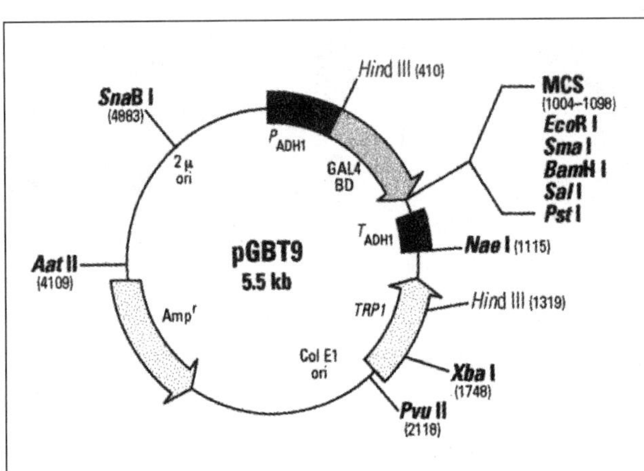

Figure 3. pGBT9 map. Unique sites are in bold. pGBT9 (3) is a cloning vector used to generate fusions of a bait protein with the GAL4 DNA-BD (a.a. 1–147). The hybrid protein is expressed at low levels in yeast host cells from a truncated *ADH1* promoter. The hybrid protein is targeted to the yeast nucleus by nuclear localization sequences (50). pGBT9 contains the *TRP1* gene for selection in Trp- auxotrophic yeast strains. GenBank Accession No. U07646.

Notes:

- BU salts are included in the medium to maintain the optimum pH for β-gal and to provide the phosphate needed for the assay.
- The X-gal should be incorporated into the medium before the plates are poured. If the X-gal is spread over the surface of the agar plates, it can result in uneven distribution and thus localized variations in X-gal concentration. Also, the extra liquid on the plate surface (from spreading the X-gal) may lead to uneven spreading of the cell suspension and will delay absorption of the liquid.
- X-gal is heat-labile and will be destroyed if added to hot (i.e., >55°C) medium.
- Prepare the required number of plates in advance. Allow plates to dry (unsleeved) at room temperature for 2–3 days or at 30°C for 3 h prior to spreading or streaking the cells. Excess moisture on the agar surface can lead to uneven spreading of cells.

1. Streak, replica plate, or spread the transformants to be assayed on selection medium containing X-gal and BU salts.

   - When performing a two-hybrid library screening where very few of the cotransformants are expected to be positive for *lacZ* expression (or where it is difficult to predict the number of interactors), plate the cells at a high density. We recommend plating at two different densities to cover a range; e.g., 0.5×10^6 cfu on some (150-mm plates) and 2×10^6 on others.
   - When performing a two-hybrid assay where most or all of the individual colonies may be LacZ$^+$, spread 200–400 cfu per 100-mm plate.

2. Incubate plates at 30°C for 4–6 days.

3. Check plates every 12 h (up to 96 h) for development of blue color.

   Note: Colonies grown on X-gal–containing medium will be somewhat smaller than those grown without X-gal.

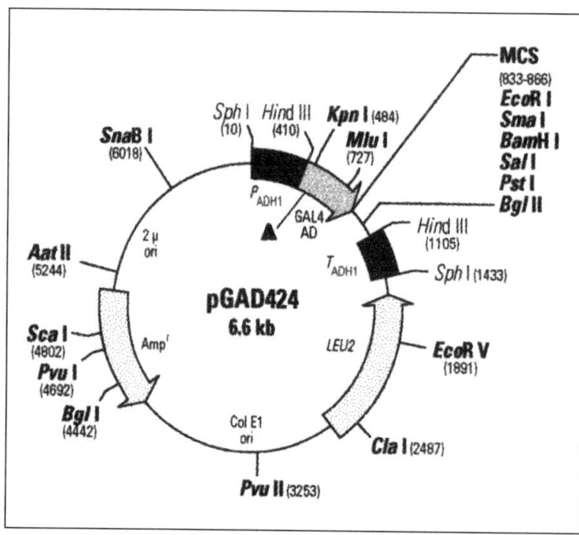

Figure 4. pGAD424 map. Unique sites are in bold. pGAD424 (3) is used to generate a hybrid containing the GAL4 AD (a.a. 768–881) and another protein of interest (or a protein encoded by a cDNA in a fusion library). The hybrid protein is expressed at low levels in yeast host cells from a truncated *ADH1* promoter and is targeted to the yeast nucleus by the SV40 T-antigen nuclear localization sequence (10). pGAD424 contains the *LEU2* gene for selection in Leu- auxotrophic yeast strains. GenBank Accession No. U07647.

7.3 Colony-Lift Filter Assay

Reagents and Materials

- Whatman #5 (Whatman LabSales, Hillsboro, OR, USA) or VWR Grade (VWR Scientific Products, West Chester, PA, USA) 410 paper filters, sterile

  Notes:
  - 75-mm filters (e.g., VWR Catalog No. 28321-055) can be used with 100-mm plates; 125-mm filters (e.g., VWR Catalog No. 28321-113) can be used with 150-mm plates

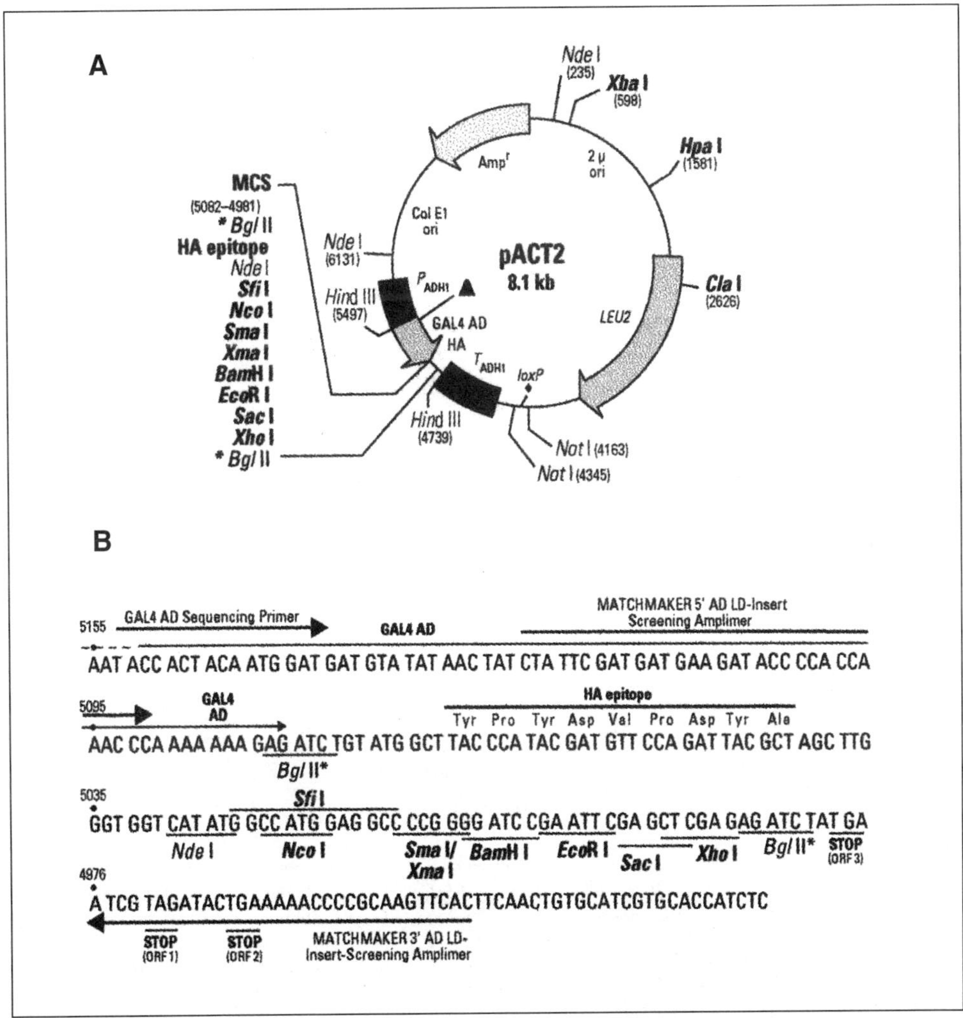

Figure 5 (A) pACT2 map and (B) MCS. Unique sites are in bold. pACT2 (37) is used to generate a hybrid containing the GAL4 AD (a.a. 768–881), an HA epitope tag (11), and another protein of interest (or a protein encoded by a cDNA in a fusion library). The hybrid protein is expressed at medium levels in yeast host cells from an enhanced, truncated *ADH1* promoter and is targeted to the yeast nucleus by the SV40 T-antigen nuclear localization sequence (10). pACT2 contains the *LEU2* gene for selection in Leu- auxotrophic yeast strains. The *Bgl*II (*) sites can be used as a unique cloning site, but this will remove the HA epitope. pACT2 libraries from CLONTECH are constructed in λACT2; recombinant plasmids are released by Cre-mediated recombination at *loxP* sites. GenBank Accession No. U29899.

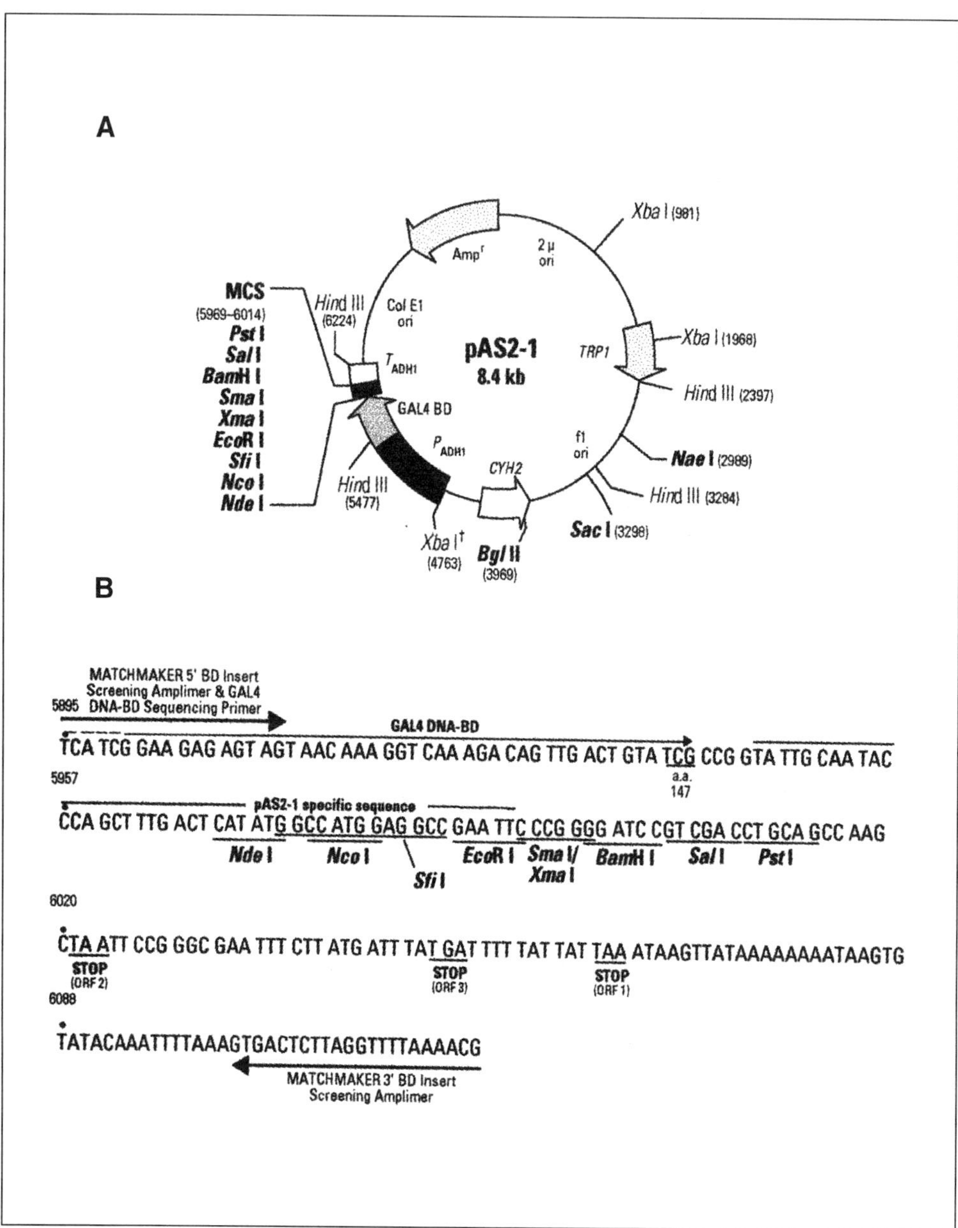

Figure 6 (A) pAS2-1 map and (B) MCS. Unique sites are in bold. pAS2-1 is a cloning vector used to generate fusions of a bait protein with the GAL4 DNA-BD (a.a. 1–147). pAS2-1 is derived from pAS2 and hence from pAS1CHY2 (25) and carries the CYH^S2 gene for cycloheximide sensitivity. The hybrid protein is expressed at high levels in yeast host cells from the full-length *ADH1* promoter. The hybrid protein is targeted to the yeast nucleus by nuclear localization sequences (50). The *Xba*I site at bp 4763 (†) is methylation sensitive. pAS2-1 contains the *TRP1* gene for selection in Trp⁻ auxotrophic yeast strains. Plasmid modification was performed at CLONTECH. GenBank Accession No. U30497. Compared with pAS2, pAS2-1 contains a neutral, short peptide instead of an HA epitope tag. Removing the HA epitope tag and converting a.a.149 from Glu to Val completely eliminates the autonomous activation activity of pAS2 (assayed in Y187 using the *lacZ* reporter; 29). pAS2-1 has a different MCS than pAS2; however, the cloning sites that they have in common are in the same reading frame. Furthermore, the *Eco*RI, *Xma*I/*Sma*I, *Bam*HI, *Sal*I, and *Pst*I sites in the pAS2-1 MCS are in the same reading frames as those in pGBT9. Thus, inserts from pAS2 or pGBT9 can be excised and moved to pAS2-1 without changing the reading frame. The *Eco*RI site in the pAS2-1 MCS is unique because the *Eco*RI site in the CYH^S2 gene in pAS2 was eliminated by site-directed mutagenesis.

- Alternatively, 85- and 135-mm filters can be specially ordered from Whatman LabSales.
- Nitrocellulose filters also can be used, but they are prone to crack when frozen.
- Forceps for handling the filters
- Z buffer

$Na_2HPO_4 \cdot 7H_2O$	16.1	g/L
$NaH_2PO_4 \cdot H_2O$	5.50	g/L
KCl	0.75	g/L
$MgSO_4 \cdot 7H_2O$	0.246	g/L

Adjust to pH 7.0 and autoclave. Can be stored at room temperature for up to 1 year.

- X-gal stock solution

Dissolve X-gal in N,N-dimethylformamide (DMF) at a concentration of 20 mg/mL. Store in the dark at -20°C.

- Z buffer/X-gal solution (prepare fresh just before use)

100	mL	Z buffer
0.27	mL	β-mercaptoethanol (β-ME; Sigma Chemical; Catalog No. M-6250)
1.67	mL	X-gal stock solution

- Liquid nitrogen

1. For best results use fresh colonies (i.e., grown at 30°C for 2–4 days), 1–3 mm in diameter.

   Notes:

   - If only a few colonies are to be assayed, streak them (or spread them in small patches) directly onto master SD selection agar plates. Incubate the plates at 30°C for an additional 1–2 days, and then proceed with the β-gal assay below.

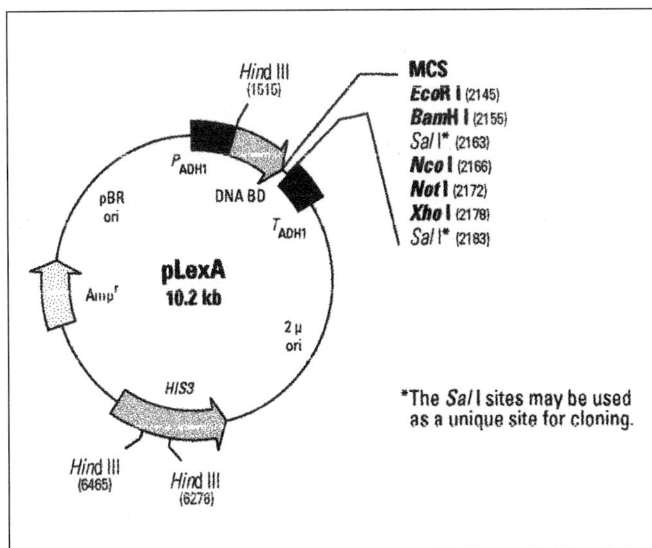

Figure 7. pLexA map. pLexA was originally published as pEG202 in Reference 24. pLexA is used to generate fusions of the BD (the 202-residue LexA protein) with a test (or bait) protein. Fusion protein expression is controlled by the strong yeast *ADH1* promoter. pLexA may be propagated and selected for in *E. coli* and yeast. The *HIS3* transformation marker is used for selection in yeast. Unique sites are in bold.

- Use the SD selection medium appropriate for your system and plasmids. When testing LexA transformants, be sure to use gal/raff induction medium.

2. Prepare Z buffer/X-gal solution as described above.

3. For each plate of transformants to be assayed, presoak a sterile Whatman #5 or VWR grade 410 filter by placing it in 2.5–5 mL of Z buffer/X-gal solution in a clean 100- or 150-mm plate.

4. Using forceps, place a clean, dry filter over the surface of the plate of colonies to be assayed. Gently rub the filter with the side of the forceps to help colonies cling to the filter.

5. Poke holes through the filter into the agar in three or more asymmetric locations to orient the filter to the agar.

6. When the filter has been evenly wetted, carefully lift it off the agar plate with forceps and transfer it (colonies facing up) to a pool of liquid nitrogen. Using the forceps, completely submerge the filters for 10 s.

   Note: Liquid nitrogen should be handled with care; always wear thick gloves and goggles.

7. After the filter has frozen completely (approx. 10 s), remove it from the liquid nitrogen and allow it to thaw at room temperature.

8. Carefully place the filter, colony side up, on the presoaked filter. Avoid trapping air bubbles under or between the filters.

9. Incubate the filters at 30°C (or room temperature) and check periodically for the appearance of blue colonies.

   Note: The time it takes colonies producing β-gal to turn blue in a filter assay varies from 0.5–8 h, depending on the plasmid(s) and host strain. For example, yeast transformed with the β-gal positive control plasmid will turn blue within 20–30 min. Most yeast reporter strains cotransformed with the positive controls for a two-hybrid interaction give a positive blue signal within 60

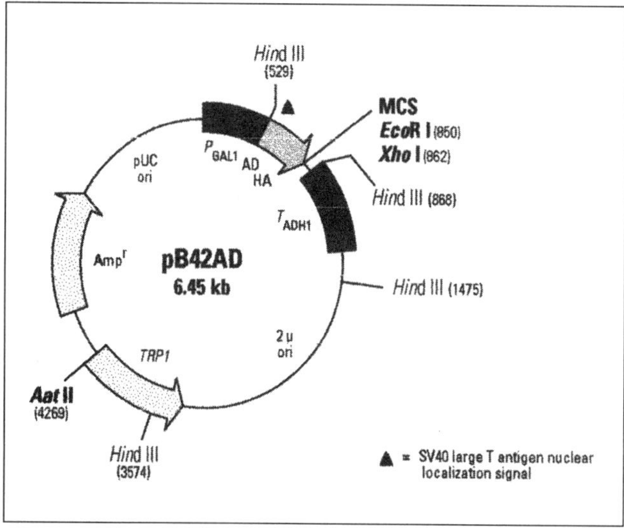

Figure 8. pB42AD map. pB42AD was originally published as pJG4-5 in Reference 24. pB42AD expresses cDNAs or other coding sequences inserted into the unique *Eco*RI and *Xho*I sites as translation fusions to a cassette consisting of the SV40 nuclear localization sequence, the 88-residue B42 acidic activator (AD), and the HA epitope tag. Fusion protein expression is under the control of the *GAL1* inducible promoter. pB42AD may be propagated and selected for in *E. coli* and yeast. The *TRP1* transformation marker is used for selection in yeast. Unique sites are in bold.

min. CG-1945 cotransformed with the control plasmids may take an additional 30 min to develop. If the controls do not behave as expected, check the reagents and repeat the assay. In a library screening, it can take up to 8 h for a positive colony to turn blue; incubating longer than 8 h tends to give false positives.

10. Identify the β-gal producing colonies by aligning the filter to the agar plate using the orienting marks. Pick the corresponding positive colonies from the original plates to fresh medium. If the entire colony was lifted onto the filter, incubate the original plate for 1–2 days to regrow the colony.

7.4 Quantitative Liquid Assay Using a Chemiluminescent Substrate

Reagents and Materials

- Appropriate liquid medium (Appendix F)
- 50-mL culture tubes
- Z buffer
- Galacton-Star reaction mixture (Provided with the Luminescent β-galactosidase Detection Kit II, CLONTECH; Catalog No. K2048-1)
- Liquid nitrogen
- Luminometer or scintillation counter with single-photon counting program
- Optional: 96-well, opaque white, flat-bottom microtiter plates (Xenopore, Hawthorne, NJ, USA; Catalog No. WBP005)
- Optional: Purified β-gal (for a standard curve)

   Note: For best results, we recommend using CLONTECH's Luminescent β-Galactosidase Detection Kit II, which includes a reaction buffer containing the Galacton-Star substrate and the Sapphire II™ accelerator, positive control bacterial β-gal, and a complete User Manual.

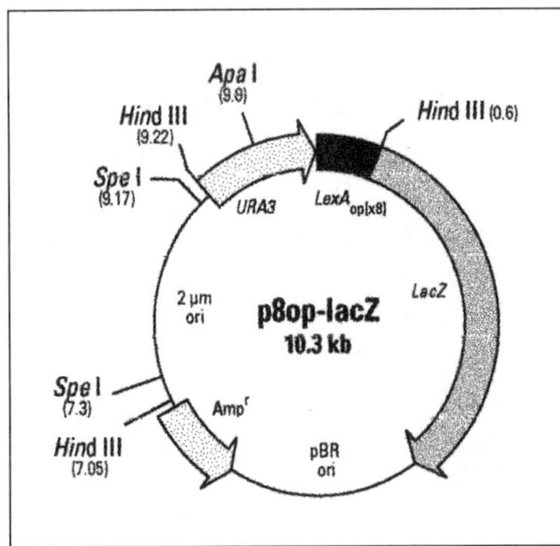

Figure 9. p8op-lacZ map. p8op-lacZ was originally published as pSH18-34 (13,20). p8op-lacZ carries the *lacZ* reporter gene under the control of eight LexA operators and the minimal TATA region from the *GAL1* promoter; the *GAL4* UAS sites were completely removed in the parent plasmid pLR11 (55). p8op-lacZ exhibits zero transcriptional activity in the absence of a LexA-fused activator. p8op-lacZ can be maintained as an autonomously replicating plasmid in strain EGY48 or can be forced to integrate into the EGY48 genome.

7.4.1 Chemiluminescent Detection of β-gal

- For the most accurate results, we recommend that you keep the samples within the linear range of the assay. High-intensity light signals can saturate the photomultiplier tube in luminometers, resulting in false low readings. In addition, low intensity signals that are near background levels may be outside the linear range of the assay. If in doubt, determine the linear range of the assay and, if necessary, adjust the amount of lysate used to bring the signal within the linear range. See Campbell et al. (8) for a chemiluminescent β-gal assay used in a yeast two-hybrid experiment.

- Use positive and negative controls. (Lines 4 and 12 from Table 2 for positive controls, and Line 7, 10, or 11 for a negative control.) Use the same host strain (untransformed with the test plasmid; if possible, grown overnight in SD/-Ura and processed in parallel with the other samples) as a blank for the assay.

- The time needed for a positive β-gal result to develop depends on the strength of the interaction. The positive control (pCL1) takes 3–5 min; certain experimental cotransformants may take several hours to develop.

1. Prepare 5-mL overnight cultures in liquid SD medium. Use the appropriate SD selection medium that will maintain selective pressure on the plasmids. See Table 1 for the nutritional markers of the plasmids. For HIS3 reporter strains harboring interacting hybrid plasmids, use the following SD media lacking His to keep selective pressure on the interaction:
   - for HF7c or CG-1945, use SD/-Leu/-Trp/-His;
   - for Y190, use SD/-Leu/-Trp/-His/+25 mM 3-AT.

   Note: For qualitative data, a whole colony, resuspended in Z buffer, may be used for the assay directly. See instructions following this section.

2. On the day of the experiment, prepare the Galacton-Star reaction mixture. Keep buffer on ice until you are ready to use it.

3. Vortex-mix the overnight culture tube for 0.5–1 min to disperse cell clumps. Immediately transfer at least 2 mL of the overnight culture to no more than 8 mL of YPD (except for the LexA System). For the LexA System, use the appropriate SD/Gal/Raff induction medium for the strains being assayed.

4. Incubate the fresh culture at 30°C for 3–5 h with shaking (230–250 rpm) until the cells are in mid-log phase (OD_{600} of 1 mL = 0.4–0.6).

5. Vigorously vortex-mix the culture tube for 0.5–1 min to disperse cell clumps. Record the exact OD_{600} when you harvest the cells.

6. Place 1.5 mL of culture into each of three 1.5-mL microcentrifuge tubes. Centrifuge at $16\,000\times g$ for 30 s.

7. Carefully remove supernatants. Add 1.5 mL of Z buffer to each tube and thoroughly resuspend the pellet.

8. Centrifuge at $16\,000\times g$ for 30 s.

9. Remove the supernatants. Resuspend each pellet in 300 μL of Z buffer. (Thus, the concentration factor is 1.5/0.3 = fivefold.)

10. Read the OD_{600} of the resuspended cells. The OD_{600} should be approx. 2.5. If the cell density is lower, repeat Steps 5–9 except resuspend the cells in <300 μL of Z buffer.

11. Vortex-mix each cell suspension and transfer 100 µL to a fresh tube.

    Note: The remaining cell suspension can be stored at -70°C to -80°C.

12. Place tubes in liquid nitrogen for 0.5–1 min to freeze the cells.

13. Place frozen tubes in a 37°C water bath for 0.5–1 min to thaw.

14. Repeat freeze/thaw cycle (Steps 12 and 13) once to ensure that cells have been cracked open.

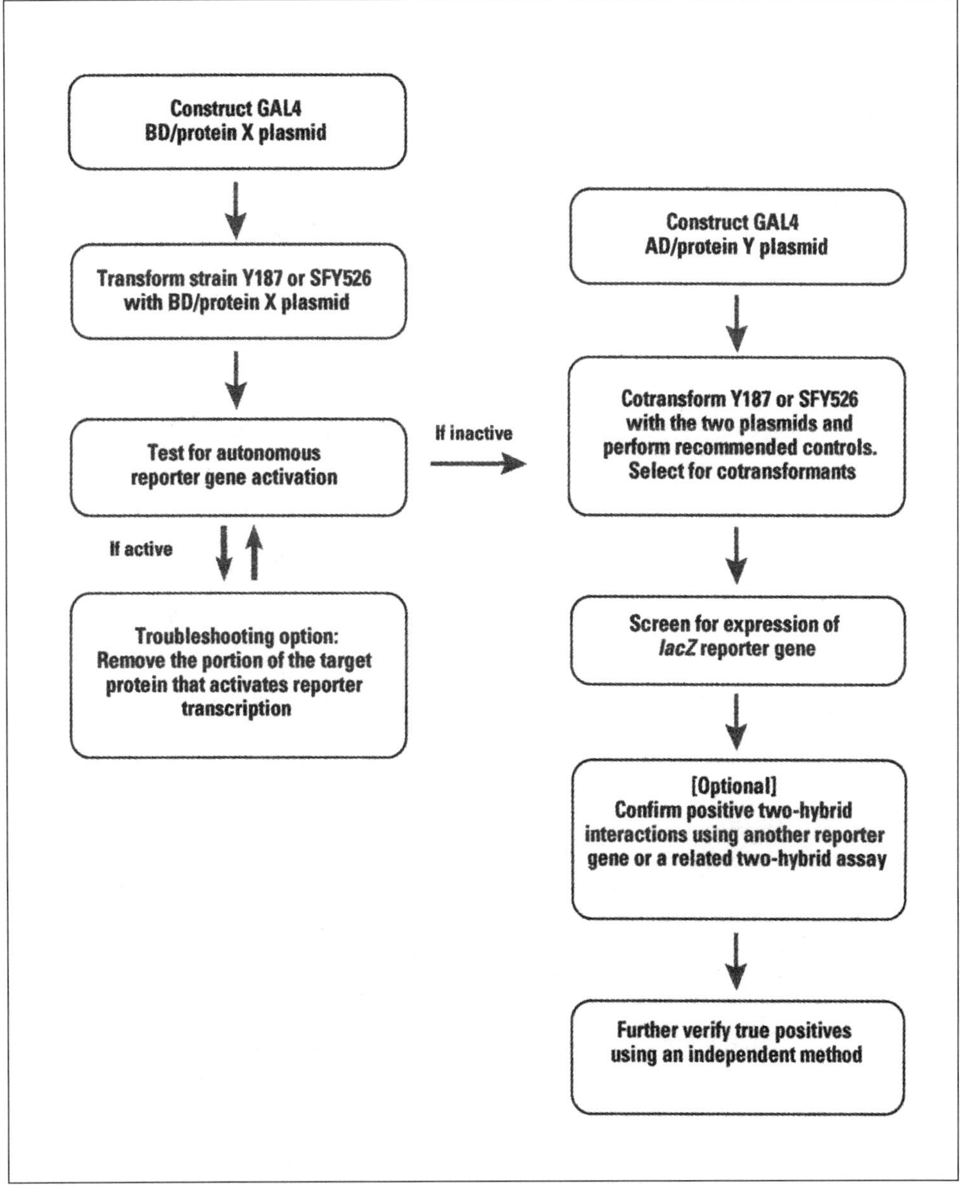

Figure 10. Overview of using the GAL4-based two-hybrid assay for protein–protein interactions. For LexA-based systems, the procedure is the same if EGY48[p8op-lacZ] is used as the host reporter strain; however, the selection media will be different.

41

15. Warm to room temperature enough reaction buffer for the entire experiment.

16. Set up a blank tube with 25 μL of Z buffer.

17. (Optional) If you wish to obtain absolute as well as relative data, set up a series of β-gal standard tubes containing 0.0005, 0.001, 0.003, 0.010, and 0.020 units of β-gal in 25 μL of Z buffer.

18. Place 20–30 μL of each cell lysate into a separate sample tube (or into wells of an opaque 96-well, flat-bottom microtiter plate suitable for plate luminometers). If you are using a sample tube, the tube should hold at least 0.5 mL. Record the exact volume used.

    Note: The amount of yeast extract required may vary depending upon the level of β-gal expression and the detection device used. Use 10–30 μL of extract for positive controls and 20–30 μL for experimental samples with potentially low levels of enzyme activity. It is important to vary the amount of extract to keep the signal within the linear range of the assay.

19. Add 200 μL of Galacton-Star reaction mixture to each sample tube or well and mix gently.

20. Incubate at room temperature (20°–25°C) for 60 min.

    Note: Light signals produced during this incubation are stable for >1 h; therefore, detection can be performed 1–2 h after the incubation.

21. Centrifuge tubes at $16\,000\times g$ for 1 min at 4°C. (If you are using microtiter plates, centrifuge plates at $1000\times g$ for 5 min in a specially adapted rotor.) Proceed directly to the appropriate detection steps for your assay: Step 22, 23, 24, or 27.

22. Detection using a tube luminometer

    a. Turn on the tube luminometer. Set the integration time for 5 s.

    b. Calibrate the luminometer according to the manufacturer's instructions.

    c. If the sample is not already in a tube suitable for luminometer readings, transfer the entire solution from (Step 21) to an appropriate tube. Do not disturb the pellet.

    d. Place one sample at a time in the luminometer compartment and record the light emission (RLU) as a 5-s integral. Use your blank sample as a reference when interpreting the data.

23. Detection using a plate luminometer

    After Step 21, simply record light signals as 5-s integrals.

24. Detection using a scintillation counter

    a. Transfer the entire solution from Step 21 to a 0.5 mL microcentrifuge tube.

       Note: Plan to use scintillation counter adaptors that keep the tubes upright.

    b. Place the tube in the washer of the scintillation counter adaptor and place the adaptor in the machine's counting rack. Set the integration time for at least 15 s.

       Note: Integration times <15 s may not produce accurate results.

    c. To detect chemiluminescent signals, use a single-photon count program. Consult your scintillation counter's manufacturer for further information about this software.

25. For detection methods described in Steps 22–24: Calculate the β-gal activity in

terms of RLU/ OD_{600} unit of cell culture. (Note that Miller unit calculations are not possible using these methods.)

26. (Optional) If you have set up β-gal standards, prepare a standard curve of RLU vs. the amount of β-gal. Estimate the quantity of β-gal in the unknown samples using the standard curve. Determine the amount of enzyme per OD_{600} unit of cell culture. The final OD_{600} units of cells assayed per sample is calculated as follows:

$$OD_{600} \text{ (from Step 5)} \times \text{vol (from Step 18)} \times \text{conc. factor (from Step 9)}$$

27. Detection by exposure of x-ray film

   a. Light emission can also be recorded by exposure of x-ray film to reaction samples in opaque 96-well flat-bottom microtiter plates. The relative intensity of the resulting spots on the film can be estimated by comparison to positive and negative controls.

   Note: x-ray film is several orders of magnitude less senstive than a luminometer or scintillation counter.

   b. Overlay the microtiter plate with x-ray film, cover the film with plastic wrap, and place a heavy object such as a book on top to hold the film in place. Expose the film at room temperature for 5–30 min.

   Note: To compare samples accurately, they must be within the linear response capability of the x-ray film. We therefore recommend that you obtain several different exposures.

7.5 Qualitative Liquid Assay Using a Chemiluminescent Substrate

This alternative cell preparation method directly detects β-gal activity in resuspended yeast colonies. It is recommended for detecting extremely weak *lacZ* transcriptional signals that cannot be detected by X-gal filter assays. For a qualitative (+/-) result, it is more labor-intensive than a filter assay. However, because of its greater sensitivity, it is less likely to give a false-negative result.

1. Grow colonies on the appropriate SD selection medium.

2. Transfer an entire large (2–3 mm), fresh (2–4-day-old) colony to a 0.5-mL tube containing 50 µL of Z buffer. If colonies are small, use several. At the same time, prepare a master or reference plate of the colonies to be assayed.

3. Completely resuspend the colony in the Z buffer by repeatedly pipetting up and down.

4. Place tubes in liquid nitrogen for 0.5–1 min to freeze the cells.

5. Continue with Step 13 of the main procedure.

6. Compare results with those of the negative control.

8. ANALYSIS OF POSITIVE TWO-HYBRID INTERACTIONS

A significant proportion of the positive candidates may contain AD fusion proteins that activate both the *lacZ* and nutritional reporters whether or not the specific BD fusion protein is present (3). To eliminate these false positives, the candidate AD fusion plasmids are subsequently tested for reporter gene activation with various BD fusion constructs or the parental BD plasmid via cotransformation. Truly positive AD fusion proteins activate both reporter genes only when the original BD fusion protein is also present (Table 2).

Detailed information on the analysis and verification of putative positive clones are in Chapter 6. In general, your strategy may involve switching cloning vectors by moving the library insert from the AD to the BD vector and vice versa, and then repeating the two-hybrid assay (10,52). Alternatively, you may wish to test the library and bait inserts using a different two-hybrid system (e.g., a GAL4 vs. LexA system). Another approach is to create a frameshift mutation just upstream of the insert in the AD plasmid by cutting at the *Mlu* I site, filling in the overhangs, and then re-ligating (5) and performing the cotransformation with the mutated AD/Y plasmid and the bait plasmid. The frameshift mutation should knock out the reporter gene expression relative to a control experiment using the unmutated AD library plasmid.

The interacting protein domains may be further characterized by generating site-specific deletion or substitution mutants and repeating the two-hybrid assay. The relative strength of the interactions can be measured using the quantitative β-gal assay. Compare the results to those obtained using the original proteins.

Most researchers choose to confirm the protein–protein interactions using independent, biochemical methods such as affinity chromatography and/or immunoprecipitation (16). For example, the hybrid plasmids can be separately transferred to an appropriate (preferably yeast) expression vector, where the encoded protein can be expressed at high levels. (Yeast expression systems are commercially available.) Purified protein can then be fixed to a chromatographic column that is then used to fractionate a cellular extract. Bound proteins are eluted, and the presence of the interacting protein in this fraction can often be detected by an immunoassay. (For examples of verification methods using coimmunoprecipitation, see References 11 and 58.)

9. CONCLUSIONS

Overall, the many advantages of the two-hybrid technology make it an attractive research tool. However, the classic yeast two-hybrid systems described in this chapter are associated with potential limitations, such as the following:

- Some test proteins may have intrinsic DNA-binding and/or transcriptional activating properties. This is especially likely if the test protein is a transcription factor; however, other proteins that are not normally involved in transcription are sometimes capable of activating transcription, particularly if they contain acidic amphipathic domains (39,44,45). It may be necessary to delete certain portions of the test proteins to eliminate the unwanted activity before the proteins can be used in a two-hybrid system, as discussed in Bartel et al. (4). Note that such deletions may also eliminate a potentially interacting domain. Alternatively, other variants of the two-hybrid system that are less sensitive to autonomously activating proteins may be used (see Chapter 8).

- Some hybrid proteins may not normally localize to the yeast nucleus. To study such proteins, it may be necessary to generate mutant forms of the protein that can be transported across the nuclear membrane. For protein–protein interactions that normally occur on the cell surface, an alternative method, such as a phage display system, may be a more appropriate choice.

- In some cases, the BD or AD fusion moiety may occlude the normal site of interaction or may impair the proper folding of the hybrid protein and thus interfere with the ability of the test proteins to interact (52). The rules defining such occlusions have not as yet been identified.
- The conditions in yeast cells may not provide the proper folding or post-translational modifications (such as glycosylation and tyrosine phosphorylation) required for interactions and/or stability of some mammalian proteins. Conversely, the detection of a specific interaction between mammalian proteins in this heterologous assay system does not necessarily indicate that there is a corresponding interaction in the proteins' native environment (16). To verify that such a protein–protein interaction occurs in native cells, we recommend using a modified yeast two-hybrid system, such as that described in Osborne et al. (43) or a mammalian two-hybrid assay. Alternatively, the positive clones can be used to generate antibodies for use in co-immunoprecipitation studies, or proteins for further biochemical analysis.

REFERENCES

1.**Allen, J.B., M.W. Wallberg, M.C. Edwards, and S.J. Elledge.** 1995. Finding prospective partners in the library: the yeast two-hybrid system and phage display find a match. TIBS *20*:511-516.
2.**Bai, C. and S.J. Elledge.** 1997. Searching for interacting proteins with the two-hybrid system I. *In* P.L. Bartel and S. Fields (Eds.), The Yeast Two-Hybrid System. Oxford University Press, New York.
3.**Bartel, P.L., C.-T. Chien, R. Sternglanz, and S. Fields.** 1993. Using the two-hybrid system to detect protein–protein interactions, p. 153-179. *In* D.A. Hartley (Ed.), Cellular Interactions in Development: A Practical Approach. Oxford University Press, Oxford.
4.**Bartel, P.L., C.-T. Chien, R. Sternglanz, and S. Fields.** 1993. Elimination of false positives that arise in using the two-hybrid system. BioTechniques *14*:920-924.
5.**Bendixen, C., S. Gangloff, and R. Rothstein.** 1994. A yeast mating-selection scheme for detection of protein–protein interactions. Nucleic Acids Res. *22*:1778-1779.
6.**Breeden, L. and K. Nasmyth.** 1985. Regulation of the Yeast HO Gene. Cold Spring Harbor Symposium Quant. Biol. *50*:643-650.
7.**Brent, R. and M. Ptashne.** 1985. A eukaryotic transcriptional activator bearing the DNA specificity of a prokaryotic repressor. Cell *43*:729-736.
8.**Campbell, K.S., A. Buder, and U. Deuschle.** 1995. Interaction of p56lck with CD4 in the yeast two-hybrid system. Ann. N.Y. Acad. Sci. *766*:89-92.
9.**Chang, E.C., M. Barr, Y. Wang, V. Jung, H.-P. Xu, and M. Wigler.** 1994. Cooperative interaction of *S. pombe* proteins required for mating and morphogenesis. Cell *79*:131-141.
10.**Chien, C.T., P.L. Bartel, R. Sternglanz, and S. Fields.** 1991. The two-hybrid system: A method to identify and clone genes for proteins that interact with a protein of interest. Proc. Nat. Acad. Sci. USA *88*:9578-9582.
11.**Durfee, T., K. Becherer, P.L. Chen, S.H. Yeh, Y. Yang, A.E. Kilbburn, W.H. Lee, and S.J. Elledge.** 1993. The retinoblastoma protein associates with the protein phosphatase type 1 catalytic subunit. Genes Devel. *7*:555-569.
12.**Ebina, Y., Y. Takahara, F. Kishi, and A. Nakazawa.** 1983. LexA protein is a repressor of the colicin E1 gene. J. Biol. Chem. *258*:13258-13261.
13.**Estojak, J., R. Brent, and E.A. Golemis.** 1995. Correlation of two-hybrid affinity data with in vitro measurements. Molec. Cell. Biol. *15*:5820-5829.
14.**Fields, S.** 1993. The two-hybrid system to detect protein–protein interactions. METHODS: A Companion to Meth. Enzymol. *5*:116-124.
15.**Fields, S. and O. Song.** 1989. A novel genetic system to detect protein–protein interactions. Nature *340*:245-247.
16.**Fields, S. and R. Sternglanz.** 1994. The two-hybrid system: an assay for protein–protein interactions. Trends Genet. *10*:286-292.
17.**Fritz, C.C. and M.R. Green.** 1992. Fishing for partners. Current Biol. *2*:403-405.
18.**Gietz, D., A. St. Jean, R.A. Woods, and R.H. Schiestl.** 1992. Improved method for high-efficiency transformation of intact yeast cells. Nucleic Acids Res. *20*:1425.

19. **Golemis, E.A. and R. Brent.** 1997. Searching for interacting proteins with the two-hybrid system III. *In* P.L. Bartel and S. Fields (Eds.), The Yeast Two-Hybrid System. Oxford University Press, New York.

20. **Golemis, E.A., J. Gyuris, and R. Brent.** 1994. Interaction trap/two-hybrid systems to identify interacting proteins Ch. 13 & 14. *In* Current Protocols in Molecular Biology, John Wiley & Sons, Inc.

21. **Golemis, E.A., J. Gyuris, and R. Brent.** 1996. Analysis of protein interactions; and Interaction trap/two-hybrid systems to identify interacting proteins. Ch. 20.0 & 20.1. *In* Current Protocols in Molecular Biology. John Wiley & Sons, Inc.

22. **Guarente, L.** 1993. Strategies for the identification of interacting proteins. Proc. Natl. Acad. Sci. USA 90:1639-1641.

23. **Guthrie, C. and G.R. Fink.** 1991. Guide to yeast genetics and molecular biology. *In* Methods in Enzymology, 194:1-932. Academic Press, San Diego.

24. **Gyuris, J., E. Golemis, H. Chertkov, and R. Brent.** 1993. Cdi1, a human G1 and S phase protein phosphatase that associates with Cdk2. Cell *75*:791-803.

25. **Harper, J.W., G.R. Adami, N. Wei, K. Keyomarsi, and S.J. Elledge.** 1993. The p21 Cdk-interacting protein Cip1 is a potent inhibitor of G1 cyclin-dependent kinases. Cell *75*:805-816.

26. **Heslot, H. and C. Gaillardin.** 1992. Molecular Biology and Genetic Engineering of Yeasts. CRC Press, Inc..

27. **Hill, J., K.A. Donald, and D.E. Griffiths.** 1991. DMSO-enhanced whole cell yeast transformation. Nucleic Acids Res. *19*:5791.

28. **Hollenberg, S.M., R. Sternglanz, P.F. Cheng, and H. Weintraub.** 1995. Identification of a new family of tissue-specific basic helix-loop-helix proteins with a two-hybrid system. Mol. Cell. Biol. *15*:3813-3822.

29. **Holtz, A.E. and L. Zhu.** 1995. CLONTECHniques X(3):20.

30. **Hope, I.A. and K. Struhl.** 1986. Functional dissection of a eukaryotic transcriptional protein, GNN4 of yeast. Cell *46*:885-894.

31. **Hopkin, K.** 1996. Yeast two-hybrid systems: more than bait and fish. J. NIH Res. *8*:27-29.

32. **Ito, H., Y. Fukada, K. Murata, and A. Kimura.** 1983. Transformation of intact yeast cells treated with alkali cations. J. Bacteriol. *153*:163-168.

33. **Iwabuchi, K., B. Li, P. Bartel, and S. Fields.** 1993. Use of the two-hybrid system to identify the domain of p53 involved in oligomerization. Oncogene *8*:1693-1696.

34. **Johnston, M., J.S. Flick, and T. Pexton.** 1994. Multiple mechanisms provide rapid and stringent glucose repression of GAL gene expression in *Saccharomyces cerevisiae*. Mol. Cell. Biol. *14*:3834-3841.

35. **Keegan, L., G. Gill, and M. Ptashne.** 1986. Separation of DNA binding from the transcription-activating function of a eukaryotic regulatory protein. Science *231*:699-704.

36. **Li, B. and S. Fields.** 1993. Identification of mutations in p53 that affect its binding to SV40 T antigen by using the yeast two-hybrid system. FASEB J. *7*:957-963.

37. **Li, L., S.J. Elledge, C.A. Peterson, E.S. Bales, and R.J. Legerski.** 1994. Specific association between the human DNA repair proteins XPA and ERCC1. Proc. Natl. Acad. Sci. USA *91*:5012-5016.

38. **Luban, J. and S.P. Goff.** 1995. The yeast two-hybrid system for studying protein–protein interactions. Curr. Opinion in Biotechnol. *6*:59-64.

39. **Ma, J. and M. Ptashne.** 1987. A new class of yeast transcriptional activators. Cell *51*:113-119.

40. **Ma, J. and M. Ptashne.** 1988. Converting a eukaryotic transcriptional inhibitor into an activator. Cell *55*:443-446.

41. **McNabb, D.S. and L. Guarente.** 1996. Genetic and biochemical probes for protein–protein interactions. Curr. Opin. Biotechnol. *7*(5):554-559.

42. **Mendelsohn, A.R. and R. Brent.** 1994. Biotechnology applications of interaction traps/two-hybrid systems. Curr. Opinion in Biotechnol. *5*:482-486.

43. **Osborne, M.A., M. Lubinus, and J.P. Kochan.** 1997. Detection of protein–protein Interactions Dependent on Post-Translational Modifications. *In* P.L. Bartel and S. Fields (Eds.), The Yeast Two-Hybrid System. Oxford University Press, New York.

44. **Ruden, D.M.** 1992. Activating regions of yeast transcription factors must have both acidic and hydrophobic amino acids. Chromosoma *101*:342-348.

45. **Ruden, D.M., J. Ma, Y. Li, K. Wood, and M. Ptashne.** 1991. Generating yeast transcriptional activators containing no yeast protein sequences. Nature *350*:250-251.

46. **Sambrook, J., E.F. Fritsch, and T. Maniatis.** 1989. Molecular Cloning: A Laboratory Manual. Cold Spring Harbor Laboratory, Cold Spring Harbor, NY.

47. **Scharf, S.J.** 1990. Cloning with PCR. p. 84-91. *In* M.A. Innis, D.H. Gelfand, J.J. Sninsky and T.J. White (Eds.), PCR Protocols: A Guide to Methods and Applications, Academic Press, Inc., San Diego.

48. **Schiestl, R.H. and R.D. Gietz.** 1989. High-efficiency transformation of intact cells using single stranded nucleic acids as a carrier. Curr. Genet. *16*:339-346.

49. **Schneider, S., M. Buchert, and C.M. Hovens.** 1996. An *in vitro* assay of β-galactosidase from yeast. BioTechniques *20*:960-962.

50. **Silver, P.A., L.P. Keegan, and M. Ptashne.** 1984. Amino terminus of the yeast GAL4 gene product is sufficient for nuclear localizaton. Proc. Natl. Acad. Sci. USA *81*:5951-5955.

51. **Stargell, L.A. and K. Struhl.** 1996. Mechanisms of transcriptional activation in vivo: Two steps forward. Trends Genet. *12*:311-315.
52. **Van Aelst, L., M. Barr, S. Marcus, A. Polverino, and M. Wigler.** 1993. Complex formation between RAS and RAF and other protein kinases. Proc. Natl. Acad. Sci. USA *90*:6213-6217.
53. **Vojtek, A., S. Hollenberg, and J. Cooper.** 1993. Mammalian Ras interacts directly with the serine/threonine kinase Raf. Cell *74*:205-214.
54. **Vojtek, A.B., J.A. Cooper, and S.M. Hollenberg.** 1997. Searching for interacting proteins with the two-hybrid system II. *In* P.L. Bartel and S. Fields (Eds.), The Yeast Two-Hybrid System. Oxford University Press, New York.
55. **West, R.W. Jr., R.R. Yoccum, and M. Ptashne.** 1984. *Saccharomyces cerevisiae* GAL1-GAL10 divergent promoter region: Location and function of the upstream activating sequence UAS$_\text{G}$. Mol. Cell. Biol. *4*:2467-2478.
56. **Yang, M., Z. Wu, and S. Fields.** 1995. Protein-peptide interactions analyzed with the yeast two-hybrid system. Nucleic Acid Res. *23*:1152-1156.
57. **Ye, Q. and H.J. Worman.** 1995. protein–protein interactions between human nuclear lamins expressed in yeast. Exp. Res. *219*:292-298.
58. **Zhang, X., J. Settleman, J.M. Kyriakis, E. Takeuchi-Suzuki, S.J. Elledge, M.S. Marshall, J. Bruder, U.R. Rapp, and J. Avruch.** 1993. Normal and oncogenic p21ras proteins bind to the amino-terminal regulatory domain of c-Raf-1. Nature *364*:308-313.

3

Screening a Yeast Two-Hybrid Library to Identify Proteins that Interact with a Bait Protein

Kristen Mayo[1], Ann E. Holtz[1], and Li Zhu[1,2]
*[1]CLONTECH Laboratories, Palo Alto,
and [2]Genetastix, San Jose, CA, USA*

1. INTRODUCTION

One of the most fruitful and exciting applications of yeast two-hybrid technology has been the identification of proteins that interact with a known protein "bait". In some cases, the results have been confirmatory, that is, the interactions were already known (or suspected) from other types of data. Typically, however, a two-hybrid library screening reveals at least one or a few totally unexpected interactions involving either previously known or unknown proteins.

The two-hybrid system utilizes a powerful growth selection—the conditional expression of a nutritional reporter gene—to screen large numbers of yeast transformed with a specially constructed fusion library for interacting proteins. When a positive clone is identified, the two-hybrid system provides immediate access to the gene encoding the interacting protein of interest. Site-directed mutagenesis or deletion analysis, combined with additional two-hybrid assays, can then be used to pinpoint the interacting domains. Purified target proteins or antibodies to the protein of interest are not required for two-hybrid library screening. However, purified proteins or antibodies may subsequently be used to confirm a two-hybrid result, as discussed in Chapter 2 and the Chapter by S.B. Hua and A. Fanidi.

For a description of the basic principles of the original GAL4 and LexA-based two-hybrid systems, see Chapter 2 and References 4, 5, 11, and 12. For additional reviews on yeast two-hybrid systems, see 1, 3, 7, 14, 16, 17, 20, 23, 26, 27, and 33.

2. OVERVIEW OF TWO-HYBRID LIBRARY SCREENING

For two-hybrid library screening, the bait protein is generally expressed as a fusion with a DNA-binding domain (BD): amino acids (aa) 1–147 of the yeast GAL4 protein in the case of the GAL4 system, or the entire LexA protein in the case of the LexA system. A library of cDNAs is cloned into an appropriate AD vector such that

Yeast Hybrid Technologies
Edited by L. Zhu and G.J. Hannon
© 2000 Eaton Publishing, Natick, MA

at least one million different hybrid proteins will be expressed as a fusion with a transcriptional activation domain (AD). In GAL4 systems, the AD is aa 768–881 of the GAL4 protein; in the LexA system, an 88-residue acidic *Escherichia coli* peptide (B42) serves as an AD. The bait/BD fusion vector and the AD fusion library are cotransformed into an appropriate yeast host strain and plated on synthetic dropout (SD) medium lacking two or more required nutrients (Figure 1). The exact type of selection medium used depends on which vectors and host strains are being used and the screening strategy. For a discussion and comparison of the classic, two-hybrid reporter genes and their promoters, see Chapter 2, Table 1. Chapter 5 describes another, more advanced variation of the GAL4-based two-hybrid system, which will not be discussed here.

In the LexA system, the cotransformants are first plated on minimal SD non-induction medium that selects for both the bait and the AD/library plasmids, but not for the two-hybrid interactions directly. After cotransformant colonies have grown, they are re-plated on SD medium containing galactose (for induction of fusion protein expression) but lacking Leu, for selection of positive colonies. Because galactose induction is not a feature of the GAL4 systems (Chapter 2), cotransformants in this system may be directly plated onto SD medium lacking His for selection of positive clones (for quickest results). Alternatively, a "jump-start" amplification protocol, similar to that used in the LexA system, may be used to obtain a broader spectrum of positive clones in a GAL4 system. See Figure 2 for an overview of the library screening protocol.

Both the GAL4 and LexA systems make use of at least one additional reporter gene (*lacZ*) whose promoter differs from that of the nutritional reporter in all but the *GAL* UAS (or the LexA operators). Thus, screening initially positive clones for expression of the second reporter eliminates many of the false positive AD/library fusions, particularly those that do not bind to the bait fusion but instead interact directly with promoter sequences flanking the UAS or LexA binding site, or to proteins bound to the flanking sequences. Nevertheless, putative true positive clones should be tested further to determine that they can activate the reporter genes only in the presence of the bait/BD protein (Eliminate False Positives and the Chapter by Y.R. Hong). After sequencing the positive clones, most researchers choose to confirm the protein–protein interactions using independent, biochemical methods such as affinity chromatography and/or immunoprecipitation (Chapters 2 and 17)

3. CHOOSING AND TESTING A BAIT PROTEIN

Many different types of proteins, including proteins involved in cellular machinery, signal transduction, and viral-host interactions, have successfully been used as bait proteins in a two-hybrid library screening. While there is no evidence to suggest that certain types of proteins make particularly good or poor two-hybrid baits, proteins with a known transcriptional activation function or a strong membrane localization peptide would be difficult to use in a two-hybrid library screening without some genetic re-engineering. The sequence encoding the bait protein is generally inserted in a DNA-binding domain (BD) vector in the correct reading frame and orientation so that a fusion of the bait protein and the BD will be expressed, as described in Chapter 2. The bait construction is typically verified by restriction digestion and/or sequencing. Recommended BD vectors for use with the GAL4 systems are pGBT9, pAS2, and pGBKT7; a recommended BD vector for use with Brent's LexA system is pLexA.

(See Chapter 2 and Appendix A for vector information and references; pGBKT7 is described in Chapter 4.) Other BD vectors may be used and are compatible with these systems if they carry the appropriate nutritional marker for selection in the yeast host strains (*TRP1* for *GAL4* systems; *HIS3* for Brent's LexA system). Many of these BD vectors are available from CLONTECH Laboratories (Palo Alto, CA, USA).

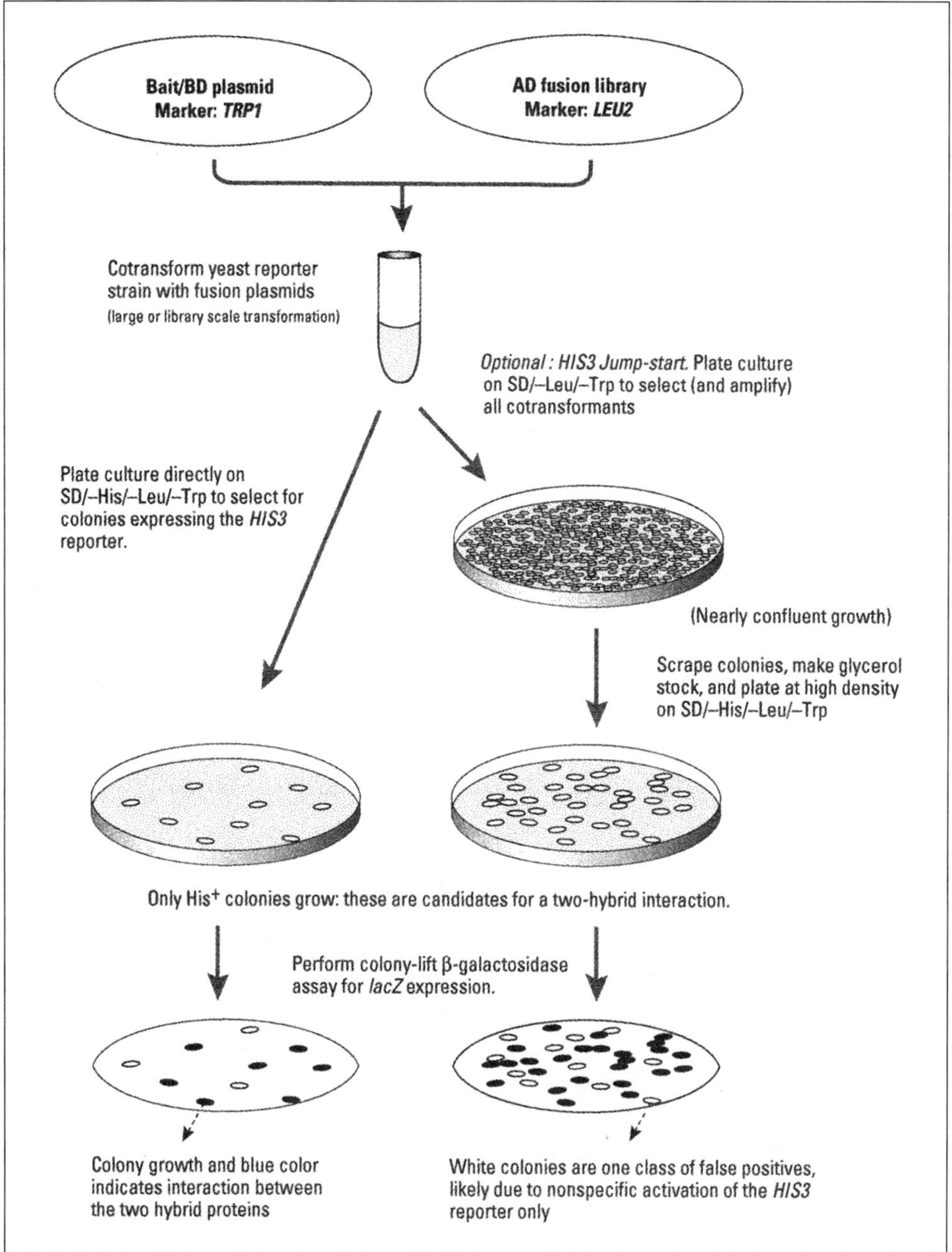

Figure 1. Using a GAL4-based two-hybrid system to screen an AD fusion library for proteins that interact with a bait protein. For LexA-based systems, the procedure is the same as the *HIS3* jump-start protocol for GAL4 systems except that the selection media will be different due to different plasmid transformation markers and interaction reporters.

Before proceeding further with a chosen bait, we recommend that you demonstrate that the bait cannot autonomously activate the reporter genes when fused to a BD, due, for example, to the presence of a cryptic transcriptional activation sequence. (Often, acidic amphipathic domains are responsible for unwanted transcriptional activation; 28,29.) If autonomous activation is observed (in a GAL4-based two-hybrid system), the use of 3-aminotriazole (3-AT) in the growth medium may sufficiently

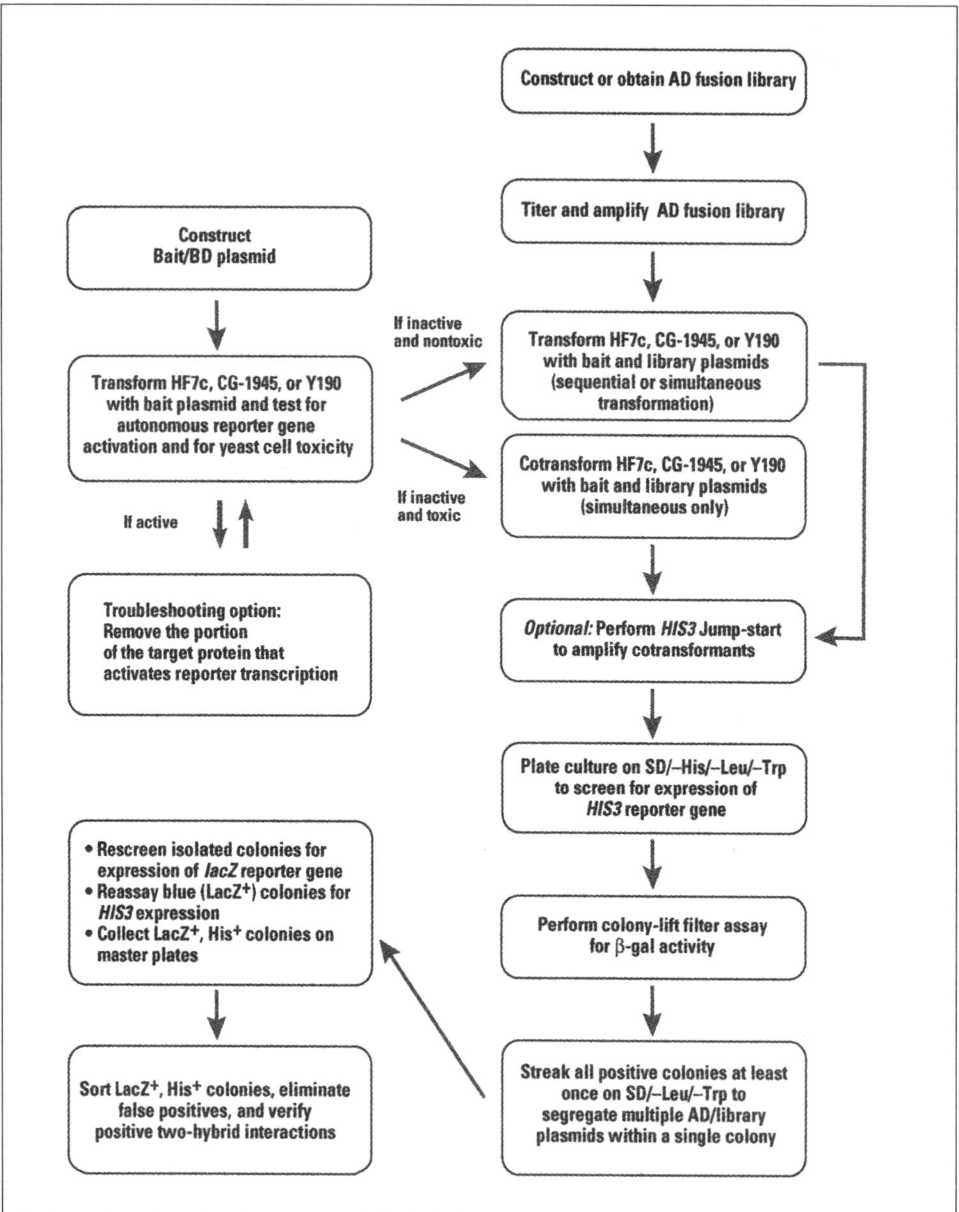

Figure 2. Overview of using a GAL4-based two-hybrid system for performing a two-hybrid screening of an AD fusion library. For the LexA-based system, the procedure is the same if EGY48[p8op-lacZ] is used as the host reporter strain; however, the jump-start step is not optional, and the selection media will be different due to different markers and interaction reporters.

suppress the background His3p signal so that the bait can be used. Alternatively, it may be possible to modify the bait molecule (by deletion) before using it in a two-hybrid screening. Note that such deletions may also eliminate a potentially interacting domain. Special two-hybrid systems have been developed to deal with this category of protein baits (Chapters 8 and 9).

To test the BD/bait protein for transcriptional activation function, the host strain you plan to use for the library screening is transformed with the hybrid construct using the small-scale yeast transformation protocol (Chapter 2). HF7c, CG-1945, Y190, and AH109 (Chapter 4) are recommended host strains for library screening using a GAL4 system; EGY48 or EGY48[p8op-lacZ] are recommended for use with the LexA system; see Table 1 for a comparison of the host strains and Appendix D for host strain genotypes and references. We recommend that you verify the phenotype of your yeast reporter strain before using it in any transformations (Chapter 2) and always include positive and negative controls. If a known protein partner for your bait protein is available, it can be used to check whether an interaction is detectable with this particular BD/bait protein before investing in a library search. Transformants are plated on the appropriate SD selection medium to separately select for individual plasmids and for colonies harboring interacting hybrid proteins. Typically, cotransformants are assayed for lacZ reporter gene expression using the colony-lift filter assay for β-galactosidase (β-gal) (Chapter 2). If transformants seem to be growing unusually slowly, your DNA-BD/bait protein may be toxic to the host cells. If this is the case, you may not be able to use sequential transformation, and simultaneous cotransformation may be a better choice.

4. CHOOSING AND PREPARING AN AD FUSION LIBRARY FOR SCREENING

4.1 General Considerations

The selection markers in the library vector must be compatible with the yeast reporter strains. Two-hybrid libraries are nearly always built into the AD vector rather than into the BD vector, because fusing random proteins to a BD will produce a much larger percentage of hybrids that function as transcriptional activators on their own than if the library-encoded proteins are fused to an AD (25). Examples of suitable AD cloning vectors for use with GAL4 two-hybrid systems are: pGAD424, pACT, pACT2, pGAD10, pGAD GL, pGAD GH, and pGADT7 (Chapter 4); pB42AD is a popular AD cloning vector used with the LexA system. (See Chapter 2 and Appendices for vector information and references). Other AD vectors may be used and are compatible with these systems if they carry the appropriate nutritional marker for selection in the yeast host strains (*LEU2* for GAL4 systems; *TRP1* for Brent's LexA system). Many of these AD vectors, as well as premade MATCHMAKER™ cDNA libraries from a variety of tissues, are available from CLONTECH.

For the best chance of finding a positive interaction, the library should express at least one million different hybrid proteins (2), and the library should be made with cDNA (polyA$^+$ RNA) from an appropriate species or tissue source. (Genomic DNA may be used if it is isolated from an organism, such as bacteria or yeast, which contains few or no introns.) Any standard method for generating cDNA may be used (2,30). The cDNA may be either directionally or nondirectionally cloned. The advan-

Table 1. Choosing a Reporter Strain for AD Library Screening

Strain	System	Reporter Genes	Advantages	Disadvantages
HF7c	GAL4 2H	*HIS3, lacZ*	• Relatively tight *HIS3* expression	• Weak *lacZ*
Y190	GAL4 2H	*HIS3, lacZ*	• Higher *lacZ* reporter expression • Most frequently cited in literature	• Leaky *HIS3* expression may result in a high background of colony growth, and the use of 3-AT in the selection medium may obscure some weak two-hybrid interactions
CG-1945	GAL4 2H	*HIS3, lacZ*	• *HIS3* expression much less leaky than in Y190, so CG-1945 provides more sensitive His3p screening	• Relatively weak *lacZ* reporter expression may result in some LacZ false negatives
AH109	GAL4 2H	*HIS3, ADE2 lacZ*	• *ADE2* reporter is very stringent, compared with *HIS3*	
EGY48	LexA 2H	*LEU2*		
EGY48 [p8op-lacZ]	LexA 2H	*LEU2, lacZ*	• Chromosomally integrated *lacZ* reporter	

tage of an oligo(dT)-primed, directionally cloned library is that one-third of all the clones will be expressed in the correct frame, thereby maximizing the chances of finding a positive interaction. The disadvantage of a directionally cloned library is that the 5′ ends of very long transcripts and, hence, the amino terminus of some library proteins, may be underrepresented. On the other hand, all portions of the transcripts, including the 5′ ends, are equally represented in a randomly primed, nondirectionally cloned library; however, only one-sixth of the clones will be expressed in the correct reading frame. For detailed information regarding construction of two-hybrid cDNA libraries, see References 9, 10, 24, 32, and 34.

Premade libraries should be obtained as *E. coli* transformants, not as purified DNA. When you have received an aliquot of a premade AD library (such as those from CLONTECH), titer the library and then amplify it to produce enough plasmid DNA to screen the library in yeast. You will need 100–500 µg of plasmid DNA to screen approximately 1×10^6 independent clones (Table 3). The following protocol has been optimized for amplification of a MATCHMAKER Library from CLONTECH. If you have obtained a library from another commercial source, follow the manufacturer's recommended procedures.

Tips and precautions for handling libraries:

- Diluted libraries are always less stable than undiluted libraries. Therefore, once the library dilutions are made, use them within the next h, before drastic reductions in titer can occur.

- Use proper sterile technique when aliquoting and handling libraries.

- Design and use appropriate controls to test for cross-contamination.

- Always use the recommended concentration of antibiotic (i.e., ampicillin) in the medium to ensure plasmid stability.

- pACT and pACT2 libraries are released from λACT and λACT2 libraries, respectively. Incubating cultures of pACT and pACT2 libraries at 37°C can result in plaques on high-density plates due to the presence of residual phage in the library. The presence of a few λ plaques on the titering plates should not affect the results. However, if the plaques are so numerous that they obscure the results, repeat the titer using a lower incubation temperature (i.e., 28°C–30°C). The lower temperature may require extending the incubation time, as noted below. Likewise, if you follow the amplification protocol exactly (through Step 7), the presence of a few λ plaques on the amplification plates should not affect the quality or yield of plasmid. If you choose to include Step 8 (an optional liquid culture step to increase yield in the amplification), be sure to perform this incubation at 28°C–30°C. Lysis is more likely to occur in liquid cultures, and you risk lysing the entire culture at temperatures over 30°C. Growth at 28°C–30°C is not necessary once the pACT or pACT2 library DNA has been transferred to a new *E. coli* host.

4.2 Amplification of Premade Libraries

Materials and Reagents

- LB/amp agar plates (30)

Notes:

- The exact number of plates required depends on the size of the library: plan on using 50 150-mm diameter plates per 1×10^6 independent clones in the library.

- Allow the agar plates to dry (unsleeved) at room temperature for 2–3 days, or at 30°C for 3 h, prior to plating cells. Moisture droplets on the agar surface can lead to uneven spreading of cells.
- Sterile glass spreading rod or bent Pasteur pipette
- LB/glycerol (1 L; LB broth containing 25% glycerol)
- *Optional*, for Step 7; LB/amp broth (2 L; Reference 30) and sterile, 50%–80% glycerol.

Procedure

1. If you have not done so already, titer the plasmid library using standard methods.

2. Plate the library directly onto selective medium (e.g., LB/amp plates) at a high enough density so that the resulting colonies will be nearly confluent (approx. 20 000–40 000 cfu per 150-mm plate). Plate enough cfu to obtain at least 2–3× the number of independent clones in the library.

   Notes:
   - The number of independent clones is the number of independent colonies present in the library before amplification. If you have purchased a library from CLONTECH, the size of the library is stated on the Product Analysis Certificate.
   - To promote even growth of the colonies, continue spreading the inoculum over the agar surface until all visible liquid has been absorbed. Then allow plates to sit at room temperature for 15–20 min.

3. Invert the plates and incubate at 37°C for 18–20 h.
   - Alternatively, for pACT or pACT2 libraries: incubate plates at 30°–31.0°C for 36–48 h, or until confluent.
   - Growing the transformants on solid medium instead of in liquid culture minimizes uneven amplification of the individual clones.

4. Add approximately 5 mL of LB/glycerol to each plate and scrape up the colonies into the liquid. Pool all the resuspended colonies in one flask and mix thoroughly.

5. Set aside one-third of the library culture. This is roughly equivalent to 3 L of overnight culture for the purpose of plasmid preparation; this portion can be stored at 4°C if you plan to use it within 2 weeks. For storage >2 weeks, divide the culture into 50-mL aliquots, and store at -70°C.

6. Prepare plasmid DNA using any standard method that yields a large quantity of highly purified plasmid. (See Reference 30 for CsCl gradient purification, if necessary.)

   Note: The cell culture from Step 4 will be very dense ($OD_{600} \gg 1$), so adjust the plasmid preparation protocol accordingly (i.e., follow the procedure as if you were processing 3 L of overnight liquid culture).

7. Expected yields of plasmid DNA per 1×10^6 cfu (cfu estimated from Step 2 above):

- pACT and pACT2 Libraries: 0.25 mg
- all other MATCHMAKER GAL4 Libraries: approximately 1 mg

8. *Optional*; To obtain higher yields, scrape the colonies into LB/amp (no glycerol) and pool the colonies in one flask. Incubate at 30°–31°C for 2–4 h with vigorous shaking (200 rpm). Add sterile glycerol to 25%, then proceed with Step 5 above.

5. STRATEGIES FOR SCREENING AN AD FUSION LIBRARY

5.1 Direct Plating vs. a "Jump-Start" Protocol

When using a GAL4 two-hybrid system, one has the choice of using either a standard library screening procedure or a *HIS3* "jump-start". The two procedures differ only in the plating and selection strategies. The so-called jump-start is the standard procedure used in LexA-based two-hybrid library screenings, but has been adapted for use in GAL4 systems also. Both procedures require about the same amount of total incubation time, although the jump-start protocol involves more steps. In the jump-start procedure, the library transformants are first plated on medium that selects only for the plasmid markers; the selection for a two-hybrid interaction occurs after a second plating, which is physically separate from the first step (Figure 1). In contrast, the conventional GAL4 procedure does not include two separate plating steps; instead, the library transformants are immediately plated onto medium that selects, simultaneously, for plasmid markers and the interaction.

Because it is simpler, most researchers prefer to try the standard procedure first and, if they have trouble picking up positive clones, will repeat the library transformation using the jump-start plating protocol. The jump-start procedure provides an initial phase of growth (or amplification) that maximizes plasmid copy number in each cell. The amplification of plasmids results in higher levels of fusion protein (and His3 protein, in the case of a positive interaction) by the time the library cotransformants are plated onto SD medium lacking His. If you suspect that your bait protein interacts only weakly or transiently with other proteins, you may choose to divide the transformation mixture into two parts and plate half of it using the standard procedure and half using a jump-start.

In a standard GAL4 library screening procedure, library transformation mixtures are plated directly on SD medium lacking Leu and Trp (to select for the plasmids) and also lacking His to select for interacting hybrid proteins (or lacking His and Ade in the case of host strain AH101; Chapter 5). At least 1.5–3 times the number of independent clones in the library are plated on SD, and Leu$^+$, Trp$^+$, His$^+$ colonies will be evident after 10–14 days. In a jump-start screening procedure, library transformants are first plated on SD medium lacking Leu and Trp to select for the plasmids, but containing His, so there is no initial selection for the two-hybrid interaction. In this procedure, at least 1.5–3 times the number of independent clones in the library is initially plated on SD/-Leu/-Trp. After cotransformants have grown, they are scraped up, pooled, and plated at high cell density on SD/-Leu/-Trp/-His to screen for *HIS3* reporter gene expression. In a jump-start procedure, approximately 10 times the number of independent clones will be screened altogether.

A *LEU2* jump-start procedure is recommended when using a galactose-inducible fusion vector, such as those typically employed in the LexA two-hybrid system. After library transformation, at least 1.5-3 times the number of independent clones in the library are plated on glucose-containing non-induction medium (SD/-Trp/-His) to select for the plasmids. Trp⁺, His⁺ cotransformants are scraped up, pooled, and re-plated at high density on SD/-Trp/-His/-Leu to select for *LEU2* reporter expression. With this procedure, 5–10 times the number of independent clones will be screened for activity. If the host strain is EGY48 pretransformed with the autonomously replicating *lacZ* reporter plasmid (p8op-lacZ), *lacZ* and *LEU2* reporter expression can be screened simultaneously by including X-gal in the SD induction medium. This in vivo, whole-plate assay for β-gal is more convenient and less labor-intensive than colony-lift β-gal filter assays when screening large numbers of cotransformants. However, whole-plate assays are more expensive (because of the amount of X-gal needed), and yeast colonies grow a bit slower on X-gal-containing medium than on medium without X-gal. Furthermore, because the whole-plate assay is much less sensitive than colony-lift filter assays, it can only be used when the cells to be assayed contain many copies of the *lacZ* reporter gene. It does not work with host strains (such as those commonly used in GAL4 systems) that contain only one copy of an integrated *lacZ* reporter.

Jump-start protocols typically result in many more positive colonies than the standard procedure and may result in a population preference for clones exhibiting stronger activation of the two-hybrid reporter. Thus, extra steps may be required to sort the positive clones into groups before you proceed with further analysis. Also if you are using a yeast reporter strain, such as Y190 or CG-1945 (GAL4 systems), which has leaky *HIS3* expression, it is important to optimize the 3-AT concentration needed to control background growth.

5.2 Choosing a Host Reporter Strain

The primary considerations when choosing a host reporter strain are *(i)* that its reporter gene promoter is compatible with your BD vector; and *(ii)* that its genotype will permit selection of both the BD and AD plasmids (Table 1 and Appendix D). In all yeast hybrid reporter strains, the nutritional reporter gene is integrated into the host genome for stability and better control of expression levels. In addition, in the LexA system, you have a choice of using the *LEU2* reporter host strain (EGY48) either with the *lacZ* reporter integrated or carried on a plasmid (EGY48[p8op-lacZ]). Keep in mind that if the reporter gene is integrated into the host genome, there will be fewer copies and the system will be less sensitive than one in which the reporter gene is maintained on an autonomously replicating plasmid. The induced level of β-gal activity depends on the strength of the promoter (Chapter 2).

In the GAL4-based systems, there are several HIS3 reporter strains available. The strains differ in their degree of leaky *HIS3* expression (Table 1) due to differences in their promoter construct and, in some cases, to the presence of transforming plasmid. Leaky *HIS3* expression can lead to bothersome background growth on SD medium lacking His unless 3-AT (3-amino-1,2,4-triazole), a competitive inhibitor of the yeast HIS3 protein, is included in the medium (10, 11). New reporter strains that carry an *ADE* reporter (e.g., PJ69-2A and AH109) offer a much tighter selection than *HIS3* (see Chapter 4).

Notes:

• Transformants derived from HF7c may demonstrate slightly leaky *HIS3* expression; the extent of *HIS3* leakiness depends primarily on the presence of transforming plasmids. A small amount of 3-AT (e.g., 1–5 mM) is generally sufficient to suppress background growth on SD/-His.

- Y190 is very leaky for *HIS3* expression and, in the absence of 3-AT, grows normally on SD/-His; 25–60 mM 3-AT is required in the medium to suppress background growth of Y190 and transformant strains derived from Y190. To optimize the 3-AT concentration, plate Y190 transformed with your DNA-BD/bait plasmid on a series of SD/-Trp/-His plates containing 3-AT in the range of 25 to 60 mM (i.e., try 25, 35, 45, 55, and 60 mM). Use the lowest concentration of 3-AT that, after one week, allows only small (<1 mm) colonies to grow.

- CG-1945 is slightly leaky for *HIS3* expression and, in the absence of 3-AT, grows slowly on SD/-His (maximum colony size approx. 1 mm); 5–15 mM 3-AT in the medium is sufficient to suppress background growth of CG-1945 and transformant strains derived from CG-1945. To optimize the 3-AT concentration, plate CG-1945 transformed with your DNA-BD/bait plasmid on a series of SD/-Trp/-His plates containing 3-AT in the range of 0 to 15 mM (i.e., try 0, 2.5, 5, 7.5, 10, 12.5, and 15 mM). Use the lowest concentration of 3-AT that, after one week, allows only small (<1 mm) colonies to grow.

- Too much 3-AT in the medium can kill freshly transformed cells. Thus, if you wish to use excess 3-AT to select only very strong two-hybrid interactions in a library screening, we recommend the *HIS3* jump-start alternative protocol.

5.3 Simultaneous vs. Sequential Transformations

The BD/bait plasmid and the AD/library plasmids can be introduced into the reporter strain either simultaneously or sequentially. In sequential transformation, the BD/bait plasmid is introduced first (small-scale transformation); selected transformants are then grown up and transformed with the AD fusion library (large-scale transformation). Simultaneous cotransformation is generally preferred because it is simpler to perform than sequential transformation—and because of the risk that expression of the BD/bait protein may be toxic to the cells. If the BD/bait protein is toxic, clones arising from spontaneous deletions in the BD/bait plasmid will have a growth advantage and will accumulate at the expense of clones containing intact plasmids. However, if there is no selective disadvantage to cells expressing the first hybrid protein, sequential transformation may be preferred because it uses significantly less plasmid DNA than simultaneous cotransformation (Table 3). We provide protocols for three library screening options (compared in Tables 2 and 3):

- library-scale, simultaneous cotransformation

- large-scale, simultaneous cotransformation

- large-scale, sequential transformation

Large-scale transformation uses the same amount of cells as the small-scale transformation (Chapter 2). However, the small-scale method is for several (up to 15) individual transformations, whereas the large-scale method is for one transformation using up to 50 µg of one type of plasmid. The library-scale method transforms over five times as many competent cells as the small- and large-scale methods and uses up to 500 µg of plasmid DNA. Common applications of the three transformation scales are summarized in Table 2.

Table 2. Guide to Yeast Transformation Protocols

<u>Small-scale (Chapter 3)</u>

• Uses 1.5 mL of competent cells for up to 15 small-scale transformations using 0.1 µg of each plasmid

• Use to:

   • Verify that BD/bait protein does not autonomously activate reporter genes

   • Look for toxicity effects of BD/bait protein on host

   • Perform control experiments (Tables 4 & 5)

   • Transform host strain with BD/bait plasmid as first step of sequential transformation in library screening

<u>Large-scale</u>

• Uses 1.5 mL of competent cells in one large-scale transformation using 10–50 µg of the limiting plasmid

   <u>Large-scale, sequential transformation:</u>

   – Introduce AD/library plasmids into host strain previously transformed with the BD/bait plasmid (second step of sequential transformation)*

   – Yields approximately 10^6 cotransformants

   <u>Large-scale, simultaneous cotransformation:*</u>

   • Yields approximately 10^5 cotransformants

<u>Library-scale simultaneous cotransformation*</u>

   • Uses 8 mL of competent cells and 100–500 µg of the limiting plasmid

   • Yields approximately 10^6 cotransformants

*Recommended if expression of the DNA-BD/target protein is toxic to the cells.

6. LIBRARY- AND LARGE-SCALE YEAST TRANSFORMATION PROTOCOLS

6.1 General Information

In this section, we provide detailed protocols for the PEG/LiAc-mediated transformation of yeast, adapted for two-hybrid library screening. The LiAc method for preparing yeast competent cells (15,18,21,31) typically results in transformation efficiencies of 10^5 per µg of DNA when a single type of plasmid is introduced into yeast. When the yeast is simultaneously cotransformed with two plasmids having different selection markers, the efficiency is usually an order of magnitude lower due to the lower probability that a particular yeast cell will take up both plasmids.

When setting up a library transformation experiment, we recommend that you include proper controls for transformation efficiencies. In the case of simultaneous cotransformation, it is important to determine the transformation efficiencies of both plasmids together, as well as of each type of plasmid independently. Example calculations are shown after the transformation protocol. It is also important to include positive controls for β-gal and for a two-hybrid interaction. The recommended setup (including controls) for a standard GAL4 two-hybrid library screening are shown in Table 4 and for a LexA library screening in Table 5.

Table 3. Comparison of Two-Hybrid Library Transformation Methods

Transformation Type	Amount of Limiting Plasmid	Transformation Efficiency[a]	No. of Indep. Clones Amplified[b]	No. of Plates (150-mm)
Library-scale Simultaneous	100–500 µg	10^3–10^4	1×10^6	50
Large-scale Simultaneous	10–50 µg	10^3–10^4	1×10^5	5
Large-scale Sequential	10–50 µg	10^4–10^5	1×10^6	50

[a]cfu per µg of limiting (in this case, AD library) plasmid DNA.
[b]Total approximate number of transformants expected on SD/-Leu/-Trp selection plates assuming, in each case: 1) the minimal amount of plasmid was used; 2) the transformed cells were resuspended in the volumes recommended in the protocol; 3) 200 µL of transformed cells were spread onto each plate; and 4) the transformation efficiencies were optimal.

6.2 Yeast Transformation Method and Controls

Materials and Reagents

- YPD or the appropriate SD liquid medium (Appendix F)
- Appropriate SD agar plates (Table 4 or 5; Appendix F)

Notes:

  - Prepare the selection media and pour the required number of agar plates in advance
  - Allow SD agar plates to dry (unsleeved) at room temperature for 2–3 days or at 30°C for 3 h prior to plating any transformation mixtures. The presence of moisture droplets on the agar surface can lead to uneven spreading of cells.

- Appropriate yeast reporter strain
- Appropriate plasmid DNA in solution (check amounts required)
- Sterile 1× TE/LiAc (Prepare immediately prior to use from 10× stocks; see Chapter 2 for stock recipes)
- TE buffer (Prepare from 10× TE buffer, Chapter 2) or sterile, distilled H_2O
- Appropriate sterile tubes for the transformation.
- Herring testes carrier DNA (10 mg/mL). See notes in Chapter 2.
- Sterile PEG/LiAc solution (Check volume needed and prepare immediately prior to use from 10× stocks; recipe in Chapter 2)
- 100% dimethylsulfoxide (DMSO); Sigma Chemical, St. Louis, MO, USA; Catalog No. D-8779
- Sterile glass rod, bent Pasteur pipette, or 5-mm glass beads for spreading cells on plates.
- 65% Glycerol/$MgSO_4$ solution (Sterile; for harvesting library transformants in a jump-start procedure):

Table 4. Setup For a Two-Hybrid Library Screening (GAL4-based systems)[a]

Transformation	Plasmid(s)	Plate on SD Minimal Medium[b]	LacZ Phenotype
Control 1	pCL1[c]	-Leu	Blue
Control 2	AD/TAg + BD/p53	-Leu[d]	White
		-Trp[d]	White
		-Leu/-Trp[e]	Blue
		-His/-Leu/-Trp/+3-AT[e]	Blue
Experimental	BD/bait + AD library	-Leu[f]	
		-Trp[f]	
		-Leu/-Trp[f]	
		-His/-Leu/-Trp/+3-AT	(to be tested)[g]

[a]The indicated plasmids and selection media are designed for a direct library screening procedure. If you are planning a *HIS3* jump-start protocol, see Strategies for Screening an AD Fusion Library for media requirements.

For simultaneous cotransformation, all transformations are performed in CG-1945 or Y190; for the second transformation in the sequential procedure, use CG-1945 or Y190 previously transformed with your BD/bait plasmid for the library transformation.

[b]See Choosing a Host Reporter Strain for guidelines on how much 3-AT to use in library screening plates to control background growth.

[c]Control for β-gal assay

[d]Controls for individual plasmid selection markers and media, and for the transformation efficiency of each plasmid type

[e]Controls for interaction of two hybrid proteins and for activation of both reporter genes

[f]Controls for transformation efficiency

[g]True interacting positives should turn blue on a β-gal colony-lift assay. Many of the His[+] positives will not turn blue; these might be false positives.

	Final concentration
Glycerol	65% vol/vol
$MgSO_4$	100 mM
Tris-HCl (pH 8.0)	25 mM

Note: The YEASTMAKER™ Yeast Transformation System (Catalog No. K1606-1) contains all the solutions (except media, H_2O, and DMSO) required for yeast transformation. YEASTMAKER reagents have been optimized for use in the MATCHMAKER Two-Hybrid Systems.

Tips for a successful transformation

• Fresh (one- to three-week-old) colonies will give best results for liquid culture inoculation. A single colony may be used for the inoculum if it is 2–3 mm in diameter. Scrape the entire colony into the medium. If colonies on the stock plate are smaller than 2 mm, scrape several colonies into the medium.

Note: To aid in resuspending the cells, first place the colony in a 1.5-mL tube containing 0.5 mL of medium and vortex-mix vigorously. Then transfer

Table 5. Set-Up For a Two-Hybrid Library Screening (LexA-based systems)[a]

Transformation	Plasmid(s)	Plate on SD Minimal Medium	LacZ Phenotype
Control 1[b]	pLexA-Pos	Gal/Raff/-His/-Ura[c]	Blue
Control 2[d]	pLexA-53 + pB42AD-T	Gal/Raff/-His/-Leu/-Trp/-Ura	Blue
Experimental	BD/bait + AD library	Gal/Raff/-His/-Leu/-Trp/-Ura[e]	(to be tested)

[a]The indicated plasmids and selection media are designed for screening library cotransformants that have been selected initially on SD/-His/-Trp(-Ura) non-induction medium (LexA system). For information on the cloning vectors, see Appendix A, and for control plasmids, see Appendix C. The media must be lacking Ura to keep selection on the p8op-lacZ reporter plasmid. Include X-gal and BU salts if you wish to perform in vivo whole-plate β-gal assays. For simultaneous cotransformation, all transformations are performed in EGY48 or EGY48[p8op-lacZ]; for the second transformation in the sequential procedure, use the same host strain for the controls and the host strain previously transformed with your BD/bait plasmid for the library transformation.

[b]Control for β-gal assay.

[c]SD containing glucose (rather than galactose + raffinose) may be used; however, colonies grown on SD/Gal/Raff are more directly comparable to those on the library screening plates.

[d]Controls for interaction of two hybrid proteins and for activation of both reporter genes.

[e]True interacting positives should turn blue on this medium (containing X-gal and BU salts). Many of the Leu[+] positives will not turn blue; these might be false positives.

the cell suspension to the complete volume of culture medium.

- If the overnight or 3-h cultures are visibly clumped, disperse the clumps with vigorous vortex-mixing before using them in the next step.
- When you are collecting cells by centrifugation, a swinging bucket rotor results in better recovery of the cell pellet.
- For the highest transformation efficiency (as is necessary for library screening), use competent cells within 1 h of their preparation. If necessary, competent cells can be stored (after Step 11) at room temperature for several hours with a minor reduction in competency.

Procedure

	Transformation Scale	
	<u>LARGE</u>	<u>LIBRARY</u>
1. Inoculate 1 mL of YPD or SD[a] with several colonies, 2–3 mm in diameter.		
2. Vortex-mix vigorously to disperse any clumps.		
3. Transfer cells to a flask containing appropriate volume of YPD or SD[a]:	50 mL	150 mL
4. Incubate at 30°C for 16–18 h with shaking (at 250 rpm) to stationary phase ($OD_{600} > 1.5$).		

5. Transfer overnight culture (enough to produce an OD_{600} = 0.2–0.3) into the appropriate volume of YPD: — 300 mL / 1 L

6. Incubate at 30°C for 3 h with shaking (230–270 rpm). The OD_{600} will be 0.5 ± 0.1.

7. Place cells in 50-mL tubes and centrifuge at 1000× g for 5 min at room temperature (20°–21°C).

8. Discard the supernatant and resuspend cell pellets by vortex-mixing in appropriate volume of sterile TE or H_2O: — 25–50 mL / 500 mL

9. Pool cells in one tube and centrifuge at 1000× g for 5 min at room temperature.

10. Decant the supernatant.

11. Resuspend the cell pellet in appropriate volume of freshly prepared, sterile 1× TE/LiAc: — 1.5 mL[b] / 8 mL[b]

12. Prepare PEG/LiAc solution — 10 mL / 100 mL

13. Add the following to each tube[c] and mix:
    - BD vector construct (bait)[d] — 20–100 µg / 0.2–1.0 mg
    - AD vector construct (library) — 10–50 µg / 0.1–0.5 mg
    - Herring testes carrier DNA — 2 mg / 20 mg

14. Add appropriate volume of yeast competent cells: — 1 mL / 8 mL
    and mix well by vortex-mixing.

15. Add appropriate volume of sterile PEG/LiAc solution: — 6 mL / 60 mL
    and vortex-mix at high speed.

16. Incubate at 30°C for 30 min with shaking (200 rpm).

17. Add appropriate volume of DMSO: — 700 µL / 7.0 mL
    Mix well by gentle inversion or swirling. Do not vortex-mix.

18. Heat shock for 15 min in a 42°C water bath. Swirl occasionally to mix.

19. Chill cells on ice for 1–2 min.

20. Centrifuge cells for 5 min at 1000× g at room temperature.

21. Remove the supernatant.

22. Resuspend cells in appropriate volume of 1× TE: — 1.0 mL or 10 mL[e] / 10 mL

23. Proceed to next section for plating.

ᵃUse SD/ Trp (GAL4 system), or SD/ His (Lex A system), when performing the second transformation in a sequential library transformation protocol.

ᵇFor simultaneous library cotransformations only: remove two 100-µL aliquots of competent cells to perform control transformations with the β-gal positive control, and the BD/p53 + AD/TAg two-hybrid positive control (Table 4).

ᶜUse a sterile 50-mL conical tube or 500-mL centrifuge tube for large-, and library-scale transformations, respectively.

ᵈFor simultaneous cotransformation, a molar ratio of 2:1 (BD vector:AD vector) is recommended for optimal efficiency. For sequential transformations, add either the BD vector construct or the AD vector construct (not both).

ᵉUse 1.0 mL for simultaneous cotransformation; 10 ml for the second transformation in a sequential transformation protocol.

6.3 Plating and Screening Transformation Mixtures

Important:

- Refer to Tables 4 and 5 for the selection media needed for library plating and screening and control plates.

- GAL4 system users: Before screening a library, be sure to titer the optimal concentration of 3-AT needed to eliminate background growth on SD-His selection plates if your host strain is leaky for *HIS3* expression.

- For best results in the *HIS3* jump-start library screening protocol, use three sets of SD/-His plates: one set with the optimal concentration of 3-AT; one set with a 10–15 mM higher 3-AT concentration (to control for background growth due to the extremely high plating densities); and one set with a 10–15 mM lower 3-AT concentration (for improved growth of weak positives).

- LexA system users: Be sure to use (Gal/Raff) induction media when screening for two-hybrid interactions and for the control plates. Also, if you wish to simultaneously assay for expression of the *lacZ* and *LEU2* reporter genes, include X-gal and BU salts in the SD induction medium (Chapter 2). Do not spread X-gal on the surface of the medium because localized variations in X-gal concentration can give inaccurate indications of relative interaction strength.

- To obtain an even growth of colonies on the plates, continue to spread the transformation mixtures over the agar surface until all liquid has been absorbed. Alternatively, use 5-mm sterile glass beads (5–7 beads per 100-mm plate; 7–9 beads per 150-mm diameter plate) to promote even spreading of the cells.

Procedure

1. Plate transformation mixtures:

   a. For small-scale cotransformations using two types of plasmids (e.g., the control cotransformations), spread 200 µL per 100-mm plate when selecting for both plasmids.

   b. For all other small-scale transformations and cotransformations, spread 100 µL on each 100-mm plate.

   c. For large- and library-scale transformations:

- Spread 100 µL of a 1:1000, 1:100, and 1:10 dilution on 100-mm plates for cotransformation efficiency controls:
  GAL4 system: SD/-Leu/-Trp
  LexA system: SD/-His/-Trp

- Spread 1 µL (diluted in 100 µL of H_2O) on 100-mm plates to check the transformation efficiency of each plasmid:
  GAL4 system: SD/-Leu and SD/-Trp
  LexA system: SD/-His and SD/-Trp

- Spread the remaining transformation suspension on 150-mm plates (200 µL per plate):
  GAL4 system (direct screening): SD/-His/-Leu/-Trp/+3-AT
  GAL4 system (HIS3 jump-start): SD/-Leu/-Trp and add Step 4 below to the procedure.
  LexA system (standard): SD/-His/-Trp and add Step 4 below to the procedure.

2. Incubate plates, upside down, at 30°C until colonies appear.

3. Calculate the transformation efficiency and estimate the number of clones screened as described in the next section.

4. For the LexA system and the GAL4 system *HIS3* jump-start procedures: Harvest the library transformants from all the 150-mm plates as follows:

   a. Place plates at 4°C for 3–4 h to harden.

   b. Add 1–5 mL of TE buffer (pH 7.0) to the surface of each plate. Carefully scrape the colonies into the liquid using a heat-bent, sterile Pasteur pipette. Combine all liquid containing scraped colonies into a single, sterile 50-mL tube and vortex-mix to resuspend the cells. (Be careful to avoid contamination.) These are the amplified, pooled library transformants.

   Note: Use the minimum amount of buffer necessary to cover the plate and scrape up the colonies. If the combined volume is too large to handle easily, you may wish to reduce the volume as follows: centrifuge the cell suspension for 5 min at 1000× *g*, remove all but 25–50 mL of the supernatant, and resuspend the cells in the remaining liquid by vortex-mixing thoroughly.

   c. To create a glycerol stock of your amplified yeast library cotransformants, add an equal volume of sterile 65% glycerol/$MgSO_4$ solution.

   d. Divide into 1-mL aliquots and store at -80°C. The glycerol stock of amplified yeast cotransformants can be stored at 4°C for a week and at -80°C up to 1 y.

   e. Titer the glycerol stock onto SD/-Leu/-Trp (GAL4 system) or SD/-His/-Trp (LexA system). Incubate plates at 30°C for 3–4 days or until colonies are easy to count. Calculate the cfu/µL of library.

   f. Plate the amplified yeast cotransformants at a high density (0.5–2 × 10⁶ cfu per 150-mm plate) on the appropriate selection medium. Plate 5–10 times the number of original cotransformants calculated from the number of colonies growing on the SD/-Leu/-Trp (GAL4 system) or SD/-His/-Trp (LexA system) control plates. To compensate for possible errors in the amplified library titer, plate 0.5 × 10⁶ cfu on some plates and 2 × 10⁶ cfu on others. Also, plate the appropriate controls for comparison (Tables 4 and 5).

GAL4 system: SD/-His/-Leu/-Trp/+3-AT (if using a host strain with *ADE2* reporter, see Chapter 5 for media recommendations)

LexA system: SD/Gal/Raff/-His/-Leu/-Trp (if using EGY48)

LexA system: SD/ Gal/Raff/-His/-Leu/-Trp/-Ura (if using EGY48[p8op-lacZ])

5. Incubate plates upside down at 30°C until colonies appear. LexA system users: If you have plated transformants on X-gal containing medium, check plates every 12 h (up to 96 h) for development of blue color (Chapter 2).

6. Compare the results on your screening plates to the positive and negative controls performed in parallel. Refer to Tables 4 and 5 for expected results. Choose positive transformant colonies for further analysis.

Notes:

- After 2–3 days, some His$^+$ (or Leu$^+$) colonies will be visible on the library screening plates, but plates should be incubated for 5–10 days to allow slower growing colonies (i.e., weak positives) to appear. Ignore the small, pale colonies that may appear after 2 days but never grow to >2 mm in diameter. True His$^+$ (or Leu$^+$) colonies are robust and can grow to >2 mm (on medium without X-gal). Colonies grown on X-gal containing medium will be somewhat smaller than those grown without X-gal.

- On plates containing X-gal (LexA system), you can distinguish variations in blue color intensity of the positive colonies by three days after plating; the color variations may or may not reflect the relative strength of the two-hybrid interaction. By five days, all positive colonies will appear to have the same color intensity. Beyond five days, there is an increased risk of false positive results due to the sensitivity of the *lacZ* reporter system.

- Not all of the transformants surviving the two-hybrid selection will be true two-hybrid positives. Screening for expression of the second reporter gene (*lacZ*) can eliminate the most common class of false positives.

7. Assay for β-gal activity.

- GAL4 system: Streak out or replica plate His$^+$ colonies on fresh SD/-Trp/-Leu /-His/+3-AT master plates and grow for 2–4 days at 30°C until colonies are at least 1 mm in diameter. At this point, you can perform a β-gal filter assay (Chapter 2) on the fresh colonies using a sterile filter. After lifting the colonies for the β-gal assay, place the master plates at 30°C for 1–2 days to allow the colonies to regrow. Then seal the master plates with Parafilm M™ and store at 4°C for up to 3–4 weeks.

- LexA system: If you have not included X-gal in the medium, assay the Leu$^+$ colonies for β-gal activity using the colony-lift filter assay after replica plating them to fresh SD/Gal/Raff/-His/-Leu/-Trp/-Ura induction plates and incubating as above. Take note of Leu$^+$, LacZ$^+$ colonies and collect samples on SD/ -His/-Trp/-Ura master plates without X-gal. SD/Gal/Raff/-His/-Leu/-Trp/-Ura induction medium may be used for the master plates, although the colonies will grow more slowly on induction medium than on SD plates containing glucose. More importantly, the continuous expression of some AD fusion proteins may be deleterious to the cells. Incubate plates at 30°C for 4–6 days, then seal the master plates with Parafilm M and store at 4°C for up to 3–4 weeks.

8. Restreak positive colonies

   Often more than one AD/library fusion plasmid will be present in each His+ (or Leu+), LacZ+ colony, which can complicate the analysis of putative positive clones. Restreak the original positive colonies on SD/-Leu/-Trp (or SD/-His/-Trp/-Ura) plates at least once (preferably 2–3 times) to allow segregation of some of the AD/library plasmids to take place while maintaining selective pressure on the two plasmid types (i.e., BD/bait and AD/library) and p8op-lacZ, in the LexA system. Incubate plates at 30°C for 4–6 days.

9. GAL4 system users: Replica plate or transfer well-isolated colonies to SD/-His/-Leu/-Trp to verify that they maintain the correct phenotype. Re-assay colonies for β-gal activity using the colony-lift filter assay.

   LexA system users: Replica plate or transfer well-isolated colonies to SD/Gal/Raff/-His/-Leu/-Trp/-Ura induction medium to verify that they maintain the correct phenotype. Include X-gal and BU salts in the medium if you wish to perform the whole-plate β-gal assay; otherwise, transfer colonies from the induction medium to filters for the colony-lift β-gal filter assay.

10. Collect the restreaked and retested His+ (or Leu+), LacZ+ colonies on appropriate master plates (without X-gal) in a grid fashion. Incubate plates at 30°C for 4–6 days. After colonies have grown, seal the plate with Parafilm M and store at 4°C for up to 4 weeks.

6.4 Calculating Cotransformation Efficiency and Number of Clones Screened

1. To calculate the cotransformation efficiency, count the colonies (cfu) growing on the SD/-Leu/-Trp (GAL4 systems) or SD/-His/-Trp (LexA system) dilution plate that has 30–300 cfu.

$$\frac{\text{cfu} \times \text{total suspension vol. (\mu L)}}{\text{Vol. plated (\mu L)} \times \text{dilution factor} \times \text{amt. DNA used (\mu g)*}} = \text{cfu/\mu g DNA}$$

   *In a cotransformation, this is the amount of limiting plasmid, i.e., the lesser of the two plasmids, not the total amount of DNA.

2. To estimate the number of clones screened (or amplified, in the case of a *HIS3* jump-start or LexA protocol):

   cfu/μg × amt. of library plasmid used (μg) = # of clones screened

   Example calculation:

   • 100 colonies grew on the 1:100 dilution transformation efficiency control plate (dilution factor = 0.01)

   • resuspension volume was 10 mL

   • amount of library plasmid used = 100 μg

$$\frac{100 \text{ cfu} \times (10 \text{ mL} \times 10^3 \text{ \mu L/mL})}{(100 \text{ \mu L} \times 0.01) \times 100 \text{ \mu g}} = 1 \times 10^4 \text{ cfu/\mu g}$$

   • In this example, 1×10^6 clones were screened (or amplified).

   Note: If, in your library transformation, you screened $<10^6$ clones, you may wish to repeat the transformation using more DNA. Calculate the amount of DNA to use in the repeat transformation as follows:

$$\frac{10^6 \text{ clones}}{(\text{observed \# of clones/ amt. DNA used})} = \mu g \text{ DNA needed}$$

6.5 Troubleshooting Yeast Transformation

A critical factor in performing a successful two-hybrid library screening is obtaining the highest possible transformation efficiency. The transformation efficiency should be at least 10^3 cfu/μg for simultaneous, and 10^4 cfu/μg for sequential, library transformations.

If you had trouble with simultaneous cotransformation, but the transformation with the AD library plasmids alone gave a transformation efficiency of $\geq 5 \times 10^4$ cfu/μg and with the bait plasmid alone gave $\geq 10^5$ cfu/μg, you may have to switch to sequential transformation. If one or both of the separate plasmids gave transformation efficiencies of $<5 \times 10^4$ cfu/μg for the AD library, or $<10^5$ for the bait plasmid, repeat the experiment using more of the plasmid that had the low transformation efficiency. Check the purity of the DNA and, if necessary, repurify it by ethanol precipitation before using it again. If you are not already doing so, we strongly recommend using the pretested and optimized YEASTMAKER Carrier DNA, which is available separately or as part of the YEASTMAKER Yeast Transformation System (CLON-TECH).

Another approach is to repeat the transformation, this time including a "recovery" period after the heat shock. To provide a recovery period, perform the simultaneous cotransformation as described, but add the following steps after step 21:

22. Resuspend cells in 1.0 L of YPD medium for a library-scale, and 100 ml for a large-scale, transformation.

23. Incubate cells for 1 h at 30°C with shaking at 230 rpm.

24. Pellet cells by centrifuging at 1000× g for 5 min at room temperature. Remove supernatant.

25. Continue protocol from step 22.

6.6 Eliminate False Positives and Verify the Two-Hybrid Interaction

A flow chart of suggested steps to manage your putative positive clones is shown in Figure 3. In cases where many positive colonies are obtained, it may be helpful to sort them using a rapid insert-screening procedure such as PCR amplification (8, 22). After the clones have been sorted into groups, a representative of each unique type is then analyzed to eliminate false-positives. A significant proportion of the positive candidates may contain AD/library proteins that activate both the *lacZ* and *HIS3* reporters whether or not the specific DNA-BD/target protein is present (4). To eliminate these false positives, the candidate AD/library plasmids are subsequently tested for reporter gene activation with various bait constructs or the parental DNA-BD plasmid via cotransformation or yeast mating. Truly positive AD/library plasmids activate both reporter genes only when the original bait plasmid is also present. In the LexA system and in a modified GAL4-based system, yeast mating may be used to significantly reduce the time—and the number of transformations—required to demonstrate specificity of the interaction when many clones are being analyzed (Chapter 5; Reference 6,13). To obtain data on the relative strength of the two-hybrid interactions, perform quantitative β-gal assays on liquid cultures using o-nitrophenyl-β-D-galactopyranoside (ONPG; Sigma Chemical; Catalog No. N-1124), chlorophenol red-β-D-galactopyranoside (CPRG; Roche Molecular Biochemicals; Catalog No. 884-308), or Galacton-Star as the substrate (Chapter 2).

7. CONCLUDING REMARKS

The classic GAL4 and LexA two-hybrid systems described in this chapter and in Chapter 2 have been in widespread use for several years. There are literally hundreds of publications describing successful two-hybrid library screens using these systems, particularly the original GAL4-based system developed by Stan Fields and coworkers. If you are inexperienced at working with yeast and two-hybrid screenings, it makes sense to use a system that is relatively simple and has repeatedly been proven to work in real-life research endeavors. If you carefully choose your fusion library and bait protein, you have an excellent chance of uncovering novel and/or known protein–protein interactions.

The most challenging aspects of using any two-hybrid system for library screening are: *(i)* the requirement for a large amount of library plasmid DNA (100–500 μg) to screen one million independent clones; *(ii)* minimizing the occurrence of false positives; *(iii)* sorting and managing the sometimes very large number of positive clones; and *(iv)* minimizing the potential to obtain false negative results. Each point is addressed separately below:

(i) The requirement for the large amount of library plasmid DNA means that you will almost certainly have to amplify your library before you prepare the plasmid. The only way we know of to get around this is to use commercially available, pre-transformed two-hybrid libraries, which are appropriate yeast host cells already transformed with the library DNA and ready to use in screening. Pretransformed libraries are convenient and easy to use because they enable you to perform library screening using yeast mating (as described in the Chapter 5), rather than a large- or library-scale transformation.

(ii) The occurrence of false positives can be an especially bothersome problem in the classic two-hybrid systems that rely on the *HIS3* interaction reporter due to leaky expression of the reporter gene in many of the common host cells. If you are using a leaky *HIS3* reporter strain, it is important to check the selection medium and make sure you have added the proper amount of 3-AT. New host strains have since been developed that utilize two nutritional reporter genes (*HIS3* and *ADE2*), which virtually eliminate the occurrence of this type of false positive (see Chapter 4).

(iii) Some library screens will produce many (>30) initially positive transformants, even if you have taken steps to reduce background growth due to leaky reporter expression. An abundance of positive clones is especially likely if you use the *HIS3* jump-start protocol (GAL4 system) or the standard LexA system for library screening, due to the nonselective amplification step. Many of the positive clones (>80%) may contain the same AD/library insert due to an apparent population bias toward clones exhibiting strong activation of the interaction reporter at the expense of those with weak or transient interactions. To save time in the analysis of the clones, you may wish first to eliminate duplicate colonies that bear the same AD/library plasmid. Many inserts can be analyzed concurrently using PCR ampli-

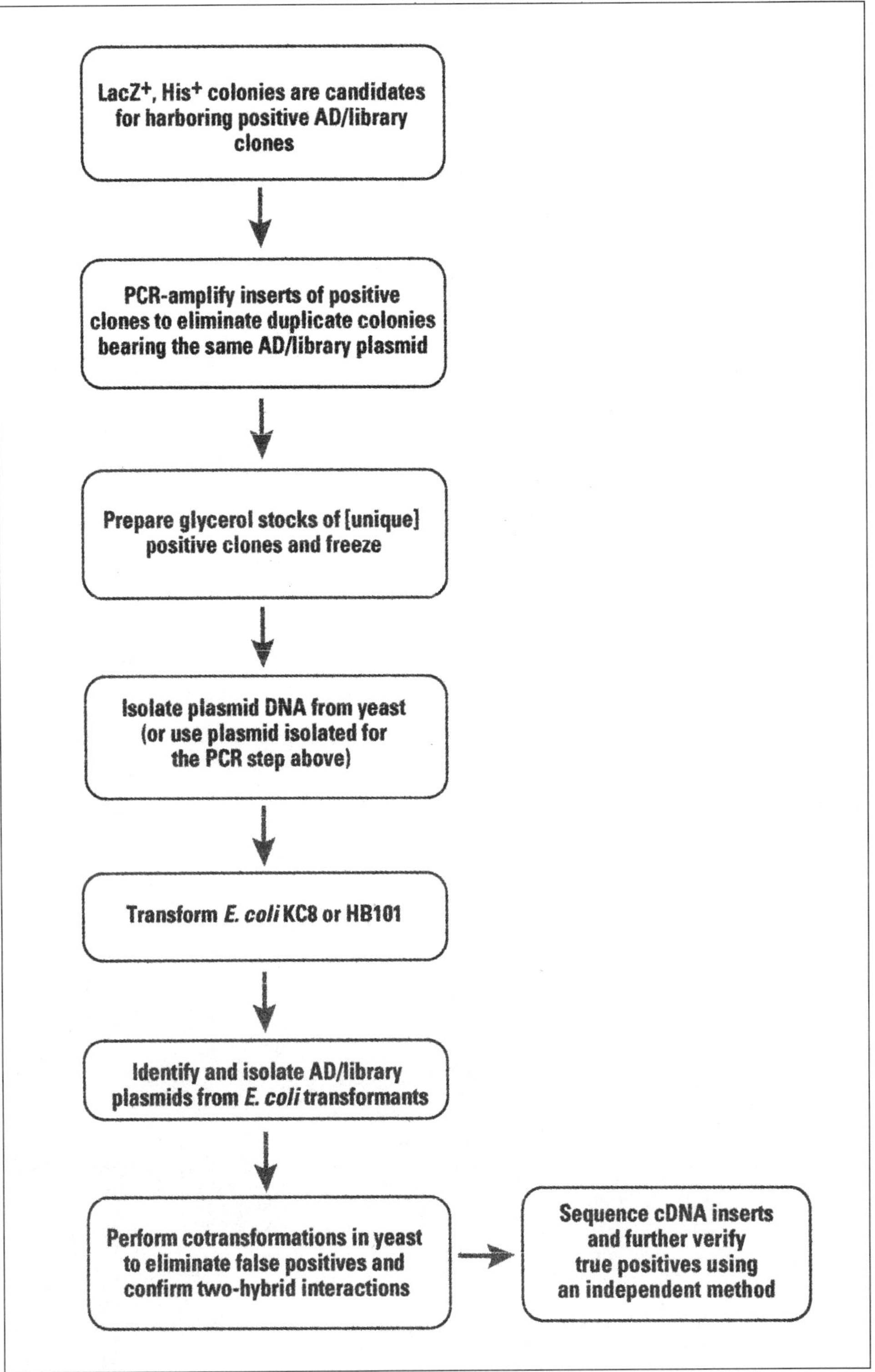

Figure 3. Suggested strategy for analyzing and verifying putative positive clones. A method for rescuing the AD/library plasmid by transforming *E. coli* KC8 or HB101 and applying a nutritional selection is described in Reference 8.

fication followed by restriction enzyme digestion and gel electrophoresis (8).

(iv) The failure to detect interactions between two proteins that normally interact in vivo (i.e., false negative results) could be due to a number of factors.

(1) If high-level expression of one or both of the hybrid proteins is toxic to the cell, transformants will not grow or will grow very slowly on the library screening plates. For this reason, we recommend that you check for cell toxicity on appropriate control plates before using a bait plasmid in a library screening. Sometimes truncation of one of the hybrid proteins will alleviate the toxicity and still allow the interaction to occur. Alternatively, you can try using vectors that express lower levels of the fusion proteins, such as pGBT9 (a DNA-BD vector), and pGAD424, pGAD GL, or pGAD10 (all AD vectors) (19).

(2) If transformation efficiency of one or both plasmids is too low, you may not be screening a sufficient number of library cotransformants. This can be critical, especially if a rare transcript in the source tissue encodes the interacting protein of interest. Chapter 2 gives basic advice for optimizing transformation efficiencies in a small-scale yeast transformation; this advice is especially important when screening a library.

(3) If one of the following situations is occurring, it may interfere with the ability of the two hybrid proteins to interact: the hybrid proteins are not stably expressed in the host cell; the fused GAL4 (or B42AD) domains occlude the site of interaction; the hybrid protein folds improperly; the hybrid protein is not posttranslationally modified as it would be in its native environment; or the hybrid protein cannot be localized to the yeast nucleus. Depending on the nature of the problem, it may help to construct hybrids containing different domains of the target protein, use an appropriately modified yeast two-hybrid system, or use a two-hybrid system adapted for screening in mammalian cells. See Chapter 2 for further information.

(4) Some types of protein–protein interaction may not be detectable in a GAL4-based system, but may be detected in a LexA system, and vice versa. Finally, it is also possible that some protein–protein interactions are not detectable using any type of two-hybrid assay.

REFERENCES

1.**Allen, J.B., M.W. Wallberg, M.C. Edwards, and S.J. Elledge.** 1995. Finding prospective partners in the library: the yeast two-hybrid system and phage display find a match. TIBS *20*:511-516.
2.**Ausubel, F.M., R. Brent, R.E. Kingston, D.D. Moore, J.G. Seidman, J.A. Smith, and K. Struhl.** 1994. Vol. 1, Chap. 5. *In* Current Protocols in Molecular Biology. John Wiley & Sons.
3.**Bai, C. and S.J. Elledge.** 1997. Searching for Interacting Proteins with the Two-Hybrid System I. *In* P.L. Bartel and S. Fields (Eds.), The Yeast Two-Hybrid System. Oxford University Press, New York.
4.**Bartel, P.L., C.-T. Chien, R. Sternglanz, and S. Fields.** 1993. Using the Two-Hybrid System to Detect Protein-Protein Interactions, p. 153-179. *In* D.A. Hartley (Ed.), Cellular Interactions in Development: A Practical Approach. Oxford University Press, Oxford.
5.**Bartel, P.L., C.-T. Chien, R. Sternglanz, and S. Fields.** 1993. Elimination of false positives that arise in using the two-hybrid system. BioTechniques *14*:920-924.
6.**Bendixen, C., S. Gangloff, and R. Rothstein.** 1994. A yeast mating-selection scheme for detection of protein-protein interactions. Nucleic Acids Res. *22*:1778-1779.
7.**Chien, C.T., P.L. Bartel, R. Sternglanz, and S. Fields.** 1991. The two-hybrid system: A method to identify and clone genes for proteins that interact with a protein of interest. Proc. Natl. Acad. Sci. USA *88*: 9578-9582.

8. **CLONTECH Yeast Protocols Handbook.** 1999. CLONTECH Laboratories, Inc. www.clontech.com.

9. **Dalton, S. and R. Treisman.** 1992. Characterization of SAP-1, a protein recruited by serum response factor to the c-fos serum response element. Cell *68*:597-612.

10. **Durfee, T., K. Becherer, P.L. Chen, S.H. Yeh, Y. Yang, A.E. Kilbburn, W.H. Lee, and S.J. Elledge.** 1993. The retinoblastoma protein associates with the protein phosphatase type 1 catalytic subunit. Genes Dev. *7*:555-569.

11. **Fields, S.** 1993. The two-hybrid system to detect protein-protein interactions. METHODS: A Companion to Meth. Enzymol. *5*:116-124.

12. **Fields, S. and R. Sternglanz.** 1994. The two-hybrid system: an assay for protein-protein interactions. Trends Genet. *10*:286-292.

13. **Finley, Jr., R.L. and R. Brent.** 1994. Interaction mating reveals binary and ternary connections between Drosophila cell cycle regulators. Proc. Natl. Acad. Sci. USA *91*:12980-12984.

14. **Fritz, C.C. and M.R. Green.** 1992. Fishing for partners. Curr. Biol. *2*:403-405.

15. **Gietz, D., A. St. Jean, R.A. Woods, and R.H. Schiestl.** 1992. Improved method for high-efficiency transformation of intact yeast cells. Nucleic Acids Res. *20*:1425.

16. **Golemis, E.A. and R. Brent.** 1997. Searching for Interacting Proteins with the Two-Hybrid System III. *In* P.L. Bartel and S. Fields (Eds.), The Yeast Two-Hybrid System. Oxford University Press, New York.

17. **Guarente, L.** 1993. Strategies for the identification of interacting proteins. Proc. Natl. Acad. Sci. USA *90*:1639-1641.

18. **Hill, J., K.A. Donald, and D.E. Griffiths.** 1991. DMSO-enhanced whole cell yeast transformation. Nucleic Acids Res. *19*:5791.

19. **Holtz, A.E. and L. Zhu.** 1995. CLONTECHniques X(3):20.

20. **Hopkin, K.** 1996. Yeast two-hybrid systems: more than bait and fish. J. NIH Res. *8*:27-29.

21. **Ito, H., Y. Fukada, K. Murata, and A. Kimura.** 1983. Transformation of intact yeast cells treated with alkali cations. J. Bacteriol. *153*:163-168.

22. **Ling, M., F. Merante, and B.H. Robinson.** 1995. A rapid and reliable DNA preparation method for screening a large number of yeast clones by polymerase chain reaction. Nucleic Acids Res. *23*:4924-4925.

23. **Luban, J. and S.P. Goff.** 1995. The yeast two-hybrid system for studying protein-protein interactions. Curr. Opin. Biotechnol. *6*:59-64.

24. **Luban, J., K.B. Alin, K.L. Bossolt, T. Humaran, and S.P. Goff.** 1992. Genetic assay for multimerization of retroviral gag polyproteins. J. Virol. *66*:5157-5160.

25. **Ma, J. and M. Ptashne.** 1987. A new class of yeast transcriptional activators. Cell *51*:113-119.

26. **McNabb, D.S. and L. Guarente.** 1996. Genetic and biochemical probes for protein-protein interactions. Curr. Opin. Biotechnol. *7*:554-559.

27. **Mendelsohn, A.R. and R. Brent.** 1994. Biotechnology applications of interaction traps/two-hybrid systems. Curr. Opin. Biotechnol. *5*:482-486.

28. **Ruden, D.M.** 1992. Activating regions of yeast transcription factors must have both acidic and hydrophobic amino acids. Chromosoma *101*:342-348.

29. **Ruden, D.M., J. Ma, Y. Li, K. Wood, and M. Ptashne.** 1991. Generating yeast transcriptional activators containing no yeast protein sequences. Nature *350*:250-251.

30. **Sambrook, J., E.F. Fritsch, and T. Maniatis.** 1989. Molecular Cloning: A Laboratory Manual. Cold Spring Harbor Laboratory Press, Cold Spring Harbor, NY.

31. **Schiestl, R.H. and R.D. Gietz.** 1989. High-efficiency transformation of intact cells using single stranded nucleic acids as a carrier. Curr. Genet. *16*:339-346.

32. **Vojtek, A., S. Hollenberg, and J. Cooper.** 1993. Mammalian *Ras* interacts directly with the serine/threonine kinase Raf. Cell *74*:205-214.

33. **Vojtek, A.B., J.A. Cooper, and S.M. Hollenberg.** 1997. Searching for Interacting Proteins with the Two-Hybrid System II. *In* P.L. Bartel and S. Fields (Eds.) The Yeast Two-Hybrid System. Oxford University Press, New York.

34. **Zhu, L., D. Gunn, and S. Kuchibhatla.** 1997. Constructing an Activation Domain Fusion Library. *In* P.L. Bartel and S. Fields (Eds.), The Yeast Two-Hybrid System. Oxford University Press, New York.

4   An Advanced Two-Hybrid System: *ADE2* and Other Advances

Nancianne Knipfer[1], Ann E. Holtz[1], Shao-bing Hua[1,2], and Li Zhu[1,2]
[1]CLONTECH Laboratories, Palo Alto, and [2]Genetastix, San Jose, CA, USA

1. INTRODUCTION

Over the last decade, many improvements have been made to the original yeast two-hybrid system (7). Two critical advancements were the development of a yeast strain that virtually eliminates false positives and the design of DNA-binding domain (BD) and activation domain (AD) vectors that facilitate downstream confirmation of protein interactions. This chapter focuses on the detection of protein interactions using an advanced two-hybrid system.

1.1 Reporter Gene Constructs

Currently, most GAL4-based two-hybrid yeast strains utilize *HIS3* and *lacZ* reporter genes to identify protein interactions. *HIS3* is a sensitive reporter. However, some reporter constructs are leaky due to the inclusion of the native *HIS3* TATA boxes. Leaky *HIS3* expression leads to false candidate clones and high background. By adding a competitive inhibitor such as 3-aminotriazole (3-AT) to the media, leaky *HIS3* expression is reduced; however, 3-AT can also reduce the sensitivity of *HIS3* (11) and kill freshly transformed cells (Holtz and Chan, unpublished). Therefore, a second reporter such as *lacZ* is needed to confirm protein interactions.

In strains such as Y190, *HIS3* and *lacZ* are under the control of the same GAL4-responsive promoter—*GAL1*. Unfortunately, the use of a single promoter leads to two major classes of false positives: those that interact directly with sequences flanking the GAL4 binding site and those that interact with transcription factors bound to specific TATA boxes (2).

To reduce the incidence of false positives, new yeast strains such as PJ69-4A (11) and AH109 (Holtz, unpublished) were constructed. These strains contain three reporters that were integrated at unrelated chromosomal locations. Furthermore, each reporter is controlled by a distinct GAL4-responsive promoter. In strain PJ69-4A, the *HIS3*, *ADE2*, and *lacZ* reporters are under control of the *GAL1*, *GAL2*, and *GAL7* promoters, respectively. In strain AH109, the *HIS3*, *ADE2*, and *lacZ* reporters are under

Yeast Hybrid Technologies
Edited by L. Zhu and G.J. Hannon
© 2000 Eaton Publishing, Natick, MA

control of the *GAL1*, *GAL2*, and *MEL1* promoters, respectively (Figure 1). In both strains, the native *HIS3* promoter was completely replaced by the tightly regulated *GAL1* promoter, which reduces the need to include 3-AT in the medium during a two-hybrid screen (5). Of the three reporters, *ADE2* is extremely stringent and virtually eliminates false positives on its own (11). Screening Ade[+]/His[+] clones for *lacZ* expression further reduces the number of false positives that may arise in a typical screen.

In addition to the *ADE2*, *HIS3*, and *lacZ* reporters, strains AH109 and PJ69-4A contain an endogenous *MEL1* gene (Figure 1). *MEL1* is a GAL4-regulated gene encoding α-galactosidase, a secreted enzyme that can be detected on medium containing the chromogenic substrate X-α-Gal (5-Bromo-4-Chloro-3-Indolyl-α-D-Galactopyranoside) (13). When GAL4 fusion proteins interact, α-galactosidase accumulates in the medium and hydrolyzes X-α-Gal, causing yeast colonies to turn blue (1). In contrast, β-galactosidase activity can only be detected on indicator plates when *lacZ* is expressed from a high-copy plasmid or the cells are lysed.

The *MEL1* gene is found in the genomes of many, but not all, yeast strains. Table 1 lists several two-hybrid yeast strains that were screened for α-galactosidase activity (Holtz, unpublished). The genotype of a two-hybrid yeast strain can be determined by transforming with a plasmid that expresses the wild-type *GAL4* gene (i.e., pCL1), and plating on X-α-Gal indicator plates.

1.2 Vectors Facilitate Downstream Confirmation

Once protein interactions have been identified in a two-hybrid screen, the interactions must be confirmed through independent biochemical methods. A fast way to confirm protein interactions is an in vitro transcription/translation reaction followed by a co-immunoprecipitation. To facilitate these assays, new BD and AD vectors containing T7 promoters and unique epitope tags were constructed (Hua and Holtz, unpublished) (15). Two such vectors, pGBKT7 and pGADT7, are illustrated in Figures 2 and 3, respectively (Holtz and Hua, unpublished).

As illustrated in the vector maps, the T7 promoters are located downstream of the

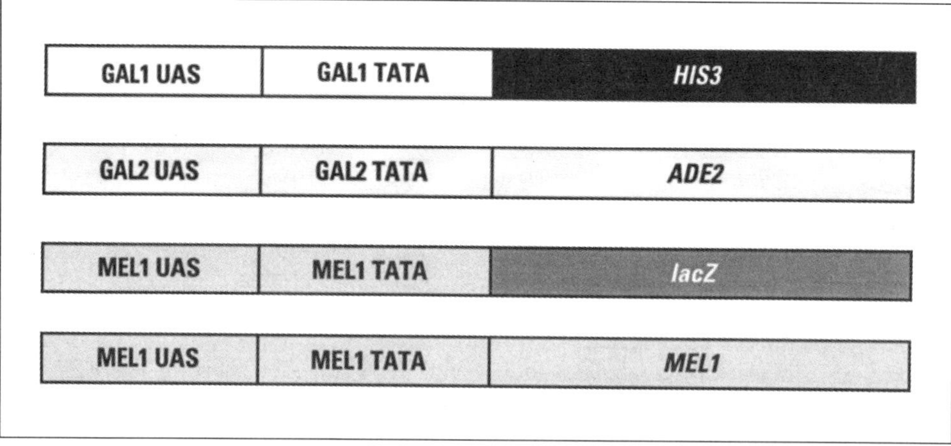

Figure 1. Reporter constructs in yeast strain AH109. Strain AH109 is a derivative of strain PJ69-2A and includes the *ADE2* and *HIS3* nutritional markers and an endogenous *MEL1* gene (11). The *lacZ* reporter gene was introduced into PJ69-2A to create strain AH109. The *HIS3*, *ADE2*, and *MEL1/lacZ* genes are under the control of three completely heterologous GAL4-responsive UAS and promoter elements—*GAL1*, *GAL2*, and *MEL1*, respectively.

Table 1. Yeast Two-Hybrid Strains Tested for α-Galactosidase Activity

Strain	α-Galactosidase Activity	Source
HF7c	–	(5), CLONTECH Laboratories
CG1945	–	CLONTECH Laboratories
SFY526	–	CLONTECH Laboratories
YRG-2	–	Stratagene
Y190	+	(8), CLONTECH Laboratories
Y187	+	(8), CLONTECH Laboratories
AH109	+	CLONTECH Laboratories
PJ69-2A	+	(11), CLONTECH Laboratories
PJ69-4A	+	(11)
J693	+	Courtesy of Dr. R. Rothstein
J692	+	Courtesy of Dr. R. Rothstein

Table 2. Increased Transformation Efficiency of Cells Carrying BD Vectors

Strain + [BD Vector]	Transformation Efficiency of pGADT7 (cfu/μg)
AH109 + [pGBKT7]	4.9×10^5
AH109 + [pAS2-1]	6.8×10^4
Y190 + [pGBKT7]	2.8×10^5
Y190 + [pAS2-1]	1.8×10^4

GAL4 coding sequences, thus the epitope-tagged bait and library proteins are transcribed and translated without the GAL4 domains. Therefore, an in vitro co-immunoprecipitation specifically detects interactions between bait and library proteins.

In pGBKT7 and pGADT7, the bait and library inserts are expressed as fusions with c-Myc and hemagglutinin (HA) epitope tags, respectively. These epitope tags are relatively small and thus unlikely to interfere with protein interactions. After an in vitro transcription/translation reaction and co-immunoprecipitation, protein interactions can be detected using antibodies against the c-Myc and HA epitopes.

1.3 Increased Transformation Efficiency

Recently, Louret and co-workers (14) constructed a BD vector—pODB8—that resulted in increased stability, expression, and transformation efficiency. pODB8 contains the HA epitope tag, which itself caused transcriptional activity (Holtz, unpublished). Therefore, the HA tag was replaced with the c-Myc tag to create pGBKT7. Table 2 demonstrates that yeast strains carrying pGBKT7 are transformed more efficiently than strains carrying other BD vectors such as pAS2-1 (Hua, unpublished). This feature is believed to occur because of the increased stability of pGBKT7 in culture (14). Overall, the higher transformation efficiency facilitates the introduction of AD fusion libraries into yeast, which maintains the complexity of the library and increases the probability of detecting two-hybrid protein interactions.

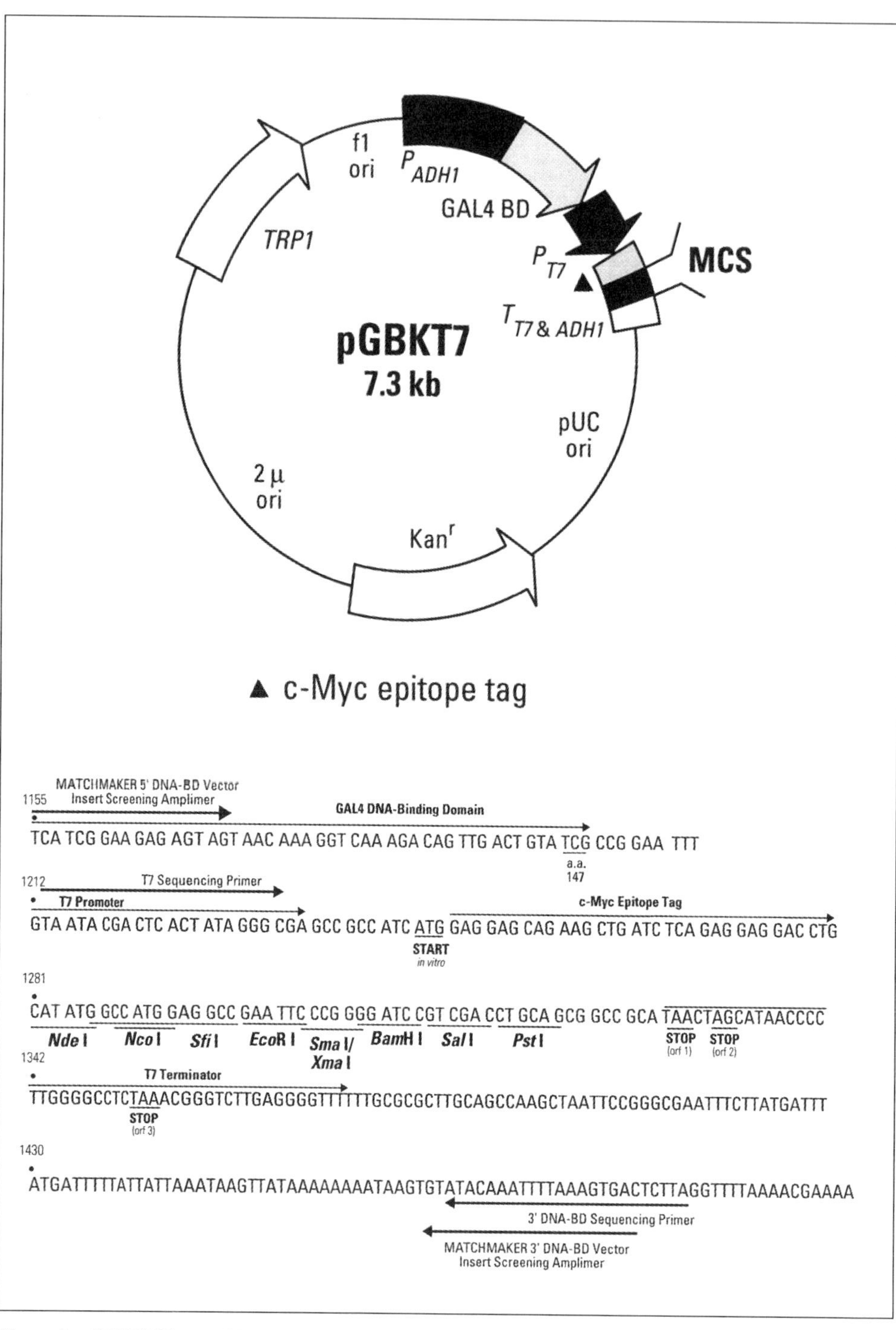

Figure 2. pGBKT7 Map and MCS. Unique restriction sites are in bold. pGBKT7 is used to generate a fusion protein between the GAL4 BD (amino acids 1–147) and another protein of interest. The fusion protein is expressed at high levels from the medium-length *ADH1* promoter. Fusion proteins are also expressed with a c-Myc epitope tag. pGBKT7 includes the T7 promoter between the GAL4 BD and the epitope tag to facilitate in vitro transcription and translation. The high-copy pUC origin of replication is included for amplification in *E. coli*. Also included is a Kan^r marker for selection in *E. coli* and a *TRP1* marker for selection in yeast. Suitable sequencing and PCR primer sites are indicated (CLONECH Laboratories).

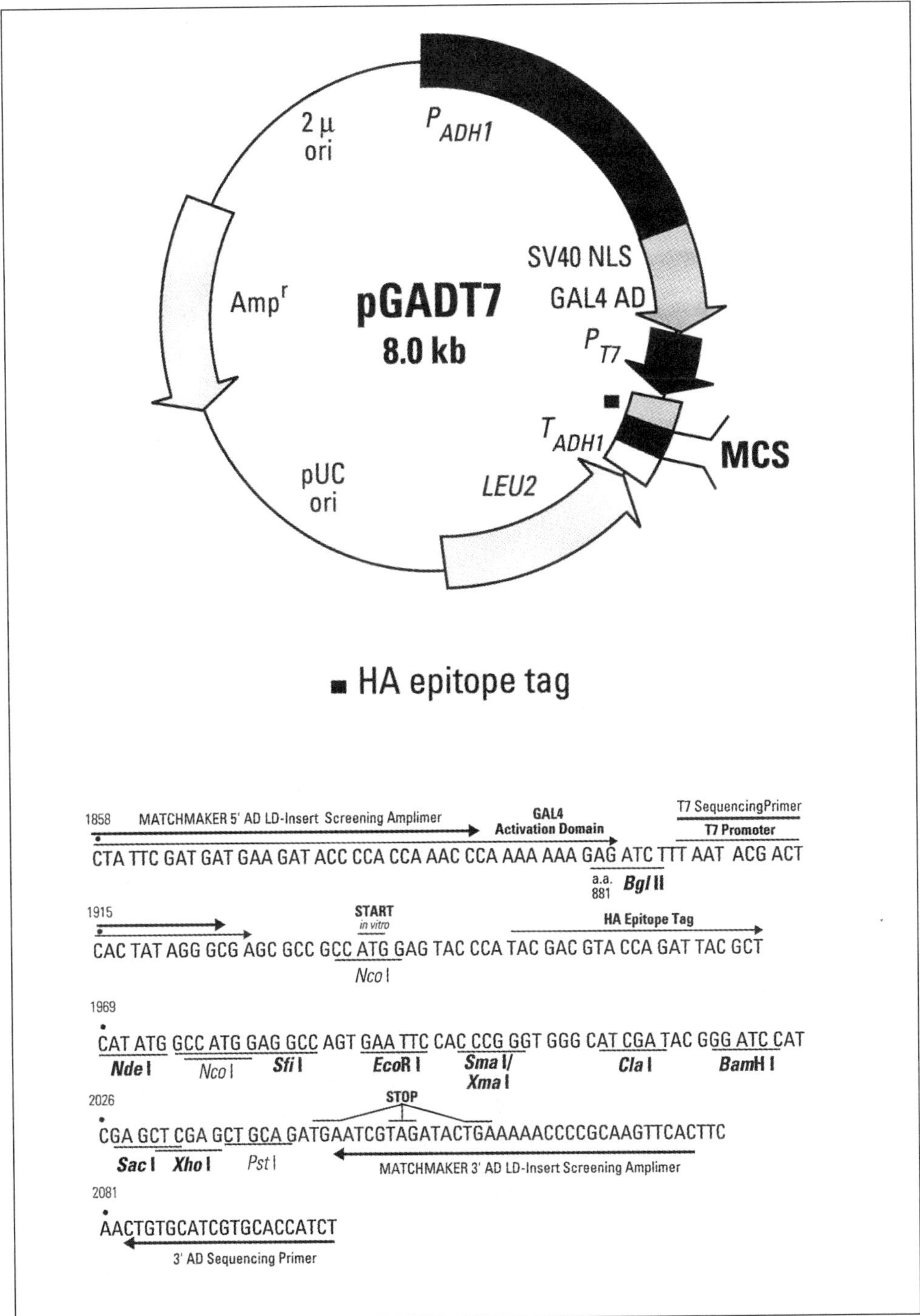

Figure 3. pGADT7 Map and MCS. Unique restriction sites are in bold. pGADT7 is used to generate a fusion protein between the GAL4 AD (amino acids 768–881) and another protein of interest or a cDNA library. The fusion protein is expressed at high levels from the full-length *ADH1* promoter. Fusion proteins are also expressed with a HA epitope tag. pGADT7 includes the T7 promoter between the GAL4 AD and the epitope tag to facilitate in vitro transcription and translation. The high-copy pUC origin of replication is included for amplification in *E. coli*. Also included is an Amp^r marker for selection in *E. coli* and a *LEU2* marker for selection in yeast and bacteria. Suitable sequencing and PCR primer sites are indicated (CLONECH Laboratories).

1.4 Overview of an Advanced Two-Hybrid Screen

As shown in Figure 4, the *ADE2* and *HIS3* reporters can control the stringency of selection during an advanced two-hybrid screen. In a high-stringency screen, library transformations are plated directly onto SD/-Ade/-His/-Leu/-Trp medium to screen for *ADE2* and *HIS3* expression. Using this selection method, virtually all false candidate clones are eliminated; however, low-affinity protein interactions may potentially be missed.

In a low-stringency screen, library transformations are plated onto SD/-His/-Leu/-Trp medium to screen for expression of only *HIS3*. His$^+$ colonies are then replica plated onto SD/-Ade/-His/-Leu/-Trp medium to screen for *ADE2* and *HIS3* expression.

An alternate plating procedure should be performed if there is difficulty identifying positive clones or if the bait protein is expected to interact only very weakly or transiently with other proteins. In this procedure, library transformants are plated onto SD/-Leu/-Trp medium to select for both the bait and the AD/library plasmids but not for the two-hybrid interaction. This selection step provides an initial phase of growth that maximizes plasmid copy number. The increased plasmid copy number results in higher fusion protein levels. This, in turn, improves the chance of detecting AD fusion proteins that interact only weakly or transiently with the bait. Candidate clones are then screened on high- or low-stringency plates.

Although *ADE2* and *MEL1* (or *lacZ*) will eliminate many classes of false positives, the alternate plating procedure typically results in many candidate colonies (up to 1000). For this reason, it is important to optimize the 3-AT concentration needed to control background growth of the reporter strain. Furthermore, this procedure may result in a population preference for clones exhibiting stronger activation of the nutritional reporters, and extra steps may be required to sort the clones into groups before proceeding to confirm protein interactions.

2. TWO-HYBRID YEAST STRAINS AND CONTROL PLASMIDS

2.2 Two-Hybrid Yeast Strains

In an advanced two-hybrid screen, yeast strain AH109 is strongly recommended because it offers four distinct reporters. Furthermore, strains AH109 and Y187 are recommended for mating assays (See Chapter 5). Both strains are *gal4*$^-$ and *gal80*$^-$; this prevents interference of native regulatory proteins with GAL4-responsive promoters during a two-hybrid screen. When a two-hybrid interaction occurs, the strains behave as though they have a *GAL4*$^+$ *GAL80*$^-$ genotype, which confers a constitutive phenotype to reporters regulated by GAL4. For the complete genotypes of strains AH109 and Y187, refer to Appendix D.

2.2.1 Nutritional Requirements

Table 3 details the nutritional requirements of strains AH109 and Y187.

2.2.2 Colony Color and Size

Y187 carries the *ade2-101* mutation. On medium with low amounts of adenine, the colonies will turn pink after a few days and then become darker as the colonies

Table 3. Growth of Two-Hybrid Yeast Strains on Various Selection Media

Strain	SD/-Ade	SD/-Met	SD/-Trp	SD/-Leu	SD/-His	SD/-Ura	YPDA
AH109	–	+	–	–	–	+	+
Y187	–	–	–	–	–	+	+

age. Colonies will grow to >2 mm in diameter. However, small (<1-mm) white colonies will form at a rate of 1%–2% due to spontaneous mutations that eliminate mitochondrial function (9). Avoid these white colonies when inoculating cultures.

In the absence of GAL4, AH109 also exhibits the *ade2-101* phenotype. However, in the presence of protein interactions, expression of *ADE2* eliminates the *ade2-101* phenotype, and thus positive clones form white colonies.

2.2.3 MEL1 *and* lacZ *Reporter Gene Expression Levels*

In a positive two-hybrid screen, strains AH109 and Y187 secrete α-galactosidase, which can be detected on X-α-Gal indicator plates (1). In contrast, the level of β-galactosidase activity in LacZ⁺ transformants is too low to be detected on X-gal indicator plates without cell lysis. Therefore, filter-lift assays or liquid cultures should be used to assay β-galactosidase activity. A modified method is described in Reference 4.

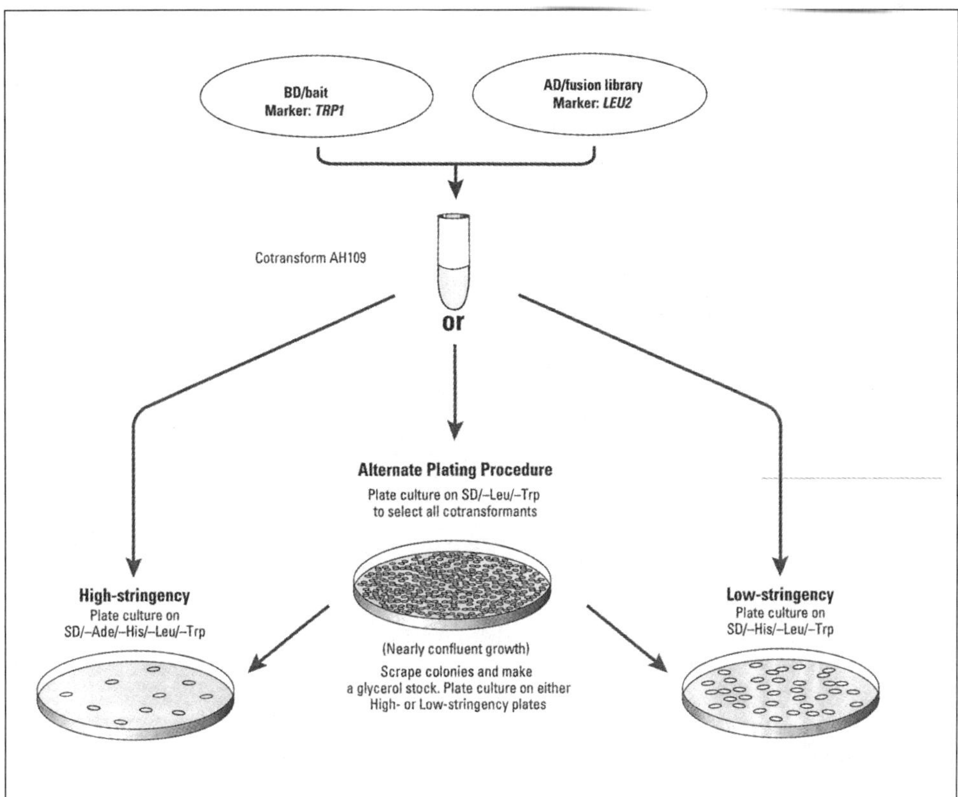

Figure 4. Screening an AD/fusion library using strain AH109.

Note: In Y187, *lacZ* is under the control of the intact *GAL1* promoter, while in AH109 *lacZ* is driven by the weaker *MEL1* promoter. As a result, Y187 exhibits a higher level of induced β-galactosidase activity than AH109 does and should be used in quantitative β-galactosidase assays. Comparisons should only be made between transformants with similar genetic backgrounds.

2.2.4 Leaky **HIS3** *Expression*

3-AT is a competitive inhibitor of the yeast His3 protein (His3p). 3-AT is used to inhibit leaky expression of His3p, and thus to suppress background growth on SD medium lacking His (3,6). In general, the *HIS3* reporter in strain AH109 does not require 3-AT. However, background growth may occur if the BD/bait exhibits weak transcriptional activation on SD/-His/-Trp plates. If background growth occurs, 3-AT should be added to the selection media. To optimize the 3-AT concentration, plate AH109 transformed with the BD/bait plasmid on SD/-His/-Trp plates containing 0, 2.5, 5.0, 7.5, 10.0, 12.5, and 15.0 mM 3-AT. Use the lowest concentration of 3-AT, which after one week allows only small (<1-mm) colonies to grow. A high concentration of 3-AT in the medium can kill freshly transformed cells (Holtz and Chan, unpublished). If you wish to use excess 3-AT to select for only very strong two-hybrid interactions, we recommend using the alternate plating procedure to increase the pool of cDNA candidates being screened.

2.3 Cloning and Control Plasmids

Table 4 describes the selectable markers and epitope tags used in the BD, AD, and control vectors. Unlike many BD and AD vectors, pGBKT7 and pGADT7 express different bacterial selection markers—kanamycin and ampicillin, respectively—which simplifies their independent isolation from *Escherichia coli*.

1. **pGBKT7** encodes a fusion between the GAL4 BD and a bait protein.

2. **pGADT7** encodes a fusion between the GAL4 AD and a library or target protein.

3. **pCL1** encodes the full-length, wild-type GAL4 protein and provides a positive control for both α-galactosidase and β-galactosidase activity.

4. **pGBKT7-53** encodes a fusion between the GAL4 BD and murine p53, which is known to interact with large T-antigen in a yeast two-hybrid screen (10,12).

5. **pGADT7-T** encodes a fusion between the GAL4 AD and SV40 large T-antigen, which is known to interact with p53 in a yeast two-hybrid screen (10,12).

6. **pGBKT7-Lam** encodes a fusion between the GAL4 BD and human lamin C and provides a control for a fortuitous interaction between an unrelated protein and either AD/T-antigen or AD/library protein. Lamin C does not interact with T-antigen and most other proteins (2,16, and S. Fields, personal communication).

3.0 PROTOCOLS

The protocols in this chapter focus on plating and screening transformations using an advanced two-hybrid system. Before beginning, read Chapters 3 and 4 for detailed

Table 4. Advanced Two-Hybrid Vectors

Type	Name	Epitope	Yeast selection	Bacterial selection
Cloning vectors				
BD	pGBKT7	c-Myc	*TRP1*	kanamycin
AD	pGADT7	HA	*LEU2*	ampicillin
Control vectors				
GAL4	pCL1		*LEU2*	ampicillin
BD/p53	pGBKT7-53	c-Myc	*TRP1*	kanamycin
AD/T-antigen	pGADT7-T	HA	*LEU2*	ampicillin
BD/lamin C	pGBKT7-Lam	c-Myc	*TRP1*	kanamycin

protocols on bait and library construction, yeast transformations, and β-galactosidase liquid and filter-lift assays.

3.1 Materials and Reagents

- MATCHMAKER™ Two-Hybrid System 3 (Catalog No. K1612-1; CLON-TECH Laboratories, Palo Alto, CA, USA)
- YPDA or the appropriate SD liquid medium (CLONTECH Laboratories)
- Sterile 1× TE/LiAc (prepare immediately prior to use from 10× stocks.)
- TE buffer or sterile, distilled H_2O
- Appropriate sterile tubes or flasks for the transformation
- Appropriate SD agar plates

    Notes:
    - Prepare the selection media and pour the required number of agar plates in advance.
    - Allow SD agar plates to dry at room temperature for 2–3 days or at 30°C for 3 h prior to plating any transformation mixtures. The presence of moisture droplets on the agar surface can lead to uneven spreading.

- Herring testes carrier DNA

    Notes:
    - Use only high-quality carrier DNA; nicked calf thymus DNA is not recommended.
    - For optimal performance, boil the carrier DNA for 20 min and quickly cool it on ice just prior to use in the transformation.

- Sterile PEG/LiAc solution (prepare immediately prior to use)
- Sterile glass rod, bent pasteur pipette, or 5-mm glass beads for spreading cells on plates
- Glycerol/$MgSO_4$ solution
- X-α-Gal (Catalog No. 8061-1; CLONTECH Laboratories)

Notes:
- Dissolve X-α-Gal at 2 mg/mL in DMSO.
- Store X-α-Gal in the dark at -20°C.

3.2 Protocol for Plating and Screening Transformation Mixtures

General Information

- Refer to Chapters 3 and 4 for tips on successful yeast transformations.
- Refer to Table 5 for library screening and control-transformation selection media.
- Before using the alternate plating procedure, be sure to titer the optimal concentration of 3-AT needed to eliminate background growth.
- To obtain an even growth of colonies on the plates, continue to spread the transformation mixtures over the agar surface until all liquid has been absorbed. Alternatively, use 5-mm sterile glass beads (5–7 beads per 100-mm plate; 7–9 beads per 150-mm diameter plate) to promote even spreading.

Procedure

1. Plate transformation mixtures as indicated in Table 5. Plate small-scale transformations on 100-mm plates and plate large- and library-scale transformations on 150-mm plates.

   For best results when using the alternate plating, use three sets of -His selection plates: one set with the optimal concentration of 3-AT; one set with a 10–15 mM-higher 3-AT concentration to control for background growth caused by extremely high plating densities; and one set with a 5 mM-lower 3-AT concentration to improve the growth of weak positives.

2. Incubate plates at 30°C until colonies appear.

3. If an AD/library was screened, calculate the transformation efficiency and estimate the number of clones screened as described in Chapter 3.

4. **Alternate plating procedure only.** Harvest the library transformants from all plates as follows:

   a. Chill plates at 4°C for 3–4 h.

   b. Add 1–5 mL of TE buffer, pH 7.0, to each plate. Carefully scrape the colonies into the liquid using a heat-bent, sterile Pasteur pipette. Combine all liquid in a single sterile 50-mL tube and vortex-mix to resuspend the cells.

      Note: If the combined volume is too large, reduce the volume as follows: centrifuge the cell suspension for 5 min at 1000× *g*, remove all but 25–50 mL of the supernatant, and vortex-mix to resuspend the cells.

   c. Create a glycerol stock by adding an equal volume of sterile 65% glycerol/MgSO$_4$ solution.

   d. Divide into 1-mL aliquots and store at -80°C. The glycerol stock can be stored at 4°C for a week and at -80°C for up to 1 year.

Table 5. Transformations in an Advanced Two-Hybrid Library Screen

Vectors	Scale[a]	SD Minimal Medium	Amount to Plate (µL)	Phenotype	
				Mel1/LacZ	His/Ade
Control					
pCL1	S	-Leu	100	Blue	+
pGADT7-T	S	-Leu	100	White	-
+ pGBKT7-53	S	-Trp	100	White	-
	S	-Leu/-Trp	200	Blue	+
	S	-Ade/-His/-Leu/-Trp[b]	200	Blue	+
pGADT7-T	S	-Leu	100	White	-
+ pGBKT7-Lam	S	-Trp	100	White	-
	S	-Leu/-Trp	200	White	-
	S	-Ade/-His/-Leu/-Trp[b]	200	White	-
Experimental					
BD/bait +	L	-Leu[c]	100	White	-
AD/library	L	-Trp[c]	100	White	-
	L	**Alternate procedure:** -Leu/-Trp[d,e,f]	100	White	-
	L	**Low:**-His/-Leu/-Trp[b,e]	200	White/Blue	-/+
	L	**High:**-Ade/-His/-Leu/-Trp[b]	200	White/Blue	-/+

[a] S = small scale; L = large scale or library scale.

[b] See text for guidelines on how much 3-AT to add.

[c] To test the transformation efficiency of each plasmid, dilute 1 µL of the transformation with 100 µL of H_2O. Spread 1 µL onto 100-mm SD/-Leu and SD/-Trp plates.

[d] To test the cotransformation efficiency, spread 100 µL of a 1:1000, 1:100, and 1:10 dilution onto 100-mm SD/-Leu/-Trp plates.

[e] Plate at least 1.5–3 times the number of independent colonies.

[f] For best results when using the alternate plate procedure, use three sets of -His selection plates: one set with the optimal concentration of 3-AT; one set with a 10–15 mM higher 3-AT concentration to control background growth caused by extremely high plating densities; and one set with a 5 mM lower 3-AT concentration for growing weak positives.

e. Titer the glycerol stock onto SD/-Leu/-Trp plates (Chapter 4). Incubate plates at 30°C for 3–4 days or until colonies are easy to count. Calculate the cfu/µL of the library.

f. Plate the amplified yeast cotransformants at 0.5–2.0×10^6 cfu per 150-mm plate on either:

   i. **High-stringency plates:** SD/-Ade/-His/-Leu/-Trp or

   ii. **Low-stringency plates:** SD/-His/-Leu/-Trp

Note: If low-stringency plates were used, replica plate resultant colonies to high-stringency plates to eliminate false positives.

To compensate for possible errors in the amplified library titer, plate 0.5×10^6 cfu on some plates and 2×10^6 cfu on others. Also, plate the appropriate controls for comparison (Table 5).

g. Incubate plates at 30°C until colonies appear.

5. Choose Ade+/His+ transformant colonies for further analysis.

Notes:

- After 2–3 days, some Ade+/His+ colonies will be visible on SD/-Ade/-Trp/-Leu/-His plates, but plates should be incubated for 5–10 days to allow weak positives to grow. Ignore the small, pale colonies that may appear after 2 days but never grow to >2 mm in diameter. True His+ colonies are robust and can grow to >2 mm. Ade+ colonies will remain white to pale pink; Ade⁻ colonies will gradually turn reddish-brown and stop growing.

- Not all of the transformants surviving this selection will be true positives. The most common class of false positives can be eliminated by screening for expression of *MEL1* or *lacZ*. Other types of false positives can be eliminated as described in Chapter 6.

6. Replica-plate or streak colonies onto fresh SD/-Ade/-His/-Leu/-Trp master plates and grow for 2–4 days at 30°C until colonies are at least 1 mm in diameter.

7. Screen Ade+/His+ colonies for either α-galactosidase or β-galactosidase activity.

a. α-galactosidase assays

   i. Add 200 µL of X-α-Gal to the top of a 15-cm plate or 100 µL to the top of a 10-cm plate and spread evenly.

   ii. Allow the plate to dry for 15 min at room temperature.

   iii. Streak cells and incubate at the appropriate temperature until the cells turn blue (1–24 h).

   Note: The color is often easier to observe through the bottom of the plate rather than the top of the plate.

b. β-galactosidase assays

   i. Perform a β-galactosidase filter-lift assay (Chapter 3).

   ii. Place the master plates at 30°C for 1–2 days to allow the colonies to regrow. Seal plates with Parafilm M™, and store at 4°C for up to 3–4 weeks.

8. Restreak and re-test positive colonies.

   Often more than one AD/library fusion plasmid will be present in each Ade+/His+/Mel1+ (or LacZ+) colony, which can complicate the analysis of putative positive clones.

a. Restreak the original positive colonies onto SD/-Leu/-Trp plates 2–3 times to allow segregation of the AD/library plasmids. Incubate plates at 30°C for 4–6 days. Reassay colonies for either α-galactosidase or β-galactosidase activity. A mixture of white and blue colonies on X-α-Gal indicator plates indicates segregation.

b. Replica plate or transfer well-isolated colonies to SD/-Ade/-His/-Leu/-Trp/X-α-Gal indicator plates to verify that they maintain the correct phenotype.

9. Transfer Ade+/His+/Mel1+ (or LacZ+) colonies to SD/-Ade/-His/-Leu/-Trp mas-

ter plates in a grid fashion. Incubate plates at 30°C for 4–6 days. After colonies have grown, seal plates with Parafilm M and store at 4°C for up to 4 weeks.

10. Sort colonies to eliminate duplicate AD/library plasmids using one or both of the insert-screening procedures described in Chapter 4.

11. Proceed to eliminate false positives (Chapter 6) and confirm protein interactions in vitro (Chapter 17).

4. DISCUSSION

Performing a yeast two-hybrid screen is a labor-intensive and lengthy task. However, using advanced strains and vectors can significantly reduce the time it requires to identify and confirm protein interactions. Two-hybrid strains, such as PJ69-4A and AH109, have incorporated two important improvements: (*i*) they use the *ADE2* marker and (*ii*) they use three reporters with distinct GAL4-responsive promoters. As a result, the number of false positive protein interactions is virtually eliminated during a two-hybrid screen.

In previous systems, β-galactosidase filter-lift assays were used to confirm protein interactions by the expression of the *lacZ* reporter gene. With strain AH109, the endogenous *MEL1* reporter can be used to quickly confirm protein interactions by screening for α-galactosidase activity on medium containing X-α-Gal.

Finally, the development of advanced two-hybrid vectors containing T7 promoters and epitope tags facilitates downstream confirmation of protein interactions.

REFERENCES

1. **Aho, S., A. Arffman, T. Pummi, and J. Uitto.** 1997. A novel reporter gene *MEL1* for the yeast two-hybrid system. Anal. Biochem. *253*:270-272.
2. **Bartel, P. L., C.-T. Chien, R. Sternglanz, and S. Fields.** 1993. Using the two-hybrid system to detect protein-protein interactions, p. 153-179. *In* D.A. Hartley (Ed.), Cellular Interactions in Development: A Practical Approach. Oxford University Press, Oxford.
3. **Durfee, T., K. Becherer, P.L. Chen, S.H. Yeh, Y. Yang, A.E. Kilburn, W.H. Lee, and S.J. Elledge.** 1993. The retinoblastoma protein associates with the protein phosphatase type 1 catalytic subunit. Genes Dev. *7*:555-569.
4. **Essers, L. and R. Kunze.** 1996. A sensitive, quick and semi-quantitative LacZ assay for the two-hybrid system. Trends Genet. *12*:449-450.
5. **Feilotter, H.E., G.J. Hannon, C.J. Ruddel, and D. Beach.** 1994. Construction of an improved host strain for two-hybrid screening. Nucleic Acids Res. *22*:1502-1503.
6. **Fields, S. and R. Sternglanz.** 1994. The two-hybrid system: an assay for protein-protein interactions. Trends Genet. *10*:286-292.
7. **Frederickson, R.M.** 1998. Macromolecular matchmaking: advances in two-hybrid and related technology. Curr. Opin. Biotechnol. *9*:90-96.
8. **Harper, J.W., G.R. Adami, N. Wei, K. Keyomarsi, and S.J. Elledge.** 1993. The p21 CdK-interacting protein Cip1 is a potent inhibitor of G1 cyclin-dependent kinases. Cell *75*:805-816.
9. **Holm, C.** 1993. A functional approach to identifying yeast homologs of genes from other species. Methods Enzymol. *5*:102-109.
10. **Iwabuchi, K., B. Li, P. Bartel, and S. Fields.** 1993. Use of the two-hybrid system to identify the domain of p53 involved in oligomerization. Oncogene *8*:1693-1696.
11. **James, P., J. Haliaday, and E.A. Craig.** 1996. Genomic libraries and a host strain designed for highly efficient two-hybrid selection in yeast. Genetics *144*:1425-1436.
12. **Li, B. and S. Fields.** 1993. Identification of mutations in p53 that affect its binding to SV40 T antigen by using the yeast two-hybrid system. FASEB J. *7*:957-963.
13. **Liljestrom, P.L.** 1985. The nucleotide sequence of the yeast *MEL1* gene. Nucleic Acids Res. *13*:7257-7268.

14.**Louret, O.F., F. Doignon, and M. Crouzet.** 1997. Stable DNA binding yeast vector allowing high bait expression for use in the two-hybrid system. BioTechniques *23*:816-819.

15.**Yavuzer, U. and C.R. Goding.** 1995. pWITCH: a versatile two-hybrid assay vector for the production of epitope/activation domain-tagged proteins both *in vitro* and in yeast. Gene *165*:93-96.

16.**Ye, Q. and H.J. Worman.** 1995. Protein-protein interactions between human nuclear lamins expressed in yeast. Exp. Cell Res. *219*:292-298.

5

Mating-Based Yeast Two-Hybrid Screening

Ann E. Holtz[1], Rene Chan[2], and Meng Sheng Qiu[3]
*[1]CLONTECH Laboratories, Palo Alto, CA; [2]Department of
Biology, University of California-San Diego, La Jolla, CA;
[3]Department of Anatomical Sciences and Neurobiology,
University of Louisville, Louisville, KY, USA*

1. INTRODUCTION

The yeast two-hybrid system has been proven to be a powerful tool to detect protein–protein interactions. It has been widely used to identify new interacting partners of proteins, to verify the protein interactions implicated by genetic or biochemical studies, and to characterize interaction domains. Many genes and gene regulatory pathways have been identified by this molecular genetic approach. The advantages of this system are apparent: it provides an easy and rapid in vivo binding assay in eukaryotic cells, and the genes encoding the interacting partners are readily available. Moreover, the two-hybrid system can be modified for rapid and high throughput screening procedures, which are becoming increasingly important in the era of functional genomics as thousands of new genes are identified and sequenced each year by the human genome project.

The traditional yeast two-hybrid screening is based on cotransformation or sequential transformation of yeast competent cells with a BD-bait fusion plasmid and an AD-cDNA fusion plasmid (for details, refer to Chapters 3 and 4). Despite the numerous successes with this method, the conventional screening procedure can be fairly labor-intensive. First, the preparation of the plasmid DNA to be used in subsequent transformations of the two-hybrid cDNA library is quite tedious and time consuming. Usually, hundreds of agar plates have to be used to amplify the library cloned in bacteria, and a large amount of plasmid DNA (hundreds of micrograms) has to be extracted for each transformation. Second, the transformation process per se involves multiple steps including preparation of competent cells, lithium acetate treatment, and heat shock. Finally, the transformation efficiency is relatively low and varies greatly with the growth conditions and strain of the yeast cells.

To circumvent the constraints of the traditional yeast two-hybrid screening, an alternative screening approach based on a yeast mating procedure has been recently developed in several labs. The rationale is that *Saccharomyces cerevisiae* cells grow persistently as either haploid cells or as diploid cells. The haploid cells are either **a** or α mating types, depending on which allele (**a** or α) occupies the transcriptionally active mating type locus, *MAT*. When cells of opposite mating types, which secrete different pheromones, detect the opposite cell type's pheromones and then make

Yeast Hybrid Technologies
Edited by L. Zhu and G.J. Hannon

physical contact with each other under nutrient-rich growth conditions, the haploid cells conjugate, or mate, with each other to form **a**/α diploid cells (for detailed mechanisms, see Reviews 14,15, and 16). This approach has been utilized to re-confirm protein–protein interactions previously isolated by conventional two-hybrid screening (8,10). Based on this observation, Bendixen et al. (3) proposed, and others (2,9,11,17) have developed mating-based screening approaches. The bait plasmid and cDNA library are first introduced separately into yeast cells of opposite mating types by transformation; the bait and cDNA plasmids are then brought together into one cell by mating. The resulting diploid cells are selected by the presence of the prototrophic markers from both plasmid types. To vary the library screening stringency, selection for the prototrophic reporters may be imposed either simultaneously or after the recovery of diploid colonies (Figure 3).

There are essentially two types of pretransformed libraries used in library screening by the yeast mating approach: pooled (9,17) or arrayed libraries (2,10,12). Pretransformed libraries are convenient and fast to use because they need only be thawed. Pretransformation provides consistency between library screens because mating rates are more consistent than transformation rates, ensuring that the same population of cDNAs are screened each time. This consistency is what makes large-scale arrayed library screening possible. In this chapter, we describe the procedures for making a pooled pretransformed library and for the subsequent library screening.

2. GENERAL MATERIALS

2.1 Yeast Strains

Most laboratory yeast strains used for conventional screening are stable in mating types and are suitable for mating screening as they carry the necessary selection markers and reporter genes. However, when selecting mating partners, several important factors should be taken into account. First, the mating partners must have non-complementing auxotrophic mutations at the same loci (e.g., *trp1*, *his3*, *leu2*); selection for both of the plasmids' prototrophic markers, which complement the mutations in the diploid cells, eliminates parental haploid cells. Second, the fusion of two yeast cells should bring together at least two independently integrated reporter genes, e.g., pGAL-HIS3 and pGAL-lacZ, ideally with different promoter structures, which will reduce false positives. Third, the mating efficiency of two mating strains should be tested in liquid culture prior to use in library screening. Table 1 lists mating strains that have been sucessfully used for two-hybrid assays.

The choice of strains enables mixing for desireable features such as cyh^R or Met+ or certain reporter constructions (6). We have studied mating between the yeast strain PJ69-2A (P. James, personal communication) and Y187 (10) (Holtz and Chan, unpublished). The strain PJ69-2A provides a tightly regulated pGAL1-HIS3 reporter and a more stringent pGAL2-ADE2 reporter (13). The Y187 provides a pGAL1-lacZ reporter, the strongest *lacZ* reporter construction of the GAL4-based systems. Both strains are also positive for *MEL1*, a naturally occurring GAL4-regulated gene that can be used as a two-hybrid reporter (1). Therefore, diploids formed between PJ69-2A and Y187 can be assayed for alpha-galactosidase, the enzyme encoded by *MEL1*. (See Chapter 4 for how to use alpha-galactosidase.) The combination of different reporters (Figure 1) also lowers the percentage of false positives that survive selection.

90

Table 1. Combinations of Haploid Hosts Used for Sorting or Screening Interactions

	Strain	Mating type	Reporters	Transformation markers	Reference
I.	**PJ69-2A**	a	*ADE2, HIS3,*	*trp1, leu2*	CLONTECH
	Y187	α	*lacZ*	*trp1, leu2*	
II.	**CG1945**	a	*HIS3, lacZ*	*trp1, leu2*	9
	Y187	α	*lacZ*	*trp1, leu2*	
III.	**Y190**	a	*HIS3, lacZ,*	*trp1, leu2*	10
	Y187	α	*lacZ*	*trp1, leu2*	
IV.	**J692**	a	*HIS3, lacZ*	*trp1, leu2*	R. Rothstein,
	J693	α	*HIS3, lacZ*	*trp1, leu2*	personal
					communication
V.	**CBY14a**	a	*HIS3, lacZ*	*trp1, leu2*	3
	CBY14α	α	*HIS3, lacZ*	*trp1, leu2*	
VI.	**CBY12a**	a	*LEU2, lacZ*	*trp1, his3*	3
	CBY12α	α	*LEU2, lacZ*	*trp1, his3*	
VII.	**EGY48**	a	*LEU2, lacZ*[1]	*trp1, his3*	17
	YPH499	α		*trp1, his3*	

[1]lacZ reporter is on a plasmid, selected by URA3 prototrophy.

2.2 Libraries

An appropriate tissue two-hybrid cDNA or genomic library should be constructed and amplified in bacteria according to the manufacturer of your library construction kit. Alternatively, one can purchase a premade two-hybrid library from a commercial source. Approximately 0.5–1.0 mg of purified library plasmid DNA will be required to transfer the library to the yeast host strain.

The quality of a yeast mating library is critical for efficient screening of protein interactions. Ideally, every independent cDNA in the original bacterial library should be represented in the yeast library. This requires a high transformation efficiency of the yeast and a large number of independent transformants. Many pretransformed libraries are commercially available (CLONTECH Laboratories, Palo Alto, CA, USA). Such pretransformed libraries will allow one to carry out library screening without going through the time consuming library construction procedures.

3. PROTOCOLS

3.1 Protocol for Construction of a Mating-Ready Yeast Library

Methods

Construct a cDNA library.

Construct a two-hybrid library according to the instructions from the manufacturer of the library construction kit. Alternatively, acquire a premade two-hybrid library from either a commercial or academic source.

Plasmid DNA isolation from bacteria.

(*a*). In bacteria, amplify at least 10^7 colonies or 2–3 times the number of original independent clones in the library on 15-cm LB agar plates, with the appropriate antibiotic selection, at an inoculum density of 20 000–40 000 cfu/plate. Amplification on plates minimizes the loss of low abundance cDNA clones. Harvest the amplified library by scraping the bacteria off the agar plates into 5 mL LB per plate. Isolate the two-hybrid library plasmids from the bacterial host according to standard plasmid isolation protocols. DNA yields of the library preparation vary depending on the replication origin of the vector. Plasmid yields of approximately 1 mg and of approximately 3–5 mg per million colonies amplified are the norm of plasmids with pBR322- and pUC-derived origins of replication, respectively.

(*b*). Alternatively, inoculate the primary library into 300 mL of LB with double the normal amount of antibiotic for that vector in a 2-L flask to an $OD_{600}=0.1$. Allow the culture to grow to a density of $OD_{600}=0.4$ and add chloramphenicol to 20–50 µg/mL to stop protein synthesis. Continue to incubate the culture for another 12–16 h and isolate plasmid. This procedure is significantly easier than a plate amplification and balances the risk of biasing the library representation in liquid culture by limiting the number of culture doublings. The culture growth is stopped in mid-log phase, but the chloramphenicol amplification yields enough DNA for a single yeast library transformation. This pro-

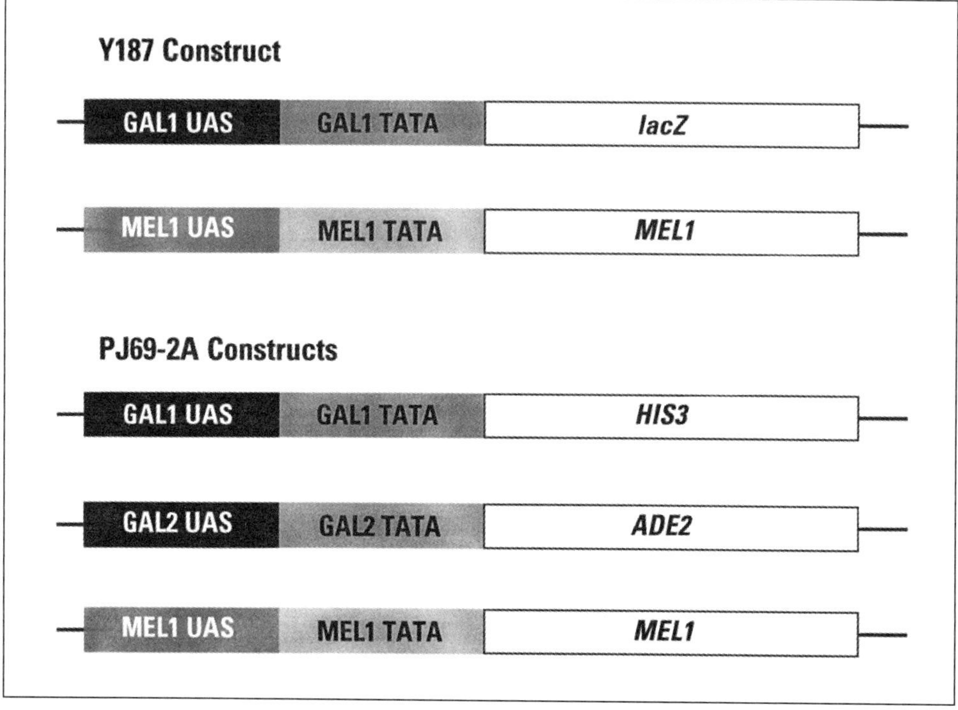

Figure 1. Combining in diploids desirable features from haploid strains. Each haploid strain provides reporter constructions that are used to screen protein–protein interactions. In the example combination PJ69-2A and Y187, the PJ69-2A provides the stringent HIS3 and more stringent ADE2 prototrophic reporters. Y187 provides a strongly induced version of the *lacZ* reporter.

cedure should be used with caution for libraries converted from lambda phage due to the risk of residual phage that could lyse the entire culture.

Transferring the library into yeast.

For an overview, see Figure 2.

(*a*). Transform MATα yeast competent cells made from 600 mL of YPDA culture (log phase: OD_{600} approximately 0.6–0.8) with 200–400 micrograms of library DNA using the LiAc/PEG method (see Chapter 3). Assuming a transformation efficiency between 1×10^4 to 1×10^5 cfu/microgram, between 2×10^6 to 4×10^7 yeast transformants should be recovered.

(*b*). After the LiAc/PEG treatment and DMSO/heat shock steps, resuspend the transformed yeast cells in synthetic dropout (SD) liquid medium and spread on SD plates (SD-Leu if using pGAD or pACT series vectors) using approximately 9–13 sterile glass beads per plate (5 mm diameter; Fisher Scientific, Pittsburgh, PA, USA; Catalog No. 11-312C). Up to 3×10^4 colonies can grow on one 15-cm plate. The number of colony forming units plated will depend on the transformation efficiency, which is estimated from empirical experience.

(*c*). Invert plates (leaving beads to drop onto the petri plate covers for use, later, in harvesting) and incubate at 30°C for 3–4 days to allow full growth of individual colonies.

(*d*). Harvest yeast cells in 3 mL of SD-Leu medium by gently rolling the glass beads around the reverted plates. It is preferable to have at least 10^7 transformants per library. Several pools may be needed if the transformation efficiency was low.

(*e*). Combine yeast cells and add glycerol to a final concentration of 20%. Make 1-mL aliquots in microcentrifuge tubes with screw caps and freeze slowly at -80°C.

(*f*). Titer the library. The resulting library should have viability of approximately 10^8–10^9 cfu/mL.

3.2 Protocol for the Preparation of the Bait Strain

Introduce a BD-bait plasmid into *MAT*a yeast cells via a small-scale LiAc transformation. Amplify transformant colonies in 50 mL of the appropriate selection media to a late log phase, then use for mating immediately or store frozen at -80°C in 20% glycerol.

3.3 Protocol for Mating and Selection

Method

Mating in liquid culture.

Mating in liquid culture is technically easier and less labor intensive than mating on plates (Figure 3). For successful screening of a million clones per mL of amplified

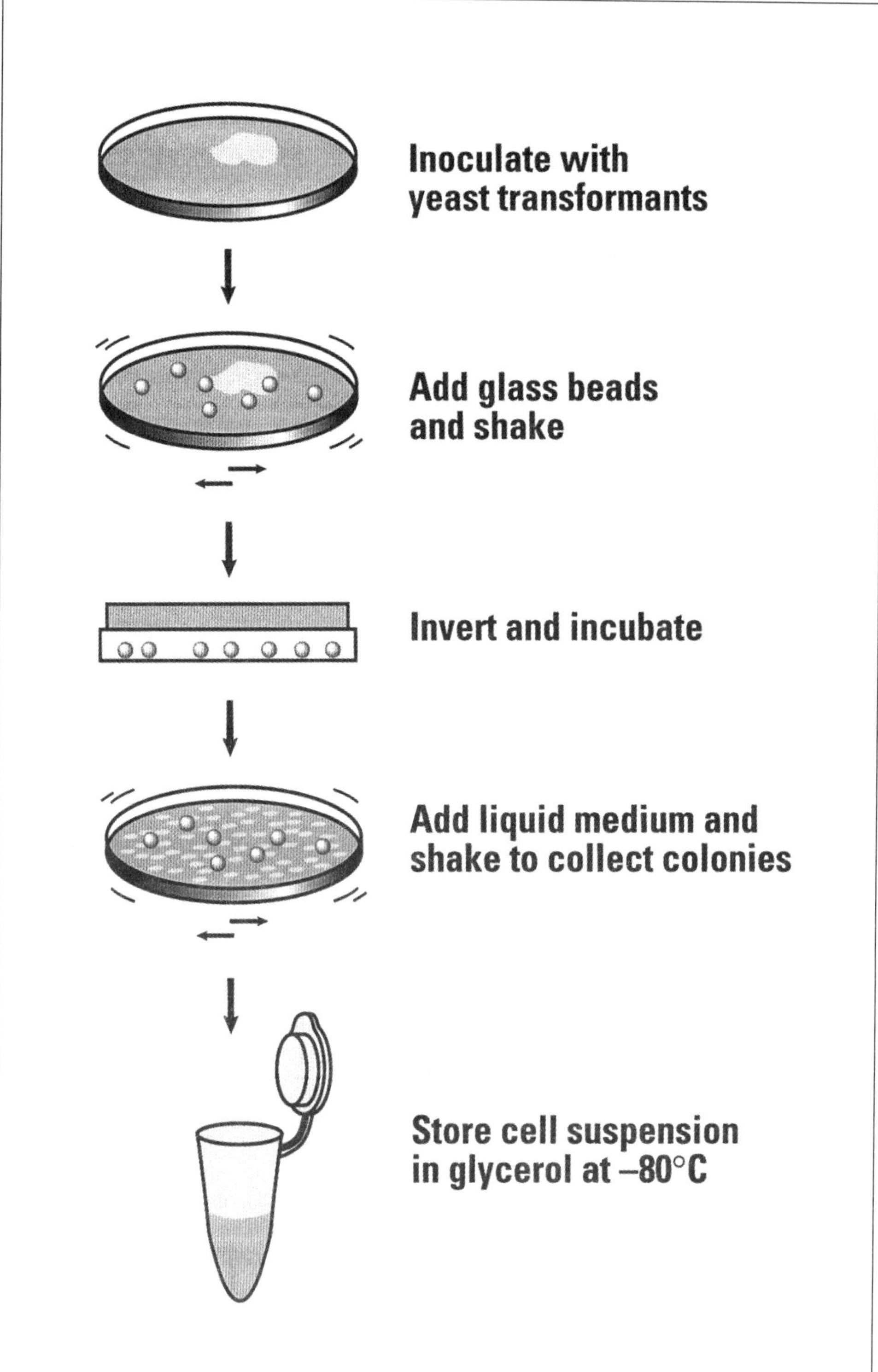

Figure 2. How to amplify a pretransformed library. This diagram emphasizes the use of glass beads to simplify the amplification of a library.

library, certain requirements must be met. The frozen, amplified library must have a viability of $>1 \times 10^8$ cfu/mL. The bait-containing strain must mate at an efficiency of greater than 1% as tested with the opposite strain containing the empty AD vector. If either requirement is not met, it is advisable to use the plate screening method.

(*a*). From above, harvest the bait-containing strain by centrifugation for 10 min at $1000 \times g$ in sterile tubes. Resuspend in 50 mL of YPDA. Inoculate into a sterile 4-L Erlenmeyer or a 2.8-L Fernbach flask.

(*b*). Thaw out an aliquot of frozen library cells in a room temperature water bath. Inoculate into the flask from Step (a), washing the remaining cells from the library tube with another 1 to 2 mL of YPDA.

(*c*). Shake gently at approximately 25 to 50 rpm at 30°C overnight to allow mating. Depending on the model of shaking incubator, the speed may vary; it should be set as slow as possible without allowing the cells to settle to the bottom of the flask. The culture should be gently swirling, which promotes oxygen exchange. Speeds faster than 50 rpm reduce mating efficiency.

(*d*). After 20–24 h, pour the cells into a sterile 50-mL microcentrifuge tube. Rinse the culture flask with an additional 50 mL of YPDA. Spin down the combined yeast cells at $1000 \times g$ for 10 min.

(*e*). Resuspend in 10 mL of the appropriate selection medium and inoculate 200 µL per 15-cm library selection plates. Reserve 100 µL to use to determine the mating efficiency. Up to 5×10^5 diploid yeast cells can be plated on each 15-cm plate. If a parental strain with a leaky HIS3 reporter, such as Y190, is used, add the appropriate amount of 3-AT to suppress the background growth.

(*f*). Serially dilute the reserved 100 µL from 1:10 to 1:10 000. Plate 100 µL of each dilution on the appropriate SD single dropout plates to separately select for the BD-bait plasmid and for the AD-cDNA library, and on SD double dropout plates to select for cells containing both types of plasmid. Calculate the mating efficiency by dividing the cfu/mL regrown on SD double dropout by the cfu/mL regrown on the limiting SD single dropout medium. (Usually, the limiting number of viable cells will be from the AD-cDNA vector selection medium.) The mating efficiency should be in the range of 1%–20%.

(*g*). Calculate the number of clones screened. Multiply the titer of the library by the mating efficiency. For example, for a library with a titer of 1×10^8 cfu/mL and with a 1% mating efficiency, 1×10^6 clones would have been screened.

Mating on plates.

One potential problem associated with the mating procedure in liquid could be biased amplification of cDNA clones during library amplification and mating. One way to prevent bias in the representation of yeast cells is to amplify the yeast library on large plates and then perform mating by replica-plating of confluent bait-containing yeast cells from separate plates (3). However, this plate-based mating is more labor and materials intensive than the bulk mating in liquid (Figure 4).

It will be necessary to use 20 of each kind of 15-cm plate per million clones to be screened.

(*a*). Thaw out frozen library cells in a room temperature water bath. From the library titer, dilute in SD medium to a concentration of 50 000 cfu/200 μL. Inoculate 200 μL per 15-cm SD plate using the single dropout amino acid mixture that selects for the AD-cDNA vector.

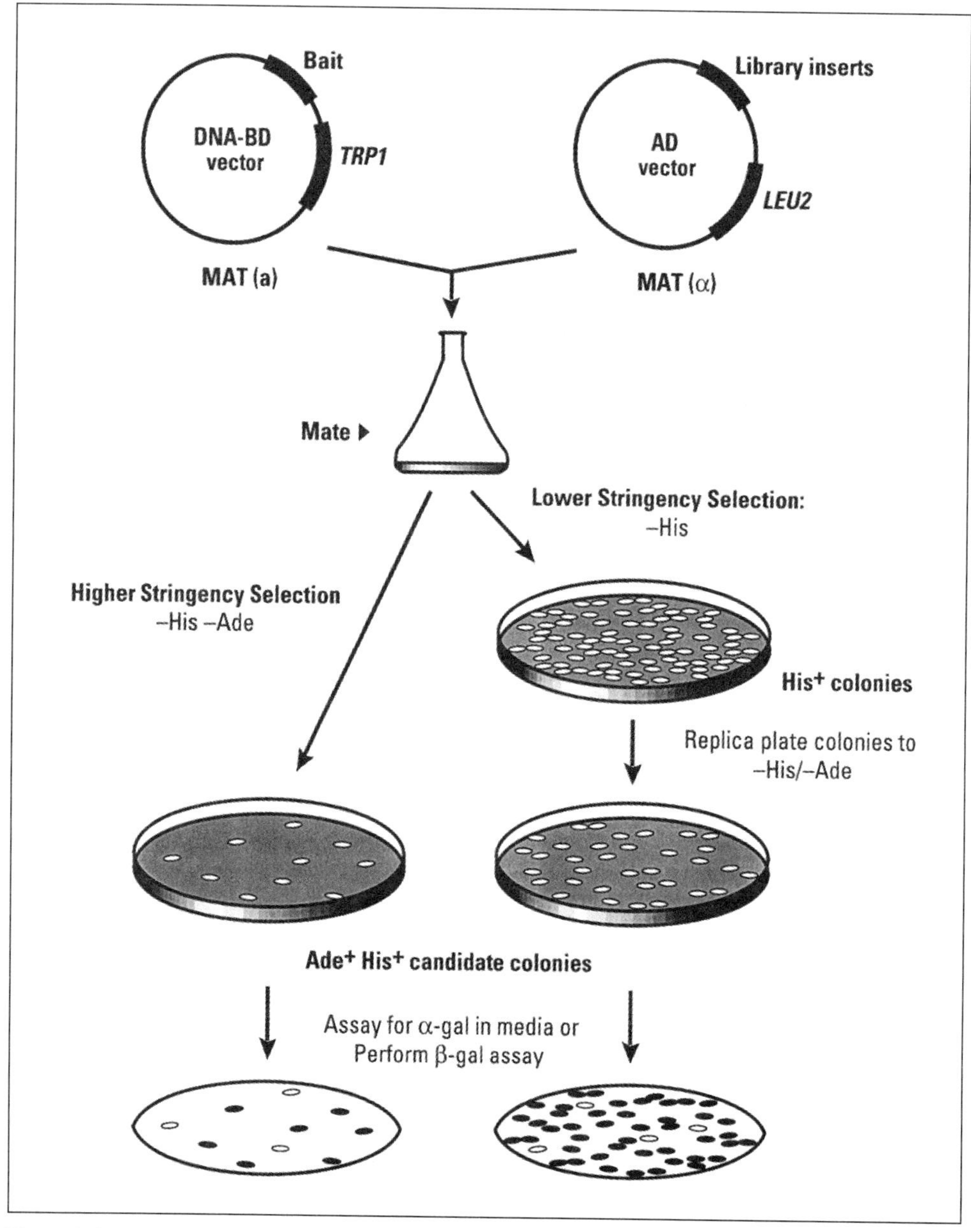

Figure 3. Screening a pretransformed library by mating in liquid culture. Using the MATa strain PJ69-2A and the MATα strain Y187 in the GAL4-based two-hybrid system, mating in liquid culture allows the use of either a higher stringency direct selection for strongly interacting clones or a lower stringency initial screen to expand the pool of possible interactors.

(*b*).　At the same time, spread at confluency (> 10^6 cfu/200 μL) yeast cells containing the bait plasmid onto an equal number of 15-cm SD single dropout plates that select for the BD-bait plasmid.

(*c*).　After 3–4 days, replica plate one of each plate onto a series of YPDA 15-cm plates.

(*d*).　After 20–24 h, replica plate the YPDA plate containing the mated yeast onto library selection plates.

Mating in arrays.

Arrayed mating can be used to study an interaction pathway or entire arrayed libraries. For a pathway, a series of bait clones and a series of AD clones are sepa-

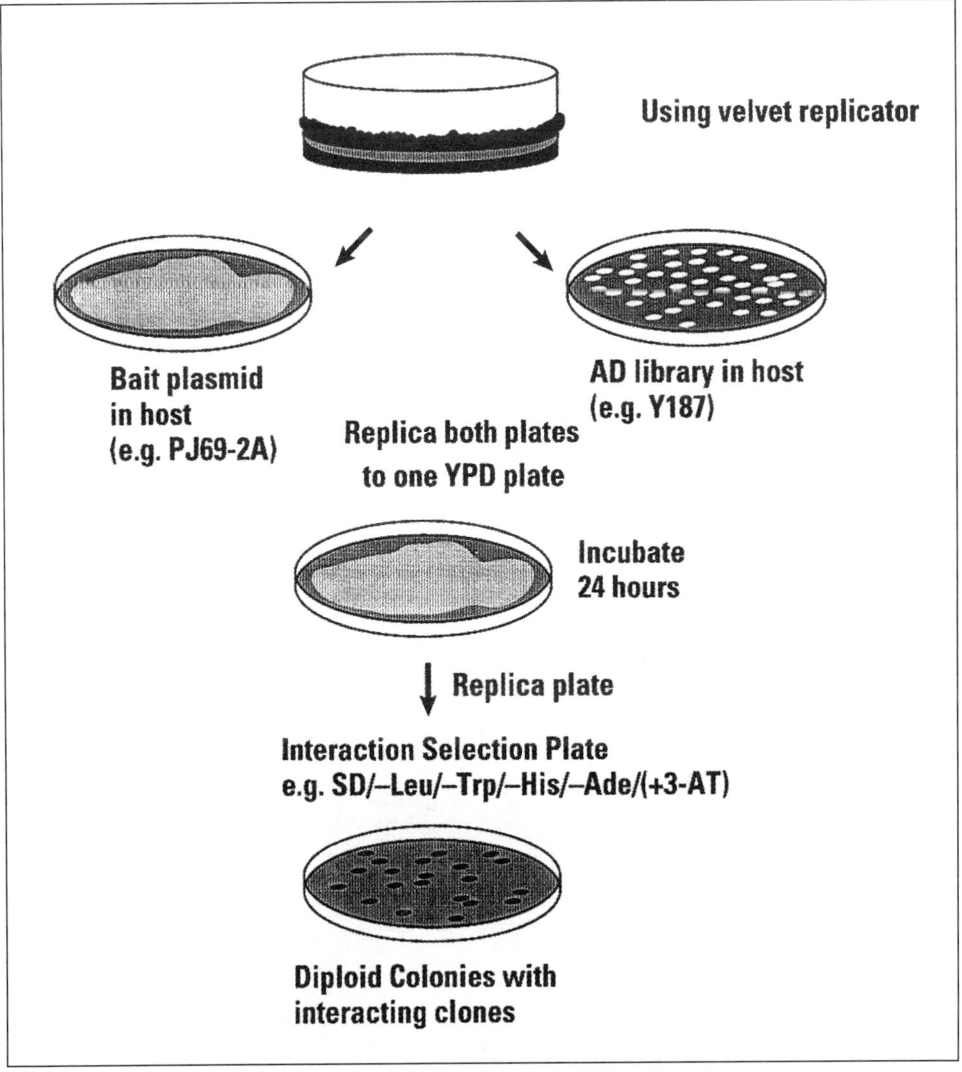

Figure 4. Screening a pretransformed library by mating on agar plates. Mating on agar plates is used to reduce library bias caused by uneven growth of the pool of interacting clones.

rately resuspended in YPDA. In a sterile microtiter plate, inoculate, pairwise, 10 μL of each suspension into 160 μL of YPDA. Shake at 200 rpm at 30°C for 4–16 h. After the mating, a 96-pin or 96-well transfer pipette (Bel-Art, Pequannock, NJ, USA) transfers 10 μL of each suspension to 2 SD 15-cm plates to determine *i*) whether mating occurred on double dropout and *ii*) whether an interaction occurred on triple or quadruple dropout.

This mating procedure essentially is scaled up for arrayed libraries. The topic of arrayed libraries is covered in detail elsewhere in this volume. (See Chapter 18).

3.4 Protocol for Screening

The procedure to screen double positive clones is essentially the same as described in the conventional method (Chapter 3). Briefly, colonies that grow on the library selection plates are individually streaked out onto a fresh plate before being assayed for β-galactosidase activity (*lacZ* assay) with X-gal or Bluo-gal as the substrate. The lacZ+/prototrophic reporter positive clones are then amplified either on the same plates or in selection liquid for plasmid DNA retrieval using commercially available kits.

Alternatively, with the GAL4-based two-hybrid systems, the X-α-gal substrate for the α-galactosidase enzyme from the *MEL1* gene may be used if either the host strain or the BD vector supplies that reporter (1). Because the enzyme is secreted, there is no need to do a lift assay if the substrate was included in the medium. With the LexA interaction trap systems, GFP can be used as an alternative reporter (4 and Chapter 10) instead of *lacZ*.

4. DISCUSSION

The mating-based two-hybrid systems provide an easy and rapid approach for detection of protein–protein interactions (8,10). It has several advantages compared to the traditional methods that involve co-transformation or sequential transformation of yeast cells. The results of library screens conducted by mating favorably compare with co-transformation results (Table 2; References 7,11).

Mating libraries greatly reduce the workload required to conduct a two-hybrid screen because there is no need for extensive library plasmid purification and exhaustive transformation. The initial library transformation is fairly labor intensive. Once a library is amplified in yeast, however, it can be stored indefinitely as frozen aliquots for subsequent multiple uses either with other baits or with the same bait under different conditions, such as with the addition of a modifying protein or drug. Premade yeast libraries have been created by academic communities (9,17) and are also available from commercial sources.

To take full advantage of the ease of mating-based two-hybrid screening, several important factors should be considered during the construction and utilization of the yeast libraries to ensure that the low-abundance cDNA clones are well represented and available for mating. For example, the quality of a mating library is an essential factor and requires that a large number of independent transformants (preferably $>10^7$) should be obtained for each library. The yeast cells should have a high rate of survival after freezing and thawing; although due to some viability loss at each freeze-thaw, it is preferable to minimize the number of cycles for each aliquot. In

Table 2. Comparing Two-Hybrid Library Screening Methods

Bait; Library	Cotransformation			Mating		
	Unique Clones	Positive[1]	Screened	Unique Clones	Positive[2]	Screened
BAD; HeLa						
Standard	1	1	4.5×10^4	4	62	3.3×10^6 (high + low stringency)
his3 Jump Start	3	54	$\sim 2 \times 10^6$			
eIF–5α; HeLa	1	1	1×10^7	2	2	6×10^6 (high + low stringency)

Two bait and library combinations were screened using either cotransformation according to CLONTECH's MATCHMAKER GAL4 Two-Hybrid User Manual or mating according to the methodology described in this chapter. For co-transformation, yeast strain HF7c was used. To mate, PJ69-2A, which contained the bait, was crossed to Y187, which contained the library. The same library DNA introduced into HF7c was used to create the pretransformed Hela library in Y187. (Chan and Holtz, unpublished).

Notes: 1) Positives are His+/lacZ+
          2) Positives are Ade+/ His+/lacZ+

addition, the cDNA libraries should be constructed in yeast strains for which highly efficient mating partners are readily available.

The mating procedure can also be used to rapidly test a large number of potential interactions in pairwise matings. As the yeast two-hybrid assay is becoming widely used, a larger number of bait genes and/or their interacting proteins in a given pathway will be available in yeast cells or can be rapidly transferred to yeast cells. These genes can be rapidly screened against one another in the 96-well format using the mating method. This matrix mating has been successfully used in the analysis of gene regulatory pathways of *Drosophila* cell cycle proteins (8).

The mating selection can be scaled up with automation to screen a large number of genes in a high throughput manner. The mating procedure was successfully used to generate a protein linkage map for the bacteria phage T7 genome (2). The advent of the human genome project has revealed tens of thousands of novel genes that await functional analyses using a variety of genetic, molecular, and biochemical approaches. The protein–protein interaction studies will play important roles in quickly gaining insights into the functional networks of new proteins to create protein linkage maps. Protein linkage maps have been created using co-transformation methodologies (5,18), but the labor intensiveness of that method precludes wider usage. The development of efficient and cost-effective large-scale cloning (11,12) and screening procedures has become increasingly important. Currently, a pilot program is being conducted using *S. cerevisiae* as the model (12). The mating-based selection is more suitable for automation than the conventional method and therefore can easily be adopted for high throughput screening. As proposed in Chapter 18, two human EST-based libraries, one of uni-gene–binding domain hybrids and another of EST-activation–domain hybrids will be constructed in yeast strains of opposite mating types. The two yeast libraries will then be used to screen against each other to construct a global human protein linkage map, which could be extremely useful in gene function studies and drug discovery.

REFERENCES

1. **Aho, S., A. Arffman, T. Pummi, and J. Uitto.** 1997. A novel reporter gene MEL1 for the yeast two-hybrid system. Anal. Biochem. *253*:270-272.
2. **Bartel, P.L., J.A. Roecklein, D. SenGupta, and S. Fields.** 1996. A protein linkage map of *Escherichia coli* bacteriophage T7. Nat. Genet. *12*:72-77.
3. **Bendixen, C., S. Gangloff, and R. Rothstein.** 1994. A yeast mating-selection scheme for detection of protein-protein interactions. Nucleic Acids Res. *22*:1778-1779.
4. **Cormack, R.S., K. Hahlbrock, and I.E. Somssich.** 1998. Isolation of putative plant transcriptional coactivators using a modified two-hybrid system incorporating a GFP reporter gene. Plant J. *14*:685-692.
5. **Cuconati, A., W. Xiang, F. Lahser, T. Pfister, and E. Wimmer.** 1998. A protein linkage map of the P2 nonstructural proteins of poliovirus. J. Virol. *72*:1297-1307.
6. **Dagher, M.C. and O. Filhol-Cochet.** 1997. Making hybrids of two-hybrid systems. BioTechniques *22*:916-922.
7. **Fernandez-Larrea, J., A. Merlos-Suarez, J.M. Urena, J. Baselga, and J. Arribas.** 1999. A role for a PDZ protein in the early secretory pathway for the targeting of proTGF-alpha to the cell surface. Mol. Cell. *3*:423-433.
8. **Finley, R.L. and R. Brent.** 1994. Interaction mating reveals binary and ternary connections between Drosophila cell cycle regulators. Proc. Natl. Acad. Sci. USA *91*:12980-12984.
9. **Fromont-Racine, M., J.C. Rain, and P. Legrain.** 1997. Toward a functional analysis of the yeast genome through exhaustive two- hybrid screens. Nat. Genet. *16*:277-282.
10. **Harper, J.W., G.R. Adami, N. Wei, K. Keyomarsi, and S.J. Elledge.** 1993. The p21 Cdk-interacting protein Cip1 is a potent inhibitor of G1 cyclin-dependent kinases. Cell *75*:805-816.
11. **Hua, S.B., Y. Luo, M. Qiu, E. Chan, H. Zhou, and L. Zhu.** 1998. Construction of a modular yeast two-hybrid cDNA library from human EST clones for the human genome protein linkage map. Gene *215*:143-152.
12. **Hudson, J.R., Jr., E.P. Dawson, K.L. Rushing, C.H. Jackson, D. Lockshon, D. Conover, C. Lanciault, J.R. Harris, S.J. Simmons, R. Rothstein, and S. Fields.** 1997. The complete set of predicted genes from *Saccharomyces cerevisiae* in a readily usable form. Genome Res. *7*:1169-1173.
13. **James, P., J. Halladay, and E.A. Craig.** 1996. Genomic libraries and a host strain designed for highly efficient two- hybrid selection in yeast. Genetics *144*:1425-1436.
14. **Marsh, L. and M.D. Rose.** 1997. The pathway of cell and nuclear fusion during mating in *S. cerevisiae*, p. 827. *In* J.R. Pringle et al. (Eds.), The molecular and cellular biology of the yeast *Saccharomyces*: cell cycle and cell biology. Cold Spring Harbor Laboratory Press, Cold Spring Harbor, New York.
15. **Sprague, G.F., Jr.** 1991. Assay of yeast mating reaction. Methods Enzymol. *194*:77-93.
16. **Sprague, G.F., Jr. and J.W. Thorner.** 1994. Pheromone response and signal transduction during the mating process of *Saccharomyces cerevisiae*, p. 657. *In* E.W. Jones et al. (Eds.), The molecular and cellular biology of the yeast *Saccharomyces*: Gene Expression. Cold Spring Harbor Press, Cold Spring Harbor, New York.
17. **Wu, X., J.X.L. Li, C.-L. Hsieh, P.M.J. Burgers, and M.R. Lieber.** 1996. Processing of branched DNA intermediates by a complex of human FEN-1 and PCNA. Nucleic Acids Res. *24*:2036-2043.
18. **Xiang, W., A. Cuconati, D. Hope, K. Kirkegaard, and E. Wimmer.** 1998. Complete protein linkage map of poliovirus P3 proteins: interaction of polymerase 3Dpol with VPg and with genetic variants of 3AB. J. Virol. *72*:6732-6741.

6

False Positives: Detection and Elimination

Yi-Ren Hong

Graduate Institute of Biochemistry, Kaohsiung Medical University, Kaohsiung, Taiwan, China

1. INTRODUCTION AND BACKGROUND

1.1 Principle of the Yeast Two-Hybrid Assay

The yeast two-hybrid assay is based on the observation that many eukaryotic transacting transcriptional regulators are composed of physically separable, functionally independent domains. Such regulators often contain a DNA-binding domain (DNA-BD), which binds to a specific promoter sequence, and an activation domain (AD) that directs the RNA polymerase II complex to transcribe the gene downstream of the DNA-binding site. Both domains are required to activate a gene and, normally (as in the case of the native yeast GAL4 protein), the two domains are part of the same protein. If physically separated by recombinant DNA technology and expressed in the same host cell, the DNA-BD and AD peptides do not directly interact with each other and thus cannot activate the responsive genes. However, if DNA-BD and AD can be brought into close physical proximity in the promoter region by two proteins that can associate with each other, the transcriptional activation function will be restored. In principle, any AD can be paired with a DNA-BD that provides the binding specificity (7,14,31).

In the past few years of its advent, the two-hybrid system has proven its utility in the study of protein interactions. It has shown a tremendous success in the charting of genetic networks of proteins involved in processes from signal transduction to transcription regulation. The two-hybrid approach may eventually be used to identify the protein–protein interaction network of a cell or an organism (Reference 13 and Chapter 18).

1.2 Definition of False-Positive Proteins

Fields and his colleagues first described different types of false positives that arose

Yeast Hybrid Technologies
Edited by L. Zhu and G.J. Hannon
© 2000 Eaton Publishing, Natick, MA

when using the two-hybrid system (2). False positive is defined as the activation of reporters unrelated to the interaction of the bait with its specific prey proteins. For example, in library screens, one can select for proteins that are initially found to activate reporter-gene expression in the presence of a specific DNA-binding–domain hybrid protein, but later observe that they interact nonspecifically with many different hybrid partners. Some proteins may contain regions with surfaces that have low affinities for many different proteins, for example, large hydrophobic surfaces, heat shock proteins, and ribosomal proteins (see survey in Reference 11). These proteins may form complexes with a number of other proteins that are stable enough to result in a detectable positive phenotype. An association with affinity higher than 10^{-6} will be detected in the yeast two-hybrid system. Therefore, any nonspecific protein interactions with affinity higher than 10^{-6} can potentially lead to a positive signal (11,33). It is true that so many protein–protein interactions as well as protein–DNA interactions exist in the cell; however, any nonspecific interactions will cause a false signal in the yeast two-hybrid system. We will analyze the false positive in detail in this chapter.

1.3 True and False Positives with Biological Consideration

One critical question in performing two-hybrid screens has been whether a positive isolate identified in the system actually represents a true in vivo interaction. Several lines of consideration have been raised. First, it is possible to identify interacting partners that never associate in vivo because they are normally expressed in different cell types, localized in distinct cellular compartments, or expressed at different developmental stages, etc. Therefore, careful consideration should be given to organism, cell type, developmental stage, and temporal patterns of gene or protein expression. When a screen is performed, target proteins and activation libraries should be carefully chosen. Moreover, it is desirable to determine whether isolated hybrid proteins that interact in the two-hybrid system correlate to biological function (18,21,37).

False positive frequency is related to the following: *(i)* bait used (bait plasmid copy number and the level of protein expressed in yeast such as PAS2 vs PGBT9), *(ii)* reporter plasmid used, *(iii)* selection stringency (e.g., 3-AT used). The protein–protein interactions identified in the yeast two-hybrid system may not be reproducible in other systems, such as the mammalian two-hybrid system (30). Others observed that certain protein–protein binding in vitro was different from that in yeast (12). The false-positive results generated by the yeast two-hybrid system cDNA library screening might be due to improper mammalian protein folding in yeast (2,5,18,29). Because the post-translational modifications in yeast are not always the same as in mammalian cells, false-positive results in two-hybrid screening may also be caused by binding of the two-hybrid proteins to a common, third endogenous yeast protein instead of binding directly to each other (30).

1.4 Types of False Positives (18)

Based on the two-hybrid system standard, three types of false positives have been classified (2). The most common, which we refer to as Type 1 false positives, are those AD:cDNA fusion products that can activate reporter gene transcription in the absence of the BD fusion protein. This type of false-positive can be eliminated by selectively "losing" the DNA-BD plasmid (see Section 2.1). Type II false positives are the ones that activate the reporter gene in the presence of any BD fusion plasmid.

They can be detected by transformation of pAD:cDNA fusion plasmid into a yeast strain containing an unrelated BD fusion plasmid (see Protocol 2.1). A positive read-out is indicative of a Type II false positive. Type III false positives are those that activate the reporter gene(s) in the presence of an 'empty' BD vector. This type of false positive has been found when using the pAS vector. It is probably the result of an interaction with the HA tag.

1.5 False Negatives

A less frequently discussed issue is false negatives that refer to the fact that a true interaction is not detected in yeast. Failure to detect expected proteins in this instance might be due to a number of situations, such as *(i)* the absence of the protein from the library used; *(ii)* the expression of hybrid proteins is toxic; *(iii)* the hybrid proteins are not stably expressed in the yeast cell; *(iv)* the folding is improper; *(v)* the hybrid protein cannot enter into the yeast nucleus; *(vi)* improper post-translational modification, such as phosphorylation, glycosylation, or acetylation; and other reasons (36). There are now a number of examples in which known interactions are not observed in the yeast two-hybrid system. There are also cases in which interaction is only observed in the one way but not in the other (i.e., when A is cloned in bait plasmid but not when it is in the AD plasmid) (11,23).

1.6 Plating Ratio and False Negatives

We have tried to measure the ratio of plating efficiency on the SD with or without selection placed on reporter gene markers (e.g., His for GAL4 system and Leu for LexA system). The ratio is measured by the number of colonies (colony forming unit, cfu) that form on SD plates that lack His or Leu over the number that form on SD plates that contain His or Leu. From our experience, the ratio of plating efficiency is usually more than 5%. In some cases, however, the ratio is less than 0.1% (our observation in p53 activation domain). Even in hybrid constructs that show the ratio at a very low level (0.1%), they may still be a true positive (Hong, unpublished data). We do not fully understand this observation, but we speculate that the qualities or quantities of the hybrid proteins may contribute to the variations of the plating ratio observed. The readout of interation in the yeast two-hybrid system may be affected by the number of binding operators (11). Harsh selective pressure of SD plates used may cause false negatives. Therefore one should carefully consider whether to use high stringency selection in library screening. High stringency of selection can help eliminate false positives but can also often miss true positives.

1.7 Survey of Frequent False-Positive Isolations

Erica Golemis' lab has made an effort to collect false positives encountered in the yeast two-hybrid system.

This survey, which included 223 investigators, revealed common sources of false positives. Their team has entered the relevant results from 100 library screenings into a database now available on the Web. Seventy-three false positives were sequenced, and the most common ones were found to be heat shock proteins (16 cases), ribosomal proteins (14 cases), cytochrome oxidase (5 cases), other mitochondrial proteins (3 cases), proteasome subunits (4 cases), ferritin (4 cases), transfer-tRNA synthase (3 cases), collagen-related proteins (3 cases), zinc-finger containing proteins (3 cases), vimentin (2 cases), inorganic pyrophosphatase (2 cases), and most surprisingly, proliferating cell nuclear antigen (PCNA) (2 cases). Elongation factors, lamin, and cytoskeletal proteins are also noted. In our experience, elongation factors, human albumin, and some repeat sequences are common false isolates in library screening.

Descriptions of the baits used and the false positives isolated in various cases can be found at **http://www.fccc.edu/research/ labs/golemis**.

2. ELIMINATION OF FALSE POSITIVES

In general, false positives can be eliminated by verifying that reporter gene expression is specific for the presence of the target protein hybrid and is not generated with unrelated DNA-binding–domain hybrids. Several methods have been designed to eliminate false positives isolated in the system. These include: swap experiments (2,23), combining plasmid (27,10), using cycloheximide plasmid (20), cotransformation (37), mating assay (4,10), putting more markers on host strains (25), and using an inducible LexA system (17). Here I describe a step-by-step procedure to eliminate false positives isolated from the yeast two-hybrid system. Some of the topics, such as mating assays, using additional selection markers, using inducible promoters (such as in the LexA system), and using cycloheximide-mediated counter selection to eliminate false positives can be found in other chapters of this book or elsewhere.

2.1 Cotransformation to Eliminate False Positives

Once one sees β-galactosidase–positive colonies from the library screening, one should immediately perform a restreaking step. This is because often more than one library plasmid can be present in a single colony. Restreaking, a very simple procedure to do, can efficiently allow plasmid segregation and only allow the "right" segregated colonies to grow further. Thus, this procedure can eliminate some false-positives right away. In many cases, one might encounter a lot of trouble downstream if this very first false-elimination step is skipped. After the restreaking step, the next step will be reconstruction, which requires the isolation of the pAD plasmid. The most common way to eliminate false positives is cotransformation. A flowchart for false positive detection is outlined in Figure 1.

AD plasmids can be isolated from yeast transformants based on the method of Hoffman and Winston (22) or Kaiser and Auser (26). Yeast-purified plasmid DNA is suitable for Long Distance (LD)-PCR (CLONTECH). It can also be used to transform an appropriate *E. coli* strain. Conventional transformation with chemical competent cells yields a low number of transformants in *E. coli*. High efficiency by electroporation yields between 500–1000 transformants (18). If one cannot get any transformants by electroporation, then these candidates should not be pursued.

Restriction enzyme analysis of cDNA clones will allow classification into groups, and the cDNA insert size should match the products from the LD-PCR. True positives are then cotransformed for confirmation. Retransform the plasmid that encodes the candidate activation domain/protein Y hybrid back into the original yeast host strain in the combinations shown in Table 1. It must be verified first that the prey (Table 1, # 1 and # 2) fusion protein does not activate the system by itself. Then, the source of bait DNA pBD-X (X stands for bait DNA) should be co-transformed with the prey to confirm the interaction. An unusual case of false positive is caused by a rare event of rearrangement of bait constructs (24). For transformations #2, #3, and #4, cotransformants are selected on SD synthetic medium lacking Trp and Leu but including His. For transformation #1, transformants are selected on SD medium lacking Leu only. Transformants are assayed for β-galactosidase activity. If *lacZ* expression is truly dependent on interaction of the two test proteins (X and Y), then the expected results are indicated in Table 1.

In an alternative approach, the positive result obtained in the two-hybrid assay can also be verified by cotransforming the two-hybrid vectors into the other yeast host strain (such as SFY526). In SFY526, the *lacZ* reporter gene is under the control of a promoter different from that in HF7c. These two promoters share only GAL4-responsive elements; the rest of the two promoter sequences differ significantly. Therefore, any positive two-hybrid interactions observed in both reporter strains are likely to require binding of the GAL4 DNA-binding domain specifically to the GAL4-responsive elements (2).

Cotransformation may be performed with any plasmid carrying an unrelated protein fused with GAL4 DNA-binding domain (not just pLAM5′ and pVA3). Yeast transformants showing His+ or LacZ+ phenotype are false positve clones. Such colonies arise in a two-hybrid assay for a number of reasons (2). In some cases, AD/library plasmids alone may activate reporter gene transcription; but in other cases, they appear to require a DNA-BD for their activity. The DNA-BD/target protein should already have been tested for autonomous activation before being used in a two-hybrid experiment, as discussed in section 1.3.

Yeast strains CG-1945 and Y190 carry two different reporter genes (*HIS3* and *lacZ*) that are under the control of different promoters. This automatically eliminates many false positives, particularly those that do not bind to the pAS2-1/target protein (CLONTECH) but instead interact with promoter sequences flanking the GAL4

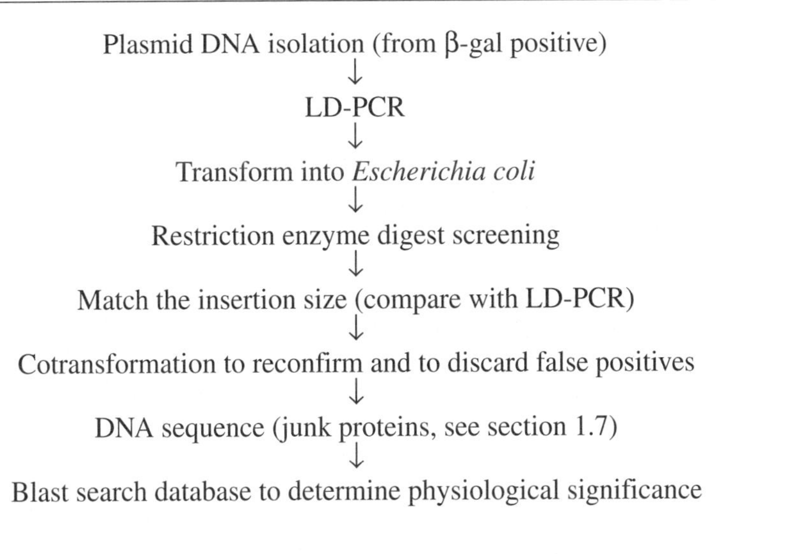

Figure 1. Flowchart describing steps employed in a typical two-hybrid screen and false-positive detection assays.

Table 1. Cotransformations to Eliminate False Positives

#	Plasmid 1	Plasmid 2 (candidate interacting protein)	Expected β-gal assay results for a true positive
1	(none)	pAD-Y hybrid	Negative[1]
2	pBD (DNA-binding domain only)	pAD-Y hybrid	Negative[2]
3	pBD-X	pAD-Y hybrid	Positive[3]
4	pLAM5′ or pVA3 (pBD/ Lamin C or p53 hybrid)	pAD-Y hybrid	Negative[4]

[1]This is to reconfirm the results obtained with the *lacZ*-Trp auxotrophs. Discard β-gal positives.

[2]Discard β-gal positives. Fortuitous interaction between the candidate protein Y and the GAL4 DNA-binding domain may lead to activation of *lacZ* expression after cotransformation with the DNA-binding–domain plasmid (pBD).

[3]The primary purpose of this test is to reconfirm the initial observation. Discard β-gal negatives. If a β-gal negative clone is found in this test, it is most likely the result of segregation of the original cotransformant due to the presence of more than one type of activation domain/Protein Y hybrid plasmid. If one of the plasmids encodes an interacting hybrid protein, while the other does not, the β-gal positive phenotype will be lost in some segregated clones. However, the loss of one of the plasmids through segregation during colony growth will result in two *lacZ* phenotypes within the same colony. If plasmid segregation is not complete at the time when the original colony is streaked out for single colonies, both phenotypes (i.e., β-gal negative and β-gal positive) will be found among the isolated colonies.

[4]Discard β-gal positives. A possible reason for this type of false positive is that the activation domain/Protein Y hybrid plasmid encodes a protein that does not bind to the test protein hybrid but instead binds to or regulates promoter sequences flanking the GAL4 binding site or to proteins already bound to the flanking sequences. In some cases, this interaction may be specific for the non-GAL 1 portion of the promoter for the reporter gene. In these cases, the presence of a GAL4 DNA-binding–domain hybrid stimulates the activation of the reporter gene with this type of false positive.

binding site or that interact with proteins bound to the flanking sequences. Nevertheless, putative positives should be tested to determine that their activity is specific for the DNA-BD/target protein. The cycloheximide counterselection and the mating assays may help to quickly identify and eliminate the great majority of false positives detected in the two-hybrid assay (20). The true positive pAD:cDNA plasmids can be sequenced to identify the open reading frames (ORFs). In most cases, library inserts contain termination codon, polyadenylation signal, and poly-A tail.

BLAST searches against Genbank database (**http://www.ncbi.nlm.nih.gov/ BLAST**) will reveal whether these sequences have homology to a known gene or not. Descriptions of the false positives isolated in previously studied cases can be found in section 1.7. In some cases, the cDNA insert isolated from the one- or two-hybrid screen are "reverse-oriented" or contain unexpected in-frame stop codon(s). We provide certain explanations for these unusual cases in section 2.3.

2.2 Swap Experiment: An Example to Eliminate False Positives and False Negatives

In this "domain swap" experiment, the bait fused with the DNA-binding domain of the GAL4 protein was inserted in place of the LEXA protein (19). The previous experiments indicate that the DNA-binding part of the GAL4 protein and the LEXA bacterial repressor protein are functionally interchangeable (32). Therefore, the methods to "swap domains" and to switch the "bait and prey" in their respective plasmids could rule out some false positives. It is noted that some true positives may not always work because of structural reasons (1). Here we use the domain-swap method to show an example of how to eliminate both false positives and negatives.

The phagocyte NADPH oxidase is a complex of membrane cytochrome b_{558} (comprised of subunits p22-*phox* and gp91-*phox*) and three cytosol proteins (p47-*phox*, p67-*phox*, and p21 *rac*) that translocate to membrane and bind to cytochrome b_{558} (35). p47-*phox* binding to both p67-*phox* and p22-*phox* is mediated by interactions between SH3 domains and proline-rich target peptides (15, 23). While studying p47-*phox*/p67-*phox* binding, we encountered several methodological problems in the two-hybrid system that are discussed in this report.

Two different two-hybrid systems were used in our studies. Whole and partial fragments of p47-*phox* and p67-*phox* were cloned into pGBT9 and pGAD 424 (MATCHMAKER™; CLONTECH Laboratories) to form GAL4 BD and AD fusion proteins. Similar constructs were made in pEG202 and pJG4-5 (19) to form LEXA BD and B42 AD fusion proteins (2). Pairs of pGBT9 and pGAD 424 plasmids and of pEG202 and pJG4-5 plasmids were co-transfected into strains of *S.cerevisiae* specific for the GAL4 and LEXA systems. GAL4 AD, GAL4 BD, and LEXA BD fusion proteins are constitutively expressed from the yeast ADH1 promoter. B42 AD fusion proteins are inducibly expressed only in galactose-containing media under control of the yeast GAL1 promoter. For the GAL4 system, p47-*phox*/p67-*phox* interactions were detected by *lacZ*. For the LEXA system, interactions were detected by a *Leu2* reporter in addition to *lacZ*.

p47-*phox* interaction with p67-*phox* is mediated by p67-*phox* carboxyl-terminal SH3$_B$ domain binding to a carboxyl-terminal SH3$_B$ domain that binds to a carboxyl-terminal p47-*phox* proline-rich peptide (23,35). Full-length p47-*phox* and p67-*phox* cDNA were cloned into pEG202 and pJG4-5, respectively (Figure 2; lane 1); or conversely into pJG4-5 and pEG202, respectively (lane 2). Interaction was seen between p47-*phox*/B42 AD and p67-*phox* LEXA BD (lane 2) but not between p47-*phox*/LEXA BD and p67-*phox*/B42 AD fusion proteins (lane 1). We further analyzed p67-*phox* fragments fused to LEXA BD and p47-*phox* fragments fused to B42 AD. It confirmed that p47-*phox*/p67-*phox* binding is mediated by an SH3 domain. This suggested that false-negative results are seen when full-length p47 *phox* is expressed as a binding domain fusion protein. We then examined whether failure of p47-*phox*/LEXA BD to bind to p67-*phox* was a characteristic of the LEXA BD or was due to constraints of the transactivator's binding domains in general. Full-length p47-*phox* and p67-*phox* were fused to GAL4 BD separately (Table 2). As shown in Table

2, p47-*phox*/GAL4 BD did not bind to either p67-*phox*/GAL4 AD or to p67-*phox*/B42 AD. In contrast, p67-*phox*/LEXA BD bound to both p47-*phox*/B42 AD and to p47-*phox*/GAL4 AD. We speculate that dimerization of the BD may prevent p47-*phox* interaction with p67-*phox*. This is possibly due to a much lower level of expression of p67-*phox*/GAL4 BD in yeast compared to that of p67-*phox*/LEXA BD.

False-positive results are common in the two-hybrid system but usually can be easily detected by control transfection experiments (2). A typical false-positive result is shown in Figure 2, lanes 3–5. Yeast transfected with p47-*phox* SH3$_{AB}$/LEXA BD were prototrophic on both galactose and glucose media, indicating transcription of the *Leu2* reporter (lanes 3 and 4) was not p67-*phox*-specific. Growth in glucose-containing media suggests that p47-*phox* SH3$_{AB}$/LEXA non-specifically interacted with an endogenous yeast protein with activation domain activity, due to the fact that B42 fusion proteins are not transcribed in glucose media. Absence of growth in lane 5 in either media showed that p47-*phox* SH3$_{AB}$ did not interact with p67-*phox*. These results indicate that false-positive problems can be circumvented by swapping the two proteins under study between LEXA BD and B42 AD.

False-negative results by their very nature are more difficult to detect. False-negative results may arise from failure to express sufficient protein in yeast. The ability of p67-*phox*/LEXA BD, but not p67-*phox*/GAL4 BD, to bind p47-*phox*/AD may be an example of this category. Expression of BD and AD fusion proteins can be determined by immunoblotting yeast cytosol extracts if anti-sera is available. False-negative results may also arise from failure of an expressed fusion protein to enter the

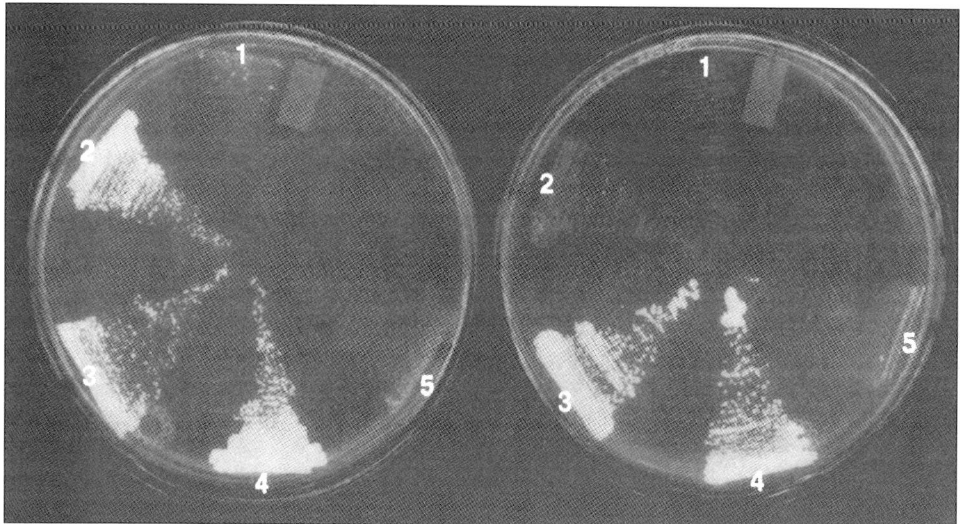

Figure 2. Interactions between p47-*phox* and p67-*phox* in the LEXA two-hybrid system. *S. cerevisiae* strain EGY48 (gift from R. Brent, Massachusetts General Hospital, Boston, MA) (*ura3, his3,trp1,lexAop:leu2*) was transformed with p47-*phox* and p67-*phox* cloned into pEG202, pJG4-5, and pSH18-34. Transformants were streaked onto galactose-containing (left plate) and glucose-containing (right plate) minimal media deficient in leucine. Minimal media deficient in uracil, histidine, and tryptophan was used to select for yeast bearing pEG202, pJG4-5, and pSH18-34 plasmids. Prototrophy on leucine-deficient minimal media indicated reconstitution of LEXA transactivator function. Expressed fusion proteins were 1) full-length p47-*phox*/LEXA BD and full-length p67-*phox*/B42 AD, 2) p67-*phox*/LEXA BD and p47-*phox*/B42 AD, 3) p47-*phox* SH3$_{AB}$ [a cDNA fragment encoding p47-*phox* SH3 domains (amino acids 154-285)]/LEXA BD and B42 AD, 4) p47-*phox* SH3$_{AB}$/LEXA BD and p67-*phox*/B42, and 5) p67-*phox*/LEXA BD and p47-*phox* SH3$_{AB}$/B42 AD (see text). Plates were incubated for two days at 30°C.

Table 2. Difference of β-galactosidase Activity in Expression of p47-*phox* and p67-*phox* Binding and Activation Domain Fusion Proteins in the GAL4 and LEXA Two-Hybrid Systems

Binding domain Activation domain	pGBT9 pGAD424[a,b]	pGBT9 pJG4-5[a,b,c]	pEG202/ pJG4-5[c,d]	pEG202/ pGAD424[b,c,d]
p47/p67-*phox*	<1[e]	<1	<1	<1
p67/p47-*phox*	<1	<1	100	35

[a]*S. cerevisiae* strain SFY526 (MATCHMAKER two-hybrid system; CLONTECH Laboratories) (*ura3,his3,trp1,leu2,ura3…gal1-lacz*) containing a β-galactosidase reporter with *gal1* upstream activation sequence was transformed with pGBT9 and either pGAD 424 or pJG4-5 plasmids.

[b]p47-*phox*/ and p67-*phox*/GAL4 BD were expressed from pGBT9, and p47-*phox*/ and p67-*phox*/GAL4 AD were expressed from pGAD 424.

[c]p47-*phox*/ and p67-*phox*/LEXA BD were expressed from pEG202, and p47-*phox* and p67-*phox*/B42 AD were expressed from pJG4-5.

[d]EGY48 yeast were transformed with pEG202 and pJG4-5 or pGAD424 as well as pSH18-34, a plasmid containing the *lacZ* reporter under LEXA transcriptional control.

[e]Representative experiment measuring hydrolysis of o-nitrophenyl-β-D-galactopyranoside (ONPG) by β-galactosidase. Results expressed as relative enzyme units according to manufacturer protocol (MATCHMAKER; CLONTECH Laboratories). Variations between replicates were less than 20%, and background activity was <1U.

nucleus. This possibility can be determined in the LEXA system by co-transfection with control plasmids (23,19).

We have identified another important source of false-negative results (23). False-negative results may occur when a polypeptide/BD fusion is presumably expressed but is nonfunctional in the two-hybrid system. The inability of either p47-*phox*/LEXA BD or p47-*phox*/GAL4 BD to interact with p67-*phox* AD fusion proteins is an example of this case. Our experience shows the value of swapping pairs of polypeptides between binding and activation domains. We recommend swapping between BD and AD when two proteins are studied in the two-hybrid system. Moreover, using more controls is always helpful for detecting false positives (10). We also recommend changing two-hybrid systems, e.g., from GAL4-based system to LEXA-based or vice versa, when an expected protein–protein interaction is not found. The two major commercially available two-hybrid systems may not always give the same results. Dramatic differences can be seen between LEXA BD and GAL4 BD or between B42 AD and GAL4 AD. In our example, p47-*phox* fused to GAL4 AD gives β-galactosidase activity one third of that of p47-*phox* fused to B42 AD (Table 2; 27).

2.3 Sequence Analysis of cDNA Inserts: Two Commonly Asked Questions

In some cases, the cDNA insert isolated from the one- or two-hybrid screen contained unexpected in-frame stop codon(s) or were "reverse-oriented". These two common cases are explained as follows:

2.3.1 Why do some isolates from the library screening contain stop codon(s) in the right ORF and only code for a short peptide?

Some people asked why could they obtain positive clones that contained in-frame stop codons? In some cases this makes the open reading frame (ORF) after the GAL4 AD only code for a very short peptide, almost impossible for any recognizable protein.

This can be explained in two ways: first, the short peptide, when fused to the GAL4 AD, may indeed cause a significant structural change of the GAL4 AD domain or create a new epitope that now becomes a bait-specific interacting fusion protein. Therefore, it is a true positive by the standard of the yeast two-hybrid system; but it is also a false negative in a biological sense. One can decide with their own judgment whether or not this type of isolate should be pursued further. Second, the yeast may actually be able to do a frame switch. This switch needs to occur right in the region between the GAL4 AD and the new fusion protein. Although there is no experimental or literature-based evidence for this, we have heard from actual users the following: the fusion partner isolated from the screening contains, despite the presence of a stop codon following the GAL4 AD, a long and meaningful ORF that could indeed interact with the bait. This means that a new ORF is used somehow in the yeast. Conceptually, the new ORF can be rendered functional either by fusing with the GAL4 AD, provided a frame switch occurred, or may act by itself if it contains an activation function by its own coding sequence. We have heard of both of these cases (Dr. Xu, Yale University, Dr. Zhu, CLONTECH, Personal communications).

2.3.2 How could one isolate positive cDNA clones with a reversed orientation?

Similar to the first rare case, the cDNA sequence does not make good sense when reading from the GAL 4 AD. It might contain a stop codon. However, the fusion partner is reasonably long and contains a good ORF in the opposite strand. The translational product from the opposite strand may encode a protein which, if transcribed and translated, makes good biological sense. How might this happen?

Stanley Fields and his colleagues reported a cryptic promoter present in the ADH terminator region. Indeed, the very first two-hybrid library (a combinatorial yeast genomic library made in 3 independent vectors) was constructed in such a way that the transcription was started from the cryptic promoter in the ADH terminator (7). They performed library screenings and isolated expected positive clones from such a library. This provided strong evidence for the function of the cryptic promoter. Recently, Dr. M. Li et al. (28) reported a first, rare case of one-hybrid screening. In their study, the reversibly oriented cDNA, which encodes a human hepatocyte cDNA in the opposite strand from that of the GAL4 AD, turned out to be a novel transcriptional factor. This factor, belonging to the nuclear receptor superfamily, binds and activates specifically human Hepatitis B virus enhancer II—the bait used in their one-hybrid screening. Since the reverse-oriented ORF cannot fuse with GAL4 AD, it must have its own activation domain. The finding by Dr. Li et al. was one of these cases. Leslie Pick et al. also describe a similar case in Chapter 19 of this book.

We expect that the number of these two types of isolates will increase in the future. Although they look odd, the cases we have encountered were not too rare. We hope one will not throw away some of these clones without careful analysis.

2.4 Additional Two-Hybrid Tests to Verify Positive Interactions

We recommend the following experiments to be performed after a two-hybrid library screening.

2.4.1 Carefully check the ORF, design a primer to exactly locate at the termination codon (in most cases, new library insert containing termination codon, polyadenylation signal, and poly-A tail), resynthesize this library insert by PCR, and insert into pAD vector again. Then repeat the two-hybrid assay.

2.4.2 Switch above (2.4.1) cloning vectors by moving the new library insert from the AD to the DNA-BD vector and vice versa, and then repeat the two-hybrid assay (7,9,36).

2.4.3 Create a frameshift mutation just upstream of the library insert in the AD plasmid by a restriction digestion, followed by fill-in and re-ligation (4). Cotransform the mutated AD/library plasmid and the DNA-BD/ target plasmid into CG-1945 or Y190, select for cotransformants, and assay for *lacZ* expression. In most cases frameshift mutation should stop the reporter gene expression (Table 1).

3. CONCLUSIONS

Although the methods described in this chapter allow several types of false positives to be eliminated, they do not address the biological significance of the interactions observed. In some instances the sequence will suggest that the interaction with the bait may have a real function. However, two-hybrid interactions can occur between proteins that normally do not interact (for example, because they are never expressed at the same time or in the same tissue or subcellular compartment). In order to show biological significance, one needs to verify the interaction by a different technique, preferably co-precipitation from a cell in which both proteins are naturally expressed. It is particularly important to do careful control experiments when dealing with chaperons or ubiquitin-conjugating enzymes, because they can interact with many proteins, some of which may be non-specific, while some may have biological significance. Although false positives were an issue, the two-hybrid system has been used for genome-wide protein linkage mapping (3,6,16,33). Finding ways to eliminate false positves from the yeast two-hybrid system will allow it to be used more widely and successfully in the future.

ACKNOWLEDGMENTS

This work was supported by NSC 87-2314-B-037-062 and NHRI DD01-87X-NG604S (Taiwan, R.O.C.) to Y.R.H. I thank Dr. Li Zhu of CLONTECH Laboratories for valuable suggestions.

REFERENCES

1.**Bai, C. and S.J. Elledge.** 1996. Gene identification using the yeast two-hybrid system. Methods Enzymol. *273*:331-347.

2. **Bartel, P.L., C.T. Chen, R. Sternglantz, and S. Fields.** 1993. Elimination of false positives that arise in using the two-hybrid system. BioTechniques *14*:920-924.

3. **Bartel, P.L., J.A. Roecklein, D. SenGupta, and S. Fields.** 1996. A protein linkage map of *Escherichia coli* bacteriophage T7. Nature Genetics *12*:72-77.

4. **Bendixen, C., S. Gangloff, and R. Rothstein.** 1994. A yeast mating-selection scheme for detection of protein-protein interactions. Nucleic Acids Res. *22*:1778-1779.

5. **Brachmann, R.K. and J.D. Boeke.** 1997. Tag games in yeast: the two-hybrid system and beyond. Curr. Opin. Biotechnol. *8*:561-568.

6. **Brent, R.** 1997. http://xanadu.mgh.harvard.edu/brentlabhomepage4.html. web site.

7. **Chien, C.T., P.L. Bartel , R. Sternglanz, and S. Fields.** 1991. The two-hybrid system: A method to identify and clone genes for proteins that interact with a protein of interest. Proc. Natl. Acad. Sci. USA *88*:9578-9582.

8. **Chuang, J.Y., C.T. Lin, C.W. Wu, and Y.S. Lin.** 1995. A movable and regulable inactivation function within the central region of a temperature-sensitive p53 mutant. J. Biol. Chem. *270*:23899-902

9. **Colas, P. and R. Brent.** 1998. The impact of two-hybrid and related methods on biotechnology. TIBTECH *16*:355-363.

10. **Dagher, M.C. and F.C. Odile.** 1997. Making hybrid of two-hybrid systems. BioTechniques *22*:916-922.

11. **Estojak, J., R. Brent, and E.A. Golemis.** 1995. Correlation of two-hybrid affinity data with in vitro measurements. Mol. Cell Biol. *15*:5820-5829.

12. **Fagan. R., K.J. Flint, and N. Jones.** 1994. Phosphorylation of E2F-1 modulates its interaction with the retinoblastoma gene product and adenoviral E4 19 kDa protein. Cell *78*:799-811.

13. **Fields, S.** 1997. The future is function. Nature Genetics *15*:325-327.

14. **Fields, S. and O. Song.** 1989. A novel genetic system to detect protein-protein interactions. Nature *340*:245-246.

15. **Finan P, Y. Schimizu, I. Gout, J. Hsuan, O. Truong, C. Butcher, P. Bennett, M.D. Waterfield, and S. Kellei.** 1994. An SH3 domain and proline-rich sequence mediate an interaction between two components of the phagocyte NADPH oxidase complex. J. Biol. Chem. *269*:13752-13755.

16. **Finley, R.L. Jr.** 1997. http://cmmg.biosci.wayne.edu/ rfinley/finlab/finlab-home.html. web site.

17. **Finley, R.L. Jr. and R. Brent.** 1995. Interaction trap cloning with yeast. p. 169-203 *In* B.D. Hames and D.M. Glover (Eds.), DNA Cloning, Expression Systems: A Practical Approach, Oxford University Press, Oxford.

18. **Geitz, R.D., B. Triggs-Raine, A. Robbins, K.C. Graham, and R.A. Woods.** 1997. Identification of proteins that interact with a protein of interest: Applications of the yeast two-hybrid system. Mol. Cell. Biochem. *172*:67-79.

19. **Golemis, E.A., I. Serebriiski, J. Gyuris, and R. Brent.** 1997. Interaction trap/two-hybrid system to identify interacting proteins. 20.1.1-20.1.35. *In* Current Protocols in Molecular Biology. John Wiley & Sons, Inc.

20. **Harper, J.W., G.R. Adami, N. Wei, K. Keyomarsi, and S.J. Elledge.** 1993. The p21 Cdk-interacting protein Cip1 is a potent inhibitor of g1 cyclin-dependent kinases. Cell *75*:805-816.

21. **Hengen, P.N.** 1997. False positives from the yeast two-hybrid system. Trends Biol. Sci. *22*:33-34.

22. **Hoffman, C.S. and F. Winston.** 1987. A ten minute DNA preparation from yeast efficiently releases autonomous plasmids for transformation of *Escherichia coli*. Gene *57*:267-272.

23. **Hong, Y.R., L. Han, J. Huang, and M.E. Kleinberg.** 1996. Elimination of false negative results in the two-hybrid system in the phagocyte NADPH oxidase. Kao Hsiung I Hsueh Ko Hsueh Tsa Chih. *12*:301-305.

24. **Housni, H.-E.L., I. Vandenbroere, D. Perez-Morga, D. Christophe, and I. Pirson.** 1998. A rare case of false positive in a yeast two-hybrid screeing: The selection of rearranged bait constructs that produce a functional Gal4 activity. Anal. Biochem. *262*:94-96.

25. **James, P., J. Halladay, and E.A. Craig.** 1996. Genomic libraries and a host strain designed for highly efficient two-hybrid selection in yeast. Genetics *144*:1425-1436.

26. **Kaiser, P. and B. Auer.** 1993. Rapid shuttle plasmid preparation from yeast cells by transfer to *Escherichia coli*. BioTechniques *14*:552.

27. **Legrain, P., M.C. Dokhelar, and C. Transy.** 1994. Detection of protein-protein interactions using different vectors in the two-hybrid system. Nucleic Acids Res. *22*:3241-3242.

28. **Li, M., Y.H. Xie, Y.Y. Kong, X. Wu, L. Zhu, and Y. Wang** 1998. Cloning and characterization of a novel human hepatocyte transcription factor, hB1F, which binds and activates enhancer II of hepatitis B virus. J. Biol. Chem. *273*:29022-29031

29. **Luban, J. and S.P. Goff.** 1995. The yeast two-hybrid system for studying protein-protein interactions. Curr. Opin. Biotechnol. *6*:59-64.

30. **Luo Y., A. Batalao, H. Zhou, and L. Zhu.** 1997. Mammalian two-hybrid system: A complementary approach to the yeast two-hybrid system. BioTechniques *22*:350-352.

31. **Ma, J. and M. Ptashne.** 1987. A new class of transcriptional activators. Cell:*51*:113-119.

32. **Ptashne, M.** 1989. How gene activators work. Sci. Am. *260*:41-47.

33. **Russell, L., R.L. Finley Jr., and R. Brent.** 1997. Two-hybrid analysis of genetic regulatory networks. *In* P.L. Bartel and S. Fields (Eds.), The Yeast Two-Hybrid System. Oxford University Press, Oxford, England.

34. **Sambrook J., E.F. Fritsch, and T. Miniatis.** 1989. Molecular cloning: A Laboratory Manual. 2nd Edn. Cold Spring Harbor Laboratory Press, Cold Spring Harbor, NY.

35. **Segal A.W. and A. Abo.** 1993. The Biochemical basis of the NADPH oxidase of phagocytes. Trends Biol. Sci. *18*:43-47.

36. **Van Aelst, L., M. Barr, S. Marcus, A. Polverino, and M. Wigler.** 1993. Complex formation between RAS and RAF and other protein kinases. Proc. Natl. Acad. Sci., USA *90*:6213-6217.

37. **Zhu, L.** 1997. Yeast GAL4 two-hybrid system. A genetic system to identify proteins that interact with a target protein. Methods Mol. Biol. *63:*173-196.

Modified Yeast Two-Hybrid Systems

7   Two-Hybrid System Applicable to Membrane Proteins

Stephan te Heesen[1,3] and Igor Stagljar[2]
[1]Microbiology Institute, ETH Zurich; [2]Institute of Veterinary Biochemistry, University Zurich-Irchel, Zurich; [3]Origin (Schweiz) AG, Zurich, Switzerland

1. INTRODUCTION

1.1 Limitations and Problems of the Yeast Two-Hybrid System

Protein–protein interactions are essential for many biological processes, including replication, transcription, splicing, secretion, signal transduction, and metabolism. Thus one central task of protein study is the determination of a protein's interaction partners. While the yeast two-hybrid system is a useful and powerful genetic tool for the in vivo analysis of protein–protein interactions (1), its application is limited. First, hybrid proteins generated in the two-hybrid assay are targeted to the nucleus. Proteins that cannot fold correctly in the cytoplasm are excluded from the method. Membrane proteins, which depend on the lipid bilayer for interaction, are also excluded. Second, interactions that are dependent on post-translational modifications that take place within the endoplasmic reticulum, such as glycosylation and disulfide bond formation, are excluded from the two-hybrid system. Third, DNA-binding protein hybrids that activate transcription in the absence of a specific activation-domain hybrid partner cannot be studied using the two-hybrid method. While the region required for inappropriate activation can sometimes be deleted, such an approach may also affect binding of the protein to its cellular partners. Finally, if the protein–protein interaction is mediated by an amino-terminal domain of either protein, the standard orientation of the hybrid constructs may not work because the interaction might be sterically hindered.

1.2 Detection of Protein–Protein Interactions Using Split Ubiquitin

The split-ubiquitin system is an alternative assay for the in vivo analysis of protein interactions. It was developed in 1994 to detect interactions between soluble proteins (3). The system was then modified and shown to work with membrane proteins (10) (Figure 1). In contrast to the two-hybrid system, where a transcription factor is cleaved and reassembled as a result of the interaction of two test proteins, the split-

Yeast Hybrid Technologies
Edited by L. Zhu and G.J. Hannon
© 2000 Eaton Publishing, Natick, MA

ubiquitin system consists of two fragments of ubiquitin. Ubiquitin is a small, 76 amino acid protein that participates in many cellular pathways involved in the degradation of proteins. Specifically, the C-terminus of ubiquitin gets linked by an amide bond to proteins that are to be degraded by a complex system of several enzymes. Ubiquitin-specific proteases (UBPs) recognize the ubiquitin part of these protein fusions and cleave the covalent bond between ubiquitin and the protein. The protein is then degraded by the proteasome (13).

To detect interactions between proteins, the ubiquitin is expressed in two separate fragments from different plasmids. The N-terminal fragment of ubiquitin (amino acids 1–37) is called NubI , with I representing isoleucine, which is the natural amino acid at position 13. The C-terminal part of ubiquitin (amino acids 35–76) is called Cub. If a reporter protein is attached to the C-terminus of Cub, and the fusion protein is coexpressed with NubI, both ubiquitin pieces rapidly reassemble and form a split-ubiquitin heterodimer. The split-ubiquitin is a substrate for UBPs, which cleave the

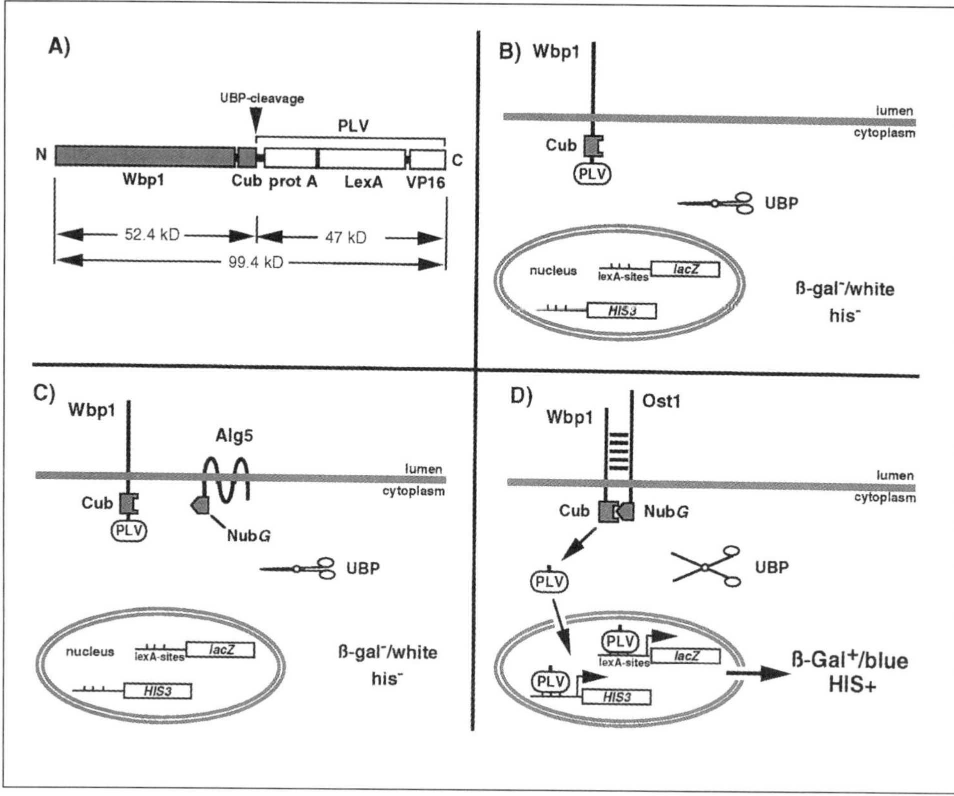

Figure 1. Membrane protein based two-hybrid system using Wbp1p, Ost1p, and Alg5p. (A) Scheme of the Wbp1-Cub-PLV fusion protein. The UBP cleaves once after the C-terminus of Cub, resulting in a 52 kDa fragment of Wbp1-Cub and the modified 47 kDa transcription factor PLV (Protein A-LexA-VP16). (B) The Wbp1-Cub-PLV protein alone is stable, with no cleavage occuring, because no split-ubiquitin can be formed. The PLV transcription factor is unable to activate genes, thus resulting in cells that lack β-galactosidase activity. (C) The coexpression of Wbp1-Cub-PLV with the non-interacting NubG-Alg5p does not lead to formation of the split-ubiquitin, hence cleavage by UBP (closed scissors) and gene activation. (D) Interaction between Wbp1 and Ost1 results in an increase in the local concentration of Cub and NubG and the formation of the split-ubiquitin heterodimer. The heterodimer is recognized and cleaved by the UBP (open scissors), liberating PLV. PLV enters the nucleus and binds to the LexA-binding sites of the *lacZ* and *HIS3* reporter genes. This results in β-galactosidase activity. Cells are blue in the presence of X-gal and grow on agar plates lacking histidine.

118

reporter from the C-terminus of Cub. NubI was modified by mutation to either NubA or NubG, which can no longer spontaneously associate with Cub. NubG and Cub-reporter reassemble only upon interaction of fused test proteins. This provides a tool to test for interactions between proteins. One protein of interest (the "bait") is fused to Cub-reporter, and the other protein (the "prey") is fused to NubG. Interaction of both proteins results in the assembly of the split-ubiquitin. The split-ubiquitin is recognized by UBPs that release the reporter cassette from the bait. The reporter cassette, which consists of protein A fused to LexA-VP16 (PLV), is now able to activate reporter genes in the nucleus. The protein A sequence serves as a tag for easy detection of the fusion protein and/or immunoprecipitation of protein complexes without the need for a specific antibody.

To test the system, we used well defined ER membrane proteins involved in N-glycosylation (Figure 1) (10). Wbp1p is an essential component of the yeast oligosaccharyltransferase complex (11) and is in close proximity to Ost1p, another essential protein of the same complex (5,8,9). Both Ost1p and Wbp1p are type I transmembrane proteins that contain cytoplasmic C-termini and serve as models for interacting proteins. Alg5p is a type II transmembrane protein that has both the N- and C-terminus in the cytoplasm and synthesizes dolichol-phosphoglucose from dicholphosphate and UDP-glucose (12). Alg5p does not interact with the oligosaccharyltransferase complex and therefore serves as a negative control for the system.

The C-terminal of Wbp1p was linked to Cub, followed by the reporter protein A-LexA-VP16 (PLV). This resulted in the fusion bait protein, Wbp1-Cub-PLV (Figure 1A). The protein A sequence from *Staphylococcus aureus* contains two IgG-binding domains that allow easy and sensitive detection of the fusion protein as well as of the cleaved product. The LexA-VP16 cassette consists of the *Escherichia coli* DNA-binding protein LexA followed by the transcriptional activation domain of VP16 (6). LexA-VP16 activates reporter genes that are under transcriptional control of LexA binding sites located in their promoter region.

The bait fusion construct Wbp1-Cub-PLV was unable to activate transcription if expressed as a single fusion protein (Figure 1B). The co-expression of a noninteracting NubG-fusion protein, like NubG-Alg5p, does not lead to gene activation because NubG does not interact with Cub and Wbp1p does not interact with Alg5p (Figure 1C). Co-expression of Ost1 protein fused to NubG, however, results in the assembly of the split-ubiquitin heterodimer. Ost1p interacts with Wbp1p, and this interaction leads to the association of Cub with NubG and to the formation of the heterodimer. The split-ubiquitin heterodimer is recognized by UBPs. The protease releases the PLV reporter protein from its membrane-bound state and is then able to activate the transcription of the *lacZ* and *HIS3* reporter genes in the nucleus (Figure 1D). In contrast to the conventional two-hybrid system, which requires nuclear localization, the interactions are detected at the natural location of the protein of interest. The prerequisite, however, is that Nub and Cub localize to the cytoplasm, since there is no UBP present in the lumen (4).

2. PROTOCOLS

To test for interactions between Wbp1p and Ost1p and between Wbp1p and Alg5p, Cub-Protein A-LexA-VP16 reporter protein is fused to the C-terminal tail of the Wbp1p, resulting in the bait construct Wbp1-Cub-PLV (Figure 1A). The other

Table 1. Vectors and Strains Used in the Membrane Protein-Applicable Two-Hybrid System

Plasmid	Characteristics	Reference
pRS304	*TRP1*, no yeast origin of replication	(7)
pRS304(Δost1-Nub)	Ost1 (aa 102–476)-Nub, *TRP1*	(10)
pRS305	*LEU2*, no yeast origin of replication	(7)
pRS305(Δwbp1-Cub-PLV)	Wbp1p (aa 251–430)-Cub-PLV*LEU2*	(10)
pRS314	CEN6/ARSH4, *TRP1*	(7)
pRS314(Nub*I*-Alg5)	Nub*I*-Alg5, CEN6/ARSH4, *TRP1*	(10)
pAS2	TRP, 2-μm origin, GAL4 DB (reference plasmid)	(2)
pOST1-Nub	OST1-Nub, 2-μm origin, *TRP1*	(10)

Yeast Strains	Genotype	Reference
L40	MAT*a*, *trp1, leu2, his3, LYS2::lexA-HIS3 URA3::lexA-lacZ*	(14)
YG0673	MAT*a*, *trp1, leu2, his3, LYS2::lexA-HIS3 URA3::lexA-lacZ, wbp1:pRS305(Δwbp1-Cub-PLV)*	(10)

candidate protein (Ost1p or Alg5p)—the prey—is fused to Nub (with Nub being Nub*I*, Nub*A*, Nub*G*) as either an N- or C- terminal fusion that creates Nub-Alg5 or Ost1-Nub. The bait plasmid encoding Wbp1-Cub-PLV is linearized and integrated into the chromosomal *WBP1* locus following the transformation of the prey construct (Alg5-Nub or Ost1-Nub). The prey construct is expressed extrachromosomally from a 2-μm plasmid, but can (e.g., in the study of yeast genes) also be integrated into the genome. In the following, both the extrachromosomal expression as well as the integrated prey fusion are described. Transformants are selected and assayed for the production of β-galactosidase. The expression of β-galactosidase indicates that the two hybrid proteins interact and reconstitute a functional ubiquitin molecule.

2.1 Protocol for Testing for an Interaction Between Wbp1p and Ost1p

Materials and Reagents

- Plasmids and Yeast strains selected from Table 1.
- Synthetic dropout (SD) plates (see Appendix F): *omit uracil and leucine*
- SD plates (see Appendix F): *omit uracil, leucine, and tryptophane*

Procedure

1. Construct two fusion genes. One is the fusion between Wbp1p and Cub-PLV (Wbp1-Cub-PLV). The other is a fusion between Ost1p and Nub*G* (Ost1-Nub*G*). As a negative control in this system, Nub*G* is fused to the Alg5p resulting in Nub*G*-Alg5. The orientation and the reading frame between the two parts of each fusion must be maintained so that the hybrid proteins containing the

Wbp1p fused to the Cub-PLV portion, and Ost1p fused to the NubG portion, can be expressed (Protocol 2.2). The native split-ubiquitin subunit, Nub*I*, fused with the protein of interest can be used as a positive control for expression and general topology. Nub*A* has an intermediate affinity to Cub, which may be useful under certain conditions. Be sure that the ubiquitin fusions localize topologically to the cytoplasm, since ubiquitin and ubiquitin-specific proteases are present only in the cytoplasm (4).

2. Transform the yeast reporter strain with the linearized bait construct that expresses the WBP1-Cub-PLV fusion. Select the transformed cells on selective (SD medium lacking uracil and leucine) agar plates. Transform the resulting strain with the prey plasmids (Ost1p-Nub or Alg5-Nub) (Protocol 2.2). Plate the transformed cells on SD media that lacks uracil, leucine, and tryptophane and incubate at 30°C for 2–3 days.

3. After colonies appear, test the transformants for β-galactosidase activity. The transformants can be streaked out in the form of patches and tested using the filter test assay (Protocol 2.3). Transformants expressing Nub*I*, Nub*A* or relevant interacting proteins will turn blue after the filter test assay. The β-galactosidase activity can be quantified using the liquid assay (Protocol 2.4).

4. Besides β-galactosidase assay, interaction can also be analyzed by assaying the cleavage of the PLV portion in vivo by Western blotting analysis and probing with peroxidase-IgG (Protocol 2.5).

2.2 Protocol for Construction of Hybrid Genes and Their Transformation in the Reporter Strain

Materials and Reagents

- SD plates: *omit leucine and uracil*
- SD plates: *omit leucine, uracil, and tryptophane*
- The available vectors and yeast strains are listed in Table 1. It is advisable to integrate the fusion gene that encodes the bait protein into the chromosome. Expression of the protein fusion from episomal or CEN/ARS plasmids might result in overexpression and false positives.

Procedure

1. Digest the bait plasmid pRS305(Δwbp1-Cub-PLV) with *Spe*I restriction enzyme. The cleavage within the *WBP1* gene results in the linearization of the plasmid. Transform the linear DNA into the yeast L40 reporter strain. The DNA integrates into the *WBP1* locus through homologous recombination with the free ends of the plasmid. As a result, an active WBP1-Cub-PLV fusion gene and a 5′-truncated *Δwbp1* gene without promotor is generated. The locus then carries the *LEU2* marker gene of the pRS305, thus allowing selection for integration events. Only the modified Wbp1-Cub-PLV, but no wild-type Wbp1p, will be present in the cell.

2. Plate the transformants on SD plates lacking leucine and uracil, and incubate

them at 30°C until colonies appear, usually after 3–5 days. This procedure results in the strain YG0673. Check transformants for correct integration by polymerase chain reaction or Southern blotting analysis.

3. Transform YG0673 with 1 µg of pOST1-Nub (Nub being Nub*I*, Nub*A*, Nub*G*) that has the prey fusion gene on a plasmid with the 2-µm origin of replication. In this case, Ost1-Nub will be overproduced and able to compete with wild type Ost1p, from the chromosomal *OST1* gene, for interaction (10).

   If the fusion protein is to be analyzed in the absence of its wild-type counterpart, it is best to integrate it into the *OST1* gene such that only the fusion protein is present in the cell. Digest the prey plasmid pRS304(Δost1-Nub) (Nub being Nub*I*, Nub*A*, Nub*G*) with the *Sph*I restriction enzyme and transform the linear DNA using standard protocols for yeast transformation (14).

   As a negative control in the interaction test, YG0673 reporter strain is transformed with the vector plasmid or 1 µg of the 2-µm plasmids pNub-Alg5 (Nub being Nub*I*, Nub*A*, Nub*G*) .

4. Plate the transformants on SD plates lacking uracil, leucine, and tryptophane and incubate them at 30°C until colonies appear.

2.3 Protocol for Filter Assay for the Detection of β-Galactosidase Activity (Figure 2)

Materials and Reagents

- SD-plates: *omit leucine, tryptophane; supplement with 0.2 mM $CuSO_4$*
- Copper(II)sulfate 0.2 mM (50 mg/L $CuSO_4.5H2O$)
- Whatman™ 3 MM filter paper, sterile
- $NaPO_4$ 0.1 M (pH 7.0)
- 20 mg/mL X-gal (5-bromo-4-chloro-β-D-galactopyranoside; dissolve 20 mg X-gal in 1 mL dimethylformamide, store in the dark at -20°C).
- Liquid Nitrogen
- Agarose

Procedure

1. Grow the yeast cells that co-express Wbp1-Cub-PLV and Nub-fusion proteins for two days at 30°C on sterile Whatman filters placed on SD agar plates lacking leucine and tryptophane and supplemented with 0.2 mM $CuSO_4$, which is needed for optimal expression of Nub-ALG5 fusion genes. The drop-out medium is used because cells tend to grow poorly in standard minimal medium. $CuSO_4$ can be omitted if no *CUP1* promotor is used.

2. Using forceps, remove the filters and dip them into liquid nitrogen for 3 min, subsequently allowing them to thaw at room temperature.

3. Overlay the filters with 1.5% agarose in 0.1 M $NaPO_4$-buffer (pH 7.0) containing 0.4 mg/mL X-gal.

4. Incubate the filters at 30°C until the blue color appears. This may require from 20 min up to 24 h.

2.4 Protocol for Quantitation of β-Galactosidase Activity

Materials and Reagents

- SD liquid Medium (see Appendix F): *omit uracil, leucine, and tryptophane, and supplemented with 0.2 mM CuSO$_4$*
- Z buffer: 113 mM Na$_2$HPO$_4$/40 mM NaH$_2$PO$_4$/10 mM KCl/1 mM MgSO$_4$ (pH 7.0) (dissolve 16.1 g Na$_2$HPO$_4$.7H$_2$O, 5.5 g NaH$_2$PO$_4$.H$_2$O, 0.75 g KCl, and 0.25 g MgSO$_4$.7H$_2$O in about 900 mL double distilled H$_2$O, adjust pH to 7.0 with H$_3$PO$_4$, and fill to 1 L with double distilled water).
- 2-Nitro-phenyl-beta-D-galactopyranoside (ONPG) 4 mg/mL in Z buffer (dissolve 40 mg ONPG in 10 mL Z buffer, which may take several h).
- β-mercaptoethanol
- 0.1 M NaCO$_3$

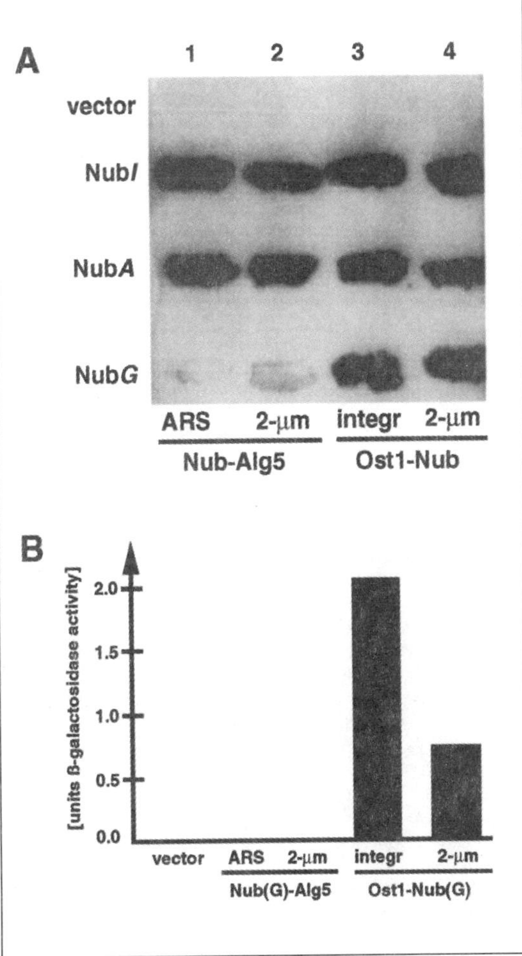

Figure 2. β-galactosidase activity of cells that co-express Wbp1-Cub-PLV and Nub-fusion proteins. (A) Filter test using X-gal as described in protocol 2.3. Cells were grown on Whatman filters, permeabilized, and incubated for 22 h at 30°C in the presence of X-gal. The blue color appears after 20 minutes of incubation. YG0673 cells expressing Wbp1-Cub-PLV and the vector pRS314 are shown (first row, vector). YG0673 cells expressing (1) NubI-Alg5, NubA-Alg5, or NubG-Alg5 from the CEN/ARS plasmid; (2) NubI-Alg5, NubA-Alg5, or NubG-Alg5 from the 2-μm plasmid; (3) Ost1-NubI, Ost1-NubA, or Ost1-NubG from an integrated fusion gene (no wild-type Ost1p present in the cell); (4) Ost1-NubI, Ost1-NubA, or Ost1-NubG from a 2-μm plasmid in the presence of wild-type Ost1p. As a negative control, YG0673 was transformed with the vector pRS314. Cells were grown on Whatman filters, permeabilized, and incubated in the presence of X-gal. Expression of β-galactosidase resulted in blue cells. **(B)** Quantitative β-galactosidase assay of cells as described in protocol 2.4. YG0673 cells expressing Wbp1-Cub-PLV together with the vector, low copy number pRS314(NubG-ALG5) (ARS), NubG-ALG5 (2-μm), OST1-NubG (integrated fusion gene, no wild-type OST1 present), OST1-NubG (2-μm, wild-type OST1 present). Shown are the results of one out of three independent experiments. The strongest activity is found when cells have only one copy of Ost1p, which is Ost1-Nub. If the Ost1-Nub protein has to compete with the wild-type Ost1p the activity is reduced, but clearly visible (2-μm).

Procedure

1. Inoculate yeast transformants that co-express Wbp1-Cub-PLV and Nub-fusion proteins into 3 mL of liquid SD medium lacking uracil, leucine, and tryptophane and supplemented with 0.2 mM $CuSO_4$.

2. Incubate at 30°C until cultures reach mid-log phase [Optical Density $(OD)_{546}$ approx. 1.0].

3. Pellet cells from 1 mL of culture, wash them once in Z buffer, and resuspend in 300 µL Z buffer.

4. Take 100 µL cells, and lyse by 3 freeze/thaw cycles.

5. Add 700 µL Z buffer containing 0.27% (vol/vol) β-mercaptoethanol and 160 µL ONPG (4 mg/mL in Z buffer) and incubate for 1–20 h at 30°C.

6. Add 400 µL 0.1 M $NaCO_3$, centrifuge the samples, and measure the OD_{420}.

7. Calculate the β-galactosidase activity using the formula:

$$\beta\text{-galactosidase units} = 1000 \times OD_{420}/(OD_{546} \times min)$$

2.5 Protocol for Western Blotting Analysis of Cells that Co-Express Wbp1-Cub-PLV and Nub-Fusions (Figure 3)

Materials and Reagents

- SD liquid medium: *omit leucine and tryptophane; supplement with 0.2 mM $CuSO_4$*

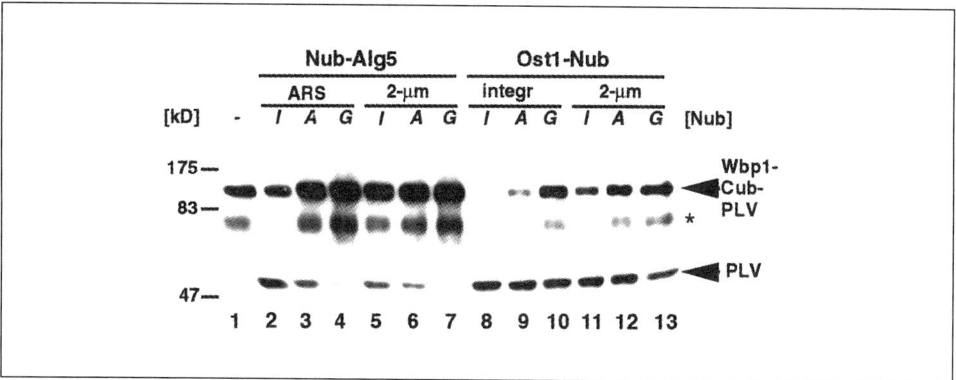

Figure 3. Interaction mediated cleavage of Wbp1-Cub-PLV in vivo. Western blotting analysis of cells that co-express Wbp1-Cub-PLV and various protein fusions to Nub as described in Protocol 2.5. Protein A carrying proteins are detected by peroxidase-IgG followed by enhanced chemiluminescence. Shown are YG0673 cells expressing Wbp1-Cub-PLV with the vector (lane 1), with either NubI-Alg5, NubA-Alg5, or NubG-Alg5 from a CEN/ARS plasmid (lanes 2–4), or from a 2-µm plasmid (lanes 5–7). YG0673 cells expressing Ost1-NubI, Ost1-NubA, or Ost1-NubG as the sole source of Ost1 in the cell (lanes 8–10, integr) or from a 2-µm plasmid in presence of wild-type Ost1p (lanes 11–13) are also depicted. The Wbp1-Cub-PLV fusion protein is visible at about 100 kDa and the cleaved PLV runs at about 52 kDa. Cleavage of Wbp1-Cub-PLV in the presence of Nub-Alg5 strongly decreases in the order NubI > NubA > NubG, both with the CEN/ARS plasmid (lanes 2–4) or with the 2-µm plasmid (lanes 5–7). Very little, if any, cleaved PLV is detected in the case of NubG-Alg5 (lanes 4,7), which is in agreement with the lack of β-galactosidase activity in Figure 2. With Ost1-Nub, the cleavage of Wbp1-Cub-PLV also decreases in the order NubI > NubA > NubG (lanes 8–10 and 11–13). However, cleaved PLV transcription factor is also present in the case of Ost1-NubG, in agreement with the observed β-galactosidase activity in Figure 3. The asterisk (*) denotes a nonspecific degradation product.

- NaCO$_3$, 0.1 M
- NaOH, 1.85 M
- Trichloroacetic acid (TCA), 50% in water
- Sodium dodecyl sulfate (SDS)-Sample buffer/8.0 M Urea: 0.0625 M Tris-HCl, 2% SDS, 5% β-mercaptoethanol, 10% glycerol, 8.0 M Urea, 0.025% bromphenol blue (mix 4.8 g urea with 0.625 mL 1 M Tris-HCl pH 6.8, 1 mL 20% SDS, 0.5 mL β-mercaptoethanol, 1.1 mL 88% glycerol, 2.5 mg bromphenol blue, fill to 10 mL with double distilled water, dissolve at 37°C).
- Peroxidase-IgG

Procedure

1. Grow yeast cells that co-express Wbp1-Cub-PLV and Nub-fusion proteins at 30°C to an OD$_{546}$ of 0.3–1.2 in SD liquid medium lacking leucine and tryptophane and supplemented with 0.2 mM CuSO$_4$.
2. Pellet the cells and resuspend them in 50 µL 1.85 M NaOH per 3 OD units of cells, and incubate on ice for 10 min.
3. Add the same volume of 50% TCA, and precipitate proteins by centrifugation for 5 min.
4. Resuspend the pellet in 50 µL of SDS sample buffer containing 8.0 M urea.
5. Add 20 µL of 1.0 M Tris base and dissolve the protein at 37°C (heating to 95°C sometimes results in clumping of membrane proteins).
6. Centrifuge the samples for 2 min, and use 10 µL extracts for SDS-PAGE/Western blotting analysis.
7. Verify the amount of protein loaded by Coomassie staining of the SDS-gels.
8. Probe the membranes with peroxidase-IgG at 1:5000 dilution. Detect Protein A-fusion proteins by enhanced chemiluminescence (Amersham Pharmacia Biotech, Arlington Heights, IL or Pierce Chemical Co., Rockford, IL, USA).

3. DISCUSSION

One of the main advantages of using the split-ubiquitin method is to leave the proteins to be tested for interactions in their natural location. For membrane proteins this is an especially important issue that, to our knowledge, no other system has addressed thus far. The method has been shown to work with proteins of the yeast oligosaccharyltransferase complex (10). Although there are no comparative studies yet available, it appears that the method is very sensitive because it is also able to detect transient interactions (3).

There are many potential applications for the split-ubiquitin-PLV system. The system can be used to test for interactions between a membrane protein of interest fused to Cub-PLV and either another membrane protein or a soluble protein fused to Nub. Nub-fusions can be made at either the N-terminus or the C-terminus of the protein of interest. The expression of PLV fusions is easily confirmed by Western blotting and probing with peroxidase-IgG. In addition, the method should also prove useful in screening for interactions between a membrane protein fused to Cub-PLV as the bait and a cDNA or genomic library that has N-terminal and C-terminal fusions to NubG.

The bait protein carrying Cub-PLV may not be overexpressed since overexpression would result in false-positives. We found that a soluble Cub-PLV results in gene activation without the need for any Nub (Stagljar and te Heesen, unpublished data). Therefore, the bait fusion protein has to be anchored to the lipid bilayer in order to test for interactions. Soluble proteins of interest might be tested by fusing them to a membrane protein anchor. In contrast, the Nub-fusion protein should be overexpressed because the fusion might have to compete with the wild-type protein for interactions within the protein complex. Ost1-Nub (even Ost1-NubI) expressed from the low copy number vector pRS314 does not give any β-galactosidase activity in the presence of wild-type Ost1p (data not shown).

The split-ubiquitin system could also be used to test for the topology of a given membrane protein. Cleavage only occurs if both Nub and Cub localize to the cytosolic side of the membrane that harbors the UBP. In this situation a membrane protein of interest would be tagged with Cub-PLV and co-expressed with a soluble Nub*I* fusion protein. For a more detailed discussion of the split-ubiquitin system, the reader is referred to (3). In summary, a new system has been generated that will be useful in the detection and the study of membrane protein interactions.

ACKNOWLEDGMENTS

The authors thank Professor Ulrich Hübscher and Professor Markus Aebi for the support of the project, Dr. Alcide Barberis for the strain L40 and the plasmid pLexA202+VP16 as well as for helpful comments. We are also greatly indebted to Dr. Robert Keller for critically reading the manuscript.

REFERENCES

1. **Bartel, P.L. and S. Fields.** 1995. Analyzing protein-protein interactions using two-hybrid system. Meth. Enzymol. *254*:241-263.
2. **Harper, J.W., G.R. Adami, N. Wei, K. Keyomarsi, and S. Elledge.** 1993. The p21 Cdk-interacting protein is a potent inhibitor of G1 cyclin-dependent kinases. Cell *75*:805-816.
3. **Johnsson, N. and A. Varshavsky.** 1994. Split ubiquitin as a sensor of protein interactions *in vivo*. Proc. Natl. Acad. Sci. USA *91*:10340-10344.
4. **Johnsson, N. and A. Varshavsky.** 1994. Ubiquitin-assisted dissection of protein transport across membranes. EMBO J. *13*:2686-2698.
5. **Karaoglu, D., D.J. Kelleher, and R. Gilmore.** 1997. The highly conserved Stt3 protein is a subunit of the yeast oligosaccharyltransferase and forms a subcomplex with Ost3p and Ost4p. J. Biol. Chem. *272*:32513-32520.
6. **Pellett, P.E., J.L. McKnight, F.J. Jenkins, and B. Roizman.** 1985. Nucleotide sequence and predicted amino acid sequence of a protein encoded in a small herpes simplex virus DNA fragment capable of trans-inducing alpha genes. Proc. Natl. Acad. Sci. USA *82*:5870-5874.
7. **Sikorski, R.S. and P. Hieter.** 1989. A system of shuttle vectors and yeast host strains designed for efficient manipulation of DNA in *Saccharomyces cerevisiae*. Genetics *122*:19-27.
8. **Silberstein, S., P.G. Collins, D.J. Kelleher, P.J. Rapiejko, and R. Gilmore.** 1995. The alpha subunit of the *Saccharomyces cerevisiae* oligosaccharyltransferase is essential for vegetative growth of yeast and is homologous to mammalian ribophorin I. J. Biol. Chem. *128*:525-536.
9. **Spirig, U., M. Glavas, D. Bodmer, G. Reiss, P. Burda, V. Lippuner, S. te Heesen, and M. Aebi.** 1997. The STT3 protein is a component of the yeast oligosaccharyltransferase complex. Mol. Gen. Genet. *256*:628-637.
10. **Stagljar, I., C. Korostenky, N. Johnsson, and S. te Heesen.** 1998. A genetic system based on split-ubiquitin for the analysis of interactions between membrane normal proteins in vivo. Proc. Natl. Acad. Sci. USA *95*:5187-5192.

11. **te Heesen, S., B. Janetzky, L. Lehle, and M. Aebi.** 1992. The yeast *WBP1* is essential for oligosaccharyl-transferase activity in vivo and in vitro. EMBO J. *11*:2071-2075.
12. **te Heesen, S., L. Lehle, A Weissmann, and M. Aebi.** 1994. Isolation of the *ALG5* locus, encoding the UDP-glucose:dolichyl phosphate glucosyltransferase from *Saccharomyces cerevisiae*. Eur. J. Biochem. *224*:71-79.
13. **Varshavsky, A.** 1997. The ubiquitin system. TIBS *22*:383-387.
14. **Vojtek, A.B., J.A. Hollenberg, and J.A. Cooper.** 1993. Mammalian ras directly interacts with the serine/threonine kinase raf. Cell *74*:205-214.

8 | CytoTrap Yeast Two-Hybrid System: Non-Transcription Activation-Based Yeast Two-Hybrid System

Hwai Wen Chang
Stratagene, Genetic Systems, and Digital Gene Technologies, La Jolla, CA, USA

1. INTRODUCTION

In many respects, the genome initiative projects have changed the way many scientists perform research. Since they began, the genome projects have yielded an enormous amount of sequence information. However, this large amount of sequence data has not been efficiently translated into biological functional data, partly due to a lack of tools useful in deciphering the role each gene product plays in the cell. One effective way to define a gene's function is to identify proteins that interact with it. Protein–protein interactions are central to many cellular processes including DNA replication, transcription, translation, splicing, secretion, cell cycle control, signal transduction, and intermediary metabolism. Methods to discover these interactions have been and will continue to be extremely useful to elucidate the in vivo function of newly discovered genes and to further develop an understanding of a known protein's function.

The first yeast two-hybrid system, developed by Dr. Stanley Fields and colleagues (12), allows a researcher to screen a cDNA library to discover whether their protein of interest interacts with any other protein in the library. The Fields two-hybrid system, and the second-generation versions that are currently used (9,10,12–14), exploit the modular nature of the DNA binding and activation domains of a transcriptional activator (18). These two-hybrid systems find interactions between a gene of interest (bait) fused to the DNA binding domain and the cDNA library (targets) fused to the activation domain. When a positive interaction occurs, the bait and target proteins reconstitute a transcriptional activator molecule in the nucleus of the cell that in turn activates transcription of a reporter gene. The two-hybrid system has been successfully used and described in papers too numerous to include.

Yeast Hybrid Technologies
Edited by L. Zhu and G.J. Hannon
© 2000 Eaton Publishing, Natick, MA

Despite these successes, there are limitations to Fields' two-hybrid system. The method of identifying protein–protein interactions in Fields' system is by the reconstitution of a transcriptional activator molecule in the nucleus. However, proteins that intrinsically modulate transcription on their own (such as transcriptional activators or repressors), cannot be studied by this method. In addition, interactions that are dependent on post-translational modifications, which take place within the endoplasmic reticulum or in the cytoplasm, may not be detected by Fields' system. Finally, proteins that interact in the cytoplasm or at the cell membrane may also not be detected by this system.

The CytoTrap yeast two-hybrid system described in this chapter provides an alternative to the conventional yeast two-hybrid screening method because it is not based on transcription activation. The CytoTrap two-hybrid system, originally called the SRS system, was developed in Michael Karin's lab at UCSD (3). Similar to Fields' system, it is a yeast-based method of detecting protein–protein interactions in vivo. This system is based upon activation of the Ras signal transduction cascade by localizing a signal pathway component, human Sos (hSos) (8), to its site of activation in the yeast plasma membrane.

The CytoTrap two-hybrid system uses a temperature-sensitive *Saccharomyces cerevisiae* yeast mutant strain cdc25H, which contains a point mutation at amino acid (aa) residue 1328 of the *cdc25* gene (19). This gene, the yeast homolog of human *sos (hsos)*, encodes a guanyl nucleotide exchange factor, which binds to and activates Ras and leads to cell growth (20,24). The mutation in the *cdc25* gene renders the protein temperature sensitive, thus preventing host growth at 37°C. At the permissive temperature, 25°C, host growth is normal. It was shown that the hSos gene product complements the cdc25 defect at 37°C and activates the yeast Ras signaling pathway when expressed and localized to the plasma membrane in yeast (2). However, if the *hsos* gene product lacks a membrane target signal it is unable to complement the cdc25 temperature-sensitive defect.

The CytoTrap system exploits this observation as follows (Figure 1). A protein of interest, or bait, is expressed as a fusion protein with hSos. The library, or target proteins, are expressed with the myristylation membrane-localization signal (22). The yeast cells are then incubated under restrictive conditions (37°C). If the bait and the target protein interact, the hSos protein is recruited to the membrane and thereby activates the Ras signaling pathway and allows the cdc25H yeast strain to grow at 37°C.

This chapter describes the use of the CytoTrap two-hybrid system to study protein–protein interactions and to identify novel dimerization partners. Experiments were carried out to validate the system with known pairs of interacting proteins. In addition, the Gal4 transcriptional activation domain was used as the bait to screen for its interacting partners using the CytoTrap system. This bait, unsuitable in the conventional two-hybrid system, yielded a number of positive clones from a yeast cDNA library.

2. PROTOCOLS

Materials

General Yeast strains and plasmids

Yeast strain:
cdc25H: MATα, *ura3, lys2, leu2, trp1, his* 200, *ade* 101, *cdc25-2, GAL⁺*.

The CytoTrap vectors:

(i) pSos and pMyr:

The pSos and pMyr vectors are designed for the construction and expression of gene fusions with hSos and Myr signal respectively (Figure 2, A and B). The pSos vector contains DNA encoding amino acids 1 to 1067 of the hSos gene (8) and unique 3′ cloning sites. It is used for the construction of bait plasmids containing a DNA insert that encodes a bait protein. The pMyr vector contains DNA encoding the myristylation membrane localization signal of v-Src (22) and unique 3′ cloning sites and is used for the construction of cDNA plasmid libraries that contain DNA inserts encoding target proteins.

Both pSos and pMyr contain the colE1 and 2 μ origins for replication in *Escherichia coli* and yeast. The pSos vector contains the *LEU2* gene and the pMyr vector contains the *URA3* gene for selection in yeast. To rapidly distinguish between the two vectors when transforming back into *E. coli*, pSos vector contains the ampicillin-resistant gene and the pMyr vector contains the chloramphenicol-resistant gene. The ADH1 promoter and GAL1 promoter govern the expression of the bait and target proteins in the pSos and pMyr vectors, respectively. Expression of the bait protein from the ADH1 promoter is constitutive. Expression of the target protein from the GAL1 promoter is repressed when yeast are grown in the presence of glucose but activated in the presence of galactose. The ADH1 and CYC1 terminators provide the signals necessary for transcriptional termination of the bait and target genes in yeast.

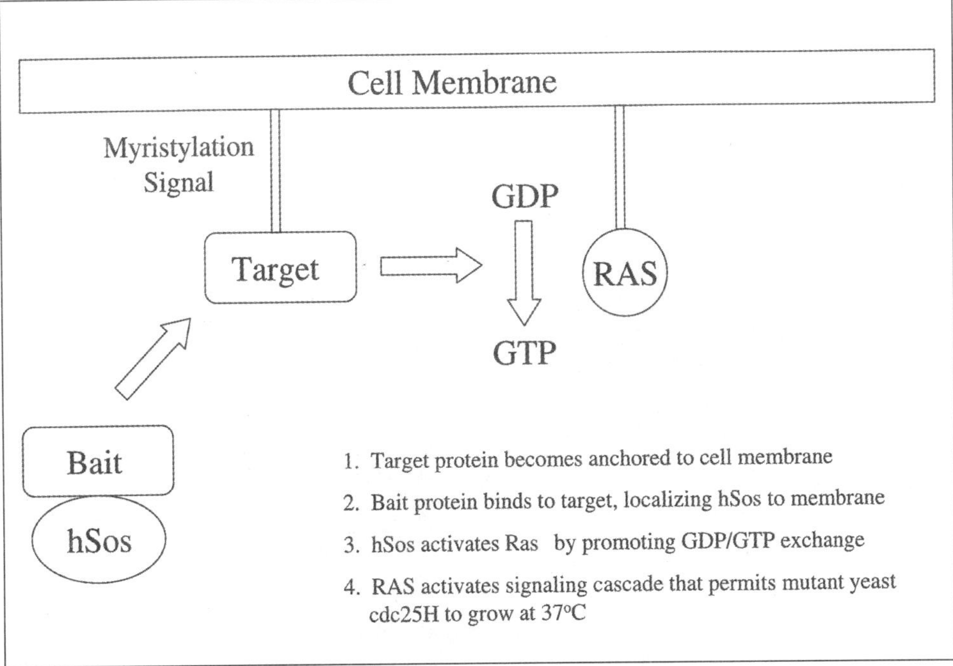

Figure 1. CytoTrap yeast two-hybrid system. Interaction between specific bait (hSos-bait) and a target (Myr-target) results in the recruitment of hSos to the plasma membrane and activation of yeast Ras. This allows the growth of cdc25H yeast strain at 37°C.

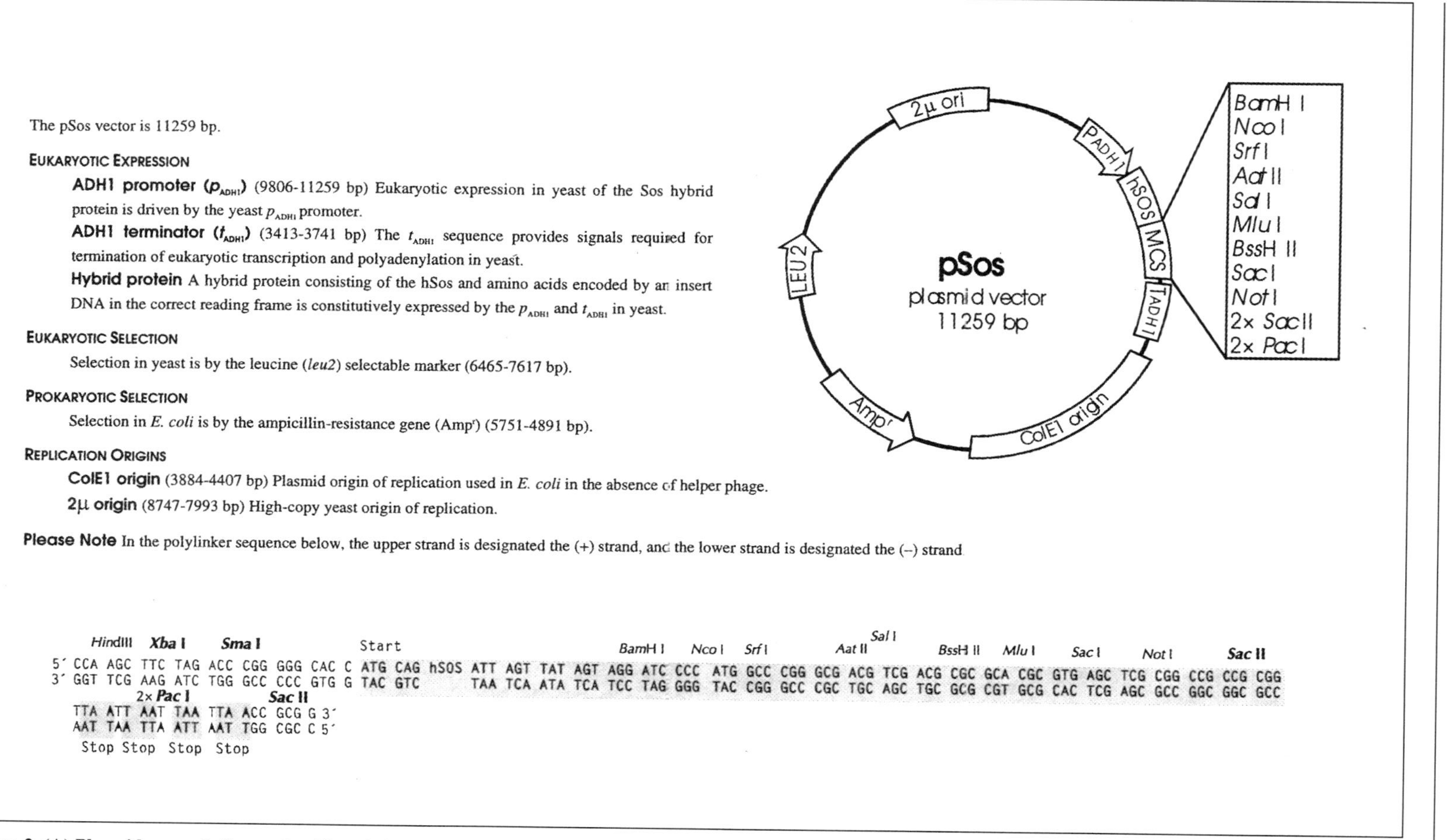

The pSos vector is 11259 bp.

EUKARYOTIC EXPRESSION

ADH1 promoter (p_{ADH1}) (9806-11259 bp) Eukaryotic expression in yeast of the Sos hybrid protein is driven by the yeast p_{ADH1} promoter.

ADH1 terminator (t_{ADH1}) (3413-3741 bp) The t_{ADH1} sequence provides signals required for termination of eukaryotic transcription and polyadenylation in yeast.

Hybrid protein A hybrid protein consisting of the hSos and amino acids encoded by an insert DNA in the correct reading frame is constitutively expressed by the p_{ADH1} and t_{ADH1} in yeast.

EUKARYOTIC SELECTION

Selection in yeast is by the leucine (*leu2*) selectable marker (6465-7617 bp).

PROKARYOTIC SELECTION

Selection in *E. coli* is by the ampicillin-resistance gene (Amp') (5751-4891 bp).

REPLICATION ORIGINS

ColE1 origin (3884-4407 bp) Plasmid origin of replication used in *E. coli* in the absence of helper phage.

2µ origin (8747-7993 bp) High-copy yeast origin of replication.

Please Note In the polylinker sequence below, the upper strand is designated the (+) strand, and the lower strand is designated the (–) strand

```
       HindIII  Xba I    Sma I                  Start                                    BamH I   Nco I  Srf I        Aat II   Sal I     BssH II  Mlu I    Sac I    Not I    Sac II
    5´ CCA AGC TTC TAG ACC CGG GGG CAC C ATG CAG hSOS ATT AGT TAT AGT AGG ATC CCC ATG GCC CGG GCG ACG TCG ACG CGC GCA CGC GTG AGC TCG CGG CCG CCG CGG
    3´ GGT TCG AAG ATC TGG GCC CCC GTG G TAC GTC      TAA TCA ATA TCA TCC TAG GGG TAC CGG GCC CGC TGC AGC TGC GCG CGT GCG CAC TCG AGC GCC GGC GGC GCC
            2× Pac I            Sac II
    TTA ATT AAT TAA TTA ACC GCG G 3´
    AAT TAA TTA ATT AAT TGG CGC C 5´
    Stop Stop Stop Stop
```

Figure 2. (A) Plasmid map of pSos vector. The relative position of the multiple cloning sites and the hSOS sequence are shown.

The pMyr XR vector is 6023 bp.

PROKARYOTIC EXPRESSION

T7 RNA promoter (5998–6017 bp) Expression in prokaryotes of the Myr hybrid protein is driven by the T7 RNA promoter.

EUKARYOTIC EXPRESSION

GAL1 promoter (p_{GAL1}) (5524–5974 bp) Eukaryotic expression in yeast of the Myr hybrid protein is driven by the yeast p_{GAL1} and induced by adding galactose to the growth media.

CYC1 terminator (t_{CYC1}) (121–387 bp) The t_{CYC1} sequence provides signals required for termination of eukaryotic transcription and polyadenylation in yeast.

Hybrid protein A hybrid protein consisting of Myr signal and amino acids encoded by an insert DNA in the correct reading frame is inducibly expressed by the p_{GAL1} and t_{CYC1} in yeast.

PROKARYOTIC SELECTION

Selection in *E. coli* is by the chloramphenicol-resistance gene (Cam') (1290–2133 bp).

EUKARYOTIC SELECTION

Selection in yeast is by the uracil (*ura3*) selectable marker (2399–3506 bp).

REPLICATION ORIGINS

f1 (+) origin (4984–5509 bp) f1 filamentous phage origin of replication allowing recovery of the sense strand of the Myr when a host strain containing the pMyr XR vector is co-infected with helper phage.

ColE1 origin (388–911 bp) Plasmid origin of replication used in *E. coli* in the absence of helper phage.

2μ origin (3512–4983 bp) High-copy yeast origin of replication.

Please Note In the polylinker sequence below, the upper strand is designated the (+) strand, and the lower strand is designated the (–) strand.

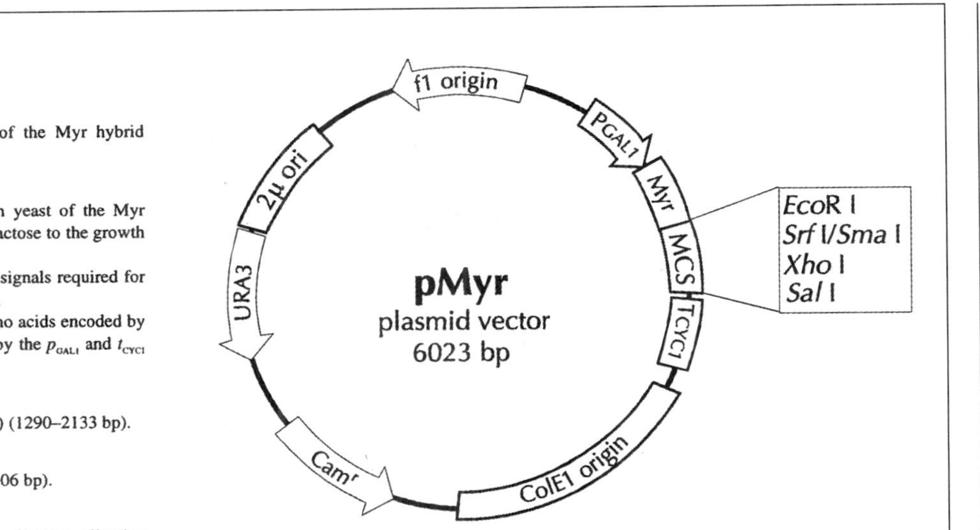


```
                  Xba I
     Hind III               Spe I  Start                                                                        Srf I/Sma I
                                                                                       Xba I   EcoR I           Xho I   Sal I   Stop Stop
5´ AAG CTT CTA GAG CTT ACT AGT ATG  GGG AGT AGC AAG AGC AAG CCT AAG GAC CCC AGC CAG CGC CGG TCT AGA GAA TTC GCC CGG GCC TCG AGG TCG ACT AAT TGA ATA
3´ TTC GAA GAT CTC GAA TGA TCA TAC  CCC TCA TCG TTC TCG TTC GGA TTC CTG GGG TCG GTC GCG GCC AGA TCT CTT AAG CGG GCC CGG AGC TCC AGC TGA TTA ACT TAT
     Stop      Xba I
   ATA AGC TCT AGA GGG CCG C  3´
   TAT TCG AGA TCT CCC GGC G  5´
```

Figure 2. (B) Plasmid map of pMyr vector. The relative position of the multiple cloning sites and the myristylation signal are shown.

(ii) Positive control plasmids:

The pSos-MafB plasmid expresses the Sos protein and full-length MafB (17) as a hybrid protein. The pMyr-MafB plasmid expresses a hybrid protein that contains the myristylation signal fused to full-length MafB. MafB belongs to a large family of bZIP transcription factors and can form homodimers with itself.

(iii) Negative control plasmids:

The pSos-Coll plasmid expresses the Sos protein and amino acids 148–357 of murine 72-kDa type IV collagenase (21). The pMyr-Lamin C plasmid expresses the myristylation signal fused to human lamin C (6).

Reagents

- Synthetic glucose minimal medium [synthetic dropout (SD)/glucose (-UL)] (per Liter)

    1.7 g yeast nitrogen base without amino acids
    5 g Ammonium sulfate
    20 g dextrose
    Add 17 g Bacto agar for SD agar plates
    Adjust the total volume to 900 mL with dH_2O
    Autoclave for 15 min at 121°C, cool to 55°C. Add 100 mL of the appropriate filter-sterilized 10× dropout solution.

- Synthetic galactose minimal medium [SD/galactose (-UL)] (per Liter)

    1.7 g yeast nitrogen base without amino acids
    5 g Ammonium sulfate
    20 g galactose
    10 g raffinose
    Add 17 g Bacto agar for SD agar plates
    Adjust the total volume to 900 mL with dH_2O
    Autoclave for 15 min at 121°C, cool to 55°C. Add 100 mL of the appropriate filter-sterilized 10× dropout solution.

- LiSORB (per Liter)

    100 mM LiOAc
    10 mM Tris-HCl (pH 8.0)
    1 mM EDTA
    1.0 M sorbitol
    Add dH_2O to a volume of 1 Liter; verify that the pH is 8.0
    Autoclave
    Store at room temperature

- PEG/LiOAc solution

    10 mM Tris-HCl (pH 8.0)
    1 mM EDTA (pH 8.0)
    100 mM LiOAc (pH 7.5)
    40% (wt/vol) PEG 3350

- Salmon Sperm DNA

    Boil 400 µL of 20 mg/mL sheared salmon sperm DNA for 5 min

Add 600 µL of LiSORB to the salmon sperm DNA and mix by pipetting
Cool the salmon sperm DNA mixture to room temperature (not below room temperature or the mixture will gel)

- Yeast Lysis solution for DNA isolation

  2.5 M LiCl
  50 mM Tris-HCl (pH 8.0)
  4% Triton X-100
  62.5 mM EDTA

Miscellaneous

- Replica block and velvets
- Sterile glass balls (4 mm)
- Acid-washed glass beads (425–600 µm)

2.1 Protocol for the Verification of the Yeast Host Strain Phenotype

Procedure

1. Prepare a fresh plate of yeast cdc25H strain on a YPAD agar plate. Incubate the plate at 25°C for 4–6 days.

2. Prepare SD agar plates using the appropriate 10× dropout solution to test the yeast host strain for the following nutritional requirements: tryptophan (Trp), leucine (Leu), histidine (His), and uracil (Ura). Streak the yeast host strain onto the agar plates with the appropriate 10× dropout solution and incubate the plates at 25°C for 4–6 days.

3. After the phenotype has been verified, use the test colony to inoculate medium for the preparation of competent yeast cells.

2.2 Protocol for the Preparation of Yeast cdc25H Competent Cells

Procedure

1. Inoculate 100 mL of YPAD broth in a 250-mL flask with a single cdc25H yeast colony. Inoculate 5 × 100 mL cultures. Incubate the yeast cultures at 25°C with vigorous shaking until the culture reaches $OD_{600} = 0.7$. (Approximately 24 h).

2. Plate 71.4 µL of the culture (approximately 1×10^6 cells) on a YPAD plate. Incubate at 37°C for at least 4 days. If more than 30 colonies appear on the plate, discard the culture that contains a high number of temperature sensitive revertants or a yeast strain that is not cdc25H.

3. Pellet the yeast culture from step 1 by spinning the culture at 4000× *g* for 10 min at 4°C. Resuspend the yeast cell pellet in 100 mL of dH_2O by repeated pipetting with a 10-mL pipette. Spin the yeast cells at 4000× *g* for 10 min at 4°C.

4. Resuspend the yeast cell pellet in 50 mL of LiSORB and incubate the cell suspension at room temperature for 30 min.

5. Pellet the yeast cells by spinning the cells at 4000× *g* for 10 min at 4°C. Resuspend the yeast cell pellet in 500 µL of LiSORB.

6. Prepare the salmon sperm DNA by boiling 400 µL of 20 mg/mL sheared salmon sperm DNA for 10 min. Add 600 µL of LiSORB to the salmon sperm DNA and mix by pipetting. Cool the salmon sperm DNA mixture to room temperature.

7. Add the 1 mL of salmon sperm DNA mixture from step 6 to the 500 µL of yeast cells from step 5. Mix thoroughly but gently by pipetting.

8. Add 9 mL of the PEG/LiOAc solution and 880 µL of DMSO. Mix thoroughly but gently by pipetting.

9. Aliquot the prepared yeast competent cells into separate microcentrifuge tubes. The yeast competent cells can be used immediately or can be frozen at -80°C for later use.

10. To estimate the number of revertants or to determine possible contamination in the yeast competent cells, add 2 µg of pSOS and 2 µg of pMyr combined with 10 µL of β-mercaptoethanol to 500 µL of yeast competent cells in a microcentrifuge tube. Mix the contents of the microcentrifuge thoroughly but gently by inversion or tapping. Incubate the transformation at room temperature for 30 min with occasional mixing. Heat shock the transformations at 42°C for 20 min. Place the transformations on ice for 5 min.

11. Centrifuge cells for 30 s at 10 000× *g* at room temperature. Remove the supernatant. Resuspend cells in 0.5 mL 1 M sorbitol.

12. Plate 10 µL and 100 µL of the cells onto 150-mm SD/glucose (-UL) agar plates. These platings will be used to determine cotransformation efficiency.

13. Plate the remainder of the transformation reaction on a 150-mm SD/glucose (-UL) agar plate with 4-mm glass beads. Incubate the inverted plate at 37°C for >4 days. No colonies should form. If colonies from, discard the plates.

14. To calculate the cotransformation efficiency, use the following equation. The number of colony forming units (cfu) is taken from the cells plated in step 11. For accurate counting, there should be at least 30 and no more than 300 cfu/plate. The transformation efficiency should be at least 0.5×10^3–1×10^4 cfu/µg.

$$\frac{\text{Number of cfu} \times \text{Total suspension volume (500 µL)}}{\text{Volume of transformation plated (µL)} \times \text{Amount of DNA used (2 µg)}}$$

$$= \text{cfu/µg DNA (Transformation efficiency)}$$

2.3 Protocol for Yeast Transformation and Detection of Protein–Protein Interaction

Procedure

1. Use standard methods to clone the gene that encodes the proteins to be tested for interaction into the pSos vector (bait vector) and the pMyr vector (target vector).

2. Transform the following combination of plasmids into cdc25H yeast strain.

    a. pSos + pMyr
    b. pSos-MafB + pMyr-MafB
    c. pSos-MafB + PMyr-Lamin C
    d. pSos-Coll + pMyr-MafB
    e. pSos-bait + pMyr-Lamin C
    f. pSos-Coll + pMyr-target
    g. pSos-bait + pMyr-target

3. Add 0.3 µg of each plasmid to 100 µL of yeast competent cells in separate microcentrifuge tubes.

4. Add 2 µL of β-mercaptoethanol to each tube. Mix the contents of each microcentrifuge tube thoroughly but gently by inversion or tapping.

5. Incubate the transformations at room temperature for 30 min with occasional tapping. Heat shock the transformations at 42°C for 20 min. Place the transformations on ice for 5 min.

6. Centrifuge the cells for 30 s at $10\,000\times g$ at room temperature. Remove the supernatant. Resuspend cells in 0.5 mL 1 M sorbitol.

7. For each transformation, plate the entire transformation reaction on a 100-mm SD/glucose (-UL) agar plate with 4-mm glass beads. Incubate the inverted plates at room temperature until colonies are visible (4–6 days).

8. For each colony to be picked, aliquot 25 µL of autoclaved H_2O to wells of sterile 96-well plates (one colony/well). For each transformation, pick at least 3 colonies into the wells (one colony/well).

9. Spot 2.5 µL of the yeast/H_2O suspension onto two SD/galactose (-UL) agar plates and two SD/glucose (-UL) agar plates.

10. Incubate one plate each at 37°C. Keep the other plates at room temperature.

11. Score the growth at 37°C after at least 5 days of incubation time.

12. The Myr-target fusion protein is expressed under the control of the GAL1 promoter. Therefore it is made at high levels in yeast grown on medium containing galactose as a carbon source but not on medium containing glucose as a carbon source. If there is a positive, specific interaction between the Sos-bait and Myr-target fusion proteins, then colonies deriving from transformation between the pSos-bait and the pMyr-target plasmids will grow on SD/galactose (-UL) plates but not on SD/glucose (-UL) plates at 37°C.

    Note: In some cases, if an interaction occurs with very high affinity and Sos-bait and Myr-target fusion proteins are very stable proteins in yeast, a weak growth will occur on glucose medium at 37°C. This is because the very low levels of transcription from GAL1 promoter on glucose medium allow accumulation of sufficient levels of stable Myr-target fusion proteins for an interaction to be detected.

13. One potential problem to be cognizant of is that the Sos-bait fusion protein may have intrinsic ability to target the Sos protein to the membrane. This would result in growth on either galactose or glucose plates at 37°C, which if substantial enough may block the ability to detect an interaction. Therefore, it

is sometimes necessary to truncate the Sos-bait fusion protein to remove membrane-targeting sequences.

14. Another potential problem to be considered is that the Myr-target fusion protein may interact with the Sos protein or activate the Ras signaling pathway itself. This would result in growth on galactose plates at 37°C. These proteins cannot be studied using the CytoTrap system.

15. If an interaction is not detected, it is important to exclude the possibility that one or both of the fusion proteins are not expressed. Expression of the Sos-fusion proteins can be determined by performing a Western blotting experiment of crude yeast lysate, using an antibody that immunoreacts with either the protein expressed from the DNA insert or the Sos protein.

Results

We verified the ability of the CytoTrap yeast two-hybrid system to detect protein–protein interaction via expression MafB proteins. MafB belongs to large family of bZIP transcription factors that are expressed in a wide variety of tissues and, reportedly, forms homodimers with itself or heterodimers with Fos and ATFs through the leucine zipper motif (17). However, due to the inherent transcriptional activation ability of MafB, it could not be studied in the conventional yeast two-hybrid system.

The full-length MafB was fused in-frame to the C-terminus of hSos (pSos-MafB) in the pSos vector or to the C-terminus of the myristylation signal (pMyr-MafB) in the pMyr vector. As a negative control, the MafB without the leucine zipper region was fused to the C-terminus of hSos [pSos-MafB (-LZ)]. Pairs of plasmids were introduced into cdc25H cells, and transformants were selected at 25°C on SD/glucose (-UL) plates. Colonies were then patched onto SD/galactose (-UL) plates or SD/glucose (-UL) plates and transferred to 37°C. Only cells containing the pSos-MafB plus pMyr-MafB constructs grew at 37°C (Figure 3). Cells that contained

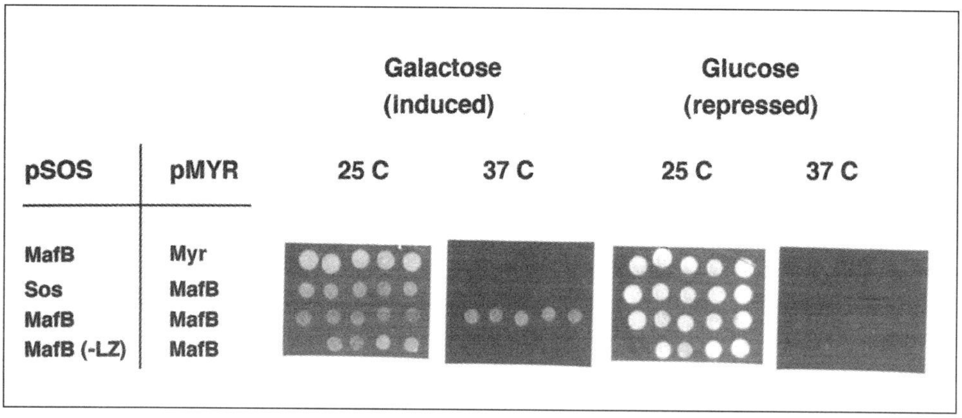

Figure 3. Complementation of the cdc25H mutation through MafB homodimerization. Cdc25H cells were transformed with the indicated plasmids and plated on glucose minimal medium supplemented with the appropriate amino acids and bases as described in Methods. Five independent transformants were patched onto glucose or galactose minimal plates and grown at either 25°C or 37°C. Only transformants expressing both Sos-MafB and Myr-MafB fusion proteins grew efficiently on galactose plates at 37°C. Expression of the Sos-MafB fusion protein was confirmed by Western blotting analysis using anti-SOS monoclonal antibody (Signal Transduction; data not shown). MafB (-LZ): MafB protein without the leucine zipper region.

pSos-MafB plus the pMyr vector, or the pSos vector plus pMyr-MafB did not grow at 37°C. In addition, no growth was observed on SD/glucose (-UL) plates at 37°C. We expected this, since the expression of the Myr-MafB fusion is under control of the GAL1 promoter and is repressed in the presence of glucose.

Other interactions demonstrated in GAL4 or LexA two-hybrid systems were also in the CytoTrap system: the interaction between CD-40 and TRAF2 (23), PI3 kinase subunits p110 and p85 (15,16), and the lambda cI repressor protein with itself (11) (data not shown).

2.4 Protocol for Library Screening

Procedure

This protocol is initially similar to that for studying specific interactions. First, Sos is fused to the protein of interest, and a series of control experiments are performed to establish whether it is suitable or must be further modified. These controls will determine whether the bait is made as a stable protein in yeast, is localized in the cytoplasm, and will not activate the system.

1. Insert the gene encoding the protein of interest (bait) into the polylinker of pSos to make an in-frame protein fusion.

2. Determine whether the Sos-bait fusion protein alone can rescue the temperature-sensitive phenotype of cdc25H yeast strain. Transform cdc25H yeast strain using the following combinations of plasmids.

   a. pSos + pMyr
   b. pSos-MafB + pMyr-MafB
   c. pSos-MafB + PMyr-Lamin C
   d. pSos-Coll + pMyr-MafB
   e. pSos-bait + pMyr-Lamin C

3. Once bait has been made and tested, one can proceed to a library screen.

4. Yeast competent cells were prepared as described in **Preparation of yeast cdc25H competent cells**. Add 40 µg of pSos-bait plasmid, 40 µg of pMyr-cDNA plasmid library, and 200 µL of β-mercaptoethanol to 10 mL of yeast competent cells in a 50-mL conical tube. Mix the contents thoroughly but gently. Transfer the contents into 20 microcentrifuge tubes.

5. In a separate microcentrifuge tube, add 2 µg of pSos plasmid with 2 µg of pMyr-cDNA plasmid library and 10 µL of β-mercaptoethanol to 500 µL of yeast competent cells.

6. Incubate the transformations at room temperature for 30 min with occasional mixing. Heat shock the transformations at 42°C for 20 min. Place the transformations on ice for 5 min.

7. Centrifuge cells for 30 s at $10\,000 \times g$ at room temperature. Remove the supernatant. Resuspend cells in 0.5 mL 1 M sorbitol.

8. For each transformation, plate the entire transformation reaction on a 150-mm SD/glucose (-UL) agar plate using 4-mm glass beads (It is very important to use glass beads to achieve even distribution of yeast colonies on the plates).

9. Incubate the inverted plates at room temperature for 48 h. Increasing incubation time will increase the sensitivity of detection but will also greatly increase background levels of temperature-sensitive revertants. These revertants can be screened at step 11. Do not incubate the plates for more than 4 days.

10. Replica plate the transformants onto SD/galactose (-UL) agar plates and incubate the plates at 37°C. Although no colonies are visible at this point, try to press evenly and hard while replica plating. Colonies should start to appear after 3 days. Keep the SD/glucose plates at room temperature to determine the transformation efficiency. Approximately 1×10^4–2×10^4 colonies should form on each plate.

11. Pick the colonies after 6 days, but keep the plates at 37°C since some colonies may appear much later (10 days). Colonies arising from the pSos and pMyr-cDNA transformation will provide an estimate of the numbers of false positive clones from the cDNA library and the temperature sensitive revertants.

12. Patch the colonies that grow at 37°C onto two SD/glucose (-UL) and one SD/galactose (-UL) agar plates. Transfer one SD/glucose (-UL) and the SD/galactose (-UL) agar plate to 37°C. Keep the other SD/glucose (-UL) agar plate at room temperature for further patching. Repatch colonies that grew at 37°C on SD/galactose (-UL) agar plates, but not on SD/glucose (-UL) agar plates. Only those colonies that grow at 37°C on SD/galactose (-UL) agar plates but not on SD/glucose (-UL) agar plates both times should be considered "putative positive" clones and analyzed further.

13. For "putative positive" clones, isolate library plasmid DNA from yeast by a rapid miniprep protocol such as the following:

    (a) Inoculate 5 mL of SD/glucose (-UL) media in a 50-mL conical tube for each of the putative positive clones.

    (b) Incubate the culture at room temperature with vigorous shaking (250 rpm) until the culture is saturated (OD_{600} >1.0).

    (c) Pellet the yeast culture at 4000× g for 5 min at 4°C.

    (d) Resuspend the yeast pellet in 0.3 mL of Yeast Lysis Solution for DNA Isolation. Transfer resuspended yeast culture to a 1.5 mL microcentrifuge tube.

    (e) Add 50 µL of acid-washed glass beads (0.5 mm) and 300 µL of phenol/chloroform to the microcentrifuge tubes. Vortex-mix vigorously for 1 min.

    (f) Spin the suspension at 14 000× g for 5 min at room temperature.

    (g) Transfer the top aqueous phase containing the DNA to a new microcentrifuge tube.

    (h) Precipitate the DNA with 600 µL of 100% (vol/vol) ice-cold ethanol at -20°C overnight or at -80°C for 15 min. Spin the suspension at 14 000× g for 10 min at 4°C. Decant the supernatant.

    (i) Wash the DNA pellet with 1 mL of 70% (vol/vol) ethanol and re-spin the pellet at 14 000× g for 5 min at room temperature. Decant the supernatant and dry the DNA pellet under vacuum.

    (j) Resuspend the DNA pellet in 40 µL of dH$_2$O. Precipitate the DNA with 4.8

μL of 3 M NaOAc (pH 5.2) and 100 μL of ethanol.

(k) Repeat steps (f)-(I).

(l) Resuspend the DNA pellet in 20 μL of dH$_2$O.

(m) Transform high-efficiency electroporation competent cells and select for pMyr cDNA plasmid by plating on LB-chloramphenicol agar plates.

(n) Identify colonies that contain the pMyr cDNA plasmid by preparing miniprep DNA from isolated colonies from the LB-chloramphenicol agar plates and by restriction digest analysis.

Verification of the specificity of protein–protein interactions

To verify the specificity of the interaction between the bait and target proteins, transform cdc25H yeast strain with the following combinations of plasmids and assay the transformants for the ability to grow at 37°C on SD/galactose (-UL) agar plates.

a. pSos + pMyr

b. pSos-bait + pMyr-target

c. pSos-Coll + pMyr-target

d. pSos-bait + pMyr-Lamin C

1. Prepare and transform the yeast competent cells by cotransformation as described in **Yeast Transformation and Detection of Protein–Protein Interaction**.

2. Patch the transformants that grow on the SD/glucose (-UL) agar plates to SD/glucose and SD/galactose agar plates. Incubate the plates for >4 days at 25°C and 37°C.

3. Determine the growth phenotype of the cotransformants.

4. Those cDNAs that meet the criteria for specificity noted above are ready to be sequenced or otherwise characterized.

Results

Gal4 has been frequently used to study eukaryotic transcriptional activators, yet few direct interactions between Gal4 and its potential transcriptional targets have been reported. To identify proteins that physically interact with the Gal4 transcriptional activation domain (Gal4AD, aa 81–874), it was fused to hSos (pSos-Gal4AD). This plasmid was used as bait to screen an expression library of yeast cDNAs fused to the myristylation signal. Approximately 5×10^5 transformants containing library plasmids and the pSos-Gal4AD plasmid were grown on SD/glucose (-UL) agar plates at 25°C, replica plated onto SD/galactose (-UL) agar plates, and transferred to 37°C. 344 colonies that grew at 37°C were isolated and tested for growth on SD/glucose (-UL) and SD/galactose (-UL) agar plates at 37°C (Figure 4A). Library plasmids were isolated from 53 clones that exhibited galactose-dependent growth at 37°C and were re-transformed into cdc25H cells with either the original bait (pSos-Gal4AD) or pSos (Figure 4B). Twenty-five isolates did not exhibit Gal4-dependent growth and thus represent the background of the cDNA library. These plasmids include members of the Ras family and components of the Ras signaling pathway

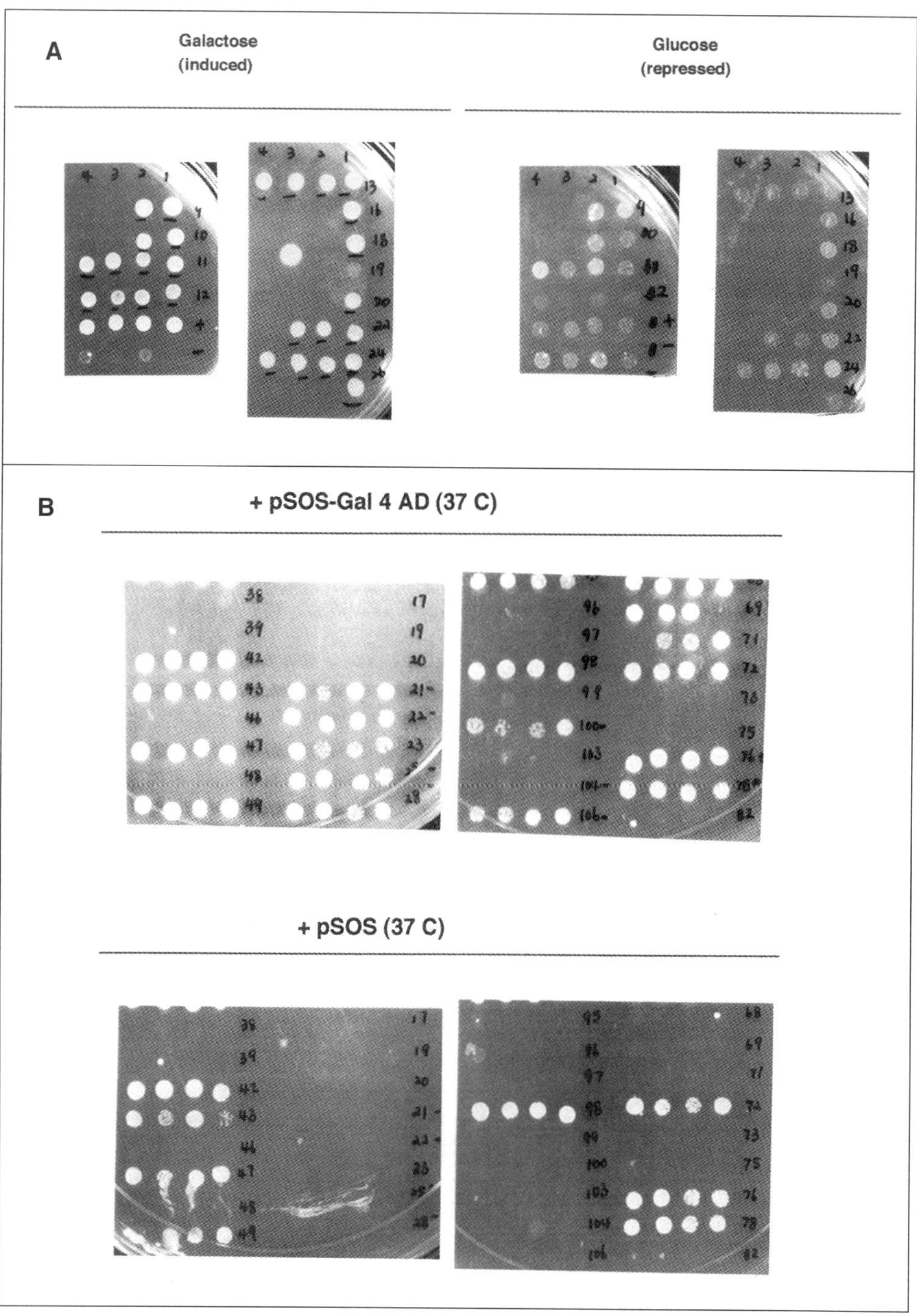

Figure 4. Screening for Gal4 transcriptional activation domain (AD) interacting proteins. (A) A yeast cDNA library in which the cDNAs are fused to a 5′ myristylation signal was cotransformed into cdc25H cells along with the pSOS-Gal4 AD expression vector. Colonies expressing both plasmids were selected at 25°C (approximately 5×10^5 transformants), then replica plated and incubated at 37°C. Colonies that grew at 37°C were then patched onto either glucose or galactose minimal plates and incubated at 37°C. Colonies that show galactose-dependent growth [colonies that were underlined in **(A)**] were chosen for further analysis. **(B)** Library plasmids were isolated from clones that exhibited galactose-dependent growth at 37°C in **(A)** and re-transformed into cdc25H cells with either pSOS-Gal4 AD or pSOS. Individual colonies were patched onto galactose plates and grown at 37°C.

such as protein kinase C, etc. Fourteen plasmids suppressed the cdc25H phenotype only in the presence of pSos-Gal4 and were sequenced. These isolates represent unique interactions to be further evaluated (manuscript in preparation).

3. DISCUSSION

The CytoTrap yeast two-hybrid system provides an alternative to the conventional yeast two-hybrid system. Because it is not based on transcriptional activation, this system is the best method to study interactions that involve transcriptional activators or repressors. In addition, certain proteins may function more physiologically when expressed in the cytoplasm rather than in the nucleus. For example, the CytoTrap system should be more suitable to examine interactions between proteins that require modification by cytoplasmic or membrane-associated enzymes. Finally, several proteins were toxic to yeast in the conventional two-hybrid system (1). Toxicity may be alleviated when these proteins are fused to hSOS and expressed in the cytoplasm.

As with all methods, the CytoTrap yeast two-hybrid system has limitations. First, this system is based on the recruitment of hSos protein to the yeast membrane. It will not be suitable for studying interactions with an integral membrane protein or a protein that interacts with a membrane component. Proteins that are localized at the membrane cannot be used as baits without modification. Second, the cytoplasmic localization requirement for interactions may be sub-optimal for baits that have nuclear localization signals. Third, the CytoTrap system is based on activation of the Ras signaling pathway. Components of the Ras pathway (such as Ras, Sos, PKA, etc.) that can activate the system when over-expressed, or proteins that can interact with hSos, constitute the majority of the false positive clones selected from library screenings.

Recently, several improvements to the CytoTrap system have been made to broaden its applicability and utility. First, it has been demonstrated that the expression of mGAP (mammalian GTPase activated protein) in cdc25H strain can reduce the number of Ras "false positives" in CytoTrap screens of mammalian cDNA expression libraries (4). Second, the bait vector, pSos was modified to include a nuclear export signal to maximize the chance of nuclear proteins to localize in the cytoplasm (Chang, H.W., unpublished result). Third, interaction mating using the two mating types of cdc25H strain has been developed to greatly facilitate the assessment of interaction specificity and false positives (Chang, H.W., unpublished result). Finally, a recently developed variation of the CytoTrap system, the Ras recruitment system (RRS), eliminates the isolation of Ras false positives during library screenings (5,7). However, the effectiveness of this system remains to be determined.

In summary, we have used the CytoTrap system to verify the interactions between various proteins. In addition, the CytoTrap system has been used to isolate novel Gal4AD interacting proteins. This underscores the utility of this system since it has been impossible to use transcriptional activation domains as baits in a yeast two-hybrid screen prior to this. Future generations of CytoTrap plasmid constructs and yeast strains will provide more efficient and powerful tools to identify interacting proteins.

ACKNOWLEDGMENT

The author would like to thank Tanya Hosfield, Ning Jiang, and Marie Callahan for excellent technical assistance, Alan Greener for critical reading of this manuscript, and Tom Kodadek, Ami Aronheim, and Makodo Nishizawa for reagents, discussions, and suggestions.

REFERENCES

1. **Allen, J.B., M.W. Walberg, M.C. Edwards, and S.J. Elledge.** 1995. Finding prospective partners in the library: the two-hybrid system and phage display find a match. Trends Biochem. Sci. *20*:511-516.
2. **Aronheim, A., D. Engelberg, N. Li, N. Al-Alawi, J. Schlessinger, and M. Karin.** 1994. Membrane targeting of the nucleotide exchange factor Sos is sufficient for activating the Ras signal pathway. Cell *78*:949-961.
3. **Aronheim, A., E. Zandi, H. Hennemann, S. Elledge, and M. Karin.** 1997. Isolation of an AP-1 repressor by a novel method for detecting protein-protein interactions. Mol. Cell. Biol. *17*:3094-3102.
4. **Aronheim, A.** 1997. Improved efficiency Sos recruitment system: expression of the mammalian GAP reduces isolation of Ras GTPase false positive. Nucl. Acids Res. *25*:3373-3374.
5. **Aronheim, A., Y.C. Broder, A. Cohen, A. Fritsch, B. Belisle, and A. Abo.** 1998. Chp, a homologue of the GTPase Cdc42Hs, activates the JNK pathway and is implicated in reorganizing the actin cytoskeleton. Curr. Biol. *8*:1125-1128.
6. **Bartel, P.L., C.-T. Chien, R. Sternglanz, and S. Fields.** 1993. p. 153-179. *In* D.A. Hartley (Ed.), Cellular Interactions in Development: A Practical Approach. Oxford Univ. Press, Oxford, England.
7. **Broder, Y.C., S. Katz, and A. Aronheim.** 1998. The Ras recruitment system, a novel approach to the study of protein-protein interactions. Curr. Biol. *8*:1121-1124.
8. **Chardin, P., J.H. Camonis, N.W. Gale, L. van Aeist, J. Schlessinger, M.H. Wigler, and D. Bar-Sagi.** 1993. Human Sos1: a guanine nucleotide exchange factor for Ras that binds to Grb2. Science *260*:1338-1343.
9. **Chien, C.-T., P.L. Bartel, R. Sternglanz, and S. Fields.** 1991. The two-hybrid system: a method to identify and clone genes for proteins that interact with a protein of interest. Proc. Natl. Acad. Sci. USA *88*:9578-9582.
10. **Durfee, T., K. Becherer, R.-L. Chen, S.H. Yeh, Y. Yang, A.E. Kilburn, W.H. Lee, and S.J. Elledge.** 1993. The retinoblastoma protein associates with the protein phosphatase type I catalytic subunit. Genes Dev. *7*:555-569.
11. **Estojak, J., R. Brent, and E.A. Golemis.** 1995. Correlation of two-hybrid affinity data with in vitro measurements. Mol. Cell. Biol. *15*:5820-5829.
12. **Fields, S. and O.-K. Song.** 1989. A novel genetic system to detect protein-protein interaction. Nature (London) *340*:245-246.
13. **Fields, S. and R. Sternglanz.** 1994. The two-hybrid system: an assay for protein-protein interactions. Trends in Genetics *10*:286-292.
14. **Gyuris, J., E.A. Golemis, H. Chertkov, and R. Brent.** 1993. Cdi 1, a human G1 and S phase protein phosphatase that associates with Cdk2. Cell *75*:791-803.
15. **Hu, P. and J. Schlessinger.** 1994. Direct association of p110 beta phosphatidyinositol 3-kinase with p85 is mediated by an N-terminal fragment of p110 beta. Mol. Cell. Biol. *14*:2577-2583.
16. **Hu, P., A. Mondino, E.Y. Skolnik, and J. Schlessinger.** 1994. Cloning of a novel, ubiquitously expressed human phosphatidylinositol 3-kinase and identification of its binding site on p85. Mol. Cell. Biol. *14*:7677-7688.
17. **Kataoka, K., K.T. Fujiwara, M. Noda, and M. Nishizawa.** 1994. MafB, a new Maf family transcription activator that can associate with Maf and Fos but not with Jun. Mol. Cell. Biol. *14*:7581-7591.
18. **Ma, J. and M. Ptashne.** 1987. A new class of yeast transcriptional activators. Cell *51*:113-119.
19. **Petitjean, A., F. Hilger, and K. Tatchell.** 1989. Comparison of thermosensitive alleles of the cdc25 gene involved in the cAMP metabolism of *Saccharomyces cerevisiae*. Genetics *124*:797-806.
20. **Quilliam, L.A., R. Khosravi-Far, S.Y. Huff, and C.J. Der.** 1995. Gaunine nucleotide exchange factors: activators of the RAS superfamily of proteins. BioEssays *17*:395-404.
21. **Reponen, P., C. Sahlberg, P. Huhtala, T. Hurskainen, I. Thesleff, and K. Truggvason.** 1992. Molecular cloning of murine 72 kDa type IV collagenase and its expression during mouse development. J. Biol. Chem. *267*:7856-7862.
22. **Resh, M.D.** 1994. Myristylation and palmitylation of Src family members: the fats of the matter. Cell *76*:411-413.
23. **Rothe, M., V. Sarma, V.M. Dixit, and D.V. Goeddel.** 1995. TRAF2-mediated activation of NF-kappa B by TNF receptor 2 and CD40. Science *269*:1424-1427.
24. **Thevelein, J.M.** 1991. Fermentable sugars and intracellular acidification as specific activators of the RAS-adenylate cyclase signalling pathway in yeast: the relationship to nutrient-induced cell cycle control. Mol. Microbiol. *5*:1301-1307.

9 | An RNA Polymerase III-Based Two-Hybrid System

Olivier Louvet[1,2] and Marie-Claude Marsolier[1]
[1]*Service de Biochimie et de Génétique Moléculaire,*
CEA-Saclay, Gif-sur-Yvette, and [2]*Quantum-Appligene,*
Illkirch, France

1. INTRODUCTION

Protein–protein interactions are critical to most biological processes, extending from the formation of cellular macromolecular structures and enzymatic complexes to the regulation of signal transduction pathways. The regulation of transcription by RNA polymerase II (Pol II) of protein-encoding genes—one of the most prominent and widely studied mechanisms that enables an organism to develop and to adapt to its environment—also occurs via protein–protein interactions. The study of interactions among transcriptional activators or repressors and various components of the basal Pol II transcription machinery cannot be done by the classical two-hybrid systems that rely on the activation of Pol II reporter genes as phenotypic readout. We present here a two-hybrid system based on yeast RNA polymerase III (Pol III) transcription. This system is particularly suited to study proteins involved in Pol II transcription or proteins that spuriously activate Pol II reporter genes.

The concept of two-hybrid systems as developed by Fields and Song (9) was based on the modularity of RNA polymerase II (Pol II) transcriptional activators in yeast. The prototypical activator Gal4 contains two independent, separable domains, a DNA-binding domain and an activating domain (4) that do not need to be covalently attached to be functional. In vivo interactions between a protein X, fused to the DNA-binding domain, and a protein Y, fused to the activating domain, reconstitute a functional transactivator whose activity can be readily monitored by a reporter gene controlled by the corresponding upstream binding sites.

We have extended this concept of functional modularity to the transcription factor C (TFIIIC) of the yeast polymerase III (Pol III) transcription system. The transcription of yeast Pol III genes (*SNR6*, tRNA, or 5S RNA genes) is much simpler than the transcription of Pol II genes. Eukaryotic cells have evolved a complex network of protein interactions to modulate the rate of transcription by Pol II. Several families of activators and coactivators or mediators intervene to influence the different steps of

Yeast Hybrid Technologies
Edited by L. Zhu and G.J. Hannon
© 2000 Eaton Publishing, Natick, MA

145

Pol II transcription such as preinitiation complex formation, promoter clearance, or elongation (33). In contrast, the activation of Pol III genes such as *SNR6* relies only on two general transcription factors: TFIIIB and TFIIIC. TFIIIC, or τ, is a large, multisubunit factor that recognizes the promoter elements (the A and B blocks) and recruits the second factor, TFIIIB, to the DNA upstream of the transcription start site. TFIIIB is the initiation factor that recruits RNA polymerase III (Figure 1A). No other transactivators are known to regulate the Pol III transcription rate in yeast.

We have previously shown that the specific interaction of TFIIIC with the B block element is dispensable if TFIIIC can be anchored to the DNA target by a different strategy. If the essential B block element of *SNR6* is replaced by Gal4-binding sites (UAS$_G$ sites), the DNA-binding function of TFIIIC can be restored artificially by fusing one of its subunits, τ138 for example, to Gal4 DNA-binding domain. This has led to a model of the one-hybrid system [(23); Figure 1B].

We then found that a direct interaction of TFIIIC with DNA via the Gal4 DNA-binding domain was not essential, and that UAS$_G$-*SNR6* could be efficiently activated when an intervening protein–protein interaction was imposed, thus providing the basis of a Pol III two-hybrid system [(24); Figure 1C]. The main advantage of this two-hybrid strategy is that it allows the analysis of proteins involved in Pol II transcriptional regulation or of proteins that spuriously activate Pol II reporter genes without any known relationship with transcription.

2. PROTOCOLS

2.1 General Materials

2.1.1 Plasmids

The UAS$_G$-*SNR6* gene used for the Northern blotting experiments corresponds to the B block-UAS$_G$ template cloned into YEp352 as described in (23). It harbors five head-to-tail repeats of Gal4-binding sites at the location of its B block and a 24 bp-insertion in its transcribed sequence—at position +73 relative to its transcription start site—so that its transcripts can be easily distinguished from the ones derived from the wild-type gene.

The UAS$_G$-*SNR6* construct used for the 5-fluoro-orotic acid (5-FOA) resistance assay is similar to the UAS$_G$-*SNR6* template used for the Northern blotting experiments except that it lacks the 24 bp insertion. It was cloned in the *LEU2* centromeric plasmid pRS315 (31) to create pRS315/UAS$_G$-*SNR6*. The *SNR6ΔB* gene that lacks both a B block and Gal4-binding sites derives from the pRS314-Δ2 plasmid described in (26) and was also cloned into pRS315.

The GAL4-(1–147)-τ138 construct has been described previously (23). The pMA424-derived GAL4-(1–147)-PRP9/11 constructs (kindly provided by P. Legrain, Institut Pasteur, Paris) have been described (15,16). The τ138-PRP21 fusion was constructed as follows: *Bam*HI sites were introduced by oligonucleotide-directed mutagenesis (14) at -6 relative to τ138 initiator codon and in front of the stop codon; the resulting *Bam*HI fragment was cloned into the Bluescript SK vector (BSSK, Stratagene) so that the *Kpn*I site was located at the 5′ end of τ138 (the BSSK-τ138 plasmid). *PRP21* coding sequence was originally cloned between the *Bam*HI and the *Pst*I sites of BSSK (plasmid pPL247 from P. Legrain). The unique *Bam*HI site of pPL247

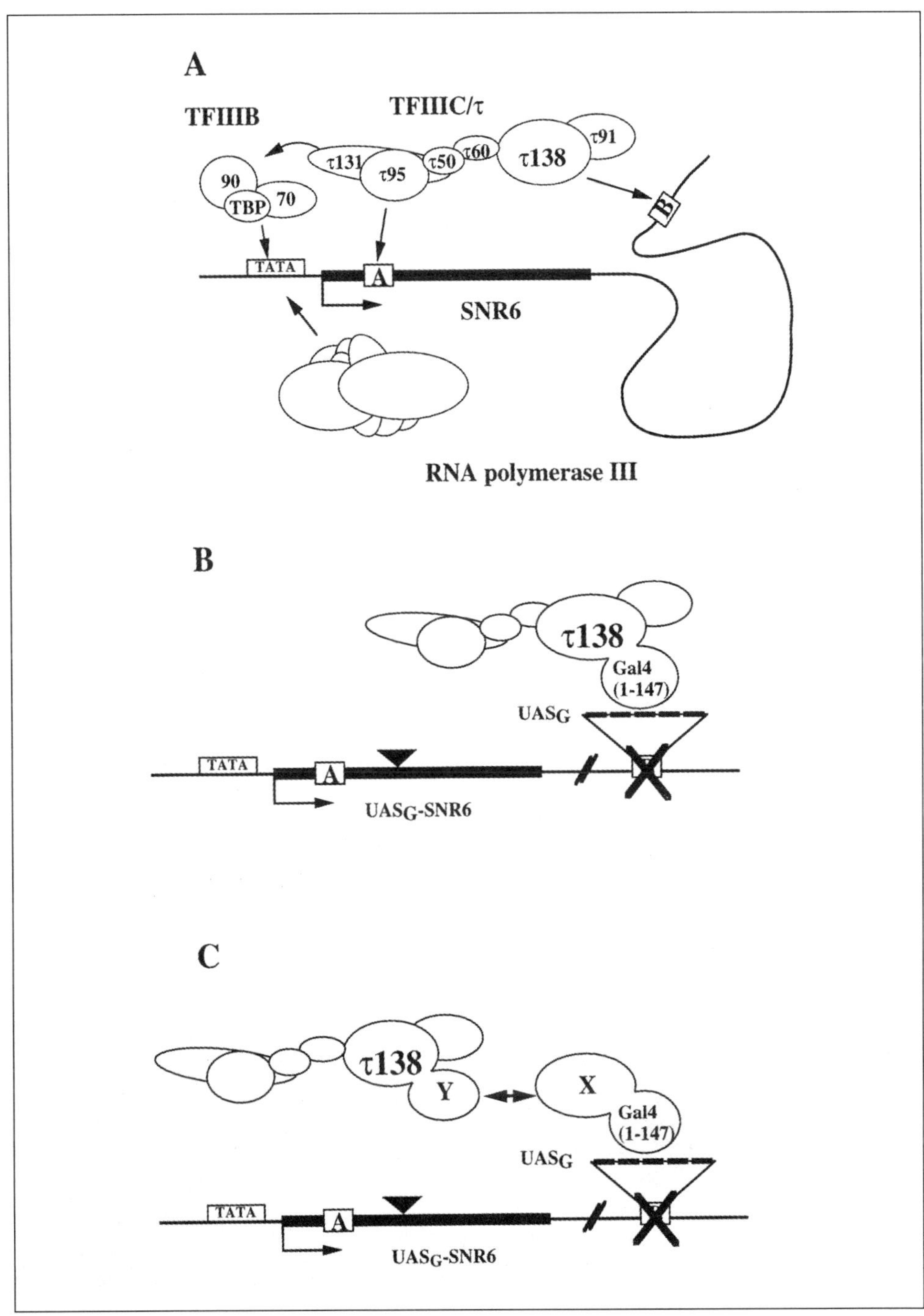

Figure 1. Models of the transcriptional activation of wild-type *SNR6* or UAS$_G$-*SNR6* genes. Critical cis-control elements such as the TATA box and A and B blocks are represented by marked, open boxes. **(A)** The TFIIIC subunit τ138 binds the downstream B block of wild-type *SNR6* as the first step of gene activation. It triggers the successive recruitment of TFIIIB and RNA polymerase III. **(B)** and **(C):** The UAS$_G$-*SNR6* templates have five head-to-tail repeats of UAS$_G$ sequence (small black rectangles) inserted at the location of the *SNR6* B block, which has been destroyed by a 2-bp deletion. In addition, the templates used for the Northern blotting experiments (maxi *SNR6*) have a 24-bp insertion at +73 (black triangle). τ138 is recruited to the UAS$_G$-*SNR6* genes either directly, via Gal4 DNA-binding domain **(B)**, or indirectly, through the interaction between X and Y **(C)**.

at the 5′ end of *PRP21* was digested, blunt-ended, and re-ligated so as to subsequently put *PRP21* in frame with τ138. The modified pPL247 plasmid was then digested by *Spe*I and *Eco*RV, and the *PRP21*-containing fragment was inserted into the *Spe*I and blunt-ended *Not*I sites of BSSK-τ138. The *Eco*RI fragment containing the τ138-PRP21 fusion was then introduced into the *Eco*RI site of the pYcDE-2 plasmid (kindly provided by B. Hall, University of Washington, Seattle). The τ138 and τ138-GAL80 constructs were made as follows: a 3.5 kb *Bam*HI-fragment containing the entire coding sequence of τ138 was first cloned into the pGEN *Bam*HI site (30). This construct (pGEN-τ138) then served as the basis for the subsequent construct in which the coding sequence of *GAL80* was fused in phase behind τ138.

pQBD1™ was derived from pODB8 (19). pODB8 was first digested by *Nhe*I followed by the Mung Bean Nuclease. The digestion product was then cut with *Eco*RI, blunt-ended, and re-ligated.

To construct pTAU2™, the *Bam*HI fragment from the pGEN-τ138 plasmid (24) that contained the entire coding sequence of τ138 was inserted into the p423GPD (28) *Bam*HI site. The *Bam*HI site located at the 5′ end of the gene was subsequently destroyed by a partial digestion followed by blunt-ending and re-ligation. Three stop codons were also introduced downstream of the multiple cloning site, in the three reading frames, and the multiple cloning site was modified so as to eliminate the *Eco*RI site and to introduce a unique *Not*I site.

The τ1-, τ3-, τ1-2-, τ1-3-, τ1-2-3-GAL4-(1–147) constructs contain the sequences encoding, respectively, amino acids 1–329 (τ1), 761–1160 (τ3), 1–676 (τ1-2), 1–329 and 676–1160 (τ1-3), and 1–1160 (τ1-2-3, the entire sequence of τ138) from τ138, cloned in front of GAL4-(1–147) into the pGEN vector (30), under control of the strong, constitutive *PGK* promoter (24).

Site-directed mutagenesis was performed to generate a GAL4-(1–147)[P26L] mutant fragment. pQBD1™ was used as a template for PCR to mutate *GAL4* nucleotides 76–78 from CCG (Pro) to CTT (Leu). First, two PCR reactions were carried out to produce two different fragments from GAL4-(1–147). In reaction 1, wild-type forward primer dbxba (5′-CTCCTTCTAGATGAAGCTACTGTCTTC-3′) and mutagenic reverse primer dblfa (5′-GCGCACTTAAGTTTTTCTTTGGAGCAC-3′) were used to introduce the mutation as well as an *Xba*I site in front of *GAL4* ATG and an *Afl*II site at the mutated site. In reaction 2, mutagenic forward primer dbafl (5′-GAAAAACTTAAGTGCGCCAAGTGTCTG-3′) and wild-type reverse primer dbbam (5′-CATGAGGATCCGCGATACAGTCAACTGTC-3′) were used to introduce the mutation and a *Bam*HI site behind the *GAL4* codon 147 and an *Afl*II site at the mutated site. PCR-derived products were then digested with *Afl*II and *Xba*I, and with *Afl*II and *Bam*HI, respectively. The resulting fragments were cloned by a three-way ligation into *Xba*I-*Bam*HI-cleaved pBSKS (to give pKS/GAL4DBP26L) for sequencing to confirm the presence of the mutant codon and the absence of other PCR-induced mutations. The *Xba*I-*Bam*HI fragment from pKS/GAL4DBP26L containing GAL4-(1–147)[P26L] was then subcloned into *Xba*I-*Bam*HI-digested p416CYC1 (28) to make p416CYC1/GAL4-(1–147)[P26L]. The *Bam*HI fragment from the pMA424/τ138 plasmid (23) containing the entire coding sequence of τ138 was inserted into the p416CYC1/GAL4-(1–147)[P26L] *Bam*HI site to create p416CYC1/GAL4-(1–147)[P26L]-τ138. Finally, to generate p416CYC1/GAL4-(1–147)[P26L]-τ1-3, the *Eco*RI and *Cla*I sites of the multiple cloning site were eliminated by a *Sal*I-*Sma*I-digestion followed by blunt-ending and re-ligation, and the resulting vector was *Eco*RI-*Cla*I-cleaved, blunt-ended, and re-ligated.

2.1.2 Strains

The Northern blotting experiments are performed using yeast strain YPH500α (31). This strain possesses the auxotrophic markers that are necessary to select the plasmids harboring the UAS$_G$-*SNR6* template and the two hybrid fusion constructs. It contains a wild-type *GAL4* gene, which has proved not to hamper the detection of protein interactions in this system.

The 5-FOA resistance assay is carried out with yeast strain yMCM616, a *snr6* mutant harboring a 2-bp deletion in the *SNR6* B block and containing a wild-type copy of the *SNR6* gene on a *URA3* plasmid. This strain is identical to FTY115 described in (26) except that it is rescued by the *URA3* centromeric plasmid pRS316, which harbors the region of the *SNR6* gene spanning bp -140 to +314 relative to the transcription start site, and not by the plasmid pRS314-U6.

The tester strain QYN1™ was constructed as follows: the UAS$_G$-*SNR6* construct previously used for the 5-FOA resistance assay was first transferred from the pRS315 plasmid to the integrative, *URA3*-marked vector pRS306 (31) to give pRS306/UAS$_G$-SNR6. The W303-1A strain was then transformed with the linearized pRS306/UAS$_G$-SNR6 vector, and a clone with pRS306/UAS$_G$-SNR6 integrated at the *SNR6* locus was selected following PCR analysis of the genomic DNA using the primers 5′snr6 (5′-ATGTCATCTTCCTGGACC-3′), 3′snr6 (5′-ATCGTACCATTGCATAGC-3′), and 5′kpnI (5′-GGTACCGGGCCCCCCCTCGAGGTC-3′). This clone was then transformed with pRS315/UAS$_G$-SNR6 and with a vector harboring the GAL4-(1–147)-τ138 construct, and recombination events at the *SNR6* locus were selected by streaking the cells on plates that contained 5-FOA. The genomic DNA of 5-FOA–resistant clones was analyzed by PCR using the primers 5′snr6 and 3′snr6, and a clone that had conserved the UAS$_G$-*SNR6* construct at the *SNR6* locus was chosen. This clone was then transformed with pQAR1™ and with pRS316/SNR6, a *URA3* plasmid harboring a wild-type copy of *SNR6*. Transformants were subsequently allowed to lose pRS315/UAS$_G$-SNR6 and the vector bearing the GAL4-(1–147)-τ138 construct, so as to create the tester strain QYN1™.

2.2 Protocol for Quantification of UAS$_G$-*SNR6* Transcripts

Procedure

Total RNA is extracted according to a modified version of the method described by (29). Small RNA species are separated by electrophoresis on a denaturing polyacrylamide gel, blotted onto a membrane, then hybridized with a ^{32}P–end-labeled oligonucleotide to probe both wild-type *SNR6* and UAS$_G$-*SNR6* transcripts.

RNA extraction:

1. Grow yeast cells overnight in 40 mL selective medium until they reach an OD$_{660}$ of 0.3–0.4. It is very important to harvest the cells at this stage. We have found that if cells are harvested later the steady-state level of the transcripts derived from the UAS$_G$-*SNR6* maxi-gene will drop too much compared to the level of the wild-type transcripts (this is probably due to the fact that the maxi-gene transcripts are less stable than the wild-type *SNR6* RNA, so that their steady state level decreases when

the cells enter the stationary phase and therefore slows down their Pol III transcription). Pellet the cells by centrifugation ($4000\times g$, 5 min, room temperature) using 50-mL Falcon tubes with conic ends so that the pellets are tightly packed.

2. Resuspend the cells in 400 µL AE buffer (50 mM sodium acetate pH 5.3, 10 mM EDTA, autoclaved) and transfer them to a 1.5-mL microcentrifuge tube (the protocol can be interrupted at that step and cells can be stored at -80°C).

3. Add 50 µL 10% sodium dodecyl sulfate (SDS) and vortex-mix.

4. Add 550 µL phenol, previously equilibrated in AE buffer, and vortex-mix again.

5. Incubate the mixture at 65°C for 4 min and then quickly chill it in a dry ice/ethanol bath.

6. After centrifugation in a microcentrifuge at $17\,000\times g$ for 6 min at room temperature, transfer the aqueous phase (about 600 µL) to a fresh tube and extract it with phenol/chloroform for 5 min at room temperature.

7. Transfer 400 µL of the aqueous phase to a fresh tube and add 40 µL 3.0 M sodium acetate pH 5.3 and 1.1 mL ethanol.

8. After centrifugation at 4°C for 30 min at $17\,000\times g$, wash the pellets once with 80% ethanol, dry them in a Speed-Vac evaporator (15 min maximum), and resuspend them in 20 µL sterile, distilled water.

Gel electrophoresis, blotting, and hybridization:

Small-size RNAs are separated on a denaturing polyacrylamide gel [8% acrylamide (acrylamide/bisacrylamide : 20/1), 8.0 M urea, 1× TBE (90 mM Tris-borate, 2 mM EDTA)]. 10–15 µg RNA are loaded per lane. Four volumes of loading buffer [90% (vol/vol) formamide, 10% (vol/vol) 10× TBE, 0.1% (wt/vol) bromophenol blue, 0.1% (wt/vol) xylene cyanol blue] are added to the RNA samples, which are then heated in boiling water for 5 min and cooled on ice for 5 min before loading.

After electrophoresis, the gel is blotted onto a positively charged Nylon membrane (Roche Molecular Biochemicals, Indianapolis, IN, USA) using the Bio-Rad Trans-Blot cell (Bio-Rad Laboratories, Hercules, CA, USA; 60 V, constant voltage, 1 h, in 0.5× TBE).

The membrane is hybridized with a probe consisting of a fragment of *SNR6* transcribed sequence. We have used either the entire *SNR6* sequence labeled by nick translation with α-^{32}P-dCTP or an oligonucleotide (5′-AATCTCTTTGTAAAACG-GTTCATCC-3′) kinased with γ-^{32}P-dATP. The hybridization protocol is derived from the method described by (6).

Hybridization:

1. Heat the hybridization buffer (0.5 M sodium phosphate pH 7.2, 10 mM EDTA, 7.0% SDS) to 65°C and add to the membrane containing the fixed nucleic acids.

2. Add the denatured probe.

3. Hybridize overnight at 65°C.

4. Wash the membrane 4–6 times with the washing buffer (40 mM sodium phosphate pH 7.2, 1% SDS) at 65°C for 5–10 min each time.

5-FOA–resistance assay:

Transformants of the tester strain containing the two hybrid constructs are selected on the appropriate media, and several independent clones are streaked onto plates containing 1.0% 5-FOA (wt/vol). The positive clones usually take 2–3 days to grow, and their growth rate (reflecting the strength of the interaction between the hybrid proteins) can be compared to that of a positive control containing the GAL4-(1–147)-τ138 construct.

3. RESULTS AND DISCUSSION

3.1 Setting Up a Pol III-Based Two-Hybrid System

Figure 1A shows a schematic representation of the *SNR6* gene with its promoter elements, a TATA box at -30, a non-consensus intragenic A block at +21, and a B block located downstream of the termination signal. Transcription complex assembly is thought to occur as for a tRNA gene (5,8,11): TFIIIC (τ) binds to the A and B blocks and helps to assemble TFIIIB to the TATA region. Once formed, the TFIIIB-DNA complex is highly stable in vitro under conditions that dissociate TFIIIC and is able to direct multiple rounds of transcription (13). TFIIIB then recruits Pol III. Previously we showed that B block-less *SNR6* genes harboring Gal4-binding sites at various locations (UAS$_G$-*SNR6* genes) could be rescued by fusing the DNA-binding domain Gal4-(1–147) to the τ138 subunit (23, Figure 1B). Other subunits of TFIIIC (τ131, τ95, τ91) were also tested for their efficiency in triggering UAS$_G$-*SNR6* transcription, but τ138 proved to be the most effective (1,23). This artificial transcription mechanism via the Gal4-(1–147)-τ138 fusion was found remarkably efficient, especially when the UAS$_G$ elements were inserted at the location of the natural *SNR6* B block (23). We therefore chose this UAS$_G$-*SNR6* template for further examination.

Both the A block and the TATA box proved essential to UAS$_G$-*SNR6* activation: UAS$_G$-*SNR6* templates whose A block had been destroyed or whose TATA box had been replaced by an unrelated sequence (GTGCACG) were unable to direct transcription in vivo via Gal4-(1–147)-τ138 (24, data not shown). These results confirmed that once TFIIIC is bound to DNA by Gal4-(1–147), a TATA box- and an A block-dependent activation pathway ensues. Could this system tolerate the insertion of an intervening protein–protein interaction step for TFIIIC recruitment? We replaced the covalent link between Gal4-(1–147) and τ138 with an interaction between two hybrid proteins, Gal4-(1–147)-protein X and τ138-protein Y (24, Figure 1C). We chose Prp9 or Prp11, and Prp21 as interacting proteins X and Y, respectively. Prp9, Prp11, and Prp21 are yeast nuclear proteins (PRP splicing factors) whose interactions were demonstrated by the classical two-hybrid system (15,16).

Plasmids harboring the UAS$_G$-*SNR6* template and the GAL4-(1–147)- and τ138-constructs were introduced into yeast cells, and the efficiency of TFIIIC recruitment, measured by UAS$_G$-*SNR6* transcription, was assessed by Northern blotting analysis. The results are shown in Figure 2 (24). Remarkably, the transcription of the UAS$_G$-*SNR6* gene was activated to the same extent when Gal4-(1–147)-τ138 was recruited directly to the Gal4-binding sites (lane 1, 100% activation), or when τ138-Prp21 was recruited indirectly via its interactions with Gal4-(1–147)-Prp9 (lane 3, 70% activation) or Gal4-(1–147)-Prp11 (lane 6, 80% activation). These results show that *SNR6* transcription can be triggered by locally recruiting TFIIIC through intermediate protein–protein interactions.

We verified the specificity of this assay by performing experiments with several

PRP9 and *PRP11* mutants (24, Figure 2). The PRP9-1 and PRP11-1 mutations decrease the Prp9/Prp21 and Prp11/Prp21 interactions about 50-fold, whereas the Prp9/Prp21 interaction is only slightly affected by the PRP9-C423 mutation, as monitored by the classical Pol II two-hybrid system (15,16). We found that the GAL4-(1–147)-PRP9-1 (lane 4) and the GAL4-(1–147)-PRP11-1 (lane 7) constructs were no longer able to promote the transcription of the UAS_G-*SNR6* template [GAL4-(1–147)-PRP11-1, lane 7] or promoted it only very weakly [GAL4-(1–147)-PRP9-1, lane 4], whereas the GAL4-(1–147)-PRP9-C423 (lane 5) fusion was almost as efficient as the wild-type GAL4-(1–147)-PRP9 construct (lane 3). The level of transcriptional activation therefore appeared to correlate with the local concentration of TFIIIC in the vicinity of the promoter, which is dependent on the strength of the intermediate protein interactions (24). Thus, our Pol III-based two-hybrid system could be validly used to monitor protein–protein interactions.

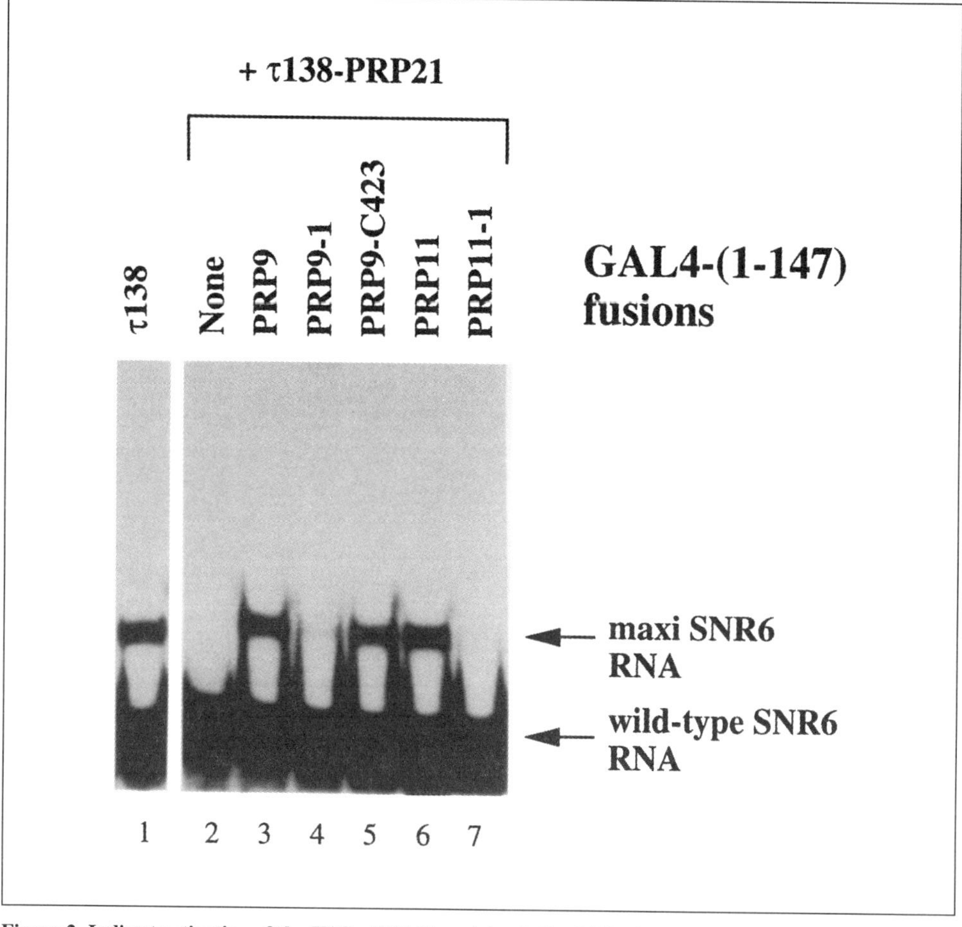

Figure 2. Indirect activation of the UAS_G-*SNR6* template via Prp21/Prp9 and Prp21/Prp11 interactions. The GAL4-(1–147)-τ138 construct alone (lane 1) or the τ138-PRP21 fusion in combination with different GAL4-(1–147) constructs as indicated (lanes 2–7) were introduced into yeast cells containing the UAS_G-*SNR6* gene. The τ138-PRP21 fusion is cloned into pYcDE-2 under control of the *ADH1* promoter. The various GAL4-(1–147) constructs are pMA424 derivatives (15,16,22). The transcripts from the UAS_G-*SNR6* template (maxi *SNR6* RNA) and the endogenous *SNR6* RNA were monitored by Northern blotting analysis.

3.2 Indirect Activation of UAS$_G$-*SNR6* Can Support Cell Growth

To derive a genetic screen, we tested whether the protein-interaction–mediated UAS$_G$-*SNR6* activation was able to support normal yeast cell growth (24, *SNR6* is a single copy, essential gene). yMCM616, a *snr6* mutant strain containing a wild-type copy of *SNR6* on a *URA3*-marked plasmid, was transformed with pRS315/UAS$_G$-SNR6—a vector harboring a UAS$_G$-*SNR6* template—and with several plasmids bearing the GAL4-(1–147)-τ138, -PRP9, -PRP11, or τ138-PRP21 constructs. These transformants were streaked onto plates containing 5-FOA, a chemical that is metabolized into a toxic compound by the *URA3* product (3). All transformants containing either GAL4-(1–147)-τ138 alone (Figure 3, panel 1, top-right) or combinations of GAL4-(1–147)-PRP9/11 with τ138-PRP21 constructs (panel 2, upper) were able to lose the URA3 plasmid harboring the wild-type SNR6 gene and to grow on the medium that contained 5-FOA. In contrast, no cell growth was seen when either GAL4-(1–147)-PRP9/11 or τ138-PRP21 were missing (panel 1, left half), or when the GAL4-(1–147)-τ138 or the GAL4-(1–147)-PRP9/11 and τ138-PRP21 fusions were co-transformed with SNR6 genes lacking both the B block and the UAS$_G$ sequences (*SNR6*ΔB, panel 1, lower right and panel 2, lower half).

In an additional control experiment, the GAL4-(1–147)-PRP9-1, -PRP9-C423, and PRP11-1 fusions expressing various Prp9 or Prp11 mutant proteins were intro-

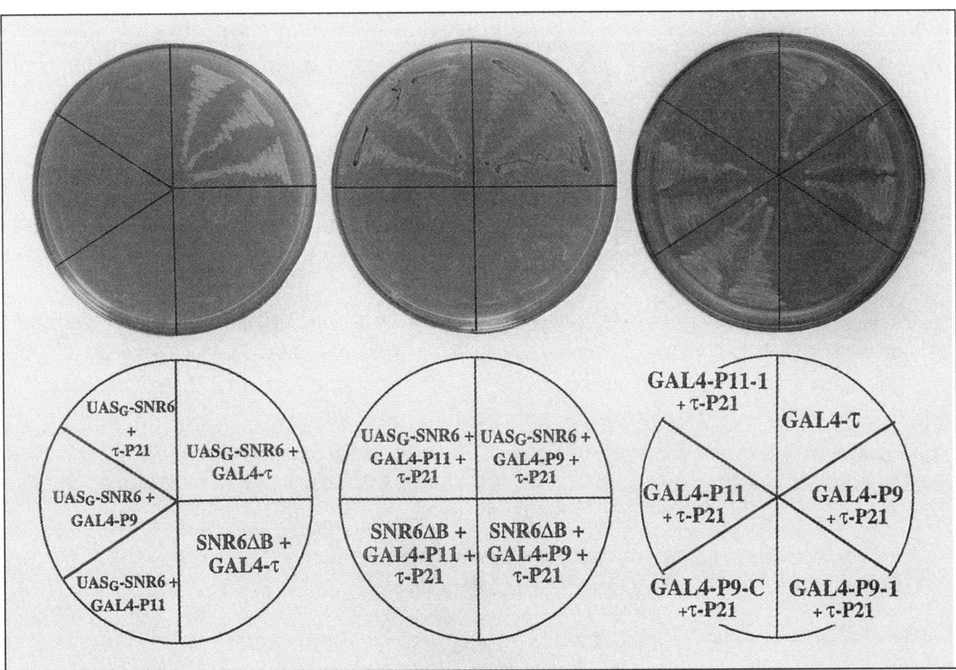

Figure 3. 5-FOA-resistance of yMCM616 transformants containing chimeric activators. yMCM616, a *snr6* mutant containing a wild-type copy of *SNR6* on a *URA3* vector, was transformed with plasmids harboring a UAS$_G$-*SNR6* template or a *SNR6* gene devoid of B block and of UAS$_G$ (*SNR6*ΔB), along with plasmids harboring the τ138-PRP21 (τ-P21), GAL4-(1–147)-τ138 (GAL4-τ), GAL4-(1–147)-PRP9 (GAL4-P9), GAL4-(1–147)-PRP9-1 (GAL4-P9-1), GAL4-(1–147)-PRP9-C423 (GAL4-P9-C), GAL4-(1–147)-PRP11 (GAL4-P11), and GAL4-(1–147)-PRP11-1 (GAL4-P11-1) constructs, as indicated. The τ138-PRP21 and GAL4-(1–147) constructs are the same as in Figure 2. Independent transformants were then streaked on 5-FOA plates. When not indicated (panel 3), the yMCM616 strain contains the UAS$_G$-*SNR6* construct.

duced into the yeast tester strain along with the τ138-PRP21 construct. Transformants were streaked on 5-FOA plates (Figure 3, panel 3) and their growth patterns were found to correlate precisely with the results shown in Figure 2 (24). As expected, no cell growth could be detected with the GAL4-(1–147)-PRP9-1 and the GAL4-(1–147)-PRP11-1 constructs, which supports the feasibility of library screening experiments using this unique Pol III design.

3.3 A Pol III-Based Two-Hybrid System to Study the Partners of Pol II Regulators: Detection of Gal4/Gal80 Interactions

This Pol III-based two-hybrid system is of special interest for the study of proteins having *bona fide* or spurious activation functions in Pol II transcription. These proteins are impossible or difficult to analyze by any classical Pol II-based two-hybrid system. Therefore, we checked whether the transcription of the UAS_G-*SNR6* reporter gene could be activated by Pol II regulators (24). In vertebrates, the same enhancers can activate Pol II and Pol III snRNA or snRNA-like genes. However, the full-length Gal4 (Figure 4), or Gal4 DNA binding domain fused with the activation domains of VP16 or p53 (27), were unable to activate the transcription of the UAS_G-*SNR6* gene by themselves (24, data not shown). Additionally, no transcriptional activation of the *SUP4* tRNA^{Tyr} gene by the yeast activator Gcn4 could be detected in vitro (18). Therefore this genetic system lent itself to the analysis of the partners of Pol II activators (or repressors).

To demonstrate that this Pol III-based two-hybrid system allowed the detection of interactions between Pol II transcription regulators and their partners, we constructed a fusion between τ138 and Gal80 and tested it against Gal4. The results are shown in Figure 4 (24). The amount of transcripts generated through the interaction between Gal4 and τ138-Gal80 (lanes 4 and 5) represented 60% of the transcript level generated with the GAL4-(1–147)-τ138 construct (lane 1). Six independent transformants overexpressing Gal4 and τ138-Gal80 were tested, and the intensity of the signal was identical (other 4 are not shown). No background signal was observed when Gal4 was co-expressed with τ138 (lane 3) or when τ138-Gal80 was co-expressed with Gal4-(1–147) (lane 2) [Gal80 interacts only with Gal4 activation region II encoded by GAL4-(768–881), (20)]. Moreover, τ138-Gal80/Gal4 interaction was able to trigger the transcription of enough UAS_G-*SNR6*–derived transcripts to support the growth of the yMCM616 strain on 5-FOA (data not shown).

Having demonstrated the capacity of a Pol III-based two-hybrid system to detect interactions between various proteins, including Pol II transcription regulators, we then developed new vectors and a more direct screening method.

3.4 New Vectors and a New Tester Strain

We developed new vectors to allow stronger expression of the fusion proteins and to make the plasmids of this Pol III system compatible with the Pol II systems.

The pQAR1™ vector

In the first version of this system, the UAS_G-*SNR6* template was harbored by pRS315/UAS_G-SNR6, a derivative from pRS315 (31)—a *LEU2* centromeric vector. To increase the sensitivity of the protein interaction detection system and to take advantage of a seldom used auxotrophic marker, the *Sst*I-*Kpn*I fragment of the pRS315/UAS$_G$-SNR6 plasmid containing the UAS$_G$-*SNR6* template was cloned into the *Sst*I and *Kpn*I sites of the *ADE2*-marked, multicopy plasmid pASZ12 (32), thus creating the pQAR1™ plasmid.

The pQBD1™ vector

To express Gal4-(1–147) fusions, we previously used several vectors that had been constructed for the Pol II two-hybrid systems, including pMA424 (22), pGBT9 (2), pAS1 (7), and pAS2 (10). These vectors contain the gene fragment encoding the Gal4 DNA-binding domain (amino acids 1 to 147) that is controlled by the *ADH1* promoter and is followed by unique restriction sites for in-frame fusions. All these vectors are multicopy plasmids and possess the *TRP1* marker, except for pMA424 that harbors the *HIS3* gene. A previous comparative study (17) indicated that expression of fusion proteins from the pGBT9 vector is much lower than from pMA424. While lower production of fusion proteins may be advantageous to avoid potential

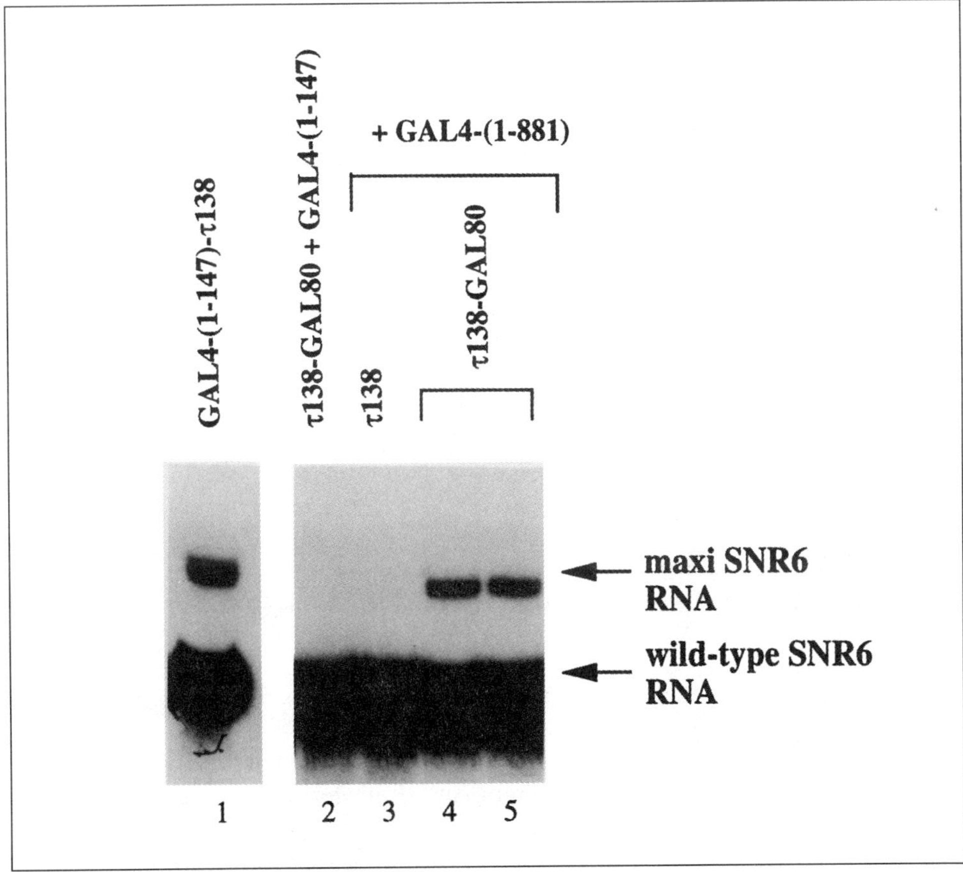

Figure 4. Gal4-directed activation of the UAS$_G$-*SNR6* template. The GAL4-(1–147)-τ138 fusion alone (lane 1), or different GAL4- and τ138- constructs as indicated (lanes 2–5), were introduced into yeast cells containing the UAS$_G$-*SNR6* template. *GAL4* [GAL4-(1–881)] is harbored by pMA210 (21). The τ138 and τ138-GAL80 constructs are cloned into pGEN (30) under control of the *PGK* promoter. The transcripts from the UAS$_G$-*SNR6* template (maxi SNR6 RNA) and the endogenous SNR6 RNA were monitored by Northern blotting analysis as in Figure 2.

toxic effects, the reduced sensitivity of pGBT9-containing two-hybrid systems could lead to overlooking weak interactions.

We have therefore constructed a new Gal4 DNA-binding domain fusion vector, pQBD1™. Its auxotrophic marker is *TRP1*, like most of the Pol II two-hybrid vectors for the expression of Gal4-(1–147) fusions, and its multiple cloning site and reading frames are compatible with many classical two-hybrid vectors including pGBT9, pAS1, pAS2, pAS2-1, pODB8, and pODB80 (2,7,10,19). Thus, any GAL4-(1–147) fusions constructed in vectors designed for the Pol II two-hybrid systems (and most probably *TRP1*-marked) can be readily tested in our system or, if needed, transferred into pQBD1™ for higher expression.

pQBD1™ was derived from pODB8 (19), a vector expressing Gal4-(1–147) fusions whose high stability in yeast was shown to lead to strong protein expression and high co-transformation efficiency. The map and polylinker sequence of pQBD1™ are shown in Figure 5.

The pTAU2™ vector

pTAU2™ was specifically designed for the efficient expression of τ138-fusion proteins. This vector is derived from a *HIS3*-marked, 2μ plasmid, p423GPD (28). It contains the full-length τ138 coding sequence under the control of the strong and constitutive promoter of GPD (glyceraldehyde-3-phosphate dehydrogenase). Because τ138 is a nuclear protein, no exogenous nuclear localization signal is required for the vector. The map and polylinker sequence of pTAU2™ are shown in Figure 6.

The QYN1™ strain

We constructed a new strain, QYN1™, which harbors a UAS_G-*SNR6* allele at the *SNR6* chromosomal locus. In the first tester strain yMCM616, the chromosomal SNR6 gene had a 2-bp deletion in its B block, giving the *snr6-ΔB* allele. However, this *snr* genotype could be reverted by a punctual mutation to produce false positives. Replacing the SNR6 mutant allele by the UAS_G-*SNR6* construct had two advantages: first, the reversion rate of UAS_G-*SNR6* is predicted to be much lower than that of the *snr6-ΔB* allele, and second, it provides one extra template for monitoring the two-hybrid interactions.

The tester strain QYN1™ is a W303-1A derivative containing the pRS316/SNR6 and the pQAR1™ plasmids and has the following genotype:

MATa *ura3-1 leu2-3, -112 trp1-1 his3-11, -15 ade2-1 can1-100 snr6-B-box::UAS$_G$×5.*

3.5 A Direct Screening Method to Detect Protein–Protein Interactions

SNR6, the reporter gene of our system, is essential to cell viability. The tester cells must therefore be supplied with *SNR6* transcripts through a process that can be turned off when protein interactions are tested. In the 5-FOA assay, *SNR6* RNAs are produced from a wild-type copy of *SNR6* present in a *URA3* plasmid that can be shuffled. Although 5-FOA assay in this design worked, we sought a simpler way to trigger the conditional activation of the *SNR6* allele.

We had previously tried to define the domains of τ138 sufficient for UAS_G-*SNR6* activation (24). The τ138 subunit of TFIIIC is a large protein (1160 amino acids) with-

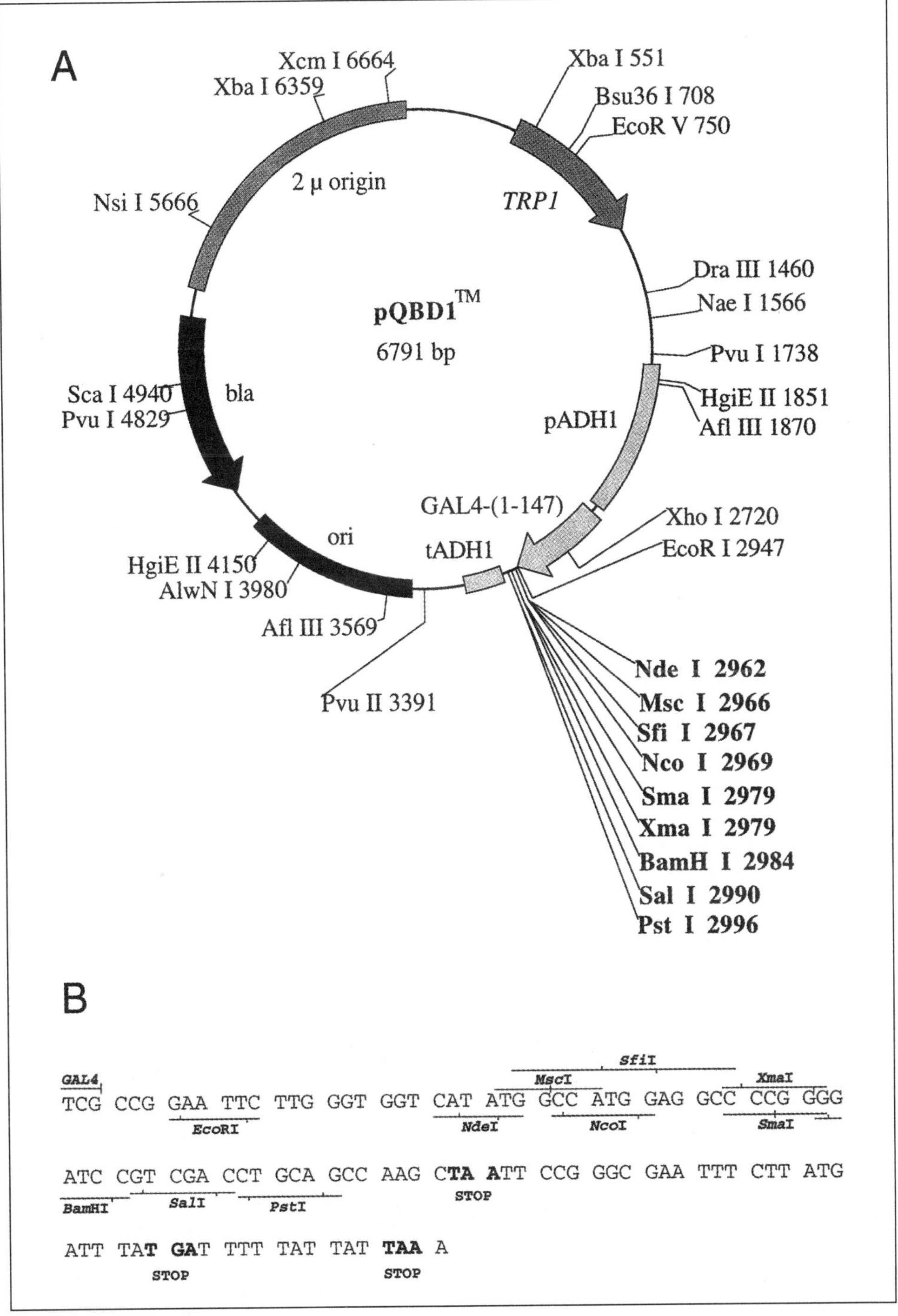

Figure 5. Map and multiple cloning site of pQBD1™. Important features include the *ADH1* promoter (pADH1), which drives transcription of the GAL4-(1–147) fragment, the *ADH1* terminator (tADH1), the yeast origin of replication (2 µ origin), the *E. coli* origin of replication (ori), the *E. coli* selectable marker for ampicillin resistance (bla), and the yeast auxotrophic marker *TRP1*. The restriction sites diagrammed below the map are unique and hence suitable for the insertion of heterologous genes. The nucleotide sequence is arranged in codons continuing in the same reading frame as the upstream GAL4-(1–147) fragment.

out apparent similarity to other eukaryotic proteins, and its DNA-binding domain is unknown. Truncated τ138 constructs with large deletions at N-terminal, C-terminal, or central parts were tested for their ability to promote the transcription of UAS$_G$-*SNR6* templates when fused to the GAL4-(1–147) moiety (24). We found that τ3-GAL4-(1–147) and τ1-3-GAL4-(1–147) constructs that code for, respectively, amino acids 761–1160 (τ3) and amino acids 1–329 and 676–1160 (τ1–3) from τ138, fused to

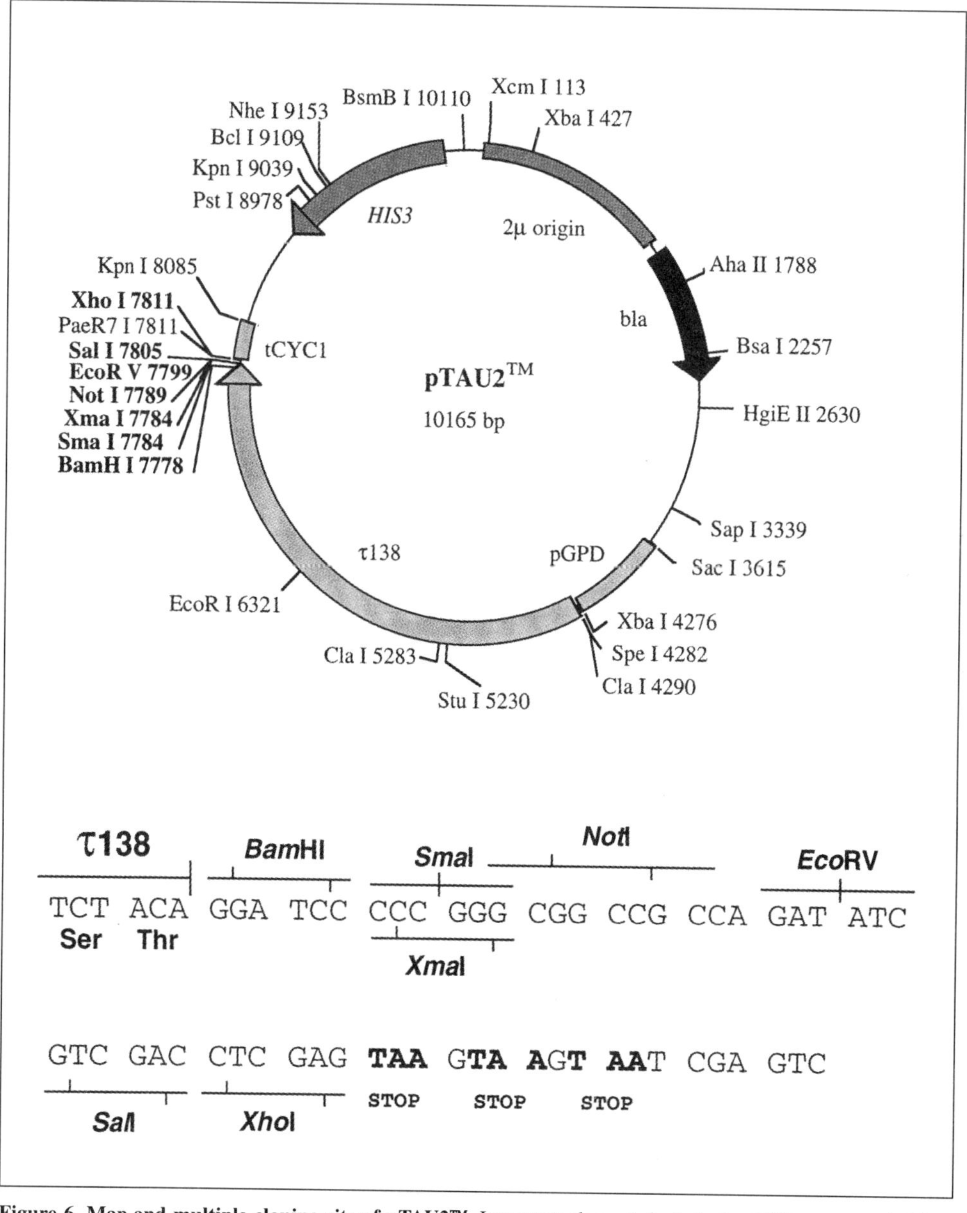

Figure 6. Map and multiple cloning site of pTAU2™. Important elements include the *GPD* promoter (pGPD), which drives transcription of the τ138 gene, the *CYC1* terminator (tCYC1), the yeast origin of replication (2 μ origin), the *E. coli* selectable marker for ampicillin resistance (bla), and the yeast auxotrophic marker *HIS3*. The restriction sites diagrammed below the map are unique and hence suitable for the insertion of heterologous genes. The nucleotide sequence is arranged in codons continuing in the same reading frame as the upstream *τ138* gene.

158

Gal4-(1–147) were able to activate the transcription of UAS$_G$-*SNR6*. They produced, respectively, 15% and 45% of the transcript level generated with the τ138-GAL4-(1–147) fusion comprising the τ138 full-length sequence [τ1-2-3, (24), Figure 7, lanes 2, 4, 5]. In contrast, deletion of the C-terminal part of τ138 (amino acids 677–1160) abolished its transcriptional activity in this system (Figure 7, lanes 1 and 3).

We then tested whether the transcript level generated from UAS$_G$-*SNR6* templates via the τ3- and the τ1-3-GAL4-(1–147) constructs was able to support normal cell growth. yMCM616 cells transformed with pRS315/UAS$_G$-SNR6 and with plasmids harboring the τ1-, τ1-2-, τ3-, τ1-3-, and τ138-GAL4-(1–147) fusions were streaked onto plates containing 5-FOA at 30°C. Only the transformants containing the τ3-, τ1-3-, and τ138-GAL4-(1–147) constructs were able to lose the URA3 plasmid harboring a wild-type copy of *SNR6* (pRS316/SNR6) and to develop colonies.

Individual transformants having lost pRS316/SNR6 were then isolated and tested for temperature sensitivity at 34°C and 38°C. Transformants containing the τ138-GAL4-(1–147) fusion were able to grow at all the temperatures tested, whereas transformants containing the τ3- and the τ1-3-GAL4-(1–147) constructs were able to grow at 34°C but not at 38°C. To quantify the thermosensitivity of the strains, cells from log phase cultures grown at 30°C were plated, incubated at 30°C and 38°C, and examined after 4 days. The results (Table 1) demonstrate that the τ3- and τ1-3-Gal4-(1–147) fusions are thermosensitive activators of UAS$_G$-*SNR6*. To our knowledge, this is the first report of conditional activation of a specific Pol III gene (thermosensitive mutations of components of the Pol III machinery do exist, but they affect the

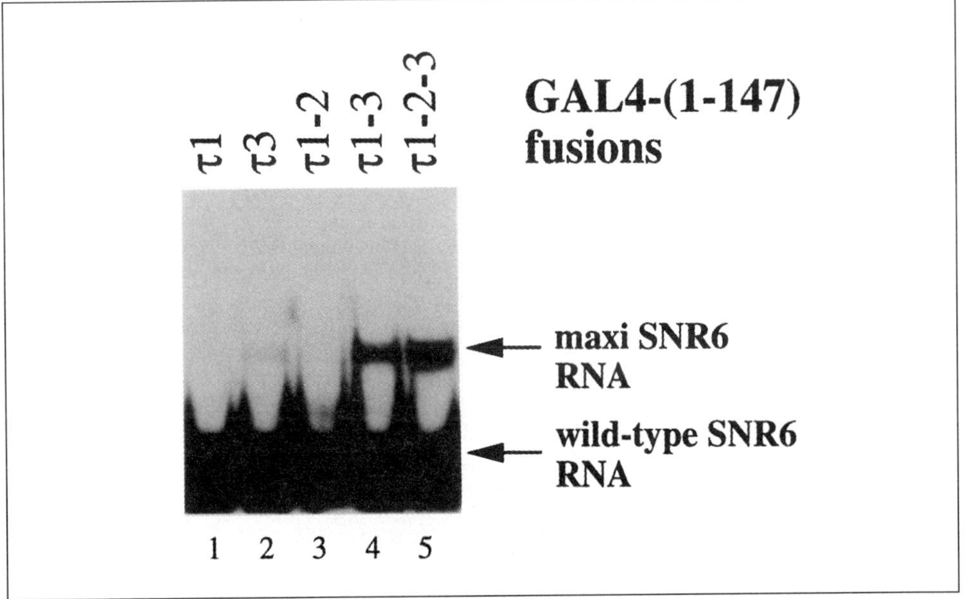

Figure 7. Activation of the UAS$_G$-*SNR6* template by truncated *τ138* constructs fused to GAL4-(1–147). The τ1-, τ3-, τ1-2-, τ1-3-, τ1-2-3-GAL4-(1–147) constructs contain the sequences encoding, respectively, amino acids 1–329 (τ1), 761–1160 (τ3), 1–676 (τ1-2), 1–329 and 676–1160 (τ1-3), and 1-1160 (τ1-2-3, the entire sequence of τ138), from τ138, cloned in front of GAL4-(1–147) into the pGEN vector (30) under control of the strong, constitutive *PGK* promoter. Northern blotting analyses were performed on YPH500α transformants harboring the UAS$_G$-*SNR6* template and various fusions between τ138 fragments and GAL4-(1–147). The transcripts from the UAS$_G$-*SNR6* gene are indicated as maxi *SNR6* RNA. The amounts of transcripts generated with the τ3 and τ1-3 fusions represent, respectively, 15% and 45% of the transcript level generated with the GAL4-(1–147)-τ138 construct.

Table 1. τ3-Gal4-(1–147) and τ1-3-Gal4-(1–147) are Thermosensitive Activators of the UAS$_G$-*SNR6* Gene

Constructs	Number of Colonies at 30°C	Number of Colonies at 38°C
τ3-GAL4-(1–147)	83	1
τ1-3-GAL4-(1–147)	100	25
τ138-GAL4-(1–147)	110	>100 000

yMCM616 transformants containing various fusions between GAL4-(1–147) and τ138 fragments were grown overnight in a medium lacking leucine and tryptophane, diluted to OD 0.2 in the same medium, and grown for 8 more hours. 100 µL of cell cultures and 100 µL of 10 000-fold dilutions were then plated and incubated at 38°C and 30°C, respectively. Plates were examined for colony numbers after 4 days.

transcription of many genes at the same time).

We were then able to set up a new strategy to screen protein interactions via this Pol III-based system. The yeast tester strain contains only one type of *SNR6* allele, the UAS$_G$-*SNR6* template, which is activated at permissive temperature by a τ1-3-Gal4-(1–147) fusion protein. After transformation by the vectors harboring the GAL4-(1–147)- and τ138-constructs, the protein interactions can be detected by plating the cells at restrictive temperature (38°C).

We then undertook to improve this UAS$_G$-*SNR6* conditional activation system. The original τ1-3-GAL4-(1–147) construct was harbored by a *TRP1*-marked, multicopy vector, pGEN (30), and was under the control of a strong, constitutive promoter from the gene encoding phosphoglycerate kinase (PGK). We aimed at reducing both the expression of the τ1-3-GAL4-(1–147) fusion and its binding to the UAS$_G$-*SNR6* template.

The P26L (Pro 26 → Leu) mutation of *GAL4* alters a proline residue located in the Gal4 DNA-binding domain, two amino acids before the second pair of cysteines that are responsible for zinc binding. This mutation was shown to make Gal4 activity dependent upon increased concentration of Zn^{2+} in vivo and in vitro (12). We therefore constructed various fusions between GAL4-(1–147)^{P26L} and τ138 fragments and tested their dependency upon Zn^{2+} to activate the UAS$_G$-*SNR6* template. These fusions were placed under the control of a weak promoter on a centromeric and *URA3*-marked vector, p416CYC1 (28). We found that the activity of the Gal4-(1–147)^{P26L}-τ138 fusion protein was Zn^{2+}-independent but moderately thermosensitive (Table 2). In contrast, Gal4-(1–147)^{P26L}-τ1-3 fusion allowed normal cell growth at 30°C and was completely thermosensitive at 38°C (Table 2). The vector harboring this construct, p416CYC1/GAL4-(1–147)^{P26L}-τ1-3, was finally chosen as the vector of our system. It combines the following advantages: first, the activation of UAS$_G$-*SNR6* template by Gal4-(1–147)^{P26L}-τ1-3 is completely temperature-dependent; second, Gal4-(1–147)^{P26L}-τ1-3 binding to the UAS$_G$ is weakened and its expression is low, which reduces the competition with Gal4-(1–147)-protein X fusions for binding to the UAS$_G$-*SNR6* template. Lastly, because p416CYC1/GAL4-(1–147)^{P26L}-τ1-3 is *URA3*-marked, it allows one to double check, using plates containing 5-FOA, that the viability of the cells at 38°C is indeed due to interacting fusion proteins.

The new vectors and strains developed for the Pol III two-hybrid system have been validated using previously characterized Prp9/Prp21 and Prp11/Prp21 interactions

Table 2. Gal4-(1–147)^{P26L}-τ1-3 is a Thermosensitive Activator of UAS$_G$-*SNR6*

Constructs	% of Thermoresistant Cells
GAL4-(1–147)-τ138	70%
GAL4-(1–147)^{P26L}-τ138	15%
GAL4-(1–147)^{P26L}-τ1-3	0.00%

QYN1 transformants containing either a GAL4-(1–147)-τ138 fusion, GAL4-(1–147)^{P26L}-τ138, or GAL4-(1–147)^{P26L}-τ1-3 constructs were grown to mid-log phase at 30°C. Serial dilutions were plated and incubated at 30°C or at 38°C. Percent survival of cells incubated at 38°C was determined relative to cells incubated at 30°C.

(data not shown). They now constitute a laboratory kit, the Q-PID™ System, distributed by Quantum-Appligene, Illkirch, France.

4. CONCLUSION

We have developed a novel, Pol III-based, two-hybrid system that can be used to analyze interactions between two known proteins or to search libraries for proteins that interact with a target protein. This system is particularly suited to the study of proteins involved in Pol II transcription or of proteins that spuriously activate Pol II reporter genes, but can be used to analyze virtually any kind of protein.

Because it is based on the interactions between two hybrid proteins in the nucleus of yeast cells, this system is subject to the same basic limitations. For instance, the two hybrid proteins must be expressed at a sufficient level, which will not occur if any are unstable or are toxic to the cell. In addition, they must be correctly folded and targeted to the nucleus, and their Gal4 or τ138 moieties must not hamper their interacting properties. Some of these problems can be solved by using only part of the proteins of interest, but it still remains possible that some physiological interactions might not be detected with this system.

However, we think that this system has a number of advantages. First, it allows the isolation and analysis of proteins that interact with *bona fide* or spurious regulators of Pol II transcription. Activation of Pol II transcription was found to be more wide-spread among proteins functionally unrelated to the transcription process than initially predicted. 1.0% of *E. coli* genomic DNA fragments randomly fused to Gal4 binding domain were found to activate Pol II gene expression (22). This promiscuity is likely due to the relative lack of specificity of Pol II activating sequences. The classical Pol II-based two-hybrid systems thus have a serious limitation. In contrast to Pol II genes, only two factors, TFIIIC and TFIIIB, are involved in the activation of *SNR6*, which makes it likely that very few sequences will activate the UAS$_G$-*SNR6* gene by themselves. Moreover, with this system interactions between proteins that play a part in the regulation of the Pol II machinery DNA and its associated proteins can be readily studied in their physiological environment.

This Pol III system is also sensitive. For two pairs of interacting proteins (Prp9/Prp21 and Prp11/Prp21), we were able to directly compare the activation of reporter genes in the Pol II and Pol III two-hybrid systems [(25), Table 3]. The UAS$_G$-*SNR6* template is activated by an indirect activation almost as strongly as it is

Table 3. Comparison of Relative Activation Level in RNA Polymerase II and III Two-Hybrid Systems

	Pol II Two-Hybrid System	Pol III Two-Hybrid System
Reporter gene	UAS_G-*lacZ*	UAS_G-*SNR6*
DNA-binding component	Gal4-(1–147)	Gal4-(1–147)
Activating component	Gal4-(768–881)	τ138
Quantification of reporter gene activation	β-galactosidase activity	*SNR6* RNA level
Reporter gene activation through direct interaction	Gal4-(1–147)-Gal4-(768–881) **100%**	Gal4-(1–147)-τ138 **100%**
Reporter gene activation through indirect interaction	Gal4-(1-147)-Prp9 and Gal4-(768-881)-Prp21 **4%**	Gal4-(1-147)-Prp9 and τ138-Prp21 **70%**
	Gal4-(1–147)-Prp11 and Gal4-(768–881)-Prp21 **6%**	Gal4-(1–147)-Prp11 and τ138-Prp21 **80%**

The transcriptional activation obtained through direct interaction between the UAS_G sites and the GAL4-(1–147)-activator construct of each system is assigned the value 100%.

by a direct one in the Pol III system (Figure 2); whereas indirect activation of the UAS_G-*lacZ* gene in the classical two-hybrid system reached only a few percent of the activation level obtained when Gal4 activating domain was fused to Gal4 DNA-binding domain (17,21).

Finally, this system can be modified for a diverse range of applications. After an interaction is identified between two proteins, their interacting domains can be mapped using the same two-hybrid search procedures. DNA-binding proteins can be identified using a one-hybrid procedure by screening a τ138-fusion library with a *SNR6* reporter gene fused to the relevant target DNA sequence. Three-hybrid systems can be developed in which expression of a third protein enhances or inhibits the interactions between two proteins.

ACKNOWLEDGMENTS

This research was supported by grants from the European Community (Science and Biotechnology programs).

REFERENCES

1.Arrebola, R., N. Manaud, S. Rozenfeld, M.C. Marsolier, O. Lefebvre, C. Carles, P. Thuriaux, C. Conesa, and A. Sentenac. 1998. Tau91, an essential subunit of yeast transcription factor IIIC, cooperates with tau138 in DNA binding. Mol. Cell Biol. *18*:1-9.

2.Bartel, P., C.T. Chien, R. Sternglanz, and S. Fields. 1993. Elimination of false positives that arise in using the two-hybrid system. BioTechniques *14*:920-924.

3.Boeke, J.D., F. LaCroute, and G.R. Fink. 1984. A positive selection for mutants lacking orotidine-5′-phosphate decarboxylase activity in yeast: 5-fluoro-orotic acid resistance. Mol. Gen. Genet. *197*:345-346.

4.Brent, R. and M. Ptashne. 1985. A eukaryotic transcriptional activator bearing the DNA specificity of a prokaryotic repressor. Cell *43*:729-736.

5. **Burnol, A.F., F. Margottin, P. Schultz, M.C. Marsolier, P. Oudet, and A. Sentenac.** 1993. Basal promoter and enhancer element of yeast U6 snRNA gene. J. Mol. Biol. *233*:644-658.

6. **Church, G.M. and W. Gilbert.** 1984. Genomic sequencing. Proc. Natl. Acad. Sci. USA *81*:1991-1995.

7. **Durfee, T., K. Becherer, P.L. Chen, S.H. Yeh, Y. Yang, A.E. Kilburn, W.H. Lee, and S.J. Elledge.** 1993. The retinoblastoma protein associates with the protein phosphatase type 1 catalytic subunit. Genes Dev. *7*:555-569.

8. **Eschenlauer, J.B., M.W. Kaiser, V.L. Gerlach, and D.A. Brow.** 1993. Architecture of a yeast U6 RNA gene promoter. Mol. Cell Biol. *13*:3015-3026.

9. **Fields, S. and O. Song.** 1989. A novel genetic system to detect protein-protein interactions. Nature *340*:245-246.

10. **Harper, J.W., G.R. Adami, N. Wei, K. Keyomarsi, and S.J. Elledge.** 1993. The p21 Cdk-interacting protein Cip1 is a potent inhibitor of G1 cyclin-dependent kinases. Cell *75*:805-816.

11. **Joazeiro, C.A., G.A. Kassavetis, and E.P. Geiduschek.** 1994. Identical components of yeast transcription factor IIIB are required and sufficient for transcription of TATA box-containing and TATA-less genes. Mol. Cell Biol. *14*:2798-2808.

12. **Johnston, M.** 1987. Genetic evidence that zinc is an essential co-factor in the DNA binding domain of GAL4 protein. Nature *328*:353-355.

13. **Kassavetis, G.A., B.R. Braun, L.H. Nguyen, and E.P. Geiduschek.** 1990. *S. cerevisiae* TFIIIB is the transcription initiation factor proper of RNA polymerase III, while TFIIIA and TFIIIC are assembly factors. Cell *60*:235-245.

14. **Kunkel, T.A., J.D. Roberts, and R.A. Zakour.** 1987. Rapid and efficient site-specific mutagenesis without phenotypic selection. Methods Enzymol. *154*:367-382.

15. **Legrain, P. and C. Chapon.** 1993. Interaction between PRP11 and SPP91 yeast splicing factors and characterization of a PRP9-PRP11-SPP91 complex. Science *262*:108-110.

16. **Legrain, P., C. Chapon, and F. Galisson.** 1993. Interactions between PRP9 and SPP91 splicing factors identify a protein complex required in prespliceosome assembly. Genes Dev. *7*:1390-1399.

17. **Legrain, P., M.C. Dokhelar, and C. Transy.** 1994. Detection of protein-protein interactions using different vectors in the two-hybrid system. Nucleic Acids Res. *22*:3241-3242.

18. **Leveillard, T., G.A. Kassavetis, and E.P. Geiduschek.** 1993. Repression and redirection of *Saccharomyces cerevisiae* tRNA synthesis from upstream of the transcriptional start site. J. Biol. Chem. *268*:3594-3603.

19. **Louvet, O., F. Doignon, and M. Crouzet.** 1997. Stable DNA-binding yeast vector allowing high-bait expression for use in the two-hybrid system. BioTechniques *23*:816-818.

20. **Ma, J. and M. Ptashne.** 1987. The carboxy-terminal 30 amino acids of GAL4 are recognized by GAL80. Cell *50*:137-142.

21. **Ma, J. and M. Ptashne.** 1987. Deletion analysis of GAL4 defines two transcriptional activating segments. Cell *48*:847-853.

22. **Ma, J. and M. Ptashne.** 1987. A new class of yeast transcriptional activators. Cell *51*:113-119.

23. **Marsolier, M.C., N. Chaussivert, O. Lefebvre, C. Conesa, M. Werner, and A. Sentenac.** 1994. Directing transcription of an RNA polymerase III gene via GAL4 sites. Proc. Natl. Acad. Sci. USA *91*:11938-11942.

24. **Marsolier, M.C., M.N. Prioleau, and A. Sentenac.** 1997. A RNA polymerase III-based two-hybrid system to study RNA polymerase II transcriptional regulators. J. Mol. Biol. *268*:243-249.

25. **Marsolier, M.C. and A. Sentenac.** 1999. RNA polymerase III-based two-hybrid system. Methods Enzymol. *303*:411-422.

26. **Marsolier, M.C., S. Tanaka, M. Livingstone-Zatchej, M. Grunstein, F. Thoma, and A. Sentenac.** 1995. Reciprocal interferences between nucleosomal organization and transcriptional activity of the yeast *SNR6* gene. Genes Dev. *9*:410-422.

27. **Melcher, K. and S.A. Johnston.** 1995. GAL4 interacts with TATA-binding protein and coactivators. Mol. Cell Biol. *15*:2839-2848.

28. **Mumberg, D., R. Muller, and M. Funk.** 1995. Yeast vectors for the controlled expression of heterologous proteins in different genetic backgrounds. Gene *156*:119-122.

29. **Schmitt, M.E., T.A. Brown, and B.L. Trumpower.** 1990. A rapid and simple method for preparation of RNA from *Saccharomyces cerevisiae*. Nucleic Acids Res. *18*:3091-3092.

30. **Shpakovski, G.V., J. Acker, M. Wintzerith, J.F. Lacroix, P. Thuriaux, and M. Vigneron.** 1995. Four subunits that are shared by the three classes of RNA polymerase are functionally interchangeable between *Homo sapiens* and *Saccharomyces cerevisiae*. Mol. Cell Biol. *15*:4702-4710.

31. **Sikorski, R.S. and P. Hieter.** 1989. A system of shuttle vectors and yeast host strains designed for efficient manipulation of DNA in *Saccharomyces cerevisiae*. Genetics *122*:19-27.

32. **Stotz, A. and P. Linder.** 1990. The *ADE*2 gene from *Saccharomyces cerevisiae*: sequence and new vectors. Gene *95*:91-98.

33. **Zawel, L. and D. Reinberg.** 1995. Common themes in assembly and function of eukaryotic transcription complexes. Annu. Rev. Biochem. *64*:533-561.

Grow and Glow: GFP Reporter Plasmid for One-Step Selection in the Two-Hybrid System

Robert S. Cormack
Max-Planck-Institut für Züchtungsforschung,
Abteilung Biochemie, Köln, Germany

1. INTRODUCTION

The yeast two-hybrid system (10) or interaction trap (21) has rapidly become one of the most widely used techniques in molecular biology. It is now the method of choice to identify protein–protein interactions from either cDNA libraries or known gene sequences (1). A related procedure, the so-called one-hybrid system (23), is also becoming a routine approach to clone genes that encode sequence-specific DNA-binding factors. Both methods rely on the *trans*-activation of reporter genes in *Saccharomyces cerevisiae* to identify positive interactions. Growth selection via activation of either a *HIS3* or *LEU2* reporter gene is used in conjunction with a second reporter gene (*lacZ*) that expresses β-galactosidase (3). The use of two reporter genes enables discrimination of false positives that activate only the auxotrophic marker. Yeast colonies that survive the growth selection scheme are chosen, re-streaked, and tested for expression of *lacZ* by means of a filter assay or growth on minimal medium plates containing 5-bromo-4-chloro-3-indolyl-β-D-galactopyranoside (X-gal). This latter step can often be time-consuming, depending on the number of primary transformants obtained.

To simplify and accelerate the screening process of the two- and one-hybrid systems, a new reporter plasmid was constructed that utilizes the green fluorescent protein (GFP) from the bioluminescent cnidarian *Aequorea victoria* (8). The GFP of the jellyfish *A. victoria* is activated in vivo by an energy transfer via the Ca^{2+}-stimulation of the photoprotein aequorin (19). The blue light generated by aequorin excites GFP and results in the emission of green light. GFP itself consists of 238 amino acids (M_r = 27 kDa) and is synthesized as an apoprotein in which post-translational formation of the chromophore occurs in an O_2-dependent manner, independent of any other gene products (4,8). It maximally absorbs light at 395 nm and has an emission peak of 509 nm. The nonsubstrate requirement for GFP activity makes this protein an

Yeast Hybrid Technologies
Edited by L. Zhu and G.J. Hannon

165

attractive reporter for gene-expression studies and this utility was initially demonstrated in both prokaryotes (*Escherichia coli*) and eukaryotes (*Caenorhabditis elegans*) (4). It has subsequently been used to monitor gene expression in many organisms, including mouse (5), *Drosophila* (24), zebrafish embryos (16), *Arabidopsis* (22), and yeast (18). In addition to the non-invasiveness of GFP detection (long-wave UV light) the protein is very stable, non-toxic, and resistant to photobleaching. These properties make GFP a viable alternative to traditional reporter genes such a β-galactosidase (*lacZ*), β-glucuronidase (GUS), chloramphenicol acetyltransferase (CAT), or firefly luciferase, which require substrate for their detection. Several modifications of the wild-type *gfp* cDNA have been engineered with optimized codon usage, improved fluorescence activity, and red-shifted variants with altered excitation maxima intended for fluorescence microscopy (8).

The *A. victoria* GFP variant, GFPuv, is optimized for maximal fluorescence by UV-light excitation and was generated using an in vitro DNA shuffling technique that introduced point mutations (7). It contains three amino-acid substitutions (F99S, M153T, V163A) relative to the wild-type *gfp10* cDNA (20), making it fluoresce 18 times brighter while retaining identical excitation and emission wavelength maxima. The GFPuv gene (CLONTECH Laboratories, Palo Alto, CA, USA) was inserted downstream of a *GAL1,10* minimal promoter and upstream of the *ADH1* terminator within the yeast episomal plasmid YEp357R (17), which contains a *URA3* selectable marker and a 2 μ origin of replication. The vector pGNG1 (Figure 1) houses four copies of the LexA operator within the *GAL1,10* minimal promoter and is designed for bait proteins fused to the LexA DNA-binding domain. A reporter plasmid similar to pGNG1, but designed for the one-hybrid system pGNG2 (Figure 1), was also constructed. pGNG2 contains a multiple cloning site about 450 bp upstream of the ATG start codon of the GFPuv gene. The multiple cloning site contains unique *Not*I, *Nru*I, and *Spe*I restriction sites for the introduction of "bait" DNA sequences, which are necessary for a one-hybrid screen. A full description of the construction of pGNG1 and pGNG2 is to be published elsewhere (6).

Both pGNG1 and pGNG2 are valuable additions to the family of reporter plasmids for the two- and one-hybrid systems. They make secondary screening of yeast colonies quicker and more cost-effective than conventional *lacZ* assays. Detection of protein–protein interactions via the GFP provides global screening of colonies without bias and can be adapted to most yeast-based two-hybrid strategies, including those that use GAL4 fusion baits and preys.

2. PROTOCOLS

General Plasmids, Yeast Strains, and Libraries

- Yeast strain *S. cerevisiae* EGY48 (*MAT*α *ura3 trp1 his3 lexA operator-LEU2*) (9); EGY48 has the *LEU2* reporter gene integrated into its genome.
- Reporter plasmids pGNG1 (World patent WO 98/46789)
- Bait plasmid pEG202 (11); pEG202 constitutively expresses the LexA DNA-binding domain.
- Prey plasmid pJG4-5 (11) containing library cDNA; pJG4-5 expresses the B42 activation domain under the control of a galactose-inducible promoter.

2.1 Protocol for the Construction of Baits

Materials and Reagents

- Plasmids selected from above
- Complete medium (CM)/glucose-uracil-histidine agar plates (10-cm diameter)
- 0.1 M lithium acetate/Tris-EDTA (TE) buffer, pH 7.5
- Long-wave UV lamp (360–400 nm)

Procedure

1. Subclone cDNA sequence of interest into the bait plasmid pEG202 within the polylinker (*Eco*RI, *Xma*I, *Bam*HI, *Sal*I, *Nco*I, *Not*I, *Xho*I, *Sal*I), so that the resulting hybrid is in frame with respect to the LexA DNA-binding domain.
2. Verify the bait construct by restriction endonuclease mapping and DNA sequencing of the plasmid.
3. Co-transform the S. *cerevisiae* strain EGY48 with pGNG1 and pEG202-Bait using the lithium acetate method (2) and select on CM/glucose-uracil-histidine agar plates (10-cm). Colonies should appear within 3 days when incubated at 30°C.

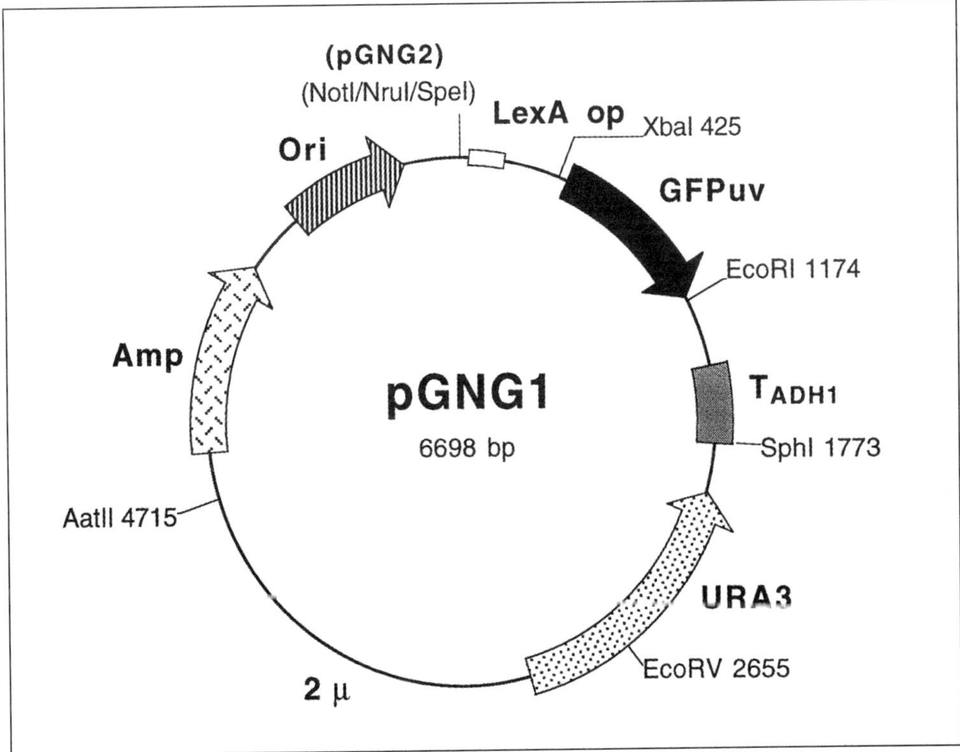

Figure 1. Reporter plasmids pGNG1 and pGNG2. The GFPuv gene is driven by a *GAL1,10* minimal promoter containing eight binding sites for LexA. Both plasmids contain the *URA3* selectable marker, 2 μ origin to allow propagation in yeast, and the ampicillin resistance gene (Amp) and origin of replication (Ori) for propagation in *E. coli*. pGNG2 has, additionally, unique restriction sites for *Not*I, *Nru*I, and *Spe*I, starting at position 7 of the plasmid sequence.

4. Restreak single colonies onto CM/glucose-uracil-histidine-leucine agar plates (10 cm) and incubate at 30°C for 1–3 days. Colonies that do not activate the *LEU2* reporter—and hence are unable to grow—express bait proteins that are suitable for a two-hybrid screen. Also, inspect the colonies for their ability to activate the GFP reporter by placing the selection plate under a long-wave UV lamp (360–400 nm).

2.2 Protocol for Library Transformation

Materials and Reagents

- Plasmids selected from **General Plasmids, Yeast Strains, and Libraries**
- CM/glucose-uracil-histidine medium
- CM/galactose/raffinose-uracil-histidine-tryptophan medium
- CM/galactose/raffinose-uracil-histidine-tryptophan-leucine agar plates (20 cm)
- 0.1 M lithium acetate/TE buffer, pH 7.5
- Sonicated salmon sperm DNA (ssDNA) (10 mg/mL)
- Dimethyl sulfoxide (DMSO)
- 40% (wt/vol) polyethylene glycol (PEG) 4000/0.1 M lithium acetate/TE buffer
- Long-wave UV lamp (360–400 nm)

Procedure

1. Grow *S. cerevisiae* EGY48/pGNG1/pEG202-Bait in 400 mL of CM/glucose-uracil-histidine medium at 30°C with shaking (200 rpm) to an OD_{600} of 1.0.
2. Harvest the yeast cells by centrifugation at $1000\times g$ for 5 min. Wash the cells in 20 mL of sterile, distilled water.
3. Resuspend the cells in 5 mL of lithium acetate/TE buffer, pellet, and discard the supernatant.
4. Resuspend the yeast in 2 mL of lithium acetate/TE buffer. Aliquot 100 μL of the competent cells into each of 10 microcentrifuge tubes. To each, add 2 μg of pJG4-5 library DNA and 60 μg of sonicated ssDNA.
5. Add 10 μL of DMSO.
6. Add 600 μL of PEG 4000/lithium acetate/TE buffer to each sample and gently invert the tubes several times to mix.
7. Incubate at 30°C for 30 min.
8. Heat-shock at 42°C for 15 min.
9. Combine all 10 samples, add 7.5 mL of CM/galactose/raffinose-uracil-histidine-tryptophan medium, and incubate for 4 h at 30°C with shaking (200 rpm) to induce the *GAL1,10* promoter of pJG4-5.
10. Pellet cells by centrifugation at $1000\times g$ for 5 min and resuspend in sterile, distilled water.
11. Spread the yeast cells onto 20–25 CM/galactose/raffinose-uracil-histidine-tryptophan-leucine agar plates (20-cm) and incubate at 30°C for 3–7 days.
12. Once colonies have appeared, examine the plates under a long-wave UV lamp

(360–400 nm) in the dark. Any colonies exhibiting green fluorescence should be restreaked onto a similar selection plate to confirm leucine auxotrophy and re-examined for GFP activation.

2.3 Protocol for Activation of the GFP Reporter Gene

The *S. cerevisiae* strain EGY48 was transformed with pGNG1 and selected on CM/glucose-uracil. The resulting strain was then transformed with different pEG202-based bait and pJG4-5-based prey plasmids and selected on the appropriate CM agar medium. Plates containing the yeast transformants were analyzed under a long-wave UV (366 nm) lamp in the dark for green fluorescence. In addition, the same bait and prey plasmids were co-transformed with pSH18-34 (9) into EGY48 and re-streaked onto CM/galactose/raffinose agar selection plates containing X-gal to visualize β-galactosidase activity (2). All strains were also restreaked onto CM medium deficient in leucine.

Yeast carrying pGNG1 alone do not visibly fluoresce when placed under a UV lamp (Table 1, line 1). However, yeast cells harboring both pGNG1 and pEG202 bait plasmids that are capable of activating *LEU2* and *lacZ* reporters, encoded by pSH18-34, activated the GFPuv gene (lines 2 and 3). Yeast cells expressing the bait human p53 (1–144) fluoresced visibly stronger under UV light than those expressing the bait PRHA(310–796) (14), suggesting that the *trans*-activation activity of the former is greater. A third bait protein, PRHA(452–796), which activated the leucine reporter gene but not *lacZ*, did not activate GFPuv (line 4).

2.3.1 Analysis of Known Protein–Protein Interactions

The standard two-hybrid system of Fields and co-workers was used previously to show that murine p53 and the large-T antigen (LTA) interact (13). Yeast cells were therefore cotransformed with plasmids expressing a LexA DNA-binding domain p53 hybrid protein and the LTA fused to the B42 activation domain. The p53–LTA interaction was observed when *LEU2* or *lacZ* was used as the reporter gene (Table 1, line 6). Additionally, the interaction between these two proteins activated the GFPuv reporter gene (line 6 and Figure 2B). Yeast cells carrying the bait plasmid alone did not activate any of the reporter genes (line 5 and Figure 2A).

Three other protein pairs were tested for their ability to activate the GFPuv reporter gene. The *Arabidopsis* cDNA clones At4, At23, and At30 were originally isolated from a standard interactor hunt (data not shown) and were shown to bind to the bait PRHA(1–270), since each protein pair activated the *LEU2* and *lacZ* reporter genes (Table 1, lines 8–10). The same bait–protein combinations activated the GFPuv reporter gene with differing degrees of intensity (lines 8–10) when they were co-expressed in the presence of pGNG1. The PRHA(1-270) bait alone could not activate any of the reporter genes (line 7).

2.3.2 Two-Hybrid Screen (Interactor Hunt)

The *Arabidopsis* cDNA clone, At23, isolated from a standard interaction trap using PRHA(1–270) as the bait (Table 1) was subcloned into the plasmid pEG202. An *Arabidopsis* cDNA library was constructed in the plasmid pJG4-5 from oligo-dT–primed total mRNA. *S. cerevisiae* EGY48 was cotransformed with pGNG1 and pEG202-At23 and selected on a CM/glucose-uracil-histidine agar plate. Approximately 20 μg of cDNA plasmid library was used to transform the resultant strain. The cells were spread onto 20 large (20-cm) CM/galactose/raffinose-uracil-histidine-tryptophan-leucine agar plates and incubated at 30°C for up to one week. Three colonies that survived the growth selection process were subsequently analyzed for GFPuv activation by placing the selection plate under a long-wave UV lamp (366 nm). Only one of the three colonies fluoresced green under these conditions. The intensity of fluorescence was comparable to that observed for the interaction between p53 and LTA (Figure 2B). The sequence of the cDNA encoding the positive prey pro-

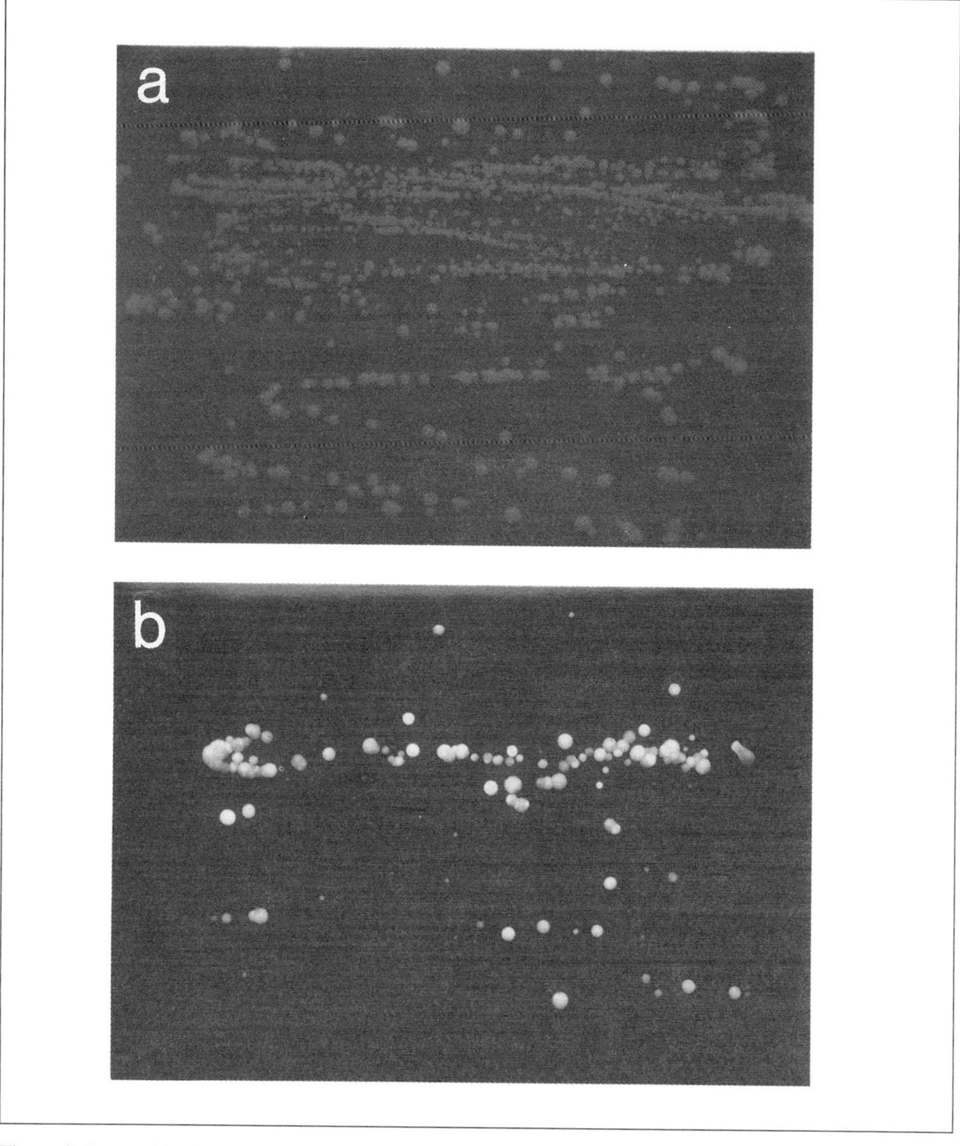

Figure 2. *S. cerevisiae* **EGY48 colonies on CM selection agar plates.** Plates contain the reporter plasmid pGNG1 and **(A)** the bait plasmid pEG202-p53(72–390) or **(B)** pEG202-p53(72–390) plus the prey plasmid pJG4-5-LTA as observed under long-wave UV light.

Table 1. Comparison of Reporter Genes Co-Transformed with Various Bait and Prey Plasmids in *S. cerevisiae* EGY48

Hybrid Plasmid	CM-Leu[a]	β-gal[b]	GFP[c]
1. None	n	-	-
2. pEG202-hp53(1–144)	y	+++	+++
3. pEG202-PRHA(310–796)	y	+++	++
4. pEG202-PRHA(452–796)	y	-	-
5. pEG202-mp53(72–390)	n	-	-
6. pEG202-mp53(72–390) pJG4-5-LTA(84–708)	y	+++	+++
7. pEG202-PRHA(1–270)	n	-	-
8. pEG202-PRHA(1–270) pJG4-5-At4	y	+++	++
9. pEG202-PRHA(1–270) pJG4-5-At23	y	+++	+
10. pEG202-PRHA(1–270) pJG4-5-At30	y	+++	++

[a]Growth after 2 days on CM/galactose/raffinose-leucine medium; y, yes; n, no.

[b]Co-transformed with pSH18-34.

[c]Co-transformed with pGNG1.

-, no signal; +, weak; ++, medium; +++, strong.

tein (At23BP) was determined following the isolation of its plasmid by standard means (2). A sequence alignment between At23 and At23BP shows that these two proteins are related homologs (Figure 3). Subsequent control transformations, including domain swapping, verified the reproducibility of the interaction to strongly activate the GFPuv reporter gene. Moreover, in vitro co-immunoprecipitation experiments confirmed the observed association (data not shown).

```
At23     1 ................................MSSRGKRKDEDVRASD 16
                                                         S
At23BP   1 MKEITEFKVRKLASEKLAIDLSEKSHKAFVRSVVEKFLDEERAREYENSQ 50


At23    17 DESETHAPAKKVAKPADDSDQSDDIVVCNISKNRRVSVRNWNGKIWIDIR 66
              E    K   K      D   D   C  S  RRV      GK    IR
At23BP  51 VNKEEEDGDKDCGKGNKEFDDDGDLIICRLSDKRRVTIQEFKGKSLVSIR 100


At23    67 EFYVKDGKTLPGKKGISLSVDQWNTLRNHAEDIEKALSDLS... 107
              E Y KDGK LP  KGISL    QW T       IE A
At23BP 101 EYYKKDGKELPTSKGISLTDEQWSTFKKNMPAIENAVKKMESRV 144
```

Figure 3. Alignment of the deduced amino-acid sequences of full-length At23 and At23BP.

3. DISCUSSION

Several refinements to the two-hybrid system/interaction trap have been developed since its initial presentation. These include the use of dual reporter genes, improved shuttle vectors and yeast strains, mating selection, lambda-based plasmid libraries, and the use of 3-aminotriazole against leaky *HIS3* expression. Many of these modifications to the original method have substantially reduced the amount of time and effort required to isolate genes deemed to encode interacting factors. The introduction of the GFP as a visible reporter gene represents a further improvement to the protocol by simplifying the screening of cDNA libraries. The reporter plasmid pGNG1 exhibits no detectable basal expression of GFP in the absence of bait and prey plasmids. However, in the presence of interacting proteins or *trans*-activating bait proteins, the GFP protein is expressed at sufficiently high levels in yeast cells to permit their identification. There is no apparent cryptic intron splicing as observed for *Arabidopsis* (12). The sensitivity of the GFP reporter gene is lower than that of *lacZ* driven by the same minimal promoter, as revealed by the intensity of fluorescence versus that of the level of blue staining. This provides a higher threshold for the determination of relevant protein–protein interactions. Nevertheless, the GFP reporter was quite capable of detecting the interaction between p53 and the large-T antigen and three protein associations originally isolated using the standard interaction trap. The variability in the intensity of fluorescence of each set of test proteins can provide a semi-quantitative measurement of the strength of interaction, a phenomenon similarly observed with β-galactosidase. It should be noted that GFP expression in yeast, like β-galactosidase, can be quantified (15,18). A standard two-hybrid screen was also performed using GFP as the secondary reporter gene. This resulted in the successful cloning of a gene encoding an interacting homologous protein, thus demonstrating the effectiveness of GFP as reporter gene in a yeast two-hybrid screen.

The practical advantages of using GFP over β-galactosidase as a reporter in the two- and one-hybrid systems are substantial. In a typical screen, individual surviving yeast colonies are picked and then tested for β-galactosidase activity either by a filter assay or growth on minimal medium plates containing X-gal. Both procedures are often labor-intensive, especially when hundreds of yeast colonies are obtained. In contrast, when pGNG1 is used as the reporter plasmid, the selection plates containing the yeast colonies are simply placed under a UV (360–400 nm) lamp and positives become immediately identifiable, thus obviating the need to restreak. More importantly, the use of GFP as the second reporter eliminates a bias in the selection of colonies for the *lacZ* assay. It is common practice to choose only those yeast colonies of sufficient size, thus ignoring some possible positives. The GFP assay is unprejudiced; every surviving yeast colony is inspected for fluorescence. Finally, the pGNG1 and pGNG2 plasmids allow a two-hybrid or one-hybrid screen to be performed using GFP as a sole reporter of protein– or DNA–protein interactions. In this approach, the entire library transformation is plated without the need for growth selection. Positive clones are subsequently identified amongst the non-fluorescing negative yeast colonies. Such an approach may be suitable for baits, which alone activate the auxotrophic selectable marker and not the GFP reporter gene.

Some comments should be made about GFPuv detection in yeast colonies on agar plates. The green fluorescence activity of GFPuv is stable for several weeks but diminishes with prolonged storage at 4°C. However, restreaking of the cells re-establishes the original fluorescence. In addition, yeast colonies from a primary transformation are generally less active with respect to GFPuv expression than those that have been restreaked onto fresh plates. Individual colonies of the same yeast clone often show variability in green fluorescence, probably attributable to copy-number effects of the bait, prey, and reporter plasmids.

In summary, a GFP reporter can be used successfully in a two-hybrid screen to detect interacting proteins. It is expressed stably in yeast cells, requires no substrate for its activity, and can be visually inspected with the unaided eye under long-wave UV light.

ACKNOWLEDGMENTS

I would like to thank Dr. Imre E. Somssich for sponsoring this work and Dr. Jeff Dangl for his generous gift of yeast vectors and strains. R.S.C. holds a Fellowship from the Alexander-von-Humboldt-Stiftung.

REFERENCES

1. **Allen, J.B., M.W. Walberg, M.C. Edwards, and S.J. Elledge.** 1995. Finding prospective partners in the library: the two-hybrid system and phage display find a match. Trends Biochem. Sci. *20*:511-516.
2. **Ausubel, F.M., R. Brent, R.E. Kingston, D.D. Moore, J.G. Seidman, J.A. Smith, and K. Struhl.** 1995. Short protocols in molecular biology, 3rd ed. John Wiley & Sons, Inc., New York, NY.
3. **Bartel, P.L., C. Chien, R. Sternglanz, and S. Fields.** 1993. Using the two-hybrid system to detect protein-protein interactions, p. 153–179. *In* D.A. Hartley (Ed.), Cellular Interactions in Development: A Practical Approach. IRL Press, Oxford, UK.
4. **Chalfie, M., Y. Tu, G. Euskirchen, W.W. Ward, and D.C. Prasher.** 1994. Green fluorescent protein as a marker for gene expression. Science *263*:802-805.
5. **Chiocchetti, A., E. Tolosano, E. Hirsch, L. Silengo, and F. Altruda.** 1997. Green fluorescent protein as a reporter of gene expression in transgenic mice. Biochim. Biophys. Acta. *1352*:193-202.
6. **Cormack, R.S., K. Hahlbrock, and I.E. Somssich.** 1998. Isolation of putative plant transcriptional co-activators using a modified two-hybrid system incorporating a GFP reporter gene. Plant J. *14*:685-692.
7. **Crameri, A., E.A. Whitehorn, E. Tate, and W.P.C. Stemmer.** 1996. Improved green fluorescent protein by molecular evolution using DNA shuffling. Nature Biotechnol. *14*:315-319.
8. **Cubitt, A.B., R. Heim, S.R. Adams, A.E. Boyd, L.A. Gross, and R.Y. Tsien.** 1995. Understanding, improving and using green fluorescent proteins. Trends Biochem. Sci. *20*:448-455.
9. **Estojak, J., R. Brent, and E.A. Golemis.** 1995. Correlation of two-hybrid affinity data with *in vitro* measurements. Mol. Cell. Biol. *15*:5820-5829.
10. **Fields, S. and O. Song.** 1989. A novel genetic method to detect protein–protein interactions. Nature *340*:245-246.
11. **Gyuris, J., E. Golemis, H. Chertkov, and R. Brent.** 1993. Cdi1, a human G1 and S phase protein phosphatase that associates with Cdk2. Cell *75*:791-803.
12. **Haseloff, J. and B. Amos.** 1995. GFP in plants. Trends Genet. *11*:328-329.
13. **Iwabuchi, K., B. Li, P. Bartel, and S. Fields.** 1993. Use of the two-hybrid system to identify the domain of p53 involved in oligomerization. Oncogene *8*:1693-1696.
14. **Korfhage, U., G.F. Trezzini, I. Meier, K. Hahlbrock, and I.E. Somssich.** 1994. Plant homeodomain protein involved in transcriptional regulation of a pathogen defense-related gene. Plant Cell *6*:695-708.
15. **Lim, C.R., Y. Kimata, M. Oka, K. Nomaguchi, and K. Kohno.** 1995. Thermosensitivity of green fluorescent protein fluorescence utilized to reveal novel nuclear-like compartments in a mutant nucleoporin NSP1. J. Biochem. *118*:13-17.
16. **Meng, A., H. Tang, B.A. Ong, M.J. Farrell, and S. Lin.** 1997. Promoter analysis in living zebrafish embryos identifies a *cis*-acting motif required for neuronal expression of GATA-2. Proc. Natl. Acad. Sci. USA *94*:6267-6272.
17. **Myers, A.M., A. Tzagoloff, D.M. Kinney, and C.J. Lusty.** 1986. Yeast shuttle and integrative vectors with multiple cloning sites suitable for construction of lacZ fusions. Gene *45*:299-310.
18. **Niedenthal, R.K., L. Riles, M. Johnston, and J.H. Hegemann.** Green fluorescent protein as a marker for gene expression and subcellular localization in budding yeast. Yeast *12*:773-786.

19.**Prasher, D.C.** 1995. Using GFP to see the light. Trends Genet. *11*:320-323.
20.**Prasher, D.C., V.E. Eckenrode, W.W. Ward, F.G. Prendergast, and M.J. Cormier.** 1992. Primary structure of the *Aequorea victoria* green-fluorescent protein. Gene *111*:229-233.
21.**Shirley, B.W. and I. Hwang.** 1995. The interaction trap: *in vivo* analysis of protein-protein interactions. Methods Cell Biol. *49*:401-416.
22.**Urwin, P.E., S.G. Moller, C.J. Lilley, M.J. McPherson, and H.J. Atkinson.** 1997. Continual green-fluorescent protein monitoring of cauliflower mosaic virus 35S promoter activity in nematode-induced feeding cells in *Arabidopsis thaliana*. Mol. Plant Microbe Interact. *10*:394-400.
23.**Wilson, T.E., T.J. Fahrner, M. Johnston, and J. Milbrandt.** 1991. Identification of the DNA binding site for NGFI-B by genetic selection in yeast. Science *252*:1296-1300.
24.**Yeh, E., K. Gustafson, and G.L. Boulianne.** 1995. Green fluorescent protein as a vital marker and reporter of gene expression in *Drosophila*. Proc. Natl. Acad. Sci. USA *92*:7036-7040.

Specialty Yeast Hybrid Systems

Yeast One-Hybrid System: A Genetic System to Identify DNA-Binding Proteins

Toshihiko Ezashi and R. Michael Roberts
Department of Animal Sciences, University of Missouri, Columbia, MO, USA

1. INTRODUCTION

The yeast single hybrid method provides a genetic screen to identify cDNAs encoding polypeptides that bind short sequences (motifs) of DNA, usually *cis*-acting regulatory elements of expressed genes (2,11,14,23). It is a modification of the yeast two-hybrid screen, which preceded it (8). Both exploit the bipartite structure of certain transcription factors, most usually the GAL4 protein, which can be engineered to provide a separate DNA-binding domain (DBD) and transcription activation domain (AD). Each cDNA in the library being explored is expressed as a fusion protein with the AD of the GAL4 protein.

In the one-hybrid screen, fusion proteins that bind the target element transactivate reporter genes (usually the yeast *His3* and the bacterial *lacZ* genes) (Figure 1, 2). The target DNA sequence has replaced the usual upstream activator sequence of these genes so that they are expressed only weakly, if at all, unless transactivated through the introduced target DNA motif within their promoters. The target DNA is usually multimerized to provide increased sensitivity to the screen. It should also be well characterized i.e., its limits well defined, and relatively short in order to minimize the number of different proteins that can potentially bind the DNA. Another important way of reducing the number of false positives, which arise from non-specific interactions of fusion proteins with the introduced DNA element, is to conduct an identical screen with a mutated target element. The latter should be minimally modified, but fail to provide an electrophoretic mobility shift with protein extracts from the tissues used to generate the cDNA library.

Yeast colonies that contain the cDNAs of interest are selected by their ability to grow in the absence of histidine and to hydrolyze X-gal (5-bromo-4-chloro-3-

Yeast Hybrid Technologies
Edited by L. Zhu and G.J. Hannon
© 2000 Eaton Publishing, Natick, MA

indolyl-β-D-galactopyranoside), i.e., turn blue. This double screen also helps reduce the number of false positives and the number of cDNAs that have to be analyzed. In the original description of the one-hybrid library screen, non-integrated reporter genes were employed (23). However, it is now standard practice to use integrated reporters to control copy numbers of the reporters and hence their expression. For example, colonies are normally selected on medium lacking histidine according to their size and appearance. The second screen for *lacZ* activity depends upon intensity of the blue color, with colonies selected on the "strength" of the reaction. Without a uniform gene copy number per cell, such semi-quantitative selection becomes much more difficult.

It is also important to note that the modified *HIS3* gene generally used in the one-hybrid screen is leaky. The native gene contains an upstream activator sequence (UAS) (the one replaced by the target DNA "bait"), which is normally recognized by the transcriptional activator GCN4. There are two TATA boxes, and GCN4 regulates only one of these TATA boxes (TR); the other TATA box (TC) drives low constitutive expressions of *HIS3*. The reporter plasmids, pHISi and pLacZi, normally used in the screen contain both these TATA boxes in their minimal promoters (Figure 2). By inserting the target element into the multi-cloning site (MCS), the regulated TR comes under the control of any fusion protein that binds the DNA "bait". There is still, however, limited expression due to TC. This leakiness is important because it is exploited in the construction of HIS3 reporter strains (Protocol for Integrating Target-Reporter Constructs into the Yeast Genome). During the genetic screen itself, growth of strains not utilizing the TR is controlled by 3-amino-1,2,4-triazole (3-AT). This compound competitively inhibits the HIS3 protein and is included in the medium at concentrations pre-determined to suppress background growth (5,9).

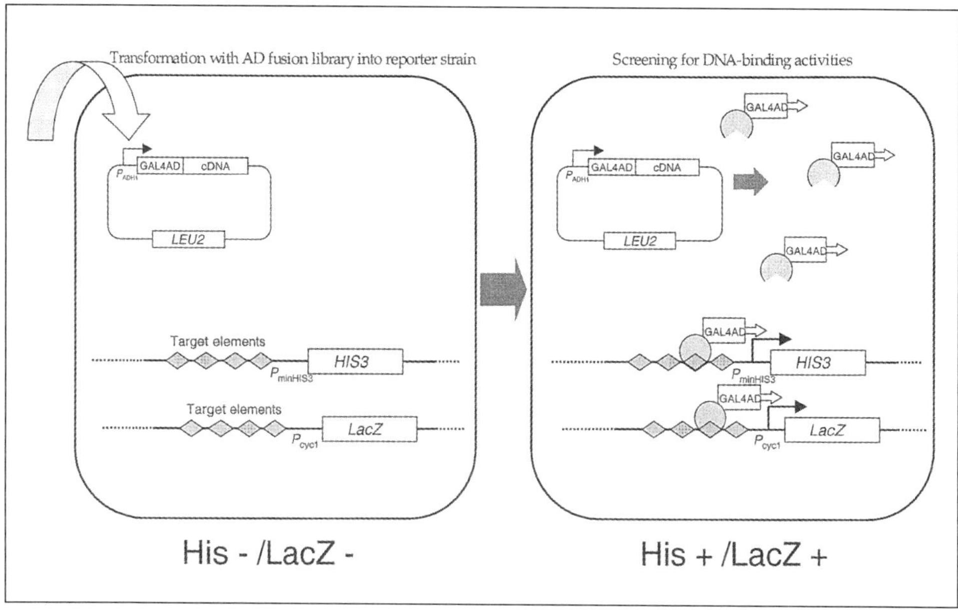

Figure 1. DNA-protein interactions in the yeast one-hybrid screen. A yeast reporter strain, integrated with the target-pHISi and target-pLacZi, is transformed with a GAL4-AD fusion cDNA library plasmid that carries a *LEU2* marker (left panel). The transformed yeast reporter cells are selected for His⁺ and LacZ⁺. Colonies capable of growth carry cDNA whose translation products can bind to the target DNA and activate transcription (right panel).

The pLacZi plasmid contains the bacterial *lacZ* gene downstream of the minimal promoter of the yeast iso-1-cytochrome *c* gene (P_{cyc1}). Target elements can be inserted into the MCS upstream of the P_{cyc1}-*lacZ* reporter (Figure 2). As mentioned previously, the extent of X-gal hydrolysis in a relatively brief (evaluation within 1 h) exposure provides some indication of the strength of transcriptional activation by the fusion protein.

In the section that follows, we have provided a detailed protocol for the one-hybrid screen. We have chosen not to describe the generation of the special libraries used for the yeast, which consist of the cDNA fused to the coding region of the GAL4 AD. Such libraries are identical to those used in the two-hybrid screen and may either be purchased, custom synthesized, or prepared in the investigator's own laboratory through use of a commercially available kit. After this step-by-step description (Figure 3), we provide an example from this laboratory of a one-hybrid screen that allowed a particular transcription factor to be identified (APPLICATIONS).

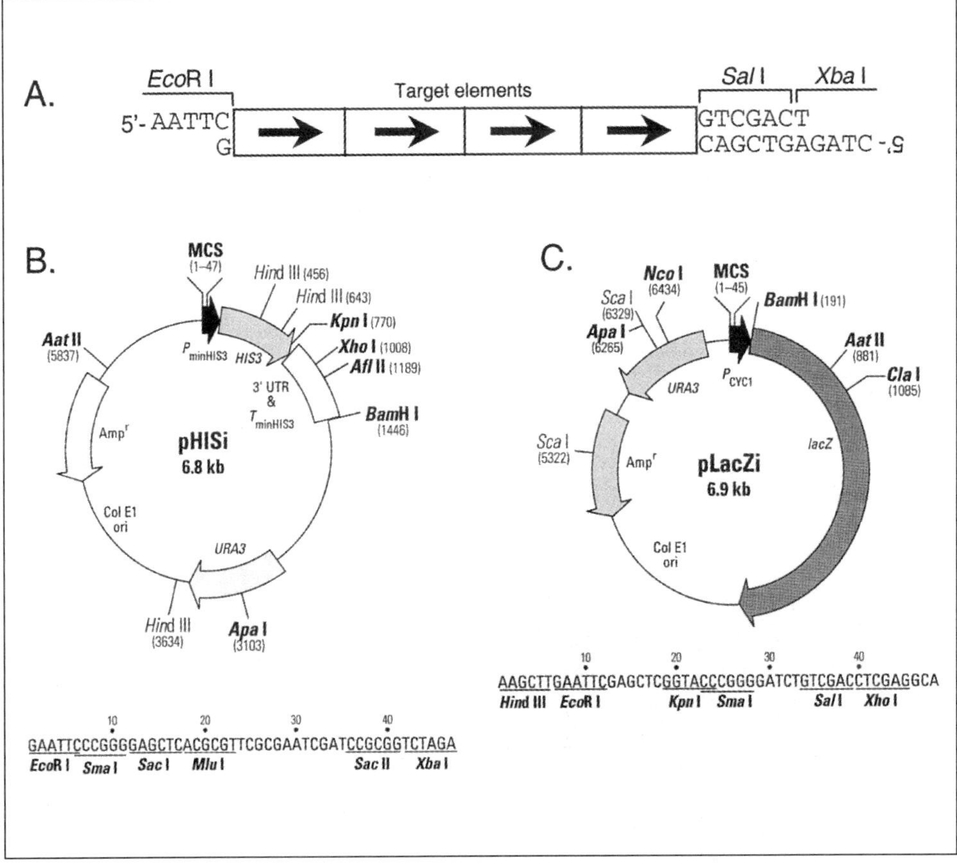

Figure 2. Maps and multi-cloning sites (MCS) of target elements (A) and yeast reporter plasmids (B) pHISi, (C) pLacZi for the one-hybrid system. (A) An annealed oligonucleotide is constructed that contains repeated target elements and *Eco*RI and *Xba*I overhangs to allow subsequent subcloning into the pHISi MCS. An *Eco*RI-*Sal*I fragment from the target-pHISi can be used for inserting into pLacZi plasmid. (B) pHISi contains the yeast *HIS3* gene downstream of the MCS and the minimal promoter of the *HIS3* locus ($P_{minHIS3}$). (C) pLacZi contains the bacterial *lacZ* gene downstream of the MCS and the minimal promoter of the yeast iso-1-cytochrome *c* gene (P_{cyc1}). (**B** and **C** are reproduced from **www.clontech.com/clontech/Vectors/Vectors.html**.)

2. PROTOCOLS

Plasmids, Yeast Strains, and Library

- The MATCHMAKER™ one-hybrid system is available from CLONTECH (Palo Alto, CA, USA). This kit provides one yeast strain (YM4271), three reporter vectors for integration (pHISi, pHISi-1, and pLacZi) and four control plasmids.
- The yeast strain, YM4271 is used for reporter vector integration. Its genotype is *MATa, ura3-52, his3-200, ade2-101, lys2-801, leu2-3, 112, trp1-901, tyr1-501, gal4-Δ512, gal80-Δ538, ade5::hisG* (16).

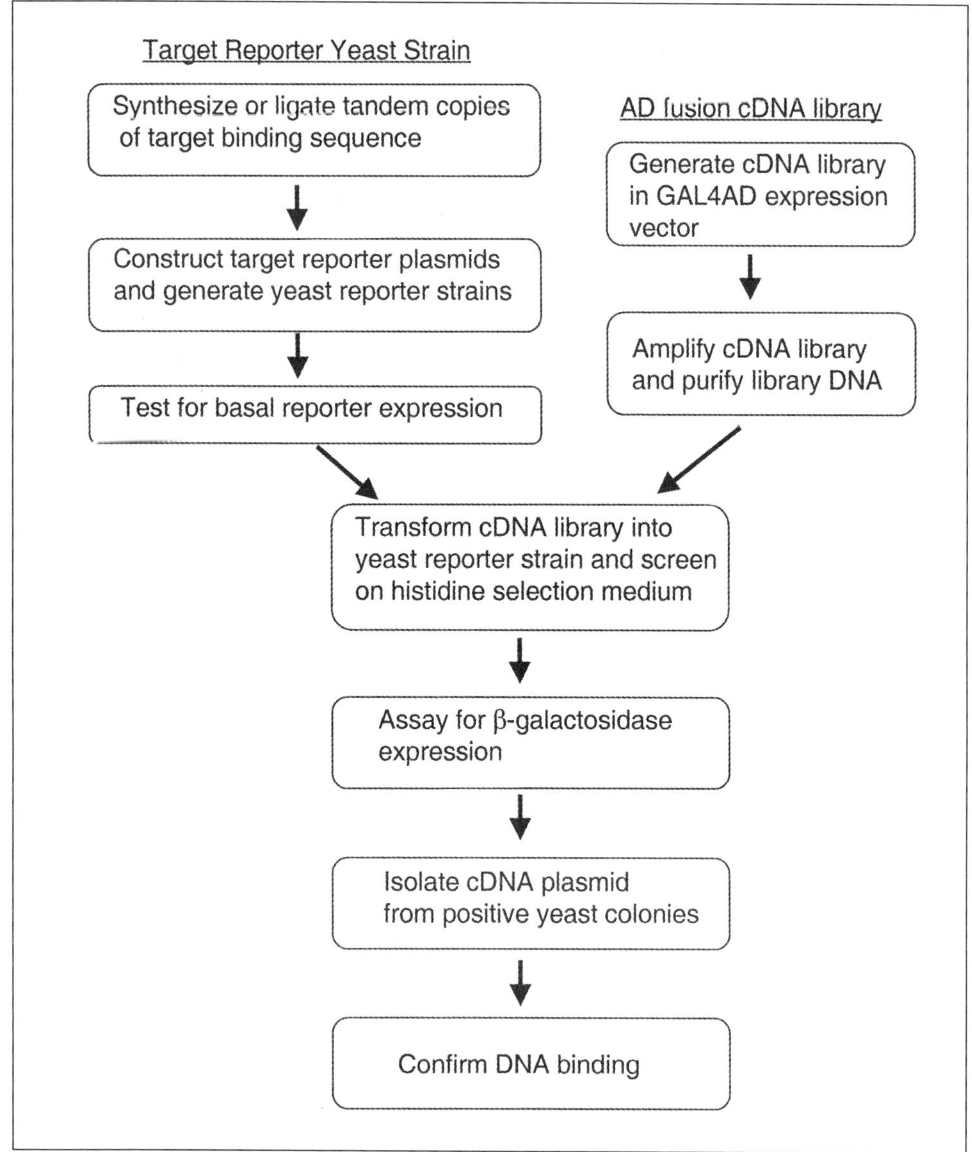

Figure 3. Schematic flow chart of the yeast one-hybrid screening procedures described here.

- The GAL4 activation domain fusion cDNA library should have at least 10^6 clones. Premade GAL4AD fusion cDNA libraries are available from some companies, for example, MATCHMAKER GAL4 libraries from CLONTECH and Two-Hybrid cDNA libraries from Stratagene (La Jolla, CA, USA). Alternatively, libraries can be constructed from isolated RNA by using kits such as the Two-Hybrid cDNA Library Construction Kit from CLONTECH and the pAD-GAL4-2.1 Library Construction Kit from Stratagene.

2.1 Protocol for Preparing Target-Reporter Constructs

Each target-reporter construct should consist of at least three tandem copies of the target element inserted upstream of the reporter gene. Although tandem copies may be generated by various methods, including ligation of a DNA fragment (15), oligonucleotide synthesis is the most convenient and reliable method for directional multimerization of a target (3). The two oligonucleotides, one representing the sense strand and the other its antisense complement, are prepared with different restriction sites on each end. When the two strands are annealed, the resulting double-stranded DNA will have different overhangs at each end for subcloning into the reporter plasmid's MCS (Figure 2). An *Eco*RI overhang for the 5′ end and an *Xba*I overhang followed by *Sal*I site for the 3′ end are often suitable (Figure 2A). A single pair of oligonucleotides can be used to construct both the target-pHISi and the target-pLacZi. After inserting the target elements (annealed oligonucleotides) into pHISi at the *Eco*RI-*Xba*I site to form target-pHISi, the *Eco*RI-*Sal*I fragment from target-pHISi can be cut out and inserted into pLacZi to form target-pLacZi.

Materials and Reagents

- Target elements: two oligonucleotides. The sense strand should have at least three tandem copies of the target element with different restriction sites on each end (Figure 2A). They should lack 5′ phosphates.
- 0.5 M NaCl
- Reporter plasmids: pHISi, pLacZi; 0.1 μg/μL
- Bovine serum albumin (BSA) 100× stock solution: 10 mg/mL BSA, for use with some restriction enzymes, including *Sal*I and *Xba*I.
- *Eco*RI, *Xba*I, *Sal*I restriction enzymes, 10–20 U/μL (New England BioLabs, Beverly, MA; Promega, Madison, WI, USA)
- T4 DNA ligase, 3 U/μL (New England Biolabs, Promega)
- 10× T4 ligation buffer [0.3 M Tris-HCl, pH 7.8, 0.1 M $MgCl_2$, 0.1 M dithiothreitol (DTT)]
- 10 mM ATP
- Competent *Escherichia coli* DH5α cells
- LB/amp plates
- Plasmid DNA purification system, e.g., Promega Wizard, QIAGEN Plasmid kits (QIAGEN, Valencia, CA, USA).
- Sequencing primers: pHISi 5′-TTCCCAGTCACGACGTTG-3′ (forward primer), pLacZi 5′-GCTACAAAGGACCTAATG-3′ (reverse primer)
- QIAEX II® Gel Extraction Kit (QIAGEN). Optional.

Procedure

1. Synthesize two oligonucleotides without 5′ phosphates, one representing the sense strand and the other its antisense strand. The sense strand consists of at least three tandem copies with different restriction sites on each end.

2. Start the annealing procedures by mixing 0.1 µg of sense-strand and 0.1 µg antisense-strand oligonucleotide in 10 µL of 50 mM NaCl.

3. Heat the oligonucleotide mixture to 70°C and maintain this temperature for 5 min. Then slowly (approx. 30 min) cool this solution to room temperature.

4. Digest 0.1 µg of pHISi by *Eco*RI and *Xba*I double digest in 20 µL for 2 h at 37°C. Confirm the complete digestion by electrophoresis of a 2 µL sample of the digest in a 1% agarose gel.

5. Incubate the digest at 65°C for 20 min for heat inactivation of restriction enzymes.

6. Mix 5 µL of digested pHISi, 1 µL of annealed oligonucleotide, 1 µL of H_2O, 1 µL of 10× T4 ligation buffer, 1 µL of 10 mM ATP, and 1 µL of T4 DNA ligase (at least 1 U) and incubate at room temperature overnight (or for at least 4 h).

   Note: Because the molar ratio of oligonucleotide to vector is 100:1 or greater, no gel purification to remove the stuffer fragment is required.

7. Transform competent DH5α, plate transformants on LB/amp plates, and incubate at 37°C overnight.

8. Prepare plasmid DNA by using Wizard Plus SV miniprep DNA Purification System or equivalent. Check for the insert by electrophoresing in a 2% agarose gel and sequence with the pHISi sequencing primer.

9. To prepare the target fragment for pLacZi, treat 5 µg of the constructed reporter plasmid (target-pHISi) with *Eco*RI and *Sal*I as a double digest in 30 µL for 2 h at 37°C. Electrophorese through a 2% low-melt agarose gel and cut out the target band with a scalpel.

10. Recover the DNA (1). For example, a QIAEX II Gel Extraction Kit is suitable.

11. To construct the target-pLacZi plasmid, follow steps 4 through 8, except pLacZi should be digested with *Eco*RI and *Sal*I.

2.2 Protocol for Integrating Target-Reporter Constructs into the Yeast Genome

The reporter plasmid target-pHISi and target-pLacZi should be integrated into the yeast genome. If they do not integrate, they will be lost when the cells replicate because these vectors do not carry a yeast replication origin. Linearization either by restriction digestion within the 3′ untranslated region—immediately after the *HIS3* marker in pHISi or within the *URA3* marker of pLacZi—significantly increases the efficiency of homologous recombination at the corresponding locus in the yeast genome (3). Because target-pHISi (or target-pLacZi) integration into the mutated *his3* (or *ura3*) locus of YM4271 confers a His⁺ (or Ura⁺) phenotype on the transformants, they can be selected on synthetic dropout (SD)/-His (or SD/-Ura) medium. The mini-scale LiAc yeast transformation procedure described here can be used for up to 6 parallel transformations. The method can also be scaled up (24).

Materials and Reagents

- Restriction enzymes: *Xho*I (10–20 U/µL, New England Biolabs or Promega) or *Afl*II (10 U/µL, New England Biolabs) for target-pHISi and *Nco*I (10 U/µL, New England Biolabs or Promega) or *Apa*I (10–20 U/µL, New England Biolabs or Promega) for target-pLacZi.
- Yeast extract Peptone Dextrose (YPD) Medium
- Sterile 250-mL flasks, and 50-mL conical tubes
- SD/-His and SD/-Ura plates (100-mm diameter)
- Sterile TE (Tris-EDTA buffer, pH 7.5)/Lithium Acetate (LiAc, 0.1 M) solution and Polyethylene glycol (PEG, 40%)/LiAc solution. Prepare only the volume needed, immediately prior to use, from 10× stocks.
- Herring testes carrier DNA [10 mg/mL, (1)]
- 100% dimethyl sulfoxide (DMSO) (Catalog No. D-8779; Sigma Chemical, St. Louis, MO or Catalog No. BP231; Fisher Scientific, Pittsburgh, PA, USA)
- Sterile 1× TE buffer
- Sterile glass rod or bent Pasteur pipette for spreading cells on plates

Procedure

1. Linearize 1 µg of target-pHISi with *Xho*I or *Afl*II and 1 µg of target-pLacZi with *Nco*I or *Apa*I separately, each in a final volume of 20 µL at 37°C for 2 h. Confirm the complete digestion by electrophoresis of a 2 µL sample of the digest in a 1% agarose gel.

2. Select at least two colonies of YM4271, 2–3 mm in diameter, and place in 1 mL of YPD.

3. Vortex-mix to disperse any clumps of cells.

4. Transfer this cell suspension into a 250-mL flask containing 15 mL of YPD.

5. Incubate at 30°C for approximately 20 h with shaking at 250 rpm. The cells should reach stationary phase growth [optical density $(OD)_{600}>1.5$].

6. Add 85 mL of YPD to the overnight culture. This dilution will bring the OD_{600} down to 0.2–0.3.

7. Incubate at 30°C for 3 h with shaking at 230 rpm. After this incubation, the OD_{600} should be within the range 0.4–0.6.

8. Place cells in 50-mL conical tubes and centrifuge at $1000 \times g$ for 5 min at room temperature.

9. Discard the supernatant and add 25 mL of sterile TE buffer or distilled H_2O to the tube. Thoroughly resuspend the cell pellets by vortex-mixing.

10. Pool cells in one tube and centrifuge at $1000 \times g$ for 5 min at room temperature.

11. Decant the supernatant and resuspend the cell pellet in 0.5 mL of freshly prepared, sterile TE/LiAc. At this stage, the cells should be competent for transformation.

    Note: If necessary, competent cells can be stored at room temperature for several h with only a minor reduction in competency. However, for the highest transformation efficiency, competent cells should be used within 1 h of their preparation.

12. Add 0.1 mg herring testes carrier DNA (in approx. 10 µL volume) to at least 1 µg of the linearized reporter plasmid (not more than 20 µL volume). In addition, set up a control transformation for each reporter plasmid using 1 µg of undigested plasmid and 0.1 mg carrier DNA in a separate 1.5-mL tube.

13. Add 0.1 mL of competent yeast cells to each tube and vortex-mix thoroughly.

14. Add 0.6 mL of sterile PEG/LiAc solution to each tube and vortex-mix at high speed for 10 s.

15. Incubate at 30°C for 30 min, shaking at 200 rpm.

16. Add 70 µL of DMSO and mix well by gentle inversion. Do not vortex-mix.

17. Heat-shock for 15 min in a 42°C water bath.

18. Chill cells on ice for 1–2 min.

19. Centrifuge cells for 5 s at $10\,000\times g$ at room temperature in a microcentrifuge. Remove the supernatant.

20. Resuspend cells in 150 µL of sterile TE buffer.

21. Plate entire transformation mixture on one plate of the appropriate SD medium. Select for pHISi transformants on SD/-His plates and pLacZi transformants on SD/-Ura plates.

22. Incubate plates, upside down, at 30°C for 4–6 days.

    Note: Usually, 20–100 pHISi transformants will be obtained per µg of DNA. For linearized target-pLacZi, approximately 300–1000 transformants per µg should be observed. For uncut reporter transformants, which act as a negative control, much fewer colonies will grow per plate. The integrated colonies are easily distinguished from colonies without an integrated functional reporter gene by their larger size. Ignore tiny background colonies.

23. Restreak colonies that arise from linearized plasmid transformation on the same selection medium used in step 22.

24. Incubate plates at 30°C for 4–6 days.

25. Seal the plates with Parafilm M™. They can be stored at 4°C for up to 4 weeks. Colonies from these plates will be used for subsequent procedures.

2.3 Protocol for Testing the Reporter Strains for Background Expression

Inserting the target element may alter the level of background *HIS3* and *lacZ* expression. The native yeast HIS3 promoter contains a UAS site recognized by the transcriptional activator GCN4 and two TATA boxes. GCN4 regulates one of the TATA boxes (TR), while the other TATA box (TC) drives low-level constitutive expression of *HIS3*. The *HIS3* reporter plasmid, pHISi, has both the *HIS3* TATA boxes present in the minimal promoter. By inserting a *cis*-acting element (target element) in the MCS, the first (regulated) TATA box (TR) can now be placed under regulatory control, although there is still a significant amount of constitutive expression due to the second TATA box of the *HIS3* TC. The leaky HIS3 expression can be suppressed by a competitive inhibitor of the yeast HIS3 protein, 3-AT. pLacZi contains the bacterial *lacZ* gene downstream of the minimal promoter of the yeast iso-1-cytochrome c gene (P_{cyc1}). The target element can be inserted into the MCS upstream of the P_{cyc1}-*lacZ* reporter.

The background expression of the target-reporter should be carefully checked to determine what concentration of 3-AT should be used in His selection medium and whether the target-pLacZi construct can be used for the library screening.

Materials and Reagents

<u>For β-galactosidase colony-lift filter assay</u>

- Whatman #5 or VMR Grade 410 paper filters, 75-mm, sterile
  Note: Nitrocellulose filters can also be used, but they are prone to crack when frozen.
- Forceps for handling the filters
- 100-mm plates
- Z buffer: 8.53 g/L Na_2HPO_4, 5.5 g/L $NaH_2PO_4 \cdot H_2O$, 0.75 g/L KCl, 0.246 g/L $MgSO_4 \cdot 7H_2O$. Adjust pH to 7.0 and autoclave. Can be stored at room temperature for up to 1 year.
- X-gal stock solution: Dissolve 5-bromo-4-chloro-3-indolyl-β-D-galactopyranoside (X-gal) in N,N-dimethylformamide (DMF) at a concentration of 20 mg/mL. Store in the dark at -20°C.
- Z buffer/X-gal solution: Prepare fresh daily as needed. 100 mL Z buffer, 0.27 mL β-mercaptoethanol (Sigma Chemical), 1.67 mL X-gal stock solution.
- Liquid nitrogen

<u>For *HIS3* background expression assay</u>

- 1.0 M 3-AT (3-amino-1,2,4-triazole) (Catalog No. A-8056; Sigma Chemical); prepare in deionized H_2O and filter sterilize. Store at 4°C.
- SD/-His plates containing 0, 15, 30, 45, and 60 mM 3-AT on each: After autoclaving the SD/-His medium, cool to approximately 55°C, add the appropriate amount of 1.0 M 3-AT stock solution and swirl to mix well.
  Note: 3-AT is heat-labile and will be destroyed if added to medium hotter than 55°C.
- Sterile 1× TE buffer
- Sterile glass rod or bent Pasteur pipette for spreading cells on plates

Procedure

1. **Test yeast colonies with integrated target-pLacZi construct in a β-galactosidase colony-lift filter assay.**

   See *Procedure* in Section 2.5.

   When the basal expression of target-pLacZi is low, dual *HIS3-lacZ* target reporters (yeast reporter strain containing both target-pHISi and target-pLacZi reporter plasmids) can be used for the screening.

a. If the control colony lift turns blue in the presence of X-gal in less than 1 h, then background *lacZ* expression is high and the pLacZi reporter should not be used.

b. If the colony lift turns blue in more than 1 h, then background *lacZ* expression is considered low. In this case, use the target-pLacZi reporter strain to construct a dual reporter strain for the library screening by integrating the target-pHISi construct into the target-pLacZi strain (Section 2.2).

2. **Test yeast colonies with integrated target-pHISi construct for background expression.**

The leaky *HIS3* expression from target-pHISi allows just enough colony growth on SD/-His medium (without 3-AT) to permit its use as a selectable marker. This background level of growth can be controlled by 3-AT in a library screening. Pick a single colony and suspend it in 1 mL of sterile TE buffer. Plate 5 μL of the suspension on each of the SD/-His plates containing 0, 15, 30, 45, and 60 mM 3-AT, respectively. Incubate plates at 30°C for 2–4 days. As the 3-AT concentration increases from 0 to 60 mM, the size of the colonies should become progressively smaller. The 3-AT concentration at which colonies fail to grow at all is sufficient to permit stringent selection of His⁺ transformants on medium lacking His. If 45–60 mM 3-AT is still not sufficient to suppress the reporter strain growth, then the target-pHISi reporter plasmid should be remade.

2.4 Protocol for Screening a cDNA Library for DNA-Binding Protein Genes

The protocol has been scaled up for screening $>1 \times 10^6$ independent clones. Fresh (no older than 3-weeks) colonies will give good growth for liquid culture inoculation. A single colony may be used as the inoculum if it is 2–3 mm in diameter. If colonies on the stock plate are smaller than 2 mm, scrape several colonies into the medium.

Materials and Reagents

- YPD medium
- Sterile 1-L and 250-mL flasks, and 50-mL conical tubes
- 15 150-mm SD/-Leu/-His plates containing the optimal selection concentration of 3-AT, several 100-mm SD/-Leu/-His/+ optimal (3-AT) plates, and one 100-mm SD/-Leu plate.
- Yeast reporter strain for making competent cells that was selected in Section 2.3.
- Sterile TE/LiAc and PEG/LiAc (Prepare only the volume needed, immediately prior to use, from 10× stocks).
- 20–50 μg of GAL4AD fusion cDNA library plasmid in solution: The library may need amplification to provide enough plasmid for the yeast transformation. If one has been purchased, consult instructions for more information on amplication.
- 2 mg herring testes carrier DNA
- 100% DMSO (Catalog No. D-8779; Sigma Chemical or Catalog No. BP231; Fisher Scientific)
- 50 mL SD/-His medium
- Sterile 1× TE buffer
- Sterile glass rod or bent Pasteur pipette for spreading cells on plates

Procedure

1. Inoculate several colonies of the appropriate yeast reporter strain, 2–3 mm in diameter, into 1 mL of YPD.

2. Vortex-mix to disperse any cell clumps.

3. Transfer this cell suspension into a 250-mL flask containing 50 mL of YPD.

4. Incubate at 30°C for approximately 20 h with shaking at 250 rpm so that the cultures reach stationary phase (OD_{600} >1.5). If the growth does not reach this OD in this time, the incubation time may be extended.

5. Transfer enough of the above overnight culture to produce an OD_{600} = 0.2–0.3 into 300 mL YPD.

6. Incubate at 30°C for 3 h with shaking at 230 rpm. After this incubation, the OD_{600} should reach 0.4–0.6.

7. Place cells in 50-mL conical tubes and centrifuge at 1000× g for 5 min at room temperature.

8. Discard the supernatant and add 25 mL of sterile TE buffer to the tube. Thoroughly resuspend the cell pellets by vortex-mixing.

9. Pool cells in one tube and centrifuge at 1000× g for 5 min at room temperature.

10. Decant the supernatant and resuspend the cell pellet in 1.5 mL of freshly prepared, sterile 1× TE/LiAc. Mix well by vortex-mixing.

    Note: Use competent cells within 1 h of their preparation for the highest transformation efficiency.

11. Add 20–50 µg of library plasmid (typically 1 µg/µL in 20–50 µL volume) to a sterile 50-mL tube with 2 mg of carrier DNA (approx. 10 µg/µL in approx. 200 µL volume), and mix well.

12. Add 1.0 mL of competent yeast cells to the DNA mixture and vortex-mix thoroughly.

    Note: If the cells are not mixed well, transformation efficiency may be declined.

13. Add 6 mL of sterile PEG/LiAc to the transformation mixture and vortex-mix at high speed for 10 s.

14. Incubate at 30°C for 30 min with shaking at 200 rpm.

15. Add 700 µL of DMSO and mix well by gentle inversion. Do not vortex-mix.

16. Heat shock for 15 min in a 42°C water bath. Swirl occasionally to mix.

17. Chill cells on ice for 2 min.

18. Centrifuge at 1000× g for 5 min at room temperature and remove the supernatant.

19. Resuspend cells in 2 × 25 mL of SD/-His/-Leu liquid medium in two 50-mL tubes.

20. Incubate at 30°C for 1 h with shaking at 200 rpm.

21. Centrifuge at 1000× g for 5 min at room temperature and remove the supernatant.

22. Resuspend cells in 7 mL of sterile TE buffer to give a final volume of approximately 7.5 mL.

23. Dilute 10 μL of well-suspended cells in 1 mL of sterile TE buffer, and plate 200 μL of the diluted sample onto a 100-mm SD/-Leu plate to determine the transformation efficiency.

    Note: To calculate the transformation efficiency, count the colonies, i.e., colony forming units (cfu), growing on the dilution plate after 2–4 days incubation. Total transformed number = cfu × 100 × 5 × 7.5 (1-5 × 10^6 clones should be screened)

24. Plate 500 μL of transformed cells on 150-mm SD/-Leu/-His plates containing the optimal selection concentration of 3-AT. Spread the cells immediately after pipetting the suspension onto the plate to avoid localized dilutions in the 3-AT concentration. To obtain an even growth of colonies after plating, continue to spread the transformation mixture over the agar surface until all liquid has been absorbed.

25. Incubate plates, upside down, at 30°C for 4–6 days.

    Note: Generally, three types of colonies will appear: First, all of the yeast in the transformation will grow a little because of the His and Leu in the YPD medium used to grow the yeast prior to transformation. This limited growth is seen as a hazy background on the plates. Second, numerous distinct colonies will slowly begin to appear on top of the hazy background on the plates. These colonies likely represent yeast that have received the library plasmids and are thus Leu+, but the majority will be His-. These colonies will soon stop growing. Third, colonies that grow to several mm in diameter over the next week are Leu+ and His+ (2).

26. Pick the largest colonies and restreak them onto 100-mm SD/-Leu/-His plates containing the optimal selection concentration of 3-AT. If a dual reporter strain is used, streak the colonies onto duplicate SD/-Leu/-His/+ optimal (3-AT) plates, and use one set of plates for a β-galactosidase colony-lift filter assay.

2.5 Protocol for Confirmation of Positive Clones: β-Galactosidase Colony-Lift Filter Assay

Assaying the target-*lacZ* reporter activity is a good independent test for transcriptional activation because it does not rely on *HIS3* growth selection, which can be leaky. Typically, positive colonies producing β-galactosidase take 0.5–2 h to turn blue. For instance, ts transcription factors (Ets-2 and GABPα) in the target-reporter strain shown in Figure 4A, turned blue in about 30 min at room temperature (Figure 4B) (7).

Materials and Reagents

See *Materials and Reagents* listed in Section 2.3; β-galactosidase colony-lift filter assay.

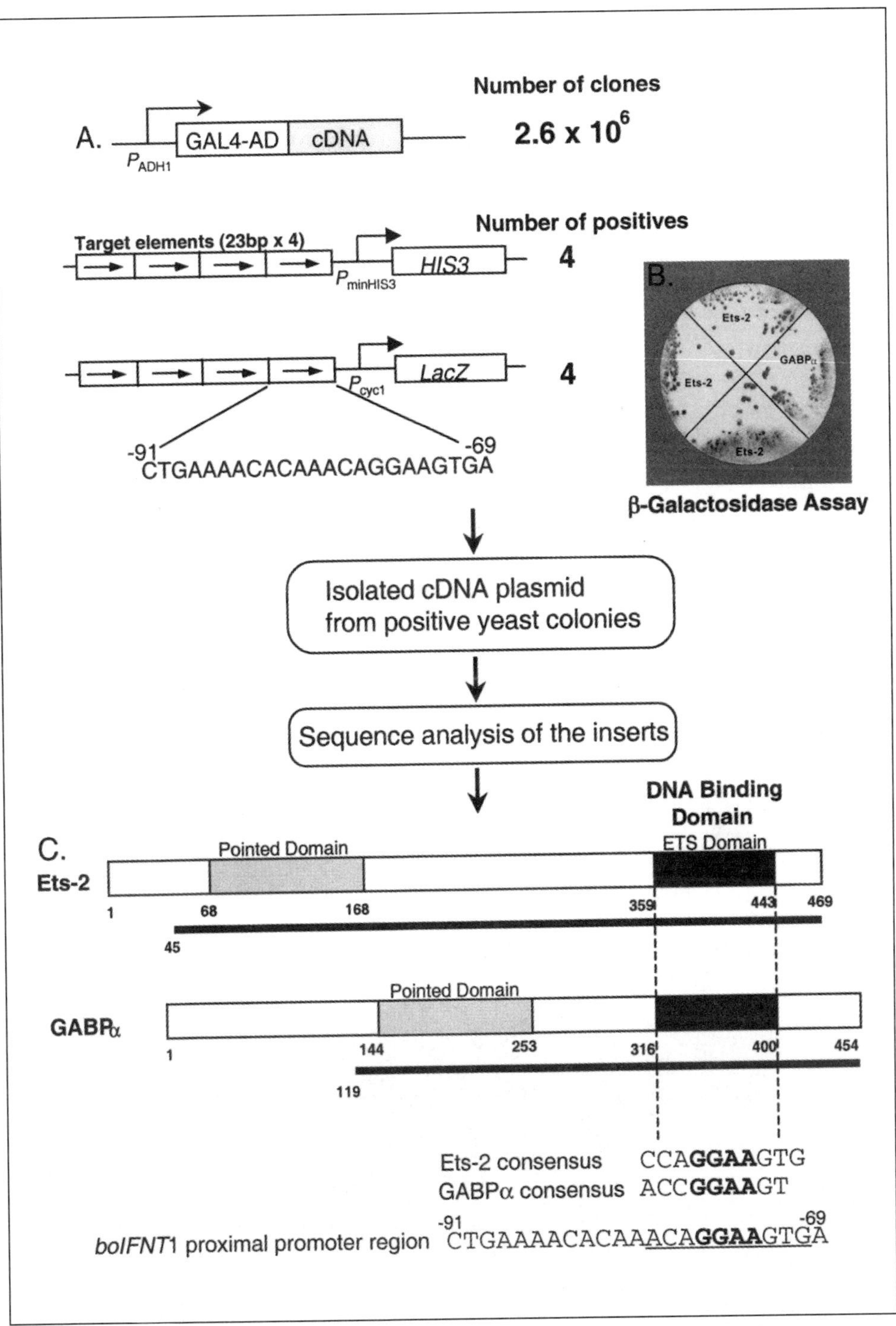

Figure 4. The one-hybrid screen for DNA-binding protein that associate with the bovine *IFNT1* promoter (-91 to –69). Out of 2.6 × 10⁶ library plasmid transformants, four colonies grew on medium lacking Leu and His (A). All four were capable of transactivating the target-LacZ reporter gene (B). Three of these cDNAs encoded the Ets-2 transcription factor (aa 45-469, shown by black bar). The fourth cDNA encoded GABPα (aa 119-454, shown by black bar), which is also a member of the Ets family that shares a highly conserved ETS DNA binding domain (C). The Ets binding motif within the *IFNT* promoter is underlined with the core motif highlighted in bold type.

Procedure

1. For best results, use fresh colonies, i.e., grown at 30°C for 2–4 days, 1–3 mm diameter.

2. Prepare Z buffer/X-gal solution only in the volume needed.

3. For each plate of transformants to be assayed, presoak a sterile filter by placing it in 1.5 mL of Z buffer/X-gal solution on a clean 100-mm plate.

4. Using forceps, place a clean, dry filter over the surface of the plate of colonies to be assayed. Gently stroke the filter with the side of the forceps to help colonies cling to the filter.

5. Poke holes through the filter into the agar in three or more asymmetric locations to orient the filter to the agar.

6. When the filter has been evenly wetted, carefully lift it from the agar plate with forceps and transfer it, with the colonies facing upwards, into a pool of liquid nitrogen. Again, by using the forceps, completely submerge the filters for 10 s.

7. After the filter has become completely frozen, remove it from the liquid nitrogen and allow it to thaw at room temperature.

   Note: This freeze/thaw treatment is to permeabilize the cells.

8. Carefully place the filter, colony side up, on the presoaked filter (step 3). Avoid trapping air bubbles under or between the filters.

9. Incubate the filters at 30°C or room temperature and check periodically for the appearance of blue colonies.

2.6 Protocol for Plasmid Isolation from Yeast by Lyticase Digestion

Isolating plasmid DNA from yeast is not trivial, primarily because of the tough cell wall. Furthermore, the relatively large size (>6 kb) and low copy number (approx. 50/cell) of some yeast plasmids provides the DNA in very low yield, regardless of the plasmid isolation method used. In addition, plasmid DNA isolated from yeast is often contaminated by genomic DNA. Yeast contains approximately three-fold more genomic DNA than *E. coli*, and the isolation method breaks the yeast chromosomes and releases them from cellular material. However, the purification method still yields sufficient DNA for use either as a polymerase chain reacton (PCR) template or for transforming *E. coli*. If a large quantity of plasmid is needed or if very pure plasmid DNA is required, for example for sequencing or for restriction enzyme digestion, it will first be necessary to transform *E. coli* by electroporation (1) and prepare the plasmid DNA from the bacteria. If chemically competent *E. coli* cells were employed, it is essential that the competent cells should be able to yield a transformation efficiency of at least 10^7 cfu/µg.

Materials and Reagents

- Sorbitol buffer: 1.2 M sorbitol (Catalog No. S-6021; Sigma Chemical) in 0.1 M sodium phosphate buffer pH 7.4 (19). Prepare 20 mL of this buffer, filter sterilize, divide into 1-mL aliquots and store at -20°C.
- Lyticase/Sorbitol buffer: Add 5.0 U/µL Lyticase (Catalog No. L-5263; Sigma

Chemical) to the Sorbitol buffer. Prepare only the volume needed, immediately prior to use.

- 20% SDS
- Phenol:choroform:isoamyl alcohol (25:24:1): Prepare with neutralized pH 7.0 phenol (19).
- 10 M ammonium acetate
- 95%–100% and 75% ethanol

Procedure

1. Prepare yeast cultures for lysis.

    a. Spread a thin film of yeast cells (approx. 2-cm^2 patch) onto the SD/-Leu/-His/+ optimal (3-AT) agar medium to keep selective pressure on the library plasmids.

    b. Incubate the plate at 30°C for 3–4 days until the patch shows abundant yeast growth.

    c. Scrape up a portion of the patch (approx. 10 mm^2) and resuspend the cells in 500 µL of sterile H_2O in a 1.5-mL microcentrifuge tube. Briefly centrifuge to pellet the cells.

2. Carefully pour off the supernatant. Resuspend the cell pellet in the residual medium (approx. 50 µL) by vortex-mixing.

3. Add 10 µL of Lyticase/Sorbitol buffer to each tube. Thoroughly resuspend the cells either by vortex-mixing or by repeatedly pipetting.

4. Incubate the tubes at 37°C overnight with no shaking.

5. Add 10 µL of 20% SDS to each tube and vortex-mix vigorously for 1 min.

6. Put the samples through one freeze/thaw cycle at -20°C and vortex-mix again to ensure complete lysis of the cells.

7. Bring the volume of the sample up to 200 µL in TE buffer (pH 7.5).

8. Add 200 µL of phenol:choroform:isoamyl alcohol (25:24:1).

9. Vortex-mix at highest speed for 5 min.

10. Centrifuge at 10 000× g for 10 min.

11. Transfer the aqueous (upper) phase to a fresh tube.

12. Add 8 µL of 10 M ammonium acetate and 500 µL of 95%–100% ethanol.

13. Place the tube at -70°C or in a dry-ice/ethanol bath for 1 h.

14. Remove from freezing source and centrifuge at 10 000× g for 10 min.

15. Discard supernatant and air-dry the pellet.

16. Resuspend pellet in 20 µL of H_2O.

    Note: The amount of plasmid DNA recovered is small relative to the contaminating genomic DNA. Therefore, yield cannot be estimated by either A_{260} measurement or as a visible band on an agarose gel.

17. Use 1–2 µL of the solution for electroporation or as a template for PCR.

3. APPLICATIONS

3.1 Identification of Ets-2 in Transcriptional Control of Interferon-Tau Gene Expression

The interferon-τ (IFN-τ) are Type I IFN structurally related to the IFN-α (approximately 50%) and IFN-ω (approximately 75%) in primary structure. They evolved about 36 million years ago in ancestors of the present day cattle and sheep by duplication of an IFN-ω gene (*IFNW*) (18). Since then, they have extensively duplicated themselves. Estimates indicate from six to ten such genes, all quite similar in sequence, are found in *Bos Taurus* (unpublished data).

The expression of the IFN-τ genes (*IFNT*) is unusual in several respects. Unlike the genes of other Type I IFN, they are not inducible by virus and are not expressed as the result of a response to pathogens. Instead they are produced constitutively in the embryonic trophectoderm of cattle and their relatives and play a key role in the responses of the mother to the presence of an embryo within her uterus. Expression is massive and sustained for several days just prior to implantation and then is silenced for the rest of embryonic development.

Transcriptional control over the *IFNT*, as in other Type I IFN, appears to reside largely, if not exclusively, in sequences located within 100 bp of the TATA box of these intronless genes. Unlike other Type I genes, however, the *IFNT* lack well-defined viral response elements (VRE) and diverge markedly from their nearest relatives, the *IFNW*, approximately 80 bp distal to the transcriptional start site. On the other hand, the 5′-flanking region of the different *IFNT* genes are highly conserved both within species and across species as distant as the sheep and giraffe.

A sequence −91 to −69 (Figure 4) from the transcriptional start site provided a stable complex in electrophoretic mobility shift experiments performed on extracts from pre-implantation ovine embryos (13). This sequence, after tetramerization, was selected as the target for the yeast one-hybrid screen of a day 13 sheep conceptus cDNA library. Four yeast colonies, from 2.6×10^6 transformants plated, grew well on medium lacking leucine and histidine. All four were capable of transactivating the *lacZ* reporter. Three contained plasmids with identical inserts 2378 bp in length, which represented the ovine equivalent of the already known human and murine Ets-2 transcription factor. The fourth cDNA encoded GABPα, another ETS family member (Figure 4).

Upon examination, it became clear that the sequence used as bait contained a 10 bp nucleotide motif (−79 to −70) that was identical at nine out of ten positions to an Ets-2 consensus sequence, and was also identical at eight of nine bases to a sequence known to bind GABPα. In particular, the core Ets-binding motif C/AGGAA/T was present at positions −78 to −73. Three ovine *IFNT* genes, which are expressed poorly, if at all, have mutations within this core and cannot bind Ets-2 (7). Electrophoretic mobility shift assays showed that the target element from active genes formed a complex with Ets-2, whereas equivalent probes from inactive genes with a mutation within the core (GGAA) could not (7).

Transfection experiments performed on human choriocarcinoma (JAr) cells, with luciferase reporter constructs, confirmed that Ets-2 could strongly transactivate the *IFNT* promoter from transcriptionally active genes (Figure 5A, C). In contrast, GABPα, even in presence of its partner GABPβ, had little effect on transcription (Figure 5B, C). Ets-2 was also shown to be localized in nuclei of the embryonic trophectoderm cells that express the IFN-τ so strongly (7).

The yeast one-hybrid screen in this instance provided a surprisingly unequivocal result. Only two candidate-binding proteins were identified and only one of them was capable of strong transactivation of the promoter under study. Moreover, a promoter from a gene that was not believed to be transcriptionally active was not transactivated (Figure 5A, B).

4. DISCUSSION

The yeast single hybrid screen has begun to replace two older procedures to identify transcription factors that target particular DNA sequences. The first of these earlier methods utilized the multimerized, double-stranded DNA element immobilized on Sepharose, which acted as an affinity matrix, to pluck out binding proteins selectively from nuclear extracts of cells or tissues (12). An affinity step such as this rarely supplies a pure product and usually has to be used in conjunction with other purifi-

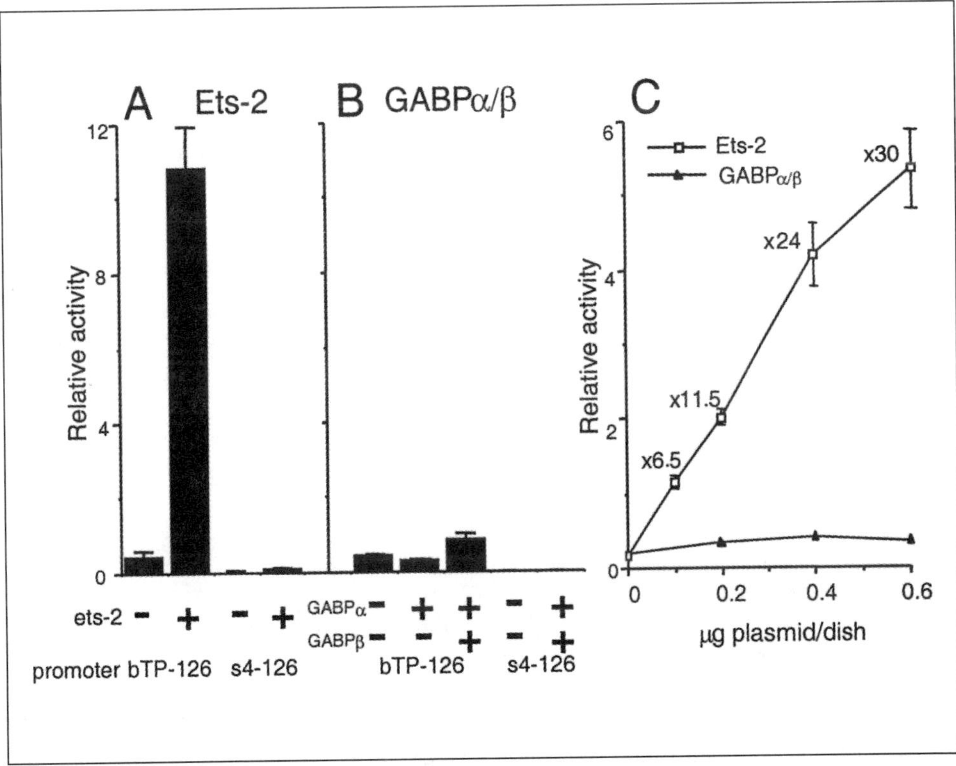

Figure 5. Transactivation of *IFNT* promoters by human Ets-2 and murine GABPα/β. A bovine *IFNT*1 promoter (bp -126 to +50, bTP-126) from a gene known to be active and the corresponding region (s4-126) from a gene expressed poorly (6) were placed ahead of luciferase-reporters. They were transfected into human JAr choriocarcinoma cells in either the absence (-) or presence (+) of the expression vectors for human Ets-2 (**A**), murine GABPα, and GABPβ (**B**). The bTP-126 promoter contains a conserved Ets binding motif whereas the s4-126 promoter contains a mutated Ets binding motif. (**C**) The dose-dependent effects of Ets-2 and GABPα/β transfection on *IFNT*-promoter activity in human JAr cells. The bTP-126 luciferase construct (3.2 µg) was co-transfected with increasing amounts (0.1 to 0.6 µg/dish) of human Ets-2 (□) or murine GABPα/β (▲) expression plasmids. Reporter activity is expressed relative to activity of a cotransfected RSVLTR-βgal plasmid. Results are means (±SE) from three independent experiments.

cation steps. As with other biochemical procedures, its success depends upon numerous confounding factors, including the rate at which the protein becomes denatured or is destroyed by proteinases. Final product yield can also be limiting. The second procedure uses radioactive double-stranded DNA probes to screen expression libraries (21). The major difficulty with this technique, as with all screens of this type, is to reduce non-specific binding. In addition, it depends upon the ability of the expressed transcription factor to fold properly as a fusion protein after being transferred to a nitrocellulose or nylon membrane from a phage plaque or bacterial colony. The one-hybrid screen, by contrast, offers maximal sensitivity because the detection of the DNA-protein interactions occurs while the proteins are in their native configurations in the yeast cytoplasm. In addition, a cDNA that encodes the DNA-binding protein is immediately available after a library screening.

Nevertheless, interactions that appear to be relatively non-specific can still create difficulties with the yeast one-hybrid screen. Endogenous yeast transcription factors that bind to the exogenous DNA sequence can, for example, cause an elevation of *IIis/lacZ* expression in the yeast and complicate the screen (17). In most cases, increasing the 3-AT concentration can offset the HIS3 reporter background to some extent, but there is no useful inhibitor for *lacZ* expression available.

Often a relatively large number of distinct cDNAs are isolated in a screen. These can often be ranked according to the robustness of colony growth or the strength of *lacZ* expression, but these two markers do not always correlate well. In any case, it is wise to follow up on all likely candidates, first by sequencing the plasmid cDNA. It is reassuring if the sequence is in frame and its translation product suggests a DNA-binding protein. A second test is to perform EMSA analysis on the translation product of the encoded cDNA. Finally, transfection experiments on permissive cells should be performed to ensure that there are effects on the intact promoter containing the DNA target element (Figure 5).

The yeast one-hybrid screen has been employed in two ways. The classic one, described here, is for detecting DNA-protein interactions. Wang and Reed (23) first described the procedure and defined a transcriptional activator Olf-1 that binds several olfactory-specific genes. Soon afterwards, Li and Herskowitz (14) independently used a similar approach to identify a protein that interacts with the yeast origin of DNA replication. Since then the screen has been employed widely. A literature search in 1998 identified 30 papers in which "yeast one-hybrid" appeared as a phrase. A similar search, performed in May 2000, identified 101.

A second more recent application of the technique has been to detect protein–protein interactions between transcription factors (10,20,22). Typically, the parental yeast strain carries an integrated copy of a selectable marker gene (usually *HIS3*) with multimerized DNA motifs upstream of the TATA element as described here. The cells are engineered to express the transcription factor that binds to this DNA element. Since it lacks an AD, this protein cannot activate *HIS3*. The library, which is transfected into the cells, again consists of cDNAs fused to the open reading frame of the transcriptional AD of a co-activator such as VP16 or GAL4. The association of a cDNA product with the protein that is already bound to the DNA element activates *HIS3*, allowing yeast colonies to be selected. This modified yeast one-hybrid screen (see Chapter 19 and 20) is particularly useful, as it allows transcription factors to be identified that may themselves have no DNA-binding ability. It is also a useful alternative to the two-hybrid screen, because it allows selection of proteins that associate with transcription factors when the latter are in a DNA-binding conformation.

194

ACKNOWLEDGMENT

This research was supported by NIH Grant No. R37HD21896.

REFERENCES

1. **Ausubel, F.M., R. Brent, R.E. Kingston, D.D. Moore, J.G. Seidman, J.A. Smith, and K. Struhl (Eds.).** 1994. Current Protocols in Molecular Biology. John Wiley & Sons, New York, NY.
2. **Chong, J.A. and G. Mandel.** 1997. Isolation of DNA-binding proteins using one-hybrid genetic screens, p. 289-297. *In* P.L. Bartel and S. Fields (Eds.), The Yeast Two-Hybrid System. Oxford University Press, New York, NY.
3. **CLONTECH Laboratories.** 1995. MATCHMAKER™ One-hybrid system protocol. Palo Alto, CA.
4. **CLONTECH Laboratories.** 1996. Yeast Protocols Handbook. Palo Alto, CA.
5. **Durfee, T., K. Becherer, P.L. Chen, S.H. Yeh, Y. Yang, A.E. Kilburn, W.H. Lee, and S.J. Elledge.** 1993. The retinoblastoma protein associates with the protein phosphatase type I catalytic subunit. Genes Dev. *7*:555-569.
6. **Ealy, A.D., J.A. Green, A.P. Alexenko, D.H. Keisler, and R.M. Roberts.** 1998. Different ovine interferon-tau genes are not expressed identically and their protein products display different activities. Biol Reprod. *58*:566-573.
7. **Ezashi, T., A.D. Ealy, M.C. Ostrowski, and R.M. Roberts.** 1998. Control of interferon-τ gene expression by Ets-2. Proc. Natl. Acad. Sci. USA *95*:7882-7887.
8. **Fields, S. and O. Song.** 1989. A novel genetic system to detect protein-protein interactions. Nature *340*:245-246.
9. **Fields, S.** 1993. The two-hybrid system to detect protein-protein interactions. Methods: A Companion to Methods Enzymol. *5*:116-124.
10. **Gstaiger, M., L. Knoepfel, O. Georgiev, W. Schaffner, and C.M. Hovens.** 1995. A B-cell coactivator of octarmer-binding transcription factors. Nature *373*:360-362.
11. **Inouye, C., P. Remondelli, M. Karim, and S. Elledge.** 1994. Isolation of a cDNA encoding a metal response element binding protein using a novel expression cloning procedure: the one hybrid system. DNA Cell Biol. *13*:731-742.
12. **Kadonaga, J.T.** 1991. Purification of sequence-specific binding proteins by DNA affinity chromatography. Methods Enzymol. *208*:10-23.
13. **Leaman, D.W., J.C. Cross, and R.M. Roberts.** 1994. Multiple regulatory elements are required to direct trophoblast interferon gene expression in choriocarcinoma cells and trophectoderm. Mol. Endocrinol. *8*:456-468.
14. **Li, J.J. and I. Herskowitz.** 1993. Isolation of ORC6, a component of the yeast origin of recognition complex by a one-hybrid system. Science *262*:1870-1873.
15. **Liaw, G.-J.** 1994. Improved protocol for directional multimerization of a DNA fragment. BioTechniques *17*:668-670.
16. **Liu, J., T.E. Wilson, J. Milbrandt, and M. Johnston.** 1993. Identifying DNA-binding sites and analyzing DNA-binding domains using a yeast selection system. In Methods: A Companion to Methods Enzymol. *5*:125-137.
17. **Luo, Y., S. Vijaychander, J. Stile, and L. Zhu.** 1996. Cloning and analysis of DNA-binding proteins by yeast one-hybrid and one-two-hybrid systems. BioTechniques *20*:564-568.
18. **Roberts, R.M., L. Liu, and A. Alexenko.** 1997. New and atypical families of Type I interferons in mammals: comparative functions, structures and evolutionary relationships. Prog. Nucl. Acids Res. Mol. Biol. *56*:287-325.
19. **Sambrook, J., E.F., Fritsch, and T. Maniatis.** 1989. Molecular Cloning: A Laboratory Manual. Cold Spring Harbor Laboratories, Cold Spring Harbor, NY.
20. **Sieweke, M.H., H. Tekotte, J. Frampton, and T. Graf.** 1996. MafB is an interaction partner and repressor of Ets-1 that inhibits erythroid differentiation. Cell *85*:49-60.
21. **Singh, H.** 1993. Specific recognition site probes for isolating genes encoding DNA-binding proteins. Methods Enzymol. *218*:551-567.
22. **Strubin, M., J.W. Newell, and P. Matthias.** 1995. OBF-1, a novel B cell-specific coactivator that stimulates immunoglobulin promoter activity through association with octamer-binding proteins. Cell *80*:497-506.
23. **Wang, M.M. and R.R. Reed.** 1993. Molecular cloning of the olfactory neuronal transcriptional factor Olf-1 by genetic selection in yeast. Nature *364*:121-126.
24. **Zhu, L.** 1997. Yeast GAL4 two-hybrid system, p.173-196. *In* R.S. Tuan (Ed.), Recombinant Protein Protocols. Methods in Molecular Biology. 63, Humana Press, Totowa, NJ.

12 Detecting Two-Protein Interaction with a (Third Non-Hybrid) Bridging Protein

Jie Zhang
Guilford Pharmaceuticals, Baltimore, MD, USA

1. INTRODUCTION

The development of the yeast two-hybrid system and its increasingly wide applications may serve as a paragon of how applied technology can benefit from basic science research, and in return, the novel technique enables new inquiry to be performed at a high level. From the early intensive studies of gene expression emerged a picture depicting the mechanism of transcription activation. This knowledge paved the way for engineering a genetic reporting system to detect protein interactions. The two-hybrid system is based on the findings that most eukaryotic transcription activation has a modular feature. A typical transcription factor contains two modules: a DNA binding domain (BD) that specifically binds to a DNA sequence and an activation domain (AD) that interacts with the basic transcription complex to enhance expression of genes under the control of the transcription factor. Fields and his colleagues discovered that the function of a transcription factor does not require a covalent linkage between BD and AD (4). In fact, when the BD and AD are expressed as fusion proteins to the interacting X and Y protein pair—the BD-X and AD-Y hybrids—association of X and Y in the yeast brings BD and AD together to reconstitute a functional transcription factor. Conversely, reporter gene expression in the system provides a sensitive way to detect X/Y interaction in vivo. Furthermore, the two-hybrid system offers a highly efficient tool to screen cDNA libraries to identify previously unknown binding partners for a given protein of interest (2). Numerous interacting proteins are cloned by the two-hybrid screening system. This greatly accelerates the process of defining arrays of protein associations to establish networks of interactions. Indeed, it is now conceivable that the two-hybrid system may be explored to decipher a "protein linkage map", which could link the genes now fragmented in the genome into functional groups (3; see also Chapter 18). On the molecular level, proteins and their interactions are the major phenotype manifestation of the genotype DNA information. Understanding different protein assembly mechanisms is fundamental to explain most life phenomenon in molecular terms.

Yeast Hybrid Technologies
Edited by L. Zhu and G.J. Hannon
© 2000 Eaton Publishing, Natick, MA

The original two-hybrid system was designed to test protein interactions in a pairwise fashion. Many protein complexes, however, involve multiple units. The traditional two-hybrid system is not an efficient means to analyze complex protein assemblies. Expanding the two-hybrid system to a multi-hybrid system would facilitate dissecting the protein linkage network. For example, in a ternary protein complex, there are cases in which protein X does not directly contact protein Y but rather requires a third protein, Z, to mediate the interaction (Figure 1A). In another situation, a protein Y may only bind to the composite site created as a result of protein X association with protein Z (Figure 2B). Both cases demand the introduction of a bridging protein on a third shuttle plasmid to form a ternary complex in yeast. The expanded hybrid system described here permits the introduction of a third component into the traditional two-hybrid system. It can be used to analyze ternary protein complexes and to isolate genes that encode components of the complex.

2. PROTOCOLS

2.1 General Considerations

Before applying the expanded hybrid system to a specific problem under study, it

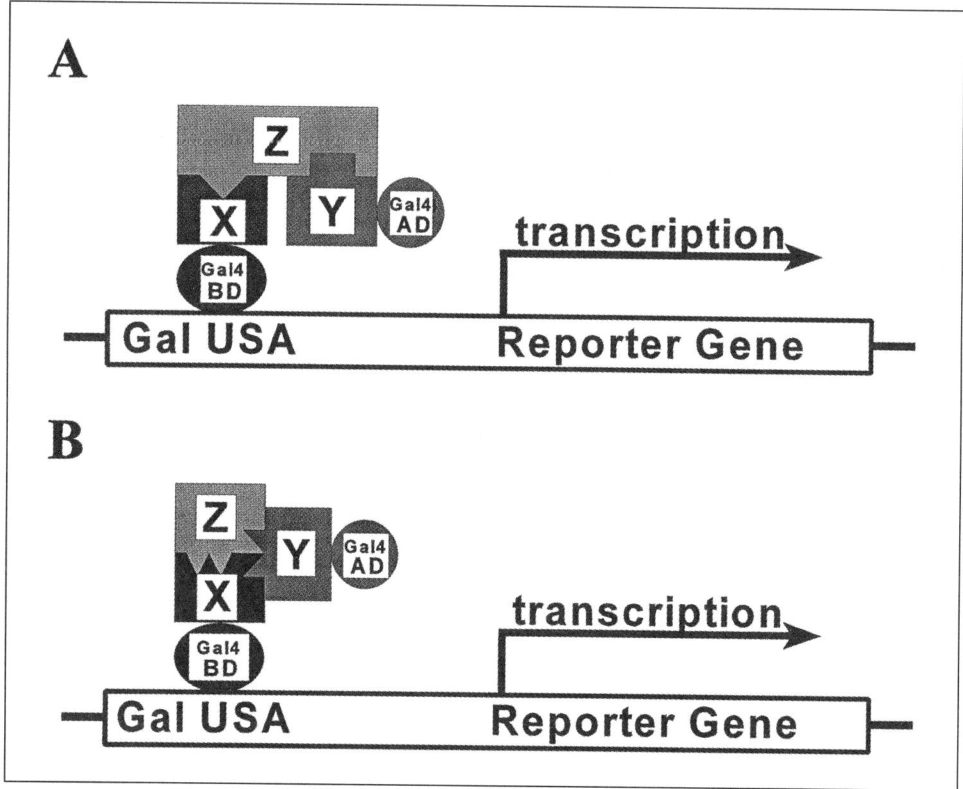

Figure 1. Uses of the expanded yeast hybrid system. The expanded hybrid system may be useful to detect ternary complex formation or for cloning the missing third component required for complex in scenarios such as in (**A**), where interaction of protein X and Y must be mediated by Z; or in (**B**), where Y only binds to a composite contour created by the combination of X and Z.

must be established that indeed more than two components are involved in forming the complex. Protein assembly initiates many key events such as DNA replication, gene transcription, gene transposition, RNA splicing, protein synthesis, gene rearrangement, cell cycle regulation, and differentiation. Some suggestive indications of multiple unit interactions are listed in Table 1. All the experimental data in the list would imply that additional factors may affect the indirectly interacting proteins

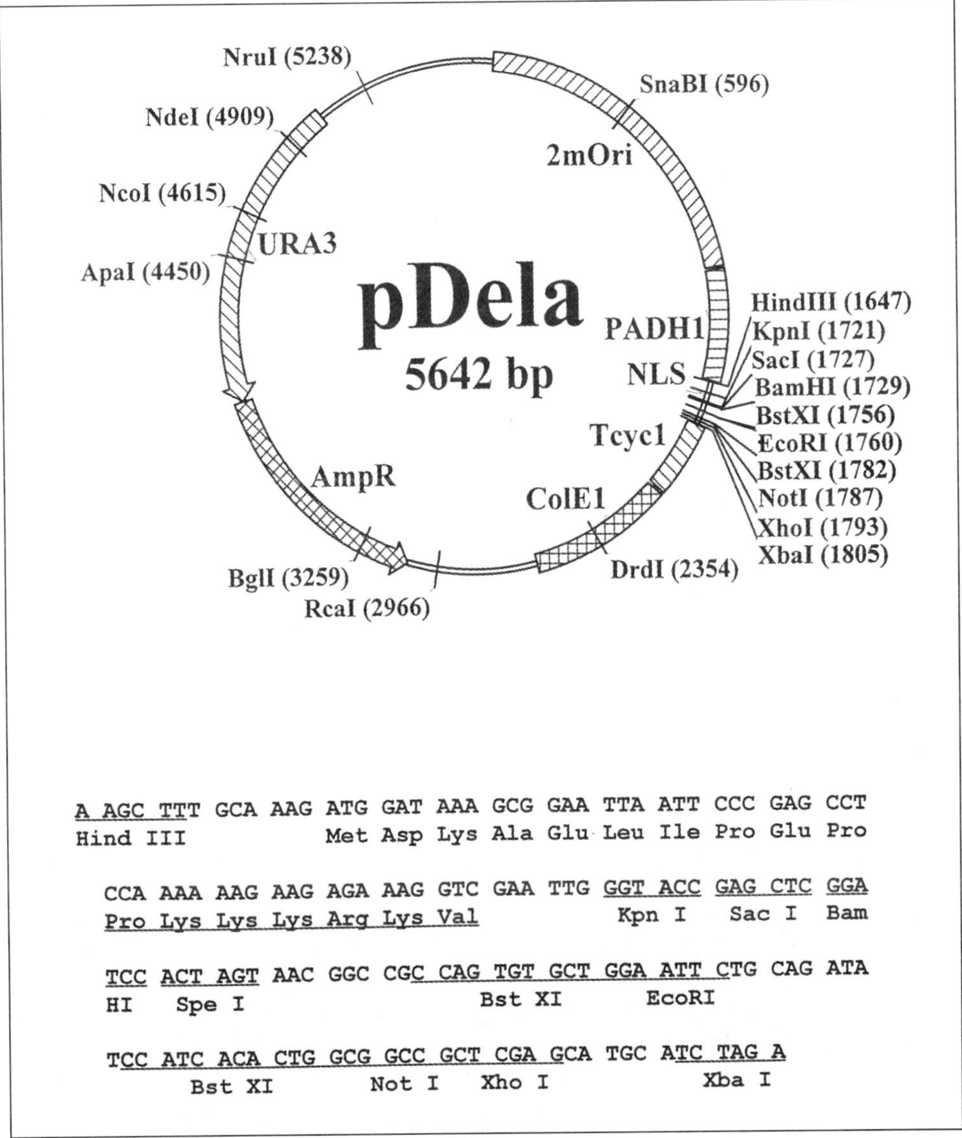

Figure 2. Map of pDela. The shuttle vector plasmid contains an *Amp*[r] gene, a Col E1 replication sequence for ampicillin-resistance selection and propagation in *E. coli*, a 2μ Ori replication sequence, and a *Ura3*[+] gene for propagation and selection in ura⁻ yeast strain to grow on uracil drop-out media. There are multiple cloning sites (MCS) for inserting proteins or protein domains of interest to be expressed in-frame as a fusion with the SV40 T-antigen nuclear localization sequence (NLS) at the amino-terminus. The DNA sequence in the MSC region is shown with the NLS and cloning sites listed. The expression of the hybrid-protein is under the control of alcohol dehydrogenase promoter (PADH1) with cyc transcription termination sequence (T cyc1).

Table 1. Typical Experimental Data That Imply Possible Multiple Protein Interactions

Experiments	Results
Immunoprecipitations	Multiple protein bands on SDS gel
Protein purifications	Co-purification of protein subunits
Enzymatic assays	Requirement of additional fractions/factors to restore the activity
Signal transductions	Missing links between up-stream and down-stream events
Genetic analysis	Multiple genes compensate/affect one phenotype

being studied. The expanded hybrid system is useful either to test if a known protein is a component of a ternary protein complex, or to screen a library to clone the component. At least two of the components in the complex should have already been cloned. These two genes, X and Y, will form the basis of the yeast two-hybrid system, the BD-X and AD-Y.

Here the EGFR/Grb2/Sos complex is used as a model system to demonstrate the principles of the expanded hybrid system. It has been established that the dimerization of epidermal growth factor receptor (EGFR) activates its tyrosine kinase in the cytoplasmic domain. As a result, autophosphorylation of tyrosines in EGFR creates high-affinity sites to bind coherent SH_2 domains. Grb2, an adaptor protein, contains a central SH_2 domain flanked by two SH_3 domains. The SH_2 domain of Grb2 binds to autophosphorylated EGFR. The SH_3 domains of Grb2 bind to the proline-rich motifs such as those in the carboxyl terminal region of Sos—a guanine-nucleotide exchange factor for Ras proteins. The interacting domains of EGFR, Grb2, and Sos have been subcloned to create pGBT9-EGFR, pDela-Grb2, and pGAD424-Sos plasmids. These plasmids can be used as positive controls for the experiments using the expanded hybrid system described.

Only the salient features for the expanded hybrid system are recapitulated here. The rest are standard general procedures for the yeast two-hybrid system and can be found in the preceeding chapters of this book.

2.2 Plasmids and Yeast Strains

Yeast strain BY3161 (MAT**a**, *leu2-3*, *trp1-901*, *his3-200*, *ura3-52*, *ade2-101*, *gal4-542*, *gal80-538*, GAL1-lacZ, GAL1-His3) was a kind gift from Dr. Jef Boeke (John Hopkins University School of Medicine). pGBT9, pGAD424, pTD1, and pVA3 were from CLONTECH Laboratories (Palo Alto, CA, USA). pDela (Figure 2) was constructed by ligating a 1.7 kb *MspA1*I-*Kpn*I fragment of pGAD424 (from nucleotide 5420–479) and a 3.9 kb *Hpa*I-*Kpn*I fragment of pYes2 [from nucleotide 2284 to 354 (pYes2 was from Invitrogen, San Diego, CA, USA)] in an orientation such that the two *Kpn*I complementary ends joined together, and the blunt ends of *MspA1*I and *Hpa*I joined together. Construction of positive control plasmids, pGBT9-EGFR, pDela-Grb2, and pGAD424-Sos, has been reported (6).

2.3 Test a Known Protein to Mediate Hybrid Interactions

The following procedures are to test if a cloned protein Z can bridge interactions between cloned proteins X and Y, which do not associate with each other directly.

200

2.3.1 Construct three fusion genes.

Standard molecular cloning techniques should be followed to insert cDNA encoding appropriate proteins or protein domains X, Y, and Z into pGBT9, pGAD424, or other compatible plasmids, and pDela, respectively, to create pGBT9-X, pGAD424-Y, and pDela-Z. To express the fusion proteins, X and Y should be in-frame with Gal4-BD and Gal4-AD. For the third protein Z, if only a partial sequence is to be expressed, its cDNA should be cloned into the multiple cloning sites in-frame with the nuclear localization sequence (NLS) in pDela. If a complete nuclear protein Z is to be expressed, it can be subcloned into the *Hin*dIII site 5′-upstream of the NLS.

2.3.2 Transform the BY3161 yeast strain with the three-hybrid plasmids, pGBT9-X/pGAD424-Y/pDela-Z.

As three negative controls (each consisting of three plasmids), also transform pGBT9/pGAD424-Y/pDela-Z, pGBT9-X/pGAD424/pDela-Z, and pGBT9-X/pGAD424-Y/pDela. Triple transformants should be selected on yeast media without Trp, Leu, and Ura.

2.3.3 Assay the transformants for β-galactosidase activity.

A positive signal for pGBT9-X/pGAD424-Y/pDela-Z but negative for pGBT9/pGAD424-Y/pDela-Z, and pGBT9-X/pGAD424/pDela-Z, pGBT9-X/pGAD424-Y/pDela would be a strong indication of a Z-protein–dependent X and Y interaction. In separate in vitro binding experiments, the X/Y/Z complex should be confirmed with the purified or recombinant components. Together, the in vivo and in vitro binding data would establish the formation of an X/Y/Z complex. A typical result for EGFR/Grb2/Sos interaction is shown in Figure 3.

2.4 Construct a cDNA Library

An educated guess must first be made of the most abundant source of mRNA encoding the putative mediator protein Z. cDNA is then made from the mRNA of the tissues or cells. Appropriate linker-adaptors compatible with the multiple cloning site (MCS) in pDela are ligated to the cDNA. Size selection for the long fragments of the cDNA is recommended. As will be discussed later, in-frame fusion between the NLS and the cDNA in pDela may not be absolutely necessary in many cases. Therefore, it is desirable to select cDNA generated from mRNA with oligo-dT primer, because this would increase the complexity of the library. This is different from an AD-cDNA fusion library, in which size fractionation for full-length cDNA could be counter-productive because potential stop codons in the 5′-untranslated sequence would prematurely terminate the translation of AD-fusion protein.

Used as an example here is a pDela-cDNA library synthesized from adult mouse brain poly(A)-selected mRNA using random and oligo(dT) primers. *Bst*XI linkers were added to the cDNA. The cDNA was sized, and fragments longer than 500 bp were inserted into two *Bst*XI sites in pDela. The primary library consisted of 6×10^5 independent colonies. More than 90% of plasmids contained inserts with an average size of 2 kb.

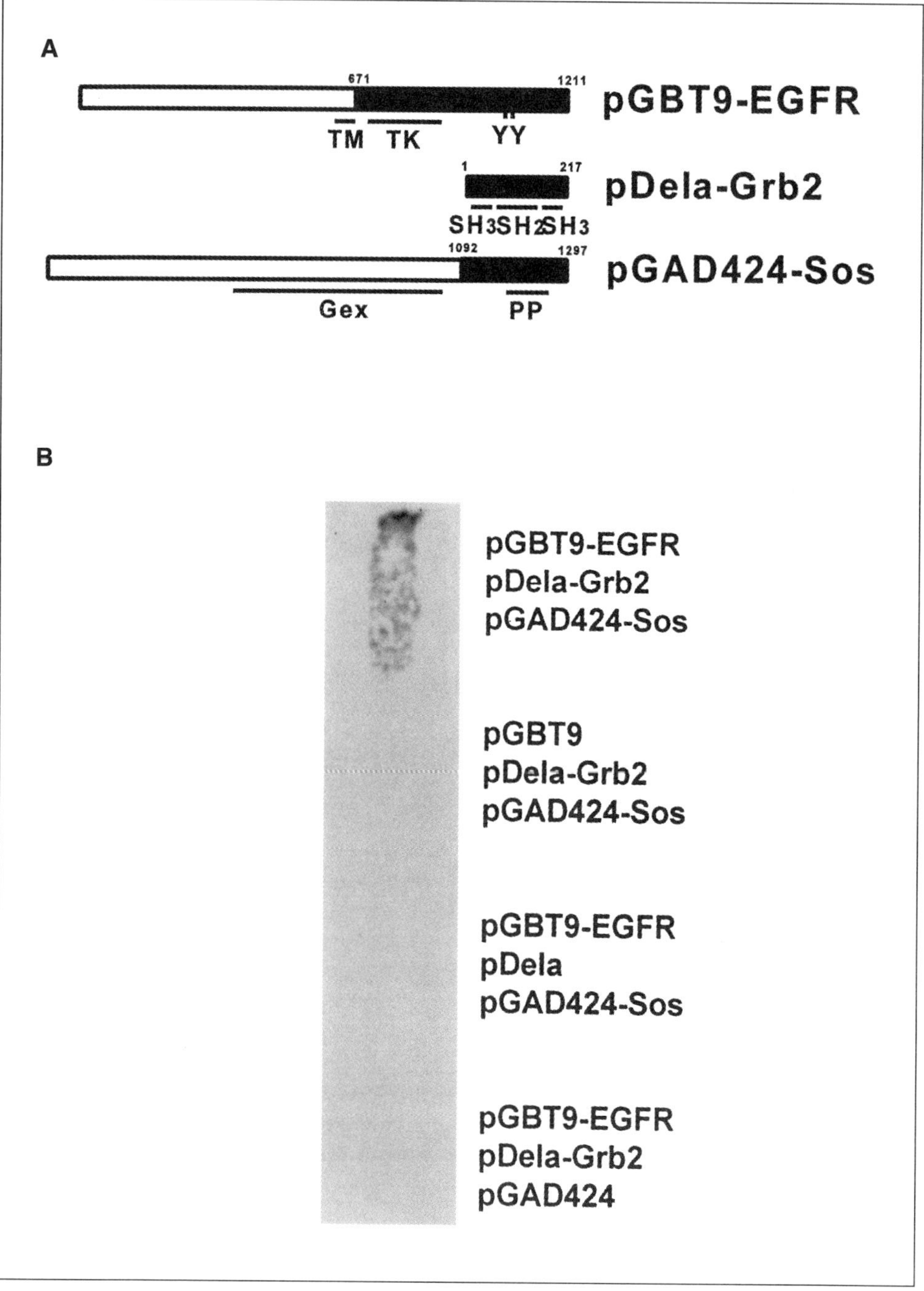

Figure 3. Detecting the formation of the ternary complex of EGFR, Grb2, and Sos in the expanded hybrid system. (A) Construction of three-hybrid expressing plasmids. DNA fragments encoding part of or whole protein (filled bars) were obtained to make pGBT9-EGFR to express GAL4-BD and the cytoplasmic domain of EGFR fusion protein, pDela-Grb2 to express the SV40 T-antigen nuclear localization domain and Grb2 fusion protein, and pGAD424-Sos to express GAL4-AD and the carboxyl terminal domain of Sos2 fusion protein. TM, transmembrane domain; TK, tyrosine kinase domain; Gex, guanine-nucleotide exchange catalytic domain; PP, proline rich domain. **(B)** Expression of β-galactosidase as a result of interactions among EGFR, Grb2, and Sos. Yeast transformed with different combinations of plasmids as indicated were grown on SD plates without tryptophan, uracil, and leucine, and their β-galactosidase activities were determined by filter X-gal assay.

2.5 Sequential Library Screening

For library screening, 500 µg of the plasmid library DNA are used to transform BY3161 yeast harboring pGBT9-X and pGAD424-Y by the PEG/lithium method. Yeast transformants are selected on SD plates with 50 mM 3-amino-1,2,4-triazole (AT) and without histidine, tryptophan, leucine, and uracil at 30°C for two weeks. His$^+$ transformants are tested for β-galactosidase activity with the filter X-gal assay. X-gal positive colonies are restreaked to obtain single colonies for further characterization. Restreaking is usually performed on +AT, -his, -trp, -leu, -ura plates to maintain his-selection pressure on weak or transient ternary complex formation.

2.6 Identify Positive Clones

2.6.1 Prepare plasmid DNA from positive yeast colony.

Refer to protocol in Chapter 11, Section 2.6.

2.6.2 Transform E. coli with plasmid DNA from step 2.6.1.

2.6.3 Pick a dozen of ampicillin-resistant E. coli transformants from step 2.6.2 and use polymerase chain reaction (PCR) to identify pDela-clone containing transformants.

A pair of primers, 5Dela3007 and 3Dela3551, can be used to specifically amplify a 545-bp fragment from the pDela-clone. 5Dela 3007 has the sequence of 5′-CCTTCATCTCTTCCACCCACACACC-3′, and 3Dela3551 has the sequence of 5′-CATCCTAGTCCTGTTGCTGCCAAGC-3′.

2.6.4 Make a large preparation of pDela-clone plasmid from the PCR positive colony in step 2.6.3.

2.7 Confirm Positive Clones

2.7.1 Transform BY3161 with pGBT9-X/pGAD424-Y/pDela-clone.

As two negative controls, transform the pGBT9/pGAD424-Y/pDela-clone and the pGBT9-X/pGAD424/pDela-clone.

2.7.2 Perform β-galactosidase assay on the triple transformants.

2.7.3 Sequence pDela-clone. Express and purify the recombinant cloned protein and test its in vitro interaction with X and Y.

As an example, a clone named E9 was isolated when the pDela–mouse-brain cDNA library was screened with BY3161 containing pGBT9-EGFR and pGAD424-Sos to search for protein links in the EGFR/Sos signal transduction pathway (Figure 4). DNA sequence revealed that the 1.3-kb insert contained the entire coding sequence for mouse *Grb2* as well as a 279-bp noncoding sequence in the 5′ and a 346-bp non-coding sequence at the 3′-end.

3. DISCUSSION

The expanded hybrid system described here should facilitate screening cDNA libraries to clone bridge proteins in ternary complex. The pDela constructs were initially designed to be compatible with the Gal4-based two-hybrid system, such as pGAD424/pGBT9. The ura-selection offered by pDela makes it easy to introduce the bridge protein into the yeast containing the other two-hybrids, the BD-fusion protein, and the AD-fusion proteins. Because many of the widely used two-hybrid systems employ trp-, leu-, his-, or ziocin-selections in the shuttle vectors, the ura+-based pDela-constructs can be used directly in those systems.

There is accumulating evidence that regulation of BD- and AD-hybrid interaction may take place not only in nuclei but also in yeast cytoplasm, even though the final activation of reporter genes by the reconstituted BD/AD transcription factor is a nuclear event. In the example of the EGFR/Grb2/Sos complex used as positive control here, the association between Grb2 and Sos apparently occurs in the cytoplasm. This is supported by sequencing data of the E9 plasmid retrieved from the library screening. It encodes Grb2 with untranslated sequences at both the 5′- and 3′-ends. A stop codon in the 5′-untranslated region prematurely terminates the translation of a

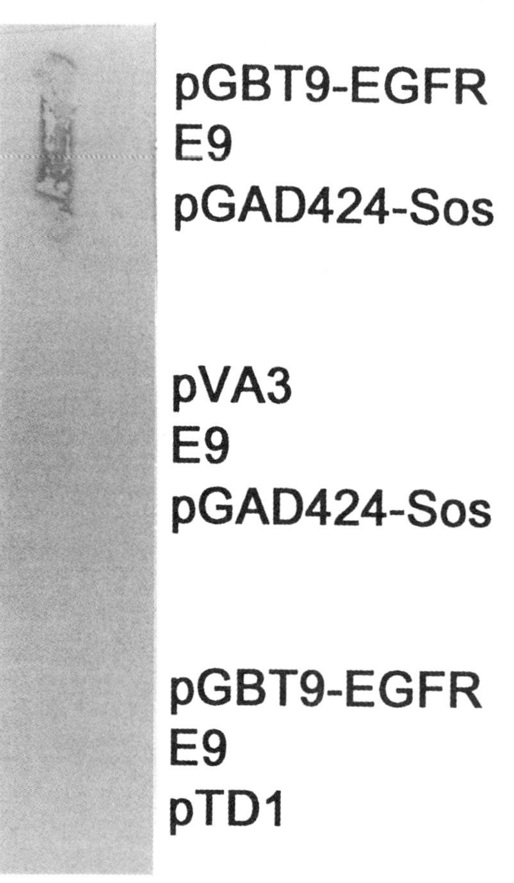

Figure 4. Isolating the E9 clone that mediates EGFR and Sos interaction. The yeast BY3161 strain containing pGBT9-EGFR and pGAD424-Sos was used to screen a mouse brain cDNA library in pDela. The pDela plasmid, retrieved from the positive clone E9, was used to retransform yeast BY3161 together with pGBT9-EGFR, pGAD424-Sos, or their cognate control plasmids, pVA3 and pTD1. Shown here are X-gal filter assay results of the triple transformants.

NLS-Grb2 fusion protein. Instead, translation seems to re-start with the initiating ATG from Grb2. It has been found that Grb2 is constitutively associated with Sos in the cytoplasm. This association may also happen in yeast cytoplasm between Grb2 and Gal4-AD-Sos. The NLS in Gal4-AD may be able to direct Grb2/Sos to nuclei where final interaction with EGFR occurs. In another example, Chaudhuri and Stephan used the yeast two-hybrid system to show that the interaction between a truncated calcineurin A subunit (CnA) and its autoinhibitory domain (AID) is inhibited by either cyclosporin A (Cys A) or FK506 (1). The inhibition is dependent on the presence of cognate immunophilins, i.e., cyclophilin A for Cys A and FKBP12 for FK506. Cys A or FK506 fails to inhibit CnA/AID interaction in the yeast two-hybrid system if the host gene for cyclophilin A or FKBP12, respectively, is deleted. It is well-established that Cys A/cyclophilin A or FK506/FKBP12 complex binds and inhibits calcineurin, a protein phosphatase, to mediate the immunosuppressive effects (5). Both cyclophilin A and FKBP12 are cytoplasmic proteins. In the yeast two-hybrid system, it appears that in the cytoplasm the Cys A/cyclophilin A complex or the FK506/FKBP12 complex binds to CnA-hybrid. As a result of this binding, CnA- and AID-hybrids can no longer form the complex to reconstitute the functional transcriptional factors to turn on reporter genes. An implication from these examples is that with minimum modification, numerous cDNA libraries previously made for screening proteins expressed in cytoplasm (e.g., the pYEUra3 cDNA libraries made by CLONTECH and the pYes2 cDNA libraries by Invitrogen) could be used in the expanded hybrid-screening system. This is because even though a cytoplasmic protein expressed from such libraries in yeast does not contain an NLS to destine it to the nuclei, its binding in cytoplasm to AD- or BD-hybrids might be sufficient to affect the reconstitution of transcription factors from AD-/BD-hybrids.

ACKNOWLEDGMENT

I am grateful to Dr. J. Boeke for the BY3161 yeast strain.

REFERENCES

1. **Chaudhuri, B. and C. Stephan.** 1995. Only in the presence of immunophilins can cyclosporin and FK506 disrupt in vivo binding of calcineurin A to its autoinhibitory domain yet strength interaction between calcineurin A and B subunits. Biochem. Biophys. Res. Commun. *215*:781-790.
2. **Chien, C.T., P.L. Bartel, R. Sternglanz, and S. Fields.** 1991. The two-hybrid system: a method to identify and clone genes for proteins that interact with a protein of interest. Proc. Natl. Acad. Sci. USA *88*:9578-9582.
3. **Evangelista, C., D. Lockshon, and S. Fields.** 1996. The yeast two-hybrid system: prospects for protein linkage maps. Trends Cell Biol. *6*:196-199.
4. **Fields, S. and O.-K. Song.** 1989. A novel genetic system to detect protein–protein interactions. Nature *340*:245-246.
5. **Liu, J., J.D. Farmer, Jr., W.S. Lane, J. Friedman, I. Weissman, and S.L. Schreiber.** 1991. Calcineurin is a common target of cyclophilin-cyclosporin A and FKBP-FK506 complexes. Cell *66*:807-815.
6. **Zhang, J. and S. Lautar.** 1996. A yeast three-hybrid method to clone ternary protein complex components. Anal. Biochem. *242*:68-72.

13 RNA-Protein Interactions Reconstituted by a Tri-Hybrid System

Ulrich Putz[1], Joachim Kremerskothen[1], Paul Skehel[2], and Dietmar Kuhl[1]

[1]*Zentrum für Molekulare Neurobiologie Hamburg (ZMNH), University of Hamburg, Hamburg, Germany;* [2]*Division of Neurophysiology, National Institute for Medical Research, London, England, UK*

1. INTRODUCTION

Interactions between RNAs and proteins are fundamental to many cellular processes. The question of how a protein recognizes a specific RNA and which proteins interact with a specific RNA are thus central to a large number of basic problems in molecular biology. Moreover, a growing body of evidence indicates the importance of RNA binding in a variety of human diseases (3,25). Current work on protein-RNA complexes can be roughly divided into two areas: *(i)* characterization of RNA target sites and *(ii)* identification of proteins and protein structures that recognize RNA.

RNA–protein interactions have been traditionally studied using extensive biochemical assays such as RNA bandshifts, footprinting, and RNA-protein cross-linking. A disadvantage of these techniques is that interacting proteins often exist in low abundance and are consequently difficult to detect. In addition, these techniques do not easily allow the identification of target RNAs recognized by a known RNA-binding protein. A further major disadvantage is that these techniques do not allow one to directly identify the genes encoding for the proteins or RNAs of interest, rather they provide additional reagents for further characterization and cloning of the cognate cDNAs. Accordingly, to alleviate some of the experimental difficulties inherent to the biochemical techniques, several experimental approaches have been employed that permit the analysis of RNA–protein interactions (13,20,27,32). Recently two genetic systems capable of rapidly identifying proteins that bind to an RNA of interest or finding RNAs that bind to a protein of interest have been developed (29,30).

In this chapter, we describe a modification of the two-hybrid system for the in vivo reconstruction of specific RNA–protein interactions (29). In this tri-hybrid system,

Yeast Hybrid Technologies
Edited by L. Zhu and G.J. Hannon
© 2000 Eaton Publishing, Natick, MA

the DNA binding and transcription activation domains of the yeast transcriptional activator GAL4 are brought together via the interaction of recombinant fusion proteins with a recombinant RNA. The method provides a system to study RNA–protein interactions that has the genetic advantages of the two-hybrid system. It may be used to detect specific RNA-binding proteins or target RNAs from a library of cDNAs, or to analyze the structural specificity of identified RNA–protein interactions.

2. THE BASIC STRATEGY OF THE TRI-HYBRID SYSTEM

The yeast two-hybrid system is a widely employed genetic screen to detect protein–protein interactions (1,9,28). We developed a modification of this system to detect RNA–protein interactions.

The two-hybrid system utilizes the modular structure of particular transcription activators that allow the DNA-binding domains and transcription activation domains to function independently. The DNA-binding and the transcription activation domains are expressed as two separate polypeptides fused to heterologous sequences. Any interaction between the heterologous polypeptides of the hybrid proteins brings together the two components of the transcription activator that induces the expression of genes under the control of the appropriate upstream activating sequences (UAS). Selectable markers such as β-galactosidase or amino acid auxotrophy can then be used to report recombinant protein–protein interactions.

To modify this approach for the detection of RNA–protein interactions, it is necessary to construct a system in which the association of the DNA-binding and transcription activation domains is dependent on an RNA–protein interaction. The basic strategy of the method is illustrated in Figure 1. We took advantage of the well characterized and specific RNA binding activity of the RevM10 mutation of the Rev protein of HIV-1. The wild-type Rev protein enhances the export of viral transcripts out of the nucleus and into the cytoplasm of the host cell. This activity is mediated in part by the specific binding of Rev protein to the RRE, a 240 nucleotide RNA sequence located in the viral *env* gene (2,6,34). This RNA–protein interaction is initiated by the binding of a Rev monomer to a Rev-binding site in the RRE (5). This high affinity binding site is located to a short stem loop (nucleotides 45–75, RRE IIb) (5). The nucleotide sequence as well as the secondary structure of the RRE is important for binding (26). Rev is a 116 amino acid protein that contains at least two functional regions. The amino terminal domain confers RNA (RRE) binding and nuclear localization, and also contributes to protein multimerization (15,24,35). The effector domain contains a signal that promotes the export of RNAs from the nucleus to the cytoplasm (10). Consequently, RevM10, which is mutated in the carboxyl-terminal region, binds with high affinity to the RRE and is localized to the nucleus but, unlike the wild-type protein, it is unable to promote the export of RNA from the nucleus to the cytoplasm (12,24,33). The three component molecules used in the tri-hybrid system are the RevM10 protein fused to the DNA-binding domain of GAL4, an RNA-hybrid containing the RRE sequence fused to a target RNA, and a second protein hybrid comprised of the activation domain of GAL4 fused to any protein of interest or to proteins expressed from a library of cDNAs. Each hybrid protein and the hybrid RNA is expressed in yeast cells that are under the control of an independent promoter and termination sequences. Upon productive interaction of the three hybrid molecules a functional GAL4 transcription factor is reconstituted. The activity of this transcription factor is then used to report the RNA–protein interaction.

3. COMPONENTS AND GENERAL PROCEDURE OF THE TRI-HYBRID SYSTEM

The general procedures of the tri-hybrid system are similar to those commonly used in the two-hybrid system. Because two-hybrid protocols are detailed in previous chapters of this book, we will mainly focus on the additional components and properties of the tri-hybrid system.

The flow chart in Figure 2 illustrates the steps involved in executing the system. In the first step two plasmids are constructed that encode three hybrid molecules (Figure 3). pDBRevM10 contains a single *Sal*I site into which hybrid RNAs can be inserted as a transcription unit cassette. The transcription unit cassette was originally derived from pPGKRRE, into which a cDNA encoding the target RNA was cloned. The pDBRevM10 plasmid was constructed in three steps. The RevM10 coding sequence

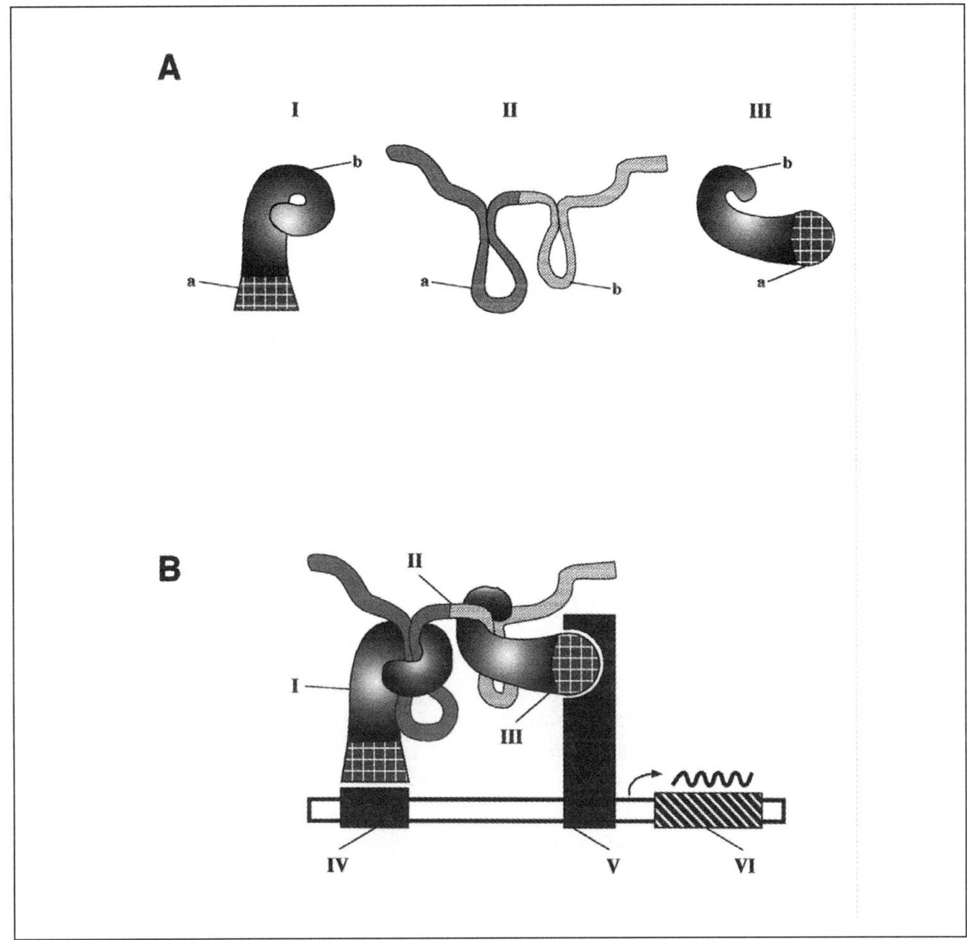

Figure 1. The basic strategy of the tri-hybrid method. (A) The first hybrid-protein (I) contains the DNA-binding domain of GAL4 (Ia) fused to the RRE-RNA-binding protein RevM10 (Ib). A hybrid-RNA (II) containing the RRE sequence (IIa) and a target RNA sequence X (IIb). The second hybrid-protein (III) contains the activation domain of GAL4 (IIIa) fused to a protein Y (IIIb) capable of recognizing the target RNA X on the RNA-hybrid. **(B)** Upon productive interaction of the three hybrids, a reconstituted GAL4 transcription factor (I+II+III) bound to a GAL4 responsive promoter (IV) stimulates the basal transcriptional machinery (V) of the *lacZ* gene and the nutritional reporter gene *HIS3* (VI). Adapted from (29).

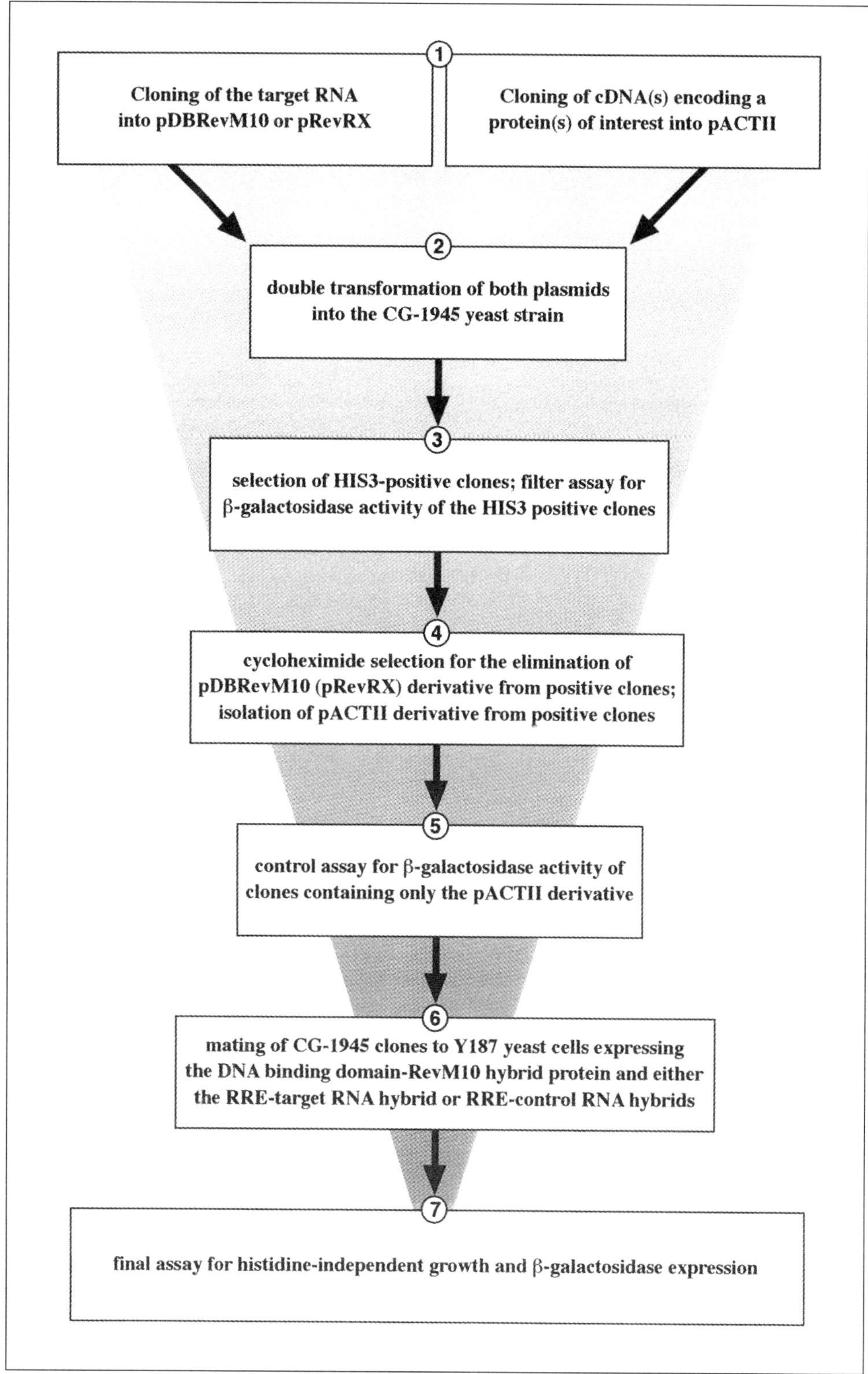

Figure 2. Flow chart of the procedures of the tri-hybrid system.

was generated in a polymerase chain reaction (PCR) using pCRev+M10 (23,24) as a template, with primer sequences: 5′-CTCGAGAAGCTTACCGCCACCATGGCAG-GAAGAAGCGGAGAC-3′, and 5′-AAGCTTATAGATCTTCTTTAGCTCCTGACT-CCA-3′. The PCR product was cloned into pCR^{TM}II (Invitrogen, Carlsbad, CA, USA). The RevM10 sequence was then removed as an *NcoI/EcoRV* fragment and

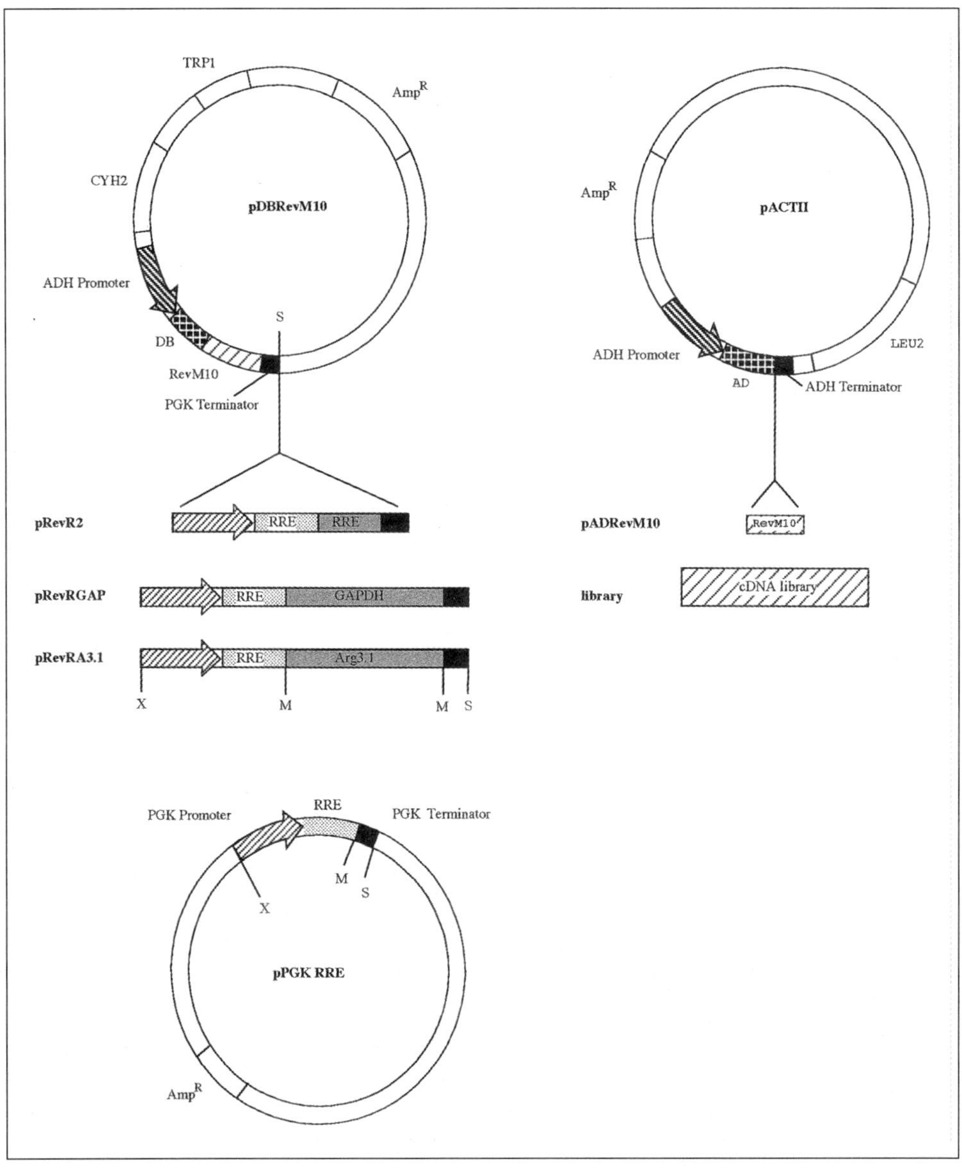

Figure 3. The plasmids used in the tri-hybrid system. The upper plasmids are *E.coli*-yeast shuttle vectors and were used in the transfection and mating experiments of Table 2 and Table 3. pPGKRRE was used as an intermediate to clone hybrid RNAs as a transcription unit cassette into pDBRevM10. The transcription units generating the hybrid proteins and the hybrid RNAs are indicated. The respective promoters and terminators are depicted. *E.coli* and yeast replication origins are not shown. *E.coli* selectable marker for ampicillin resistance, Amp^R, and the yeast selectable markers TRP1 and LEU2 are also shown. Cycloheximide sensitivity in yeast is conferred by CYH2. Relevant restriction sites are abbreviated as follows: M, *MluI*; S, *SalI*; X, *XhoI*.

cloned into the *NcoI/SmaI* sites of a pAS2 (14) derivative, in which a 300-bp *BamHI/SalI* fragment from pPGK (16) containing the PGK transcription terminator had been cloned. pPGKRRE was constructed by inserting an RRE containing *EcoRI* fragment from pGEMRREter (a derivative of pGEMRRE containing nucleotides 13–224 of the RRE (22) and has an additional *MluI* site 3′ adjacent to the RRE sequence) into the *EcoRI* site of pPGK (16). pRevR2, which contains a duplication of the RRE, was constructed by removing the PGK terminator with *MluI* and *SalI* from pPGKRRE. The *MluI* site was blunted and a blunted *SacI/SalI* fragment from pPGKRRE containing an RRE followed by the PGK terminator was inserted. From the resulting plasmid, pRRE2, the RRE-RRE transcription unit cassette was released by restriction digest with *XhoI* and *SalI* and cloned into the *SalI* site of pDBRevM10. pRevRGAP was constructed by using a 1272-bp filled in *XbaI, HincII* fragment containing the entire coding sequence of the GAPDH mRNA derived from pGPDN5 (16). This fragment was cloned into the filled *MluI* site of pPGKRRE. From the resulting plasmid, pRGAP, the RRE-GAPDH transcription unit cassette was released by cleaving with *XhoI* and *SalI* and cloned into the *SalI* site of pDBRevM10. pRevRA3.1 was constructed by using the *MluI* fragment from pSPORT1 arg3.1 (18) that contains the entire cDNA of arg3.1 (3018 bp). This fragment was inserted into the *MluI* site of pGKRRE. From the resulting plasmid pRA3.1, the RRE-Arg3.1 transcription unit cassette was released by cleaving with *XhoI* and *SalI* and cloned into the *SalI* site of pDBRevM10. pADRevM10 was constructed by inserting the *NcoI, EcoRV* fragment containing RevM10 (see above) between the *NcoI* and *SmaI* site of pACTII (7,17). The screened library was constructed by releasing > 2 × 10⁶ independent cDNAs, with an average size of 500–2000 bp from a seizure-induced hippocampal pSPORT-1 library (18), with *BamHI* and *SmaI*. These fragments were cloned between the *BamHI* and *SmaI* site of pACTII. We have now improved pDBRevM10 so that the vector already carries the RNA transcription unit cassette, and cDNAs encoding the target RNAs can be directly inserted into unique *MluI*, *SmaI*, and *NotI* sites located 3′ to the RRE (Figure 4).

Transcription of the GAL4 DNA binding domain-RevM10 protein hybrid is driven by the alcohol dehydrogenase (ADH) promoter and is terminated by the phosphoglycerol kinase (PGK) transcription terminator. The hybrid RNA is encoded by the same vector but is harbored in an independent transcription unit cassette. Transcription of the RNA hybrid is controlled by the strong PGK promoter and is terminated by the PGK terminator. The use of an RNA polymerase II promoter for transcription of the hybrid RNA allows the unrestricted analysis of long sequences. Northern blotting analysis of transformed yeast shows that the hybrid RNA is abundantly expressed. Having the hybrid RNA and one protein hybrid encoded by the same plasmid offers the advantage of using only one marker to select for their presence in yeast. As a consequence the tri-hybrid system can be used in the same yeast strains as the two-hybrid system. The second vector (pACTII) is commonly used in two-hybrid systems (7,17). cDNAs that encode a protein of interest or libraries encoding a multiplicity of proteins are cloned into the polylinker site and expressed as GAL4 activation domain fusion proteins. Both protein hybrids encoded by the two plasmids contain nuclear localization signals. The DNA binding-domain RevM10 fusion harbors two localization signals. One is contributed by the DNA-binding domain of GAL4 (31), the other by RevM10 (21). The second protein hybrid harbors a SV40-derived nuclear localization signal that has been added to the activation domain of GAL4 (4). Interaction of the hybrid RNA with the two hybrid proteins could either occur directly in the nucleus or, alternatively, a trimeric protein RNA complex is formed in the cytoplasm and then localized to the nucleus by virtue of the nuclear localization signals present on the protein hybrids. The promoter of the reporter gene can be occupied by several DNA-binding domain hybrids. Moreover, a single RRE in the RNA hybrid is capable of binding to several Rev proteins. Such multimeric complex formation is likely to significantly enhance the sensitivity of the method.

In the second step, all three hybrids (encoded by two plasmids) are introduced into the yeast strain CG-1945 (Reference 8) (Table 1) by double-transformation. CG-1945 carries inactivating mutations in the endogenous *GAL4* and *GAL80* genes so that GAL4 function is dependent on exogeneously introduced components. In addition, the strain carries inactivating mutations in the nutritional marker genes *TRP1* and *LEU2*, which are the selectable genes carried on the plasmids. The yeast strain also carries the nutritional reporter gene *HIS3*. The expression of the *HIS3* gene is under the control of a GAL4 responsive promoter that contains the upstream activating sequence of GAL1 and the TATA box of the GAL1 promoter (8). As a second reporter gene the strain carries the *E. coli lacZ* gene encoding β-galactosidase. Expression of the *lacZ* reporter gene is controlled by three copies of a 17-mer consensus sequence derived from a GAL4 regulated promoter and the TATA box of the cycloheximide promoter (8). Thus growth in medium lacking histidine as well as β-galactosidase

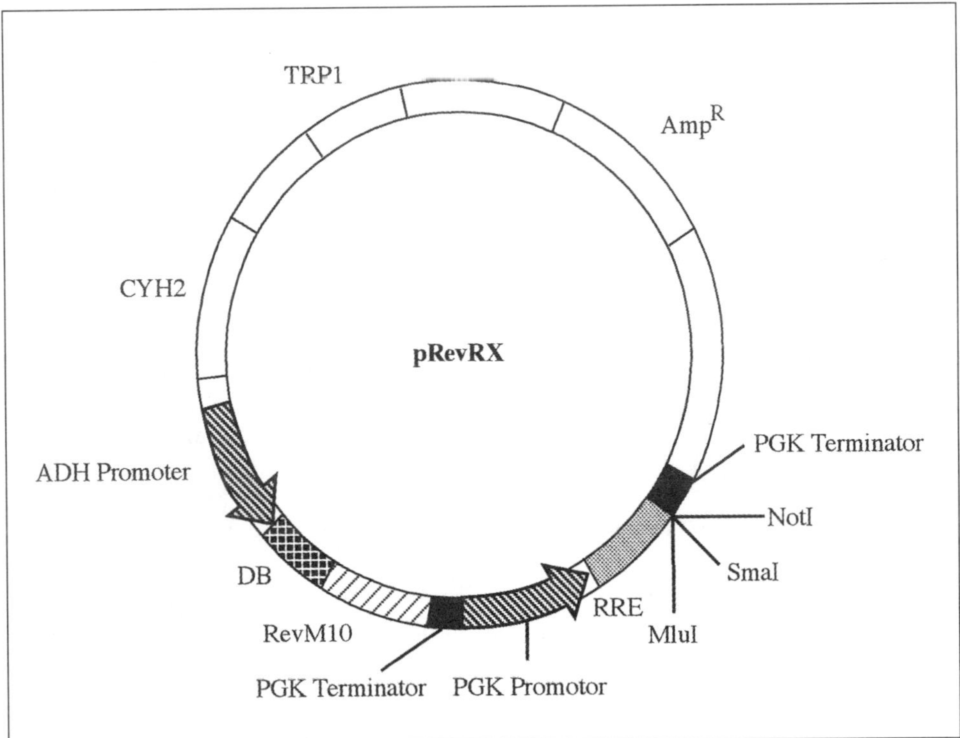

Figure 4. An improved tri-hybrid plasmid for the cloning of target RNAs. pRevRX is an *E.coli*-yeast shuttle vector and can be used in transfection and mating experiments. cDNAs encoding the target RNAs can be directly inserted into unique *MluI*, *SmaI*, and *NotI* sites located 3′ to the RRE . Abbreviations are as in Figure 3. Methods: A double-stranded oligonucleotide containing the restriction sites for *SmaI* and *NotI* was inserted between the *MluI* and *HindIII* site of pPGKRRE (see Figure 3). The transcription unit cassette containing the PGK promotor, the RRE, and the PGK terminator was released from the resulting plasmid by restriction digest with *XhoI* and *SalI* and cloned into the *SalI* site of pDBRevM10 (Figure 3), in which the *MluI* site had been destroyed, previously.

Table 1. Yeast Strains Used in the Tri-hybrid System

Strain	Genotype	Reference
CG-1945	MATa, ura3-52, his3-200, lys2-801, trp1-901, ade2-101, leu2-3, 112, gal4-542, gal80-538, LYS::GAL1-HIS3, cyh^{r}2, URA3::(GAL4 17mers) 3-CYC1-*lacZ*	(8)
Y187	MATα, ura3-52, his3-200, ade2-101, trp1-901, leu2-3, 112, gal4-542, gal80-538, URA::GAL1-*lacZ*	(14)

activity are measures of GAL4 function. If a library screen is performed, positives are identified by their ability to grow in the absence of histidine (step 3). We find it useful to include 10–30 mM 3-amino-1,2,4-triazole (3-AT), a competitive inhibitor of the HIS3-encoded enzyme. Yeast exhibiting stable interactions between a protein and the hybrid RNA will still be able to grow at this 3-AT concentration. However, to detect low affinity interactions it might be necessary to omit the 3-AT selection. Positive clones are then selected for their ability to activate the GAL4 promoter, which controls *lacZ* expression, in a filter assay for β-galactosidase activity.

In step 4 these clones are cured of the plasmid that expresses the DNA-binding domain–RevM10 protein hybrid and the recombinant RNA. This can be easily achieved by growing the clones in medium containing cycloheximide, because pDBRevM10 carries CYH2, which confers cycloheximide sensitivity to the yeast (Figure 3).

False positive clones that enhance transcription independent of a tri-hybrid interaction can be identified in a subsequent β-galactosidase assay (step 5). Only clones that are white in this assay are considered positive and are subjected to a mating test (step 6). The yeast strain Y187 (14) is of opposite mating type to CG -1945 (Table 1) and can therefore be used to reintroduce the plasmid encoding the DNA-binding domain-RevM10 protein and the RRE-target RNA hybrid into CG-1945 clones from which this plasmid has been cured in step 4. Moreover, in this step the sequence specificity of RNA binding can be assessed by mating to Y187 that expresses the DNA-binding domain RevM10 hybrid in combination with various control hybrid RNAs.

In step 7 the resulting diploids are tested for histidine independent growth and β-galactosidase expression. Clones expressing non-specific RNA binding proteins are discarded. Clones in which growth and *lacZ* expression are dependent on the presence of the target RNA and the two protein hybrids are then subjected to further analysis that might include traditional biochemical assays such as UV cross-linking or mobility shift assays.

4. RECONSTITUTION OF A SPECIFIC RNA–PROTEIN INTERACTION IN YEAST

To test the system we expressed both the DNA-binding domain and transcription activation domain of GAL4 as fusion proteins with RevM10. The yeast strain CG-1945 was transformed with plasmids expressing hybrid proteins that contained the

Table 2. Tri-Hybrid Interaction of RevM10 Protein with the RRE RNA Sequence

	Galactosidase Activity	
Expressed hybrid-protein	-RRE	+RRE
DB-RevM10 + AD-RevM10	-	+++
DB-RevM10	-	-
AD-RevM10	-	-

Tri-Hybrid interaction of RevM10 protein with the RRE RNA sequence. Relevant expressed proteins are indicated. +RRE indicates the presence of an RNA containing two copies of the RRE. -RRE indicates the absence of the hybrid RNA. DB, GAL4 DNA-binding domain; AD, GAL4 activation domain. Clones that developed visible blue color within 4 h were registered as +++. If no color developed after 16 h, clones were recorded as -. Transformants were grown on media selective for the given plasmids. When growth was a measure of GAL 4 activity the medium lacked histidine and contained 30 mM 3-amino-1,2,4-triazole. Adapted from (29).

DNA-binding domain of GAL4 fused to RevM10, the transcription activation domain of GAL4 fused to RevM10, and a recombinant RNA containing two copies of the RRE. The maps and details of construction of the expression plasmids used are given in Figure 3. Transformants were analyzed for histidine independent growth and β-galactosidase expression. Functional GAL4 activity was detected only when all three recombinant molecules were co-expressed. Results from these experiments are shown in Table 2. No GAL4 activity was detected in yeast expressing these fusion proteins alone or in combination (Table 2). When, however, a recombinant RNA carrying two copies of the RRE was co-expressed with the two RevM10 fusion proteins, functional GAL4 activity was reconstituted (Table 2). A trimeric ribonuclear protein complex, therefore, may be formed in the nucleus, thus bringing together the GAL4 DNA binding and transcriptional activation domains and recreating functional GAL4 activity. These results demonstrate the ability of a recombinant RNA to interact with two fusion proteins in a tri-hybrid system. The methodology can be extended to detect specific RNA binding proteins from a library of cDNAs.

5. SCREENING OF A cDNA LIBRARY FOR SPECIFIC RNA BINDING PROTEINS

Arg3.1 is an immediate, early gene induced by synaptic stimulation in the mammalian brain (18,19). The induced transcripts are localized to the soma and dendrites of neurons. Dendritic localization is believed to be achieved by specific RNA–protein interactions, although the identities of such proteins have not yet been reported. We have used the full-length arg3.1 transcript fused to an RRE to screen a library of cDNAs that were expressed as transcription activation domain fusion proteins. The arg3.1 mRNA was tethered to the DNA binding domain of GAL4 via the interaction between the RevM10 fusion and the RRE present on the recombinant RNA. Any polypeptide capable of interacting with the arg3.1 message will recruit the transcription activation domain into the ribonuclear protein complex and reconstitute GAL4 activity.

The yeast strain CG-1945 was co-transfected with pRevRA3.1 (Figure 3) that

Table 3. Control Transformations for Optimum Protein–Protein Interaction

	Galactosidase Activity		
Expressed hybrid-protein	RRE	GAPDH	arg3.1
DB-RevM10 + AD-protein 1	-	-	+++

Detection of polypeptides specifically interacting with the dendritically localized transcript arg3.1. Expressed hybrid-proteins are indicated. DB, GAL4 DNA-binding domain; AD, GAL4 activation domain. Arg, RRE, and GAPDH indicate the presence of the corresponding hybrid transcript in the yeast. The identified polypeptide (protein 1) interacts specifically with the arg3.1 transcript; no interaction is seen with recombinant RNAs containing two RREs or with an RRE fused to the glyceraldehyde-3-phosphate-dehydrogenase (GAPDH) mRNA. Adapted from (29).

expressed both the DNA-binding domain–RevM10 hybrid protein and the RRE-arg3.1 recombinant RNA and was also co-transfected with a library of pACTII-derived plasmids that harbored brain-derived cDNAs expressed as fusions with the GAL4 transcription activation domain. From approximately 10^5 clones analyzed, 110 were initially able to grow on medium lacking histidine and containing 30 mM 3-AT. Of these His$^+$ clones, 89 scored positive in a subsequent filter assay for β-galactosidase activity. These clones were cured of pRevRA3.1 and re-examined for β-galactosidase activity. Twelve library clones were able to activate transcription independently and did not depend on the interaction with the other two hybrids. For twenty-nine clones, β-galactosidase activity could be reconfirmed by mating to a Y187 strain carrying the pRevRA3.1 plasmid. The specificity of the arg3.1 interaction was also tested against recombinant RRE RNAs that contained an additional RRE, (pRevR2), or the glyceraldehyde-3-phosphate dehydrogenase (GAPDH) transcript (pRevRGAP, 11). In these tests, 21 clones showed interaction with the arg3.1 transcript but also showed interaction with control transcripts and therefore presumably represent general RNA binding proteins. For the eight remaining clones, β-galactosidase expression was specifically dependent on the presence of the arg3.1 transcript. As an example, the data for one clone are shown in Table 3. The results exemplify that the method allows the simultaneous testing of a large number of proteins for specific RNA-binding activity, as is required for cDNA library screening. The genes encoding the identified proteins are readily available for further analysis. Sequence analysis revealed that the primary structures of these arg3.1 RNA-binding proteins have not been previously reported; however, several of the encoded polypeptides contain motifs found in other RNA binding proteins. These clones are the subject of continued analysis.

6. FURTHER APPLICATIONS AND CONSIDERATIONS

In the previous paragraphs we described how a tri-hybrid system can be used to reconstitute a specific RNA–protein interaction and to identify components of RNA–protein complexes. In addition, because the method is a genetic screen, numerous molecular techniques can subsequently be used in combination with the tri-hybrid system to further characterize the structural specificity of the RNA–protein interactions. For example, to determine the minimal RNA sequence required for binding to the newly identified arg3.1 RNA-binding proteins, yeast clones expressing these proteins were mated to yeast strains expressing deleted arg3.1 RNA targets. The identified *cis*-acting RNA elements could be further analyzed by the introduction of point mutations that potentially disrupt secondary or tertiary structures of the target RNA. Similarly, protein motifs important for RNA binding could be dissected. The method could be used to design pharmacologically active peptides or RNAs. For example, peptides or RNAs can be tested for their ability to bind to specific RNAs or RNA binding proteins. Such peptides or RNAs could then be further tested for their ability to inhibit the function of the RNA or protein of interest. Another common application of the tri-hybrid system might lie in the identification and cloning of new target RNAs that bind to known RNA-binding proteins. Here the tri-hybrid system must be re-configured such that a library of cDNAs is expressed as RRE fusion RNAs, and the known RNA binding protein is expressed as a GAL4 activation domain fusion protein.

Obviously, ribonuclear protein complexes can contain more than a single RNA and interacting protein. In principle this tri-hybrid method could be extended to any number of interacting subunits by the co-expression of additional recombinant RNA or protein molecules.

The tri-hybrid system brings to the analysis of RNA–protein interactions the genetic advantages of the two-hybrid systems. As is the case for the two-hybrid system, several requirements have to be met. The interaction of the two hybrid proteins and the hybrid RNA must be capable of occurring in the nucleus. None of the hybrids alone or in any combination with a second hybrid may give rise to activation of reporter gene expression. The domains of each hybrid must be accessible to allow proper interaction, and secondary and tertiary structures of the target RNA must be able to form when expressed as part of an RNA hybrid. One of the major advantages of the method is that a multiplicity of proteins or RNAs can be tested simultaneously for the interaction with an RNA or protein of interest. The interaction will rely solely on the biophysical properties of the recombinant molecules. Consequently, the method is amenable to the analysis of a wide variety of ribonuclear protein complexes. However, this also implies that a tri-hybrid interaction does not ascertain a physiological function or relevance. Additional experiments are therefore mandatory.

ACKNOWLEDGMENTS

We thank Oliver Sperl and Julia Adlivankina-Kuhl for preparation of the figures.

REFERENCES

1.**Allen, J.B., M.W. Walberg, M.C. Edwards, and S.J. Elledge.** 1995. Finding prospective partners in the library: the two hybrid system and phage display find a match. Trends Biochem. Sci. *20*:511-516.
2.**Battiste, J.L., R. Tan, A.D. Frankel, and J.R. Williamson.** 1994. Binding of an HIV Rev peptide to Rev response element RNA induces formation of purine-purine base pairs. Biochemistry *33*:2741-2747.
3.**Burd, C.G. and R.B. Darnell.** 1994. Conserved structures and diversity of functions of RNA-binding proteins. Science *265*:615-621.
4.**Chien, C.T., P.L. Bartel, R. Sternglanz, and S. Fields.** 1991. The two-hybrid system: a method to identify and clone genes for proteins that interact with a protein of interest. Proc. Natl. Acad. Sci. USA *88*:9578-9582.

5.**Cook, K.S., G.J. Fisk, J. Hauber, N. Usman, and J.R. Rusche.** 1991. Characterization of HIV-1 REV protein: binding stoichiometry and minimal RNA substrate. Nucleic Acids Res. *19*:1577-1583.

6.**Daly, T.J., K.S. Cook, G.S. Gray, T.E. Maione, and J.R. Rusche.** 1989. Specific binding of HIV-1 recombinant Rev protein to the Rev-responsible element in vitro. Nature *342*:816-819.

7.**Elledge, S.J., J.T. Mulligan, S.W. Ramer, M. Spottswood, and R.W. Davis.**1991. Lambda YES: a multifunctional cDNA expression vector for the isolation of genes by complementation of yeast and *Escherichia coli* mutations. Proc. Natl. Acad. Sci. USA *88*:1731-1735.

8.**Feilotter, H.E., G.J. Hannon, C.J. Ruddell, and D. Beach.** 1994. Construction of an improved host strain for two hybrid screening. Nucleic Acids Res. *22*:1502-1503.

9.**Fields, S. and O. Song.** 1989. A novel genetic system to detect protein-protein interactions. Nature *340*:245-246.

10.**Fischer, U., J. Huber, W.C. Boelens, I.W. Mattaj, and R. Lührmann.** 1995. The HIV-1 Rev activation domain is a nuclear export signal that accesses an export pathway used by specific cellular RNAs. Cell *82*:475-483.

11.**Fort, P., L. Marty, M. Piechaczyk, S. el Sabrouty, C. Dani, P. Jeanteur, and J.M. Blanchard.** 1985. Various rat adult tissues express only one major mRNA species from the glyceraldehyde-3-phosphate-dehydrogenase multigenic family. Nucleic Acids Res. *13*:1431-1442.

12.**Fritz, C.C., M.L. Zapp, and M.R. Green.** 1995. A human nucleoporin-like protein that specifically interacts with HIV Rev. Nature *376*:530-533.

13.**Harada, K., S.S. Martin, and A.D. Frankel.** 1996. Selection of RNA-binding peptides in vivo. Nature *380*:175-179.

14.**Harper, J.W., G.R. Adami, N. Wei, K. Keyomarsi, and S.J. Elledge.** 1993. The p21 Cdk- interacting protein Cip1 is a potent inhibitor of G1 cyclin-dependent kinases. Cell *75*:805- 816.

15.**Hope, T.J., X.J. Huang, D. McDonald, and T.G. Parslow.** 1990. Steroid-receptor fusion of the human immunodeficiency virus type 1 Rev transactivator: mapping cryptic functions of the arginine-rich motif. Proc. Natl. Acad. Sci. USA *87*:7787-7791.

16.**Kang, Y.S., J. Kane, J. Kurjan, J.M. Stadel, and D.J. Tipper.**1990. Effects of expression of mammalian Gα and hybrid mammalian-yeast Gα proteins on the yeast pheromone response signal transduction pathway. Mol. Cell. Biol. *10*:2582-2590.

17.**Li, L., S.J. Elledge, C.A. Peterson, E.S. Bales, and R.J. Legerski.** 1994. Specific association between the human DNA repair protein XPA and ERCC1. Proc. Natl. Acad. Sci. U.S.A. *91*:5012-5026.

18.**Link, W., U. Konietzko, G. Kauselmann, M. Krug, B. Schwanke, U. Frey, and D. Kuhl.** 1995. Somatodendritic expression of an immediate early gene is regulated by synaptic activity. Proc. Natl. Acad. Sci. USA *92*:5734-5738.

19.**Lyford, G.L., K. Yamagata, W.E. Kaufmann, C.A. Barnes, L.K. Sanders, N.G. Copeland, D.J. Gilbert, N.A. Jenkins, A.A. Lanahan, and P.F. Worley.** 1995. Arc, a growth factor and activity-regulated gene encodes a novel cytoskeleton-associated protein that is enriched in neuronal dendrites. Neuron *14*:433-445.

20.**MacWilliams, M.P., D.W. Celander, and J.F. Gardner.** 1993. Direct genetic selection for a specific RNA-protein interaction. Nucleic Acids Res. *21*:5754-5760.

21.**Malim, M.H., D.F. McCarn, L.S. Tiley, and B.R. Cullen.** 1991. Mutational definition of the human immunodeficiency virus type 1 Rev activation domain. J. Virol. *65*:4248-4254.

22.**Malim, M.H., L.S. Tiley, D.F. McCarn, J.R. Rusche, J. Hauber, and B.R. Cullen.** 1990. HIV-1 structural gene expression requires binding of the Rev trans-activator to its RNA target sequence. Cell *60*:675-683.

23.**Malim, M.H., J. Hauber, R. Fenrick, and B.R. Cullen.** 1988. Immunodeficiency virus rev trans-activator modulates the expression of the viral regulatory genes. Nature *335*:181-183.

24.**Malim, M.H., S. Boehnlein, J. Hauber, and B.R. Cullen.** 1989. Functional dissection of the HIV-1 Rev trans-activator-derivation of a trans-dominant repressor of Rev function. Cell *58*:205-214.

25.**Okano, H.J. and R.B. Darnell.** 1997. A hierachy of Hu RNA binding proteins in developing and adult neurons. J. Neurosci. *17*:3024-3037.

26.**Olsen, H.S., P. Nelbock, A.W. Cochrane, and C.A. Rosen.** 1990. Secondary structure is the major determinant for interaction of HIV rev protein with RNA. Science *247*:845-848.

27.**Paraskeva, E., A. Atzberger, and M. Hentze.** 1998. A translational repression assay procedure (TRAP) for RNA-protein interactions in vivo. Proc. Natl. Acad. Sci. USA *95*:951-956.

28.**Phizicky, E.M. and S. Fields.** 1995. Protein-protein interactions: methods for detection and analysis. Microbiol. Rev. *59*:94-123.

29.**Putz, U., P. Skehel, and D. Kuhl.** 1996. A tri-hybrid system for the analysis and detection of RNA-protein interactions. Nucleic Acids Res. *24*:4838-4840.

30.**SenGupta, D.J., B. Zhang, B. Kraemer, P. Pochart, S. Fields, and W. Wickens.** 1996. A three-hybrid system to detect RNA-protein interactions in vivo. Proc. Natl. Acad. Sci. USA *93*:8496-8501.

31.**Silver, P.A., L.P. Keegan, and M. Ptashne.** 1984. Amino terminus of yeast GAL4 gene product is sufficient for nuclear localization. Proc. Natl. Acad. Sci. USA *81*:5951-5955.

32.Stripecke, R., C.C. Oliveira, J.E. McCarthy, and M.W. Hentze. 1994. Proteins binding to 5′ untranslated region sites: a general mechanism for translational regulation of mRNAs in human and yeast cells. Mol. Cell. Biol. *14*: 5898-5909.

33.Stutz, F. and M. Rosbash. 1994. A functional interaction between Rev and yeast pre-mRNA is related to splicing complex formation. EMBO J. *13*:4096-4104.

34.Zapp, M.L. and M.R. Green. 1989. Sequence-specific RNA binding by the HIV-1 Rev protein. Nature *342*:714-716.

35.Zapp, M.L., T.J. Hope, T.G. Parslow, and M. Green. 1991. Oligomerization and RNA binding domains of wild type 1 human immunodeficiency virus Rev protein: a dual function for an arginine-rich binding motif. Proc. Natl. Acad. Sci. USA *88*:7734-7738.

14 Reverse Two-Hybrid System: Detecting Critical Interaction Domains and Screening for Inhibitors

Candice A. Leanna and Mark Hannink
Biochemistry Department, University of Missouri-Columbia, Columbia, MO, USA

1. INTRODUCTION

In the commonly utilized standard two-hybrid system (3), one of the proteins (A) is bound to the Gal1 promoter via its fusion to the Gal4 binding domain (Gal4BD), while the other protein (B) is fused to the Gal4 activation domain (Gal4AD, Figure 1A). Association of the two fusion proteins (Gal4BD:A and Gal4AD:B) brings the activation domain of Gal4 to the Gal1 promoter. Localization of the Gal4 activation domain to the Gal1 promoter can be used to activate expression of either a reporter gene such as *lacZ*, or a gene necessary for cell viability, such as *HIS3*, in a yeast strain containing *his3* in its genotype (2,3). Thus the standard two hybrid system is useful for detecting an association between two proteins.

In contrast to the standard two-hybrid system, the reverse two-hybrid system selects against the association of two proteins such that if the two proteins associate, yeast growth is impeded. In the reverse two-hybrid system described in this chapter (5), expression of the *CYH2* allele is driven by the Gal1 promoter.

Note: In addition to the reverse two-hybrid system that utilizes the *CYH2* reporter as described here, Vidal et al. (11) have reported a reverse two-hybrid system that utilizes the *URA3* reporter. Both systems function under the same principles and may be useful in the study of protein–protein interactions.

Expression of *CYH2*, resulting from association of Gal4BD:A with Gal4AD:B, confers cycloheximide sensitivity in the otherwise cycloheximide-resistant *cyh2* CL9 yeast strain (Figure 1B). The CL9 yeast strain also contains the Gal1-driven *lacZ* reporter gene as a secondary assay to confirm loss of association.

Following cotransformation of the Gal4BD and Gal4AD expression vectors into

Yeast Hybrid Technologies
Edited by L. Zhu and G.J. Hannon
© 2000 Eaton Publishing, Natick, MA

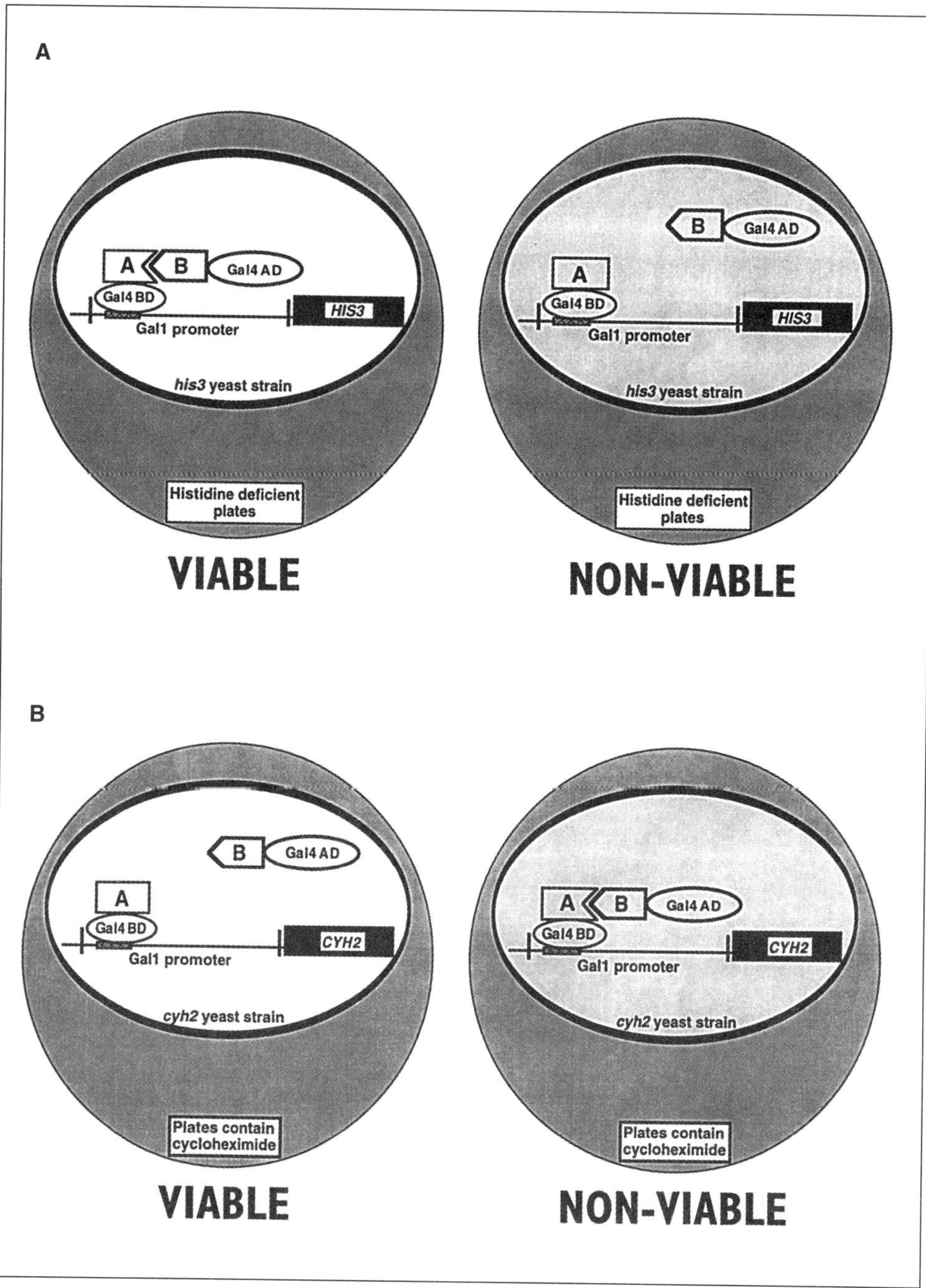

Figure 1. Schematic representation of the standard two-hybrid system and the reverse two-hybrid system. (A) The standard Gal4-dependent two-hybrid system is a genetic selection scheme that selects for the association of protein A fused to the DNA-binding domain of Gal4 (Gal4BD:A) and protein B fused to the Gal4 activation domain (Gal4AD:B). Association of Gal4BD:A and Gal4AD:B results in the localization of the Gal4 activation domain to the Gal1 promoter and in expression of *HIS3*. Survival of the *his3* auxotrophic yeast strain grown on histidine-deficient plates is dependent upon association of Gal4BD:A with Gal4AD:B. **(B)** The Gal4-dependent reverse two-hybrid system is a genetic selection scheme that selects against the association of Gal4BD:A and Gal4AD:B. Association of Gal4BD:A and Gal4AD:B results in localization of the Gal4 activation domain to the Gal1 promoter and in expression of *CYH2*. Survival of the CHX-resistant *cyh2* yeast strain grown on plates containing CHX is dependent upon the lack of association of Gal4BD:A with Gal4AD:B. (Modified from Reference 5. Used by permission.)

the CL9 yeast strain, the yeast are plated onto minimal media lacking histidine (His), leucine (Leu), and tryptophan (Trp), and containing cycloheximide (CHX). The media is lacking in His and Leu to select for the cotransformed expression vectors, while the media lacks Trp to ensure the presence of the *CYH2* gene driven by the Gal1 promoter. The media contains cycloheximide so that when *CYH2* is expressed, as a result of the association of Gal4BD:A and Gal4AD:B, growth of yeast expressing the associating proteins is impeded (Figure 2).

This chapter describes protocols for the use of the *CYH2*-based reverse two-hybrid system to identify proteins that no longer associate in a protein–protein complex. The gene encoding one of the proteins is ligated into the Gal4BD vector, while the other

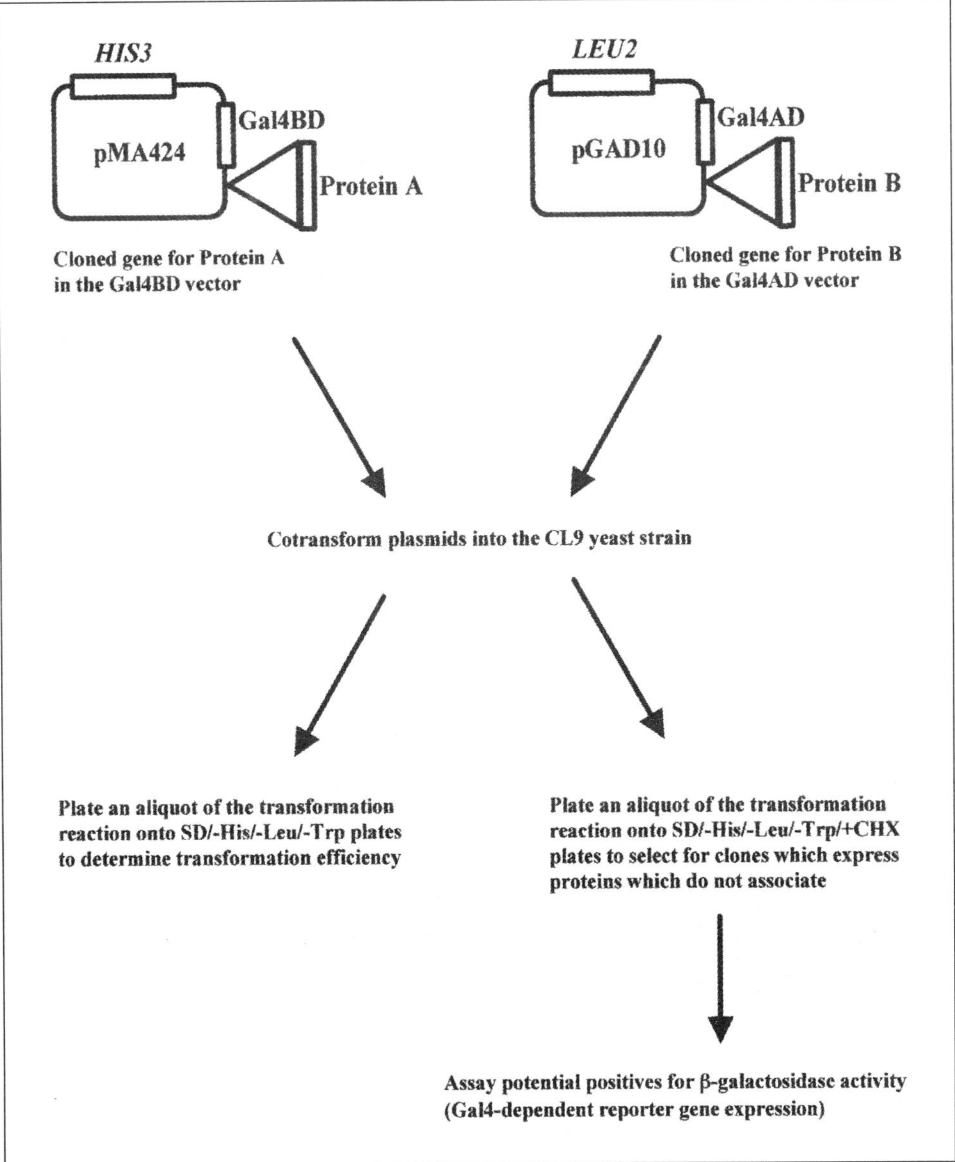

Figure 2. Selection for mutant proteins that no longer associate with a partner protein.

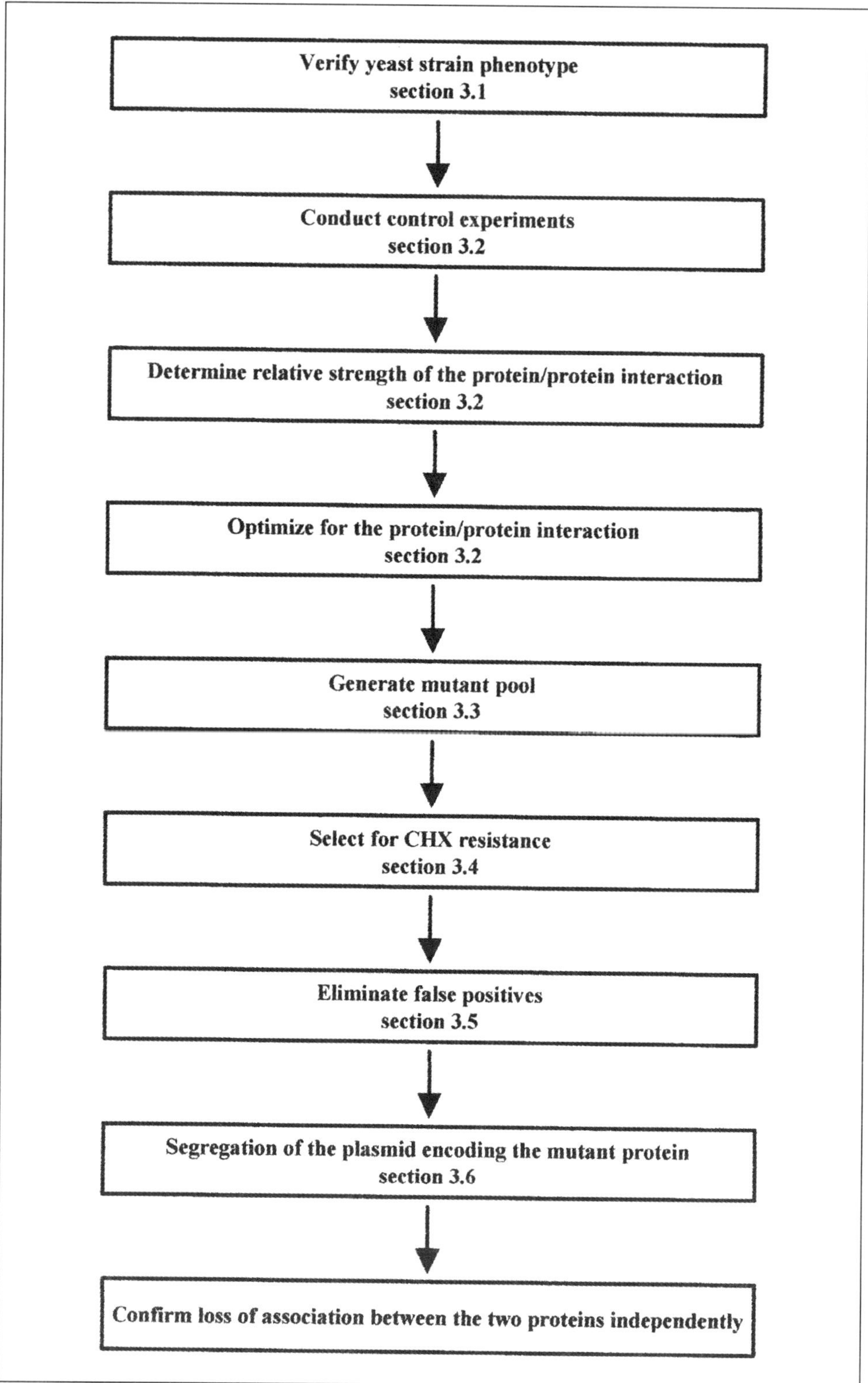

Figure 3. Guide to the reverse two-hybrid system PROTOCOLS.

gene is ligated into the Gal4AD vector. Prior to subcloning into either of the Gal4 vectors, one of the proteins has been subjected to mutagenesis such that the mutant protein can be selected for the inability to associate in a protein–protein complex. The hybrid vectors are cotransformed into the CL9 yeast strain to select for yeast clones encoding mutant proteins that lack the ability to form a protein–protein complex (Figure 3).

While the reverse two-hybrid system was originally developed to identify domains within proteins that are critical for protein–protein interactions, it can potentially be used to identify inhibitors of protein–protein interactions. These inhibitors could be regulatory proteins or exogenously added drugs. Outlined in this chapter is the procedure for use of the *CYH2*-based reverse two-hybrid system to identify domains and/or residues within a protein that are critical for a specific protein–protein interaction. The last section of this chapter provides a discussion on derivatives of the reverse two-hybrid system to identify inhibitors of protein–protein interactions and protein–DNA interactions.

2. PROTOCOLS

2A. Materials and Reagents: Preparation

2A.1 Plasmids and Yeast Strains

1. The Gal4BD-encoding vector used in the reverse two-hybrid system described here is the pMA424 vector (6), and the Gal4AD-encoding vector is the pGAD10 (2) vector. Other plasmids used in the reverse two-hybrid system as controls include pCL1 (2) and pRS423 (10). pCL1 encodes the full-length Gal4 and LEU2 proteins, while pRS423 provides the *HIS3* selectable marker.

2. The genotype of the *Saccharomyces cerevisiae* yeast strain CL9 is α *gal4 gal80 cyh2 his3 trp1-901 TRP1::Gal1→CYH2 ura3-52 URA3::Gal1→lacZ leu2-3,112*. The construction of the CL9 yeast strain is described in detail in Reference 5.

2A.2 Yeast Growth and Maintenance

1. YPD plates: 10 g/L yeast extract, 20 g/L peptone (Difco Laboratories, Detroit, MI, USA), 20 g/L glucose, and H_2O to appropriate volume. Adjust pH to 5.8. Add 20 g/L bacto-agar and autoclave 45 min. Let cool to 55°C prior to pouring plates.

2. YPD liquid media: 10 g/L yeast extract, 20 g/L peptone (Difco Laboratories), and H_2O to appropriate volume. Adjust pH to 5.3. Autoclave 45 min. Let cool to 55°C and add 20 g/L glucose.

3. SD media: 6.7 g/L yeast nitrogen base without amino acids (Difco Laboratories), H_2O to 900 mL, 20 g/L bacto-agar (for plates only). Autoclave 45 min. Let cool to approximately 55°C. Add 100 mL 20% filter sterilized sucrose, 10 mL appropriate Amino Acid Mix, 1 mL 1% Ade, 4 mL 1% Ura, and 5 mL 1% Tyr. Pour plates.

    Amino Acid Mix: 2 g/L arginine (Catalog No. A-5006; Sigma Chemical,

St. Louis, MO, USA), 1 g/L histidine (Catalog No. 8000; Sigma Chemical), 6 g/L isoleucine (Catalog No. I-2752; Sigma Chemical), 6 g/L leucine (Catalog No. L-8000; Sigma Chemical), 4 g/L lysine (Catalog No. L-5501; Sigma Chemical), 10 g/L methionine (Catalog No. M-9625; Sigma Chemical), 6 g/L phenylalanine (Catalog No. P-2126; Sigma Chemical), 5 g/L threonine (Catalog No. T-8625; Sigma Chemical), 2 g/L tryptophan (Catalog No. T-0254; Sigma Chemical) dissolved in H_2O. Filter sterilize. Store at 4°C for up to 2 months.

1% Ade: 10 g/L adenine (Catalog No. A-8626; Sigma Chemical), 980 mL/L H_2O, and 20 mL/L 5.0 M NaOH. Filter sterilize. Store at 4°C for up to 2 months.

1% Ura: 10 g/L uracil (Catalog No. U-0750; Sigma Chemical), 980 mL/L H_2O, and 20 mL/L 5.0 M NaOH. Filter sterilize. Store at 4°C for up to 2 months.

1% Tyr: 10 g/L tyrosine (Catalog No. T-3754; Sigma Chemical), 980 mL/L H_2O, and 20 mL/L 5.0 M NaOH. Filter sterilize. Store at 4°C for up to 2 months.

4. Cycloheximide: (Sigma Chemical) Dissolve in ethanol at 10 mg/mL. Store in the dark at -20°C for up to 1 month.

2B. Materials and Reagents: Methods

2B.1 Yeast Transformation

1. ssDNA: 10 mg/mL sheared and denatured salmon sperm DNA (Catalog No. D-1626; Sigma Chemical) prepared as described in Reference 8.
2. 50% polyethylene glycol 4000: (PEG, average 3350 kDa mol wt) (Sigma Chemical). Filter sterilize.
3. 100% dimethyl sulfoxide (DMSO) (Sigma Chemical).
4. 10× TE buffer: 0.1 M Tris-HCl, 10 mM EDTA, pH to 7.5. Filter sterilize.
5. 10× LiAc: 1.0 M lithium acetate (Sigma Chemical) pH to 7.5 with dilute acetic acid. Filter sterilize.
6. 1× PEG/LiAc/TE solution: 50% PEG 4000, 10% 10× TE buffer, 10% 10× LiAc; 8:1:1, respectively. Prepare fresh from stock solutions just prior to use.
7. 1× LiAc/TE solution: 10% 10× TE buffer, 10% 10× LiAc, bring to final volume with sterile H_2O. Prepare fresh from stock solutions just prior to use.

2B.2 β-galactosidase Assays

Chemiluminescent Method

1. Reaction buffer: dilute Galacton substrate from Tropix Galacto-Light™ Kit (Catalog No. BL100G; Tropix, Bedford, MA, USA) 100-fold with Galacto-Light Reaction Diluent from Tropix Galacto-Light Kit. Warm to room temperature.
2. Lysis buffer: 50 µL 100 mM potassium phosphate containing 0.2% Triton X-100 (pH 7.8), 0.5 µL 100 mM dithiothreitol (DTT), 2.5 µL 0.1% sodium dodecyl sulfate (SDS).

3. Chloroform

4. Protein dye: Bio-Rad Protein Assay dye (Catalog No. 500-0006; Bio-Rad Laboratories, Hercules, CA, USA).

5. Bovine serum albumin (BSA); (Sigma Chemical) fraction V.

<u>Filter-Lift Method</u>

1. Z buffer: 16.1 g/L $Na_2HPO_4 \bullet 7H_2O$, 5.5 g/L $NaH_2PO_4 \bullet H_2O$, 0.75 g/L KCl, 0.246 g/L $MgSO_4 \bullet 7H_2O$, 2.7 mL β-mercaptoethanol. pH to 7.0. Do not autoclave.

2. X-gal stock solution: Dissolve 5-bromo-4-chloro-3-indolyl-β-D-galactopyranoside (X-gal) in N,N-dimethylformamide (DMF) at 20 mg/mL. Store in the dark at -20°C for up to a month.

3. Z buffer/X-gal solution: Add 1 mL X-gal solution to 20 mL Z buffer.

4. Whatman® 540 hardened ashless 70-mm circles (Catalog No. 1540-070; Whatman LabSales, Hillsboro, OR, USA).

5. Whatman filter paper (Catalog No. 1003-917; Whatman LabSales).

2B.3 Small Scale Plasmid Isolation from Yeast

1. Solution I: (for 10 mL) 3.33 mg lyticase (Sigma Chemical), 4.5 mL 2.0 M Sorbitol, 2 mL 500 mM EDTA, 10 µL β-mercaptoethanol, 3.5 mL H_2O. Prepare fresh just prior to use.

2. Solution II: 1.5 mL 500 mM EDTA, 0.6 mL 2.0 M Tris base, 0.6 mL 10% SDS. Prepare fresh just prior to use.

3. TE (pH 8.0): 100 mM Tris-HCl (pH 8.0), 1 mM EDTA.

4. 5.0 M Potassium Acetate.

5. 95% ethanol.

6. Mini Phase Lock Gel™-Heavy (Catalog No. p1-257178; 5 prime → 3 prime®, Boulder, CO, USA)

7. Phenol:chloroform:isoamyl alcohol (25:24:1): Prepare with neutralized pH 7.0 phenol (see Reference 1 for preparation of phenol).

3. METHODS

3.1 Protocol for Yeast Maintenance

Procedure

<u>3.1.1 Yeast Phenotype Verification</u>

1. Prior to transforming the CL9 yeast strain, verify the nutritional requirements by streaking several independent colonies onto four separate SD plates: SD/-Trp, SD/-His, SD/-Leu, and SD/-Ura.

2. Incubate the plates at 30°C for 4–6 days until 2–3 mm colonies appear.

3. The CL9 yeast strain should grow on both the SD/-Trp and SD/-Ura plates, but on neither the SD/-His nor SD/-Leu plates.

3.1.2 Yeast Strain Maintenance

1. Streak the CL9 yeast strain onto a YPD plate. Incubate the plate at 30°C for 3–5 days. Yeast can be stored on the plate wrapped in Parafilm M™ at 4°C for up to 1 month.

2. Scrape a few isolated colonies from the plate and add to 500 μL YPD media. Vortex-mix briefly and transfer to 5 mL YPD media. Grow the culture at 30°C until saturation [Optical Density $(OD)_{600}$ >6.5].

3. To the saturated culture add an equal volume sterile glycerol and briefly vortex-mix. Store at -80°C in 500 μL aliquots.

Day 1

1. Inoculate several 2–3 mm colonies of CL9 into 0.5 mL YPD. Vortex-mix to disperse the clumps.

2. Transfer into a 50-mL flask of YPD and incubate at 30°C for 16–18 h with shaking at 250 rpm to stationary phase $(OD_{600}$ >1.5).

3. Make SD/-His/-Leu/-Trp (-HLT) plates with and without 25 μg/mL CHX. Add the CHX just prior to pouring plates. The plates should be poured as thick as possible to prevent drying out prior to the incubation time required for the assay. Let the plates sit at room temperature 16–18 h prior to plating out the transformation reaction. To ensure optimal results, do not use CHX-containing plates that are more than one day old. Also, do not spread CHX onto solidified -HLT plates to make CHX-containing plates.

Day 2

4. Transfer enough overnight culture to bring a 300 mL YPD culture to an $OD_{600} = 0.2$ to 0.3 and incubate at 30°C for 3 h with shaking at 230 rpm.

5. At 2.5 h, boil the ssDNA stock for 3 min and quick chill on ice to ensure that the DNA is denatured. Add plasmid DNAs for the transformation reactions to microcentrifuge tubes. Once the ssDNA has cooled, add 50 μg ssDNA to each reaction. Prepare 1.5 mL 1× TE/LiAc and 10 mL 1× PEG/TE/LiAc.

6. At 3 h, centrifuge the cells at 1000× g for 5 min at room temperature.

7. Pour off supernatant and resuspend cells in 50 mL sterile H_2O. Centrifuge at 1000× g or 5 min at room temperature.

8. Pour off the supernatant and resuspend the cells in 1.5 mL 1× TE/LiAc.

3.2 Protocol for Control Experiments

Procedure

3.2.1 Verify Reverse Two-Hybrid Function

To ensure that the CL9 yeast strain supports Gal4-dependent negative selection via expression of *CYH2*, the yeast transformation outlined in Table 1 should be performed.

> *Note*: The Yeast Transformation Protocol described here is a modified version of the transformation protocol described by Gietz et al. (4). Modifications were made to increase transformation efficiency and to decrease the appearance of false positives on the -HLT+CHX plates.

Aliquots of each transformation reaction are plated onto plates without and with cycloheximide. The plates that do not contain cycloheximide are used to determine transformation efficiency of each transformation reaction. The plates containing cycloheximide are used for the negative selection assay of the reverse two-hybrid system.

The transformation reactions outlined in Table 1 include positive and negative controls. Colonies should appear on the negative control plate (plate 2) at least two days prior to the appearance of colonies on the positive control plate (plate 1). Figure 4 illustrates the optimum results for verification of Gal4-dependent negative selection in the *CYH2*-based reverse two-hybrid system.

False positives, resulting from the failure to obtain negative selection, can arise for several reasons. For example, mutations that alter the *CYH2* gene such that the gene product no longer confers cycloheximide sensitivity may result in false positives.

Yeast Transformation Protocol

1. Add 100 µL cells to each transformation reaction.
2. Add 600 µL 1× PEG/TE/LiAc solution to each reaction. Vortex-mix briefly. Incubate at 30°C for 30 min with shaking at 200 rpm.
3. Add 70 µL 100% DMSO to each reaction. Mix by inversion. Heat shock for 15 min in 42°C water bath.
4. Chill cells for 5 min on ice.
5. Prepare 1× TE.
6. Microcentrifuge for 5 s at $16\,000\times g$. Remove supernatant and resuspend in 1 mL 1× TE.
7. Plate out enough of the transformation reaction onto -HLT and -HLT+CHX plates to produce a maximum of 200 colonies on each -HLT plate.

   Note: It may be necessary to determine the optimal amount of transformation reaction to be plated onto each plate. The optimal amount of transformation reaction to plate out should be determined by plating out dilutions of the transformation reaction onto the -HLT plates. Once the correct amount of the transformation reaction to be plated out has been determined, this amount of transformation reaction should be plated out onto the -HLT+CHX plates.

Do not increase the amount plated on each plate in an attempt to decrease the number of plates needed. Increasing the amount of reaction plated will only increase the number of false positives.

8. Place plates in 30°C incubator, colony side up, until all liquid is absorbed. Flip plates over and incubate at 30°C until colonies appear. Once colonies are visible on the positive control -HLT+CHX plate, the assay should be stopped because this indicates the time point at which background colonies appear.

9. Calculate the transformation efficiency from the -HLT plates. Colonies (2–3 mm in diameter) should appear on the -HLT plates before appearing on the -HLT+CHX plates.

3.2.2 Determine the Relative Strength of your Protein–Protein Interaction

It is important to determine the strength of your specific protein–protein interaction in the CL9 yeast strain. This section provides protocols for both a quantitative and a qualitative measure of your specific protein–protein interaction. By conducting both procedures the transactivation potential and the CHX-sensitivity of your specific protein–protein interaction can be compared with that of the full-length Gal4 protein.

The quantitative measure described here is a chemiluminescent β-galactosidase

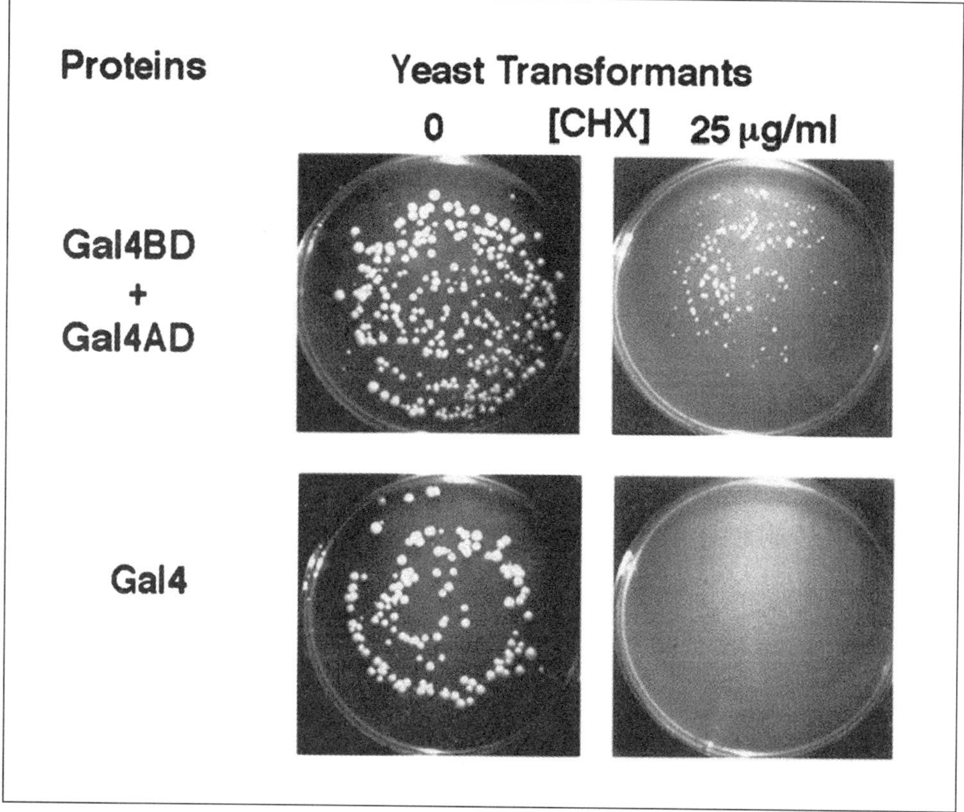

Figure 4. Gal4-dependent growth inhibition in the CL9 yeast strain. The Gal4 protein alone and Gal4BD with Gal4AD were expressed in the *S. cerevisiae* CHX-resistant *cyh2* yeast strain CL9. Equivalent aliquots of the transformed yeast were plated in the absence and presence of 25 µg/mL CHX. (Modified from Reference 5. Used by permission.)

Table 1. Growth Results of the Control Transformations for the CL9 Yeast Strain

Reaction Number	DNA BD	DNA AD	Growth on Plates Containing Cycloheximide
1	pCL1	pRS423[a]	–
2	pMA424	pGAD10	+

[a]This plasmid is actually not a DNA AD plasmid. Because the plates used in the reverse two-hybrid assay are SD/-His/-Leu/-Trp, this plasmid is used to fulfill the His requirement of the transformation reaction.

assay using yeast lysates in which association of the two fusion proteins results in expression of the β-galactosidase gene (*lacZ*).

Note: The quantitative assay includes the use of a modified version of the Tropix Galacto-Light Kit, manual version 9003-G.4. The protocol was modified to produce chemiluminescent measurements that were within the scale of the luminometer.

Note: The quantitative assay includes the determination of protein concentrations of yeast lysates. The protocol described is essentially a Bradford Assay as described in Reference 1.

When the β-galactosidase gene product cleaves its substrate, photons of light are emitted and are measured using a standard luminometer. The purpose of the quantitative assay is to provide a measure of the transactivation potential of each of the fusion proteins alone and of the protein–protein interaction. In general, the transactivation potential measured in the quantitative assay correlates with the CHX-sensitivity measured in the qualitative assay. This assay is particularly useful if optimization of the reverse two-hybrid assay is required (section 3.2.3).

The qualitative assay is simply a measure of the growth of yeast expressing different combinations of proteins on plates containing CHX. Aliquots of the transformed yeast are plated onto the appropriate minimal media plates containing CHX. After several days, colony growth from the yeast expressing the various combinations of proteins is observed and is compared to the yeast growth observed from the control reactions. The purpose of the qualitative assay is to determine the CHX-sensitivity of yeast that express the fusion proteins alone as well as with their partner proteins. The optimum result of the qualitative assay is to obtain growth of colonies that express both fusion proteins several days <u>after</u> obtaining growth of colonies that express each of the fusion proteins alone. If optimum results are not obtained for your specific protein–protein interaction in the qualitative assay, refer to section 3.2.3, which provides suggestions for modification of the reverse two-hybrid assay to obtain the optimum results.

Testing both possible combinations of fusion proteins (i.e., Gal4AD:A with Gal4BD:B and Gal4AD:B with Gal4BD:A) is recommended since one of the combinations of fusion proteins may function better than the other in the reverse two-hybrid system. For example, one of the fusion proteins may fold incorrectly when fused to either Gal4AD or Gal4BD. Therefore, subclone both cDNAs into each of the expression vectors and perform both the quantitative and qualitative assays on both combinations of fusion proteins.

Quantitative Assay to Determine Transactivation Potential

1. Perform the yeast transformations as outlined in Table 2.

2. Once colonies are 2–3 mm on the -HLT plates, perform chemiluminescent β-galactosidase assays to quantitate the transcription potential.

Chemiluminescent β-galactosidase Assay

Yeast Lysates

1. Prepare lysis solution. Add 53 µL lysis solution to microcentrifuge tubes (3 for each transformation reaction, label 1a, 1b, 1c, 2a, 2b, 2c, 3a, etc.).

2. Scrape equivalent amounts of yeast from each transformation reaction using a sterile pipette tip and add to the microcentrifuge tubes. Only use fresh colonies from -HLT plates that were incubated at 30°C.

3. Add 7.5 µL chloroform to each tube. Vortex-mix each tube for 10 s. The lysates can be stored at -80°C for 1–2 days.

β-galactosidase Assay

1. Prepare the reaction buffer and let warm to room temperature. If you are making the lysates and performing the β-galactosidase assay on the same day, prepare the reaction buffer prior to making the lysates to save time.

2. Aliquot 3 µL of each yeast lysate into a luminometer cuvette. Occasionally, the photons emitted from 3 µL of yeast lysate are too high for the luminometer to measure. If this occurs, reduce the amount of yeast lysate until the measurement is within the scale of the luminometer.

3. Add enough reaction buffer to the first cuvette to bring the total volume to 70 µL. After 30 s, add enough reaction buffer to the second cuvette to bring the total volume to 70 µL. Repeat with the remaining samples.

4. Measure the photons emitted from each lysate 1 h after addition of reaction buffer. That is, place the first cuvette into the luminometer. One h after adding reaction buffer to the first cuvette, inject 100 µL accelerator into the cuvette and record the photons emitted. Discard the first cuvette and replace with the second cuvette. At 1 h and 30 s inject 100 µL accelerator into the second cuvette and measure photons emitted. Repeat with the remaining samples.

5. Divide the photons emitted by the number of µL used in step 2 to calculate photons/µL lysate (also referred to as relative light units/µL lysate or RLU/µL lysate).

Determine Protein Concentration

Set up a standard curve:

1. Range the protein concentrations from 0 µg/mL to 12 µg/mL. Add x µL BSA (x = # µL to give appropriate µg BSA), 800 - x µL H_2O, and 200 µL protein dye to a plastic cuvette.

2. Measure the OD_{595} for each sample.

3. Set up a graph of protein concentration vs. OD_{595}.

Table 2. Control Transformations for Optimum Protein–Protein Interaction

Reaction Number	DNA BD	DNA AD
3	pMA424 + Protein A	pGAD10 + Protein B
4	pMA424 + Protein B	pGAD10 + Protein A
5	pMA424	pGAD10 + Protein B
6	pMA424	pGAD10 + Protein A
7	pMA424 + Protein A	pGAD10
8	pMA424 + Protein B	pGAD10

4. Draw the best-fit line and determine the equation of the line.

Determine the protein concentration of yeast lysates:

1. Add 30 µL yeast lysate, 770 µL H_2O, and 200 µL protein dye to a plastic cuvette.

2. Measure the OD_{595} for each yeast lysate.

3. Calculate protein concentration using the equation of the line calculated above. Divide this number by 30 to get µg protein/µL lysate.

4. Calculate the mean RLU/µg protein and the standard deviation for each transformation reaction.

5. Compare the mean RLU/µg protein for your protein combinations with the controls.

Qualitative Assay to Determine Cycloheximide Sensitivity

Use the -HLT+CHX plates from the transformation protocol for the quantitative assay to monitor the growth of colonies in the presence of CHX. The optimum result is to obtain growth of colonies that express both fusion proteins several days <u>after</u> obtaining growth of colonies expressing each of the fusion proteins alone. If you obtain the optimum result for one of the combinations of fusion proteins, proceed to section 3.3.

Note: Expression of both fusion proteins may produce viable colonies if incubated at 30°C for an extended period of time. The important condition is the time period between appearance of colonies on the positive and negative controls. The longer the period between the appearance of colonies on the negative controls (expression of each fusion protein alone) and the positive control (expression of both fusion proteins), the better. If colonies appear on the negative controls several days before the colonies appear on the positive control then the reverse two-hybrid assay will be suitable for that particular protein combination.

If you do not obtain the optimum result for either of the combinations of fusion proteins, proceed to section 3.2.3.

<u>3.2.3 Optimize the Reverse Two-Hybrid System for Your Specific Protein–Protein Interaction</u>

If neither of the protein combinations tested in section 3.2.2 result in the growth of colonies expressing both fusion proteins several days after obtaining growth of colonies that express each of the fusion proteins alone, one or a combination of the following protocols may help to optimize the reverse two-hybrid system for your protein–protein interaction. Listed below are the potential problems that may arise and the protocols to use for optimization of the reverse two-hybrid system for your protein–protein interaction. Once the reverse two-hybrid system is optimized for your protein–protein interaction proceed to section 3.3.

1. *Appearance of yeast colonies expressing both fusion proteins less than 2 days after the appearance of yeast colonies expressing each of the Gal4AD and Gal4BD fusion proteins alone: (plate 3 vs. 5 and 7 or plate 4 vs. 6 and 8).* This is most likely due to a protein–protein interaction that is too weak to be detected in the reverse two-hybrid system using the conditions as described. However, it may be possible to increase the CHX sensitivity of your protein–protein interaction in the reverse two-hybrid system by performing one or both of the optimization protocols A and B.

2. *Appearance of yeast colonies expressing both fusion proteins less than 2 days after the appearance of yeast colonies expressing the Gal4BD fusion protein alone*: (plate 3 vs. 7 or plate 4 vs. 8). This result suggests that the protein fused with the Gal4BD either contains a transactivation domain or is toxic to the yeast. Attempt to delete portions of the Gal4BD fusion protein such that growth of colonies expressing both fusion proteins is obtained several days after obtaining growth of colonies that express the fusion proteins alone. If deletion of part or all of the domain does not result in optimum results, perform one or both of the optimization protocols A and B.

3. *Appearance of yeast expressing both fusion proteins less than 2 days after the appearance of yeast colonies expressing the Gal4AD fusion protein alone*: (plate 3 vs. 5 or plate 4 vs. 6). This result suggests that the protein fused with the Gal4AD either associates with Gal4BD, binds to the Gal1 promoter, or is toxic to the yeast. Attempt to delete portions of the Gal4AD fusion protein such that growth of colonies expressing both fusion proteins is obtained several days after obtaining growth of colonies that express the fusion proteins alone. If deletion of part or all of the domain does not result in optimum results, perform one or both of the optimization protocols A and B.

OPTIMIZATION PROTOCOLS

Protocol A: Enhancement of Transactivation Potential

If your protein–protein interaction is too weak or too transient it may be helpful to increase the potential for expression of *CYH2*. This can be accomplished by addition of another transactivation domain to the Gal4AD fusion protein. For example, the transactivation domain derived from the C-terminus of the c-Rel protein can increase the transactivation potential by 3 to 12-fold (5).

1. Subclone the c-Rel transactivation domain (cTA) into your Gal4AD fusion protein vector at both the N- and C-terminus of the coding region for your protein.
2. Repeat section 3.2.2 and compare the results obtained from both the chemiluminescent β-galactosidase assays and the reverse two-hybrid assays with the results obtained following addition of the cTA to both termini of your protein.

Protocol B: Enhancement of Growth Conditions

Some interactions may require an alteration of yeast growth conditions to obtain optimum results in the reverse two-hybrid assay. It is not known how either of the alterations in growth conditions described below optimize the reverse two-hybrid system. We suggest that the alterations in growth conditions may alter the threshold level of the *CYH2* gene product, produced as a result of the protein–protein interaction, needed to overcome the resistance of the CL9 yeast strain to CHX.

Procedure

CHX Concentration Optimization

1. Repeat section 3.2.2 except make -HLT+CHX plates containing different concentrations of CHX ranging from 10 µg/mL to 50 µg/mL (in 5 µg/mL increments) and plate out aliquots of each transformation reaction onto each concentration of CHX. Plate out aliquots onto -HLT plates to determine transformation efficiency.
2. Monitor the growth on each of the CHX concentrations and determine which concentration is optimum.

Temperature Optimization

1. Repeat section 3.2.2 except make eleven -HLT+CHX plates for each reaction and plate out an aliquot of each transformation reaction onto each of the eleven plates (label A-I). Plate out aliquots on -HLT plates to determine transformation efficiency.
2. Place the sets of -HLT and -HLT+CHX plates at either 4°C or 30°C as listed on the days shown in Table 3.
3. Monitor the growth on each of the -HLT+CHX plates and determine which temperature scheme produces optimum results.

3.3 Protocol for Generation of a Mutant Pool

The pool of mutants can be generated for either of the proteins in the protein–protein interaction. Any of the commonly used methods for mutagenesis can be utilized. However, it would be best to mutagenize the cDNA that expresses your protein prior to subcloning into the expression vector. This will minimize the number of false positives arising from mutation of the expression vector itself.

Procedure

1. Perform random mutagenesis or deletion mutagenesis on the cDNA from one of the proteins.
2. Ligate the pool of mutants into the appropriate vector.

3.4 Protocol for Reverse Two-Hybrid Selection

Once the reverse two hybrid system has been optimized for your protein–protein interaction and the mutant pool has been generated, the reverse two-hybrid assay can be performed on the pool of mutants. If your pool of mutants consists of randomly generated point mutants, be aware that in addition to identifying non-interacting missense mutations, the reverse two-hybrid system may also identify nonsense mutations in which an entire domain is absent from the protein.

Table 3. Sets for Temperature Optimization of the Reverse Two-Hybrid System

Set	Days at 30°C	Days at 4°C	Days at 30°C
A	0	9	
B	1	2	
C	1	3	Until
D	1	4	colonies
E	2	2	appear on
F	2	3	the negative
G	2	4	control
H	3	2	plates
I	3	3	
J	3	4	
K	9	0	

Procedure

1. Perform a yeast transformation using the plasmids expressing the proteins outlined in Table 4 and using any of the optimized conditions. The size of the mutant pool will determine the number of -HLT+CHX plates you will need. If more than 30 -HLT+CHX plates are required to assay the entire mutant pool, more than one transformation reaction can be set up for the pool of mutants. Remember not to increase the amount plated on each plate in an attempt to decrease the number of plates needed. Increasing the amount of reaction plated will only decrease the efficiency of negative selection and will increase the number of false positives.

2. Scrape off any colonies that appear on the -HLT+CHX plates and spot in duplicate onto a new -HLT plate along with the appropriate controls from the original -HLT plates. Return the -HLT+CHX plate to the 30°C incubator. Place the new -HLT plate at 30°C for two days and eliminate false positives as described in section 3.5.

3. Repeat step 2 until colonies appear on the positive control (reaction containing both of the fusion proteins). The time necessary for colonies to become visible on the positive control plate signifies the incubation time that allowed yeast expressing both of the wild-type fusion proteins to grow on CHX.

3.5 Protocol for the Elimination of False Positive Clones

The number of false positives can be reduced by performing either the filter-lift β-galactosidase assay or the chemiluminescent β-galactosidase assay (section 3.2.2).

Note: The protocol for the Colony-Lift β-galactosidase Filter Assay is described in detail in Reference 1.

While the chemiluminescent β-galactosidase assay tends to be more accurate, it is also a more labor intensive protocol, especially when a large number of potential positives need to be assayed. Therefore, if a large number of colonies need to be further assayed, it may be worthwhile to first perform the filter-lift β-galactosidase assay to reduce the number of potential positives to be assayed in the chemiluminescent β-

Table 4. Protein Combinations for the Reverse Two-Hybrid Assay

mutant pool of fusion proteins	fusion protein partner
wild-type fusion protein	fusion protein partner
appropriate Gal4 domain	fusion protein
fusion protein	appropriate Gal4 domain
Gal4 (pCL1)	none (pRS423)
c-Rel (pER31)	p40 (pMH140)
c-Rel (pER31)	Gal4AD (pGAD10)
Gal4BD (pMA424)	p40 (pMH140)

galactosidase assay. When assaying potential positives in either assay, remember to test the potential positive clones in duplicate and include the positive and negative controls from the original -HLT plates.

Procedure

<u>Colony Lift β-galactosidase Filter Assay</u>

1. Place a mark on each of the plates using a marking pen. Place a Whatman 540 filter (Whatman Lab Sales) on each plate. Label each filter and mark the filter to correspond with the mark on the plate. The mark on the plates and the filters will help to identify which yeast lysates on the filters correspond to which colonies remaining on the -HLT plate.
2. Once the filters are wet from underneath, remove the filter and let air dry colony-side up for 5 min.
3. Place the filters colony-side up at -80°C for 10 min to lyse the yeast.
4. Place the filters on a piece of Whatman paper soaked with Z buffer/X-gal solution. Ensure that the filters become completely wet.
5. Place the filters at 30°C and monitor the color change over time until blue color appears on the negative control. Colonies from the reverse two-hybrid assay that produce a blue color at the same time as the positive control are false positives.

3.6 Protocol for the Isolation of Plasmid DNA from the Positive Clones

Once putative positive clones have been identified, the plasmid DNA encoding the mutant protein of interest will need to be isolated for further characterization.

Note: The protocol for isolation of plasmid DNA from yeast is modified from the protocol described in Reference 10. The protocol was modified to produce a plasmid preparation that resulted in a higher transformation into *E. coli*.

Following isolation of plasmid DNA, the plasmid containing the mutant cDNA should be separated from the plasmid encoding the partner protein. The easiest way to do this is to digest the plasmid that encodes the partner protein with a restriction enzyme that does not digest the plasmid encoding the mutant protein. If it is not possible to design the plasmids as such, the plasmid can be segregated by complementation in *E. coli*.

Note: The protocol for segregation of plasmids using complementation is described in Reference 1. The *E. coli* strains used for the pMA424-derived plasmid and the pGAD10-derived plasmid are BA1 and JA300, respectively.

Procedure

<u>Small Scale Plasmid Isolation from Yeast</u>

1. Scrape cells from the -HLT plates and inoculate 6 mL -HLT selective media. Incubate the cultures at 30°C on a rotator until saturated (OD_{600} >6.5).
2. Centrifuge at 1500× *g* for 5 min at room temperature. Resuspend in 1 mL sterile H_2O. Transfer to a microcentrifuge tube.
3. Microcentrifuge 10 s at 16000× *g*. Resuspend in 300 μL freshly made Solution I and incubate at 37°C for 1 h.
4. Microcentrifuge 10 s at 16000× *g*. Resuspend in 400 μL TE.
5. Add 100 μL of freshly made Solution II, mix by inverting 10 times, and incubate at 65°C for 30 min.
6. Add 100 μL 5.0 M KOAc and incubate on ice for 1 h (or overnight at 4°C).
7. Microcentrifuge at 16000× *g* for 15 min at 4°C. Transfer supernatant to a microcentrifuge tube.
8. Fill tube with 95% ethanol at room temperature. Invert 10 times to mix.
9. Microcentrifuge 5 s at 16000× *g*. Decant supernatant. Microcentrifuge 5 s at 16000× *g*. Remove remaining liquid using a micropipettor.
10. Dry pellet under vacuum.
11. Resuspend in 50 μL H_2O, add to Phase Lock Gel (5 Prime → 3 Prime, Boulder, CO, USA) tube, and add 50 μL phenol:chloroform:isoamyl alcohol. Vortex-mix well. Microcentrifuge at 16000× *g* for 3 min.
12. Transfer the aqueous phase to a new microcentrifuge tube.
13. Separate the plasmid encoding the mutant protein from the plasmid that encodes the partner protein, confirm loss of association using other methods, and further characterize the mutant protein.

4. DISCUSSION

The reverse two-hybrid system is a genetic scheme for selection against protein–protein interactions. Association of protein partners results in the expression of the dominant *CYH2* allele that confers CHX sensitivity to the CHX-resistant *cyh2* yeast strain. The reverse two-hybrid system is best suited to analyze a large number of mutants for the loss of association with a partner protein and for the identification of regulatory proteins from a library that disrupts a protein–protein interaction. However, some protein–protein interactions may not be suitable in the reverse two-hybrid system as described here. For example, the reverse two-hybrid system will not be useful if the wild-type protein–protein interaction is too weak or transient such that the expression of the *CYH2* allele is too low to confer sensitivity to CHX.

Vidal et al. (11) have described an alternative reverse two-hybrid system that utilizes the *URA3* gene as the reporter that confers negative selection. If the reverse two-hybrid system utilizing the *CYH2* reporter cannot be used with your protein-protein interaction, it may be beneficial to use this particular reverse two-hybrid system.

This chapter described the procedure for identification of domains and/or residues that are critical for a protein–protein interaction. The reverse two-hybrid system has several other potential uses. First, the reverse two-hybrid system could potentially be used to screen for drugs that disrupt the protein–protein interaction. Second, a cDNA expression library could be screened for regulatory proteins that disrupt a specific protein–protein interaction. To use the CL9 yeast strain to identify regulatory proteins, another auxotrophy would have to be generated to select for plasmids expressing the library. However, recall that the *TRP1* marker may not be used because it is required to select for *Gal1→CYH2*. Similarly, it is not recommended to use the *URA3* marker because it selects for *Gal1→lacZ*.

The reverse one-hybrid system may also be used in the *CYH2*-based negative selection system. For example, the CL9 yeast strain could be used to identify a regulatory protein of a transcription factor with transactivation potential. In the presence of the regulatory protein, the transactivation potential of the transcription factor would be inhibited such that *CYH2* was no longer expressed. Another reverse one-hybrid variation would include manipulation of the CL9 yeast strain such that the promoter driving *CYH2* would contain the DNA-binding site for a transcription factor. Upon expression of a regulatory protein that disrupts the DNA–protein interaction, *CYH2* would no longer be expressed and the yeast would be able to grow in the presence of CHX.

REFERENCES

1. **Ausubel, F.M., R. Brent, R.E. Kingston, D. Moore, J.G. Seidman, J.A. Smith, and K. Struhl.** 1994. Current Protocols in Molecular Biology. John Wiley & Sons, New York.

2. **Bartel, P.L., C. Chein, R. Sternglanz, and S. Fields.** 1993. Using the two-hybrid system to detect protein-protein interactions, p. 153-179. *In* D.A. Hartley (Ed.), Cellular Interactions in Development-A Practical Approach. Oxford University Press, Oxford.

3. **Fields, S. and O. Song.** 1989. A novel genetic system to detect protein-protein interactions. Nature *340*:245-246.

4. **Gietz, D., A. St. Jean, R.A. Woods, and R.H. Schiestl.** 1992. Improved method for high efficiency transformation of intact yeast cells. Nucleic Acid Res. *20*:1425.

5. **Leanna, C.A. and M. Hannink.** 1996. The reverse two-hybrid system: a genetic scheme for selection against specific protein/protein interactions. Nucleic Acids Res. *24*:3341-3347.

6. **Ma, J. and M. Ptashne.** 1987. A new class of yeast transcriptional activators. Cell *51*:113-119.

7. **Sambrook, J., E.F. Fritsch, and T. Maniatis.** 1989. Molecular Cloning-A Laboratory Manual. Cold Spring Harbor Laboratory Press, Cold Spring Harbor, NY.

8. **Schiestl, R.H. and R.D. Gietz.** 1989. High efficiency transformation of intact yeast cells using single stranded nucleic acids as a carrier. Curr. Genet. *16*:339-346.

9. **Sikorski, R.S. and P. Heiter.** 1992. Multifunctional yeast high-copy-shuttle vectors. Gene *110*:119-125.

10. **Strathern, J.N. and D.R. Higgins.** 1991. Recovery of plasmids from yeast into *Escherichia coli*: shuttle vectors. Methods Enzymol. *194*:319-329.

11. **Vidal, M., R.K. Brachmann, A. Fattaey, E. Harlow, and J.D. Boeke.** 1996. Reverse two-hybrid and one-hybrid systems to detect dissociation of protein-protein interactions. Proc. Natl. Acad. Sci. USA *93*:10315-10320.

Section V

Other Hybrid Systems

<table><tr><td>**15**</td><td></td></tr></table>

Mammalian Two-Hybrid Assays

Andrew Farmer
CLONTECH Laboratories, Palo Alto, CA, USA

1. INTRODUCTION

The yeast two hybrid assay has become a pervasive technique in molecular biology, as indispensable to research as the polymerase chain reaction. The principle of this assay, first described by Fields and coworkers (5,9), relies on the observation that transcription factors can be separated into two distinct domains: a DNA-binding domain (DNA-BD), which specifically binds to a promoter or other cis-regulatory element; and an activation domain (AD), which directs RNA polymerase II to transcribe the gene downstream of the DNA-binding site (14,16). While these domains may be part of the same protein (as in the case of the native yeast GAL4 protein), they can also function as part of two separate proteins as long as the AD is tethered to a DNA-BD bound to a promoter. In two-hybrid assays, that tether is the interaction between two additional proteins (X and Y) that are expressed as protein fusions with the DNA-BD and AD peptides, respectively [for review, see Fields and Sternglanz (10) and other chapters in this volume].

The yeast two-hybrid assay provides a powerful and rapid way to identify novel protein–protein interactions as well as a way to dissect the specific domain and amino-acid requirements for a known interaction. More recently, it has been realized that the assay can be adapted to help identify small molecule inhibitors of protein association with a goal toward identifying novel therapies for disease.

Despite its great power and success, there remain certain limitations to the yeast two-hybrid system that have fueled a desire to develop mammalian two-hybrid systems. Chief among these is the dependence of some interactions between human proteins on post-translational modifications. These modifications are often dependent on cellular signaling events, or occur only in a specific cell state or lineage that is impossible to replicate in yeast. Many significant interactions may therefore go undetected in the yeast assay. The desire of the pharmaceutical industry for high throughput assays in mammalian cells has also fueled the development of appropriate mammalian two-hybrid assays and related technologies, especially toward systems that utilize rapid and inexpensive methods of detection.

Yeast Hybrid Technologies
Edited by L. Zhu and G.J. Hannon
© 2000 Eaton Publishing, Natick, MA

Despite this need, mammalian two-hybrid assays have yet to reach the power and simplicity of the yeast assay. The major problem being that most methods for transferring DNA into mammalian cells are of low efficiency, making library screening cumbersome and inefficient. Moreover, the growth and maintenance of mammalian cells is both time consuming and expensive. Mammalian cells typically take 24 h to divide compared to 2–4 h for yeast. For this reason, most mammalian two-hybrid assays rely on colorimetric reporters alone to identify positive clones rather than a combination of colorimetric reporters and selective growth assays, as is typical in yeast-based systems. Use of selective growth assays in the mammalian cell setting presents an additional problem in that unless the expression plasmids encoding the proteins being tested for interaction are able to replicate autonomously (as is common in yeast plasmids), they must integrate into the genome of the cell. Since successful integration is a rare event, this further compounds the problem of library screening. Moreover, cDNAs encoded on the integrated vectors are not directly available for study, but must be isolated by PCR and subcloned before they can be studied further. In the case of colorimetric assays, current technology is not capable of readily identifying and isolating individual cells that display a positive signal. Thus, current mammalian two-hybrid assays can only enrich a population for positive clones, rather than identify specific clones, as is possible with the yeast systems. Mammalian two-hybrid assays, therefore, require more downstream analysis to identify the individual positive clones than the yeast-based assays. For these reasons, current mammalian two-hybrid assays have mostly been used to confirm interactions identified in the yeast assay, rather than to detect novel interactions. Thus the potential to detect novel interactions, which—for the reasons stated previously—might go unnoticed in the yeast-based assay, is yet to be exploited. This chapter will provide a detailed description of the basic mammalian two-hybrid assay, originally described by Dang et al. (7), which can be used to confirm interactions identified in the yeast assay (33) or to map specific binding domains in known interacting proteins (30). This assay will then provide a framework for a discussion of alternative assays that have been developed with the goal of making a mammalian cell based assay that can readily be used to screen libraries.

In the Mammalian MATCHMAKER™ Two-Hybrid Assay available from CLONTECH (Palo Alto, CA, USA) (Figure 1), test cells are co-transfected with three vectors: one expressing a DNA-BD fusion protein (bait); one expressing an AD fusion protein (prey); and a third construct encoding a reporter gene whose expression is dependent on the interaction of the proteins encoded on the first two constructs. Specifically, the pM cloning vector is used to generate fusions of a bait (X) with the GAL4 DNA-BD. Similarly, the pVP16 vector is used to construct fusions between the prey (Y) and an AD derived from the VP16 protein of herpes simplex virus. The reporter construct, pG5CAT, contains the CAT gene downstream of five consensus GAL4 binding sites and the minimal promoter of the adenovirus E1b gene. (For details of the vectors, see Figure 2). The minimal E1b promoter alone will not express significant levels of the CAT gene, so that background is low in the absence of activation from the GAL4 sites. Once the constructs are made, all three plasmids are co-transfected into a suitable mammalian host cell line, and the interaction between proteins X and Y is assayed 48–72 h later by measuring CAT gene expression. This assay cannot be used to identify novel interactions. However, it can be used to confirm interactions identified by other means (e.g., immunoprecipitation, theoretical predictions based on homology to known protein binding domains, genetic data, and interactions detected by yeast two-hybrid assays such as CLONTECH's MATCH-MAKER Two-Hybrid Systems). Such confirmation eliminates the possibility of a false-positive that is an artifact of working in yeast cells. Once an interaction between two proteins has been detected, the assay can be further used to dissect the interaction using deletional or site-directed mutagenesis [see Takacs et al. (30) for example].

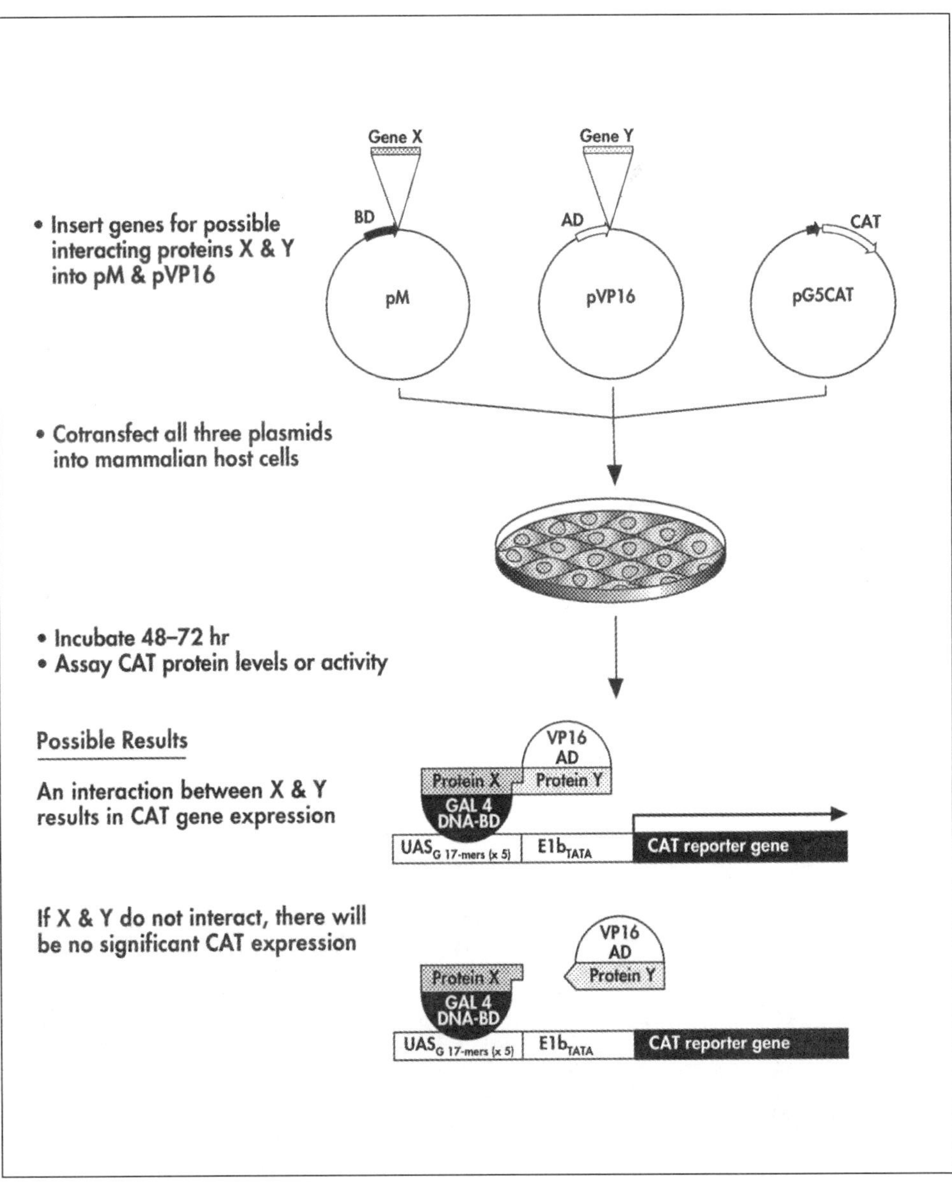

Figure 1. Schematic flow diagram outlining the mammalian two-hybrid procedure. Genes encoding potential interacting proteins are cloned into the DNA-BD and AD expression vectors, pM and pVP16. These are then co-transfected with the pG5CAT reporter into a mammalian host cell. After 48–72 h, the cells are harvested and assayed for CAT activity. If X and Y interact, this will result in activation of CAT gene transcription and CAT activity will be detected. If X and Y do not interact, the activation domain will not be brought into proximity of the promoter for the CAT gene and CAT activity will not be induced.

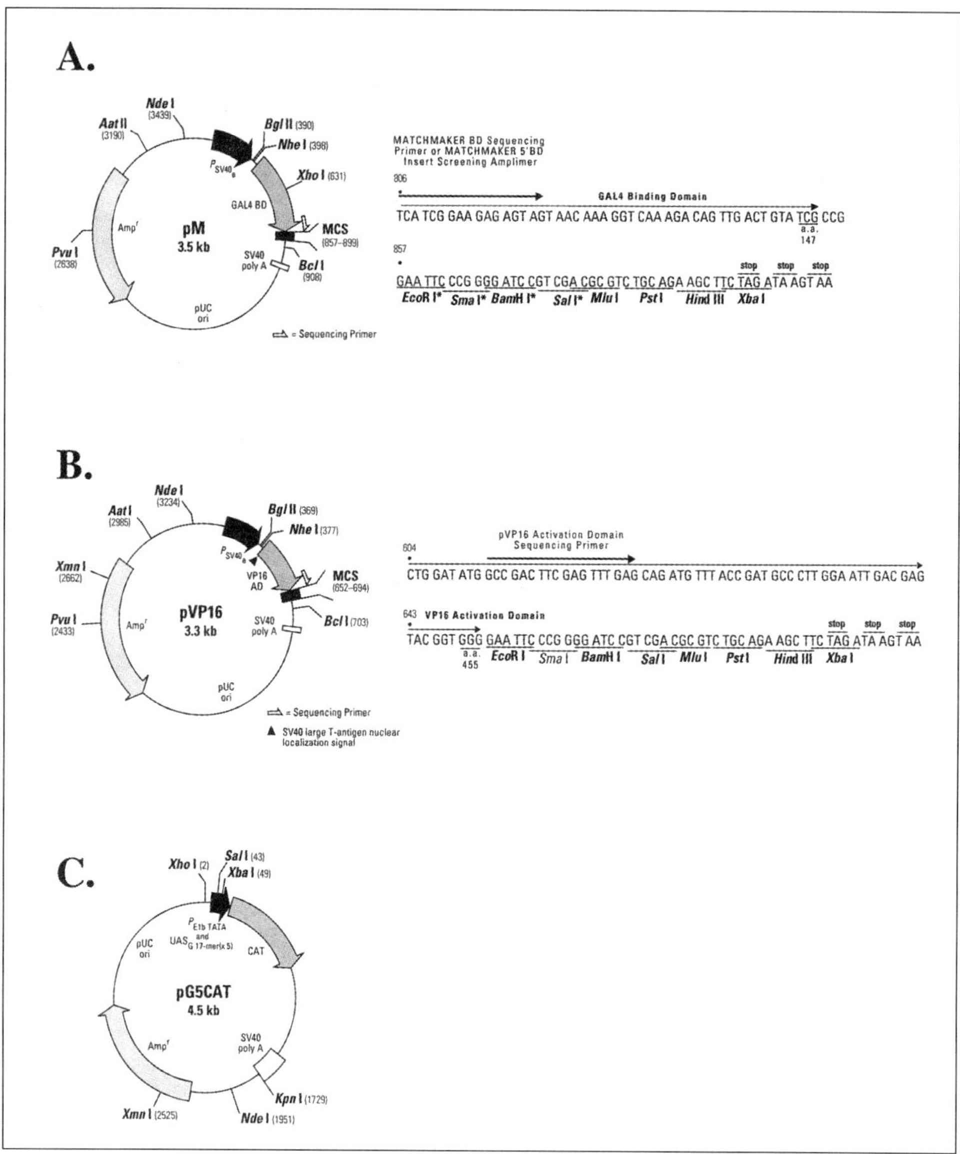

Figure 2. Maps of the expression and reporter vectors. (A) pM. pM is used to generate a fusion of the GAL4 DNA-BD (amino acids 1–147) and a protein of interest. The hybrid protein is targeted to the cell's nucleus by the GAL4 nuclear localization sequence (27). Genes encoding test proteins should be cloned, in the correct orientation and reading frame, into one of the unique restriction sites in the MCS region at the 3′ end of the GAL4 DNA-BD. Restriction sites marked with an asterisk are in the same reading frame as pGBT9 and pAS2-1 from the MATCHMAKER (GAL4) Two-Hybrid System (Catalog Nos. K1604-1 and K1605-1). Sites marked with an asterisk and *Pst*I are in the same reading frame as pLexA from the MATCHMAKER LexA Two-Hybrid System (Catalog No. K1609-1). Transcription is initiated from the constitutive SV40 early promoter (P); transcription is terminated at the SV40 poly A transcription termination signal. **(B)** pVP16. pVP16 is used to generate a fusion of the VP16 AD (amino acids 411–455) and a protein of interest. The hybrid protein is targeted to the cell's nucleus by the SV40 nuclear localization sequence. Genes encoding test proteins should be cloned, in the correct orientation and reading frame, into one of the unique restriction sites in the MCS region at the 3′ end of the VP16 AD. Transcription is initiated from the constitutive SV40 early promoter (P); transcription is terminated at the SV40 poly A transcription termination signal. **(C)** pG5CAT. pG5CAT contains five consensus GAL4 binding sites [UASG 17-mer (×5)] and an adenovirus E1b minimal promoter upstream of the chloramphenicol acetyltransferase (CAT) gene. As seen in Figure 1, the DNA-BD portion of a hybrid protein expressed from a pM-derived plasmid localizes to the GAL4 binding sites in pG5CAT. If the hybrid test protein X interacts with test protein Y (expressed as a hybrid protein from a pVP16-derived plasmid), the CAT gene will be transcribed.

2. PROTOCOLS

2.1 Plasmids

List Of Components Provided With CLONTECH's Mammalian MATCHMAKER System

(Store all components at -20°C.)

pM: GAL4 DNA-binding domain cloning vector used to express a fusion of the bait with the GAL4 DNA-BD (Figure 2A).

pVP16: activation-domain cloning vector used to express a fusion of the prey with the VP16 AD, a herpes virus protein that acts as a transcriptional activator in mammalian cells (Figure 2B).

pG5CAT: mammalian CAT reporter plasmid for co-transfection into mammalian cells with recombinant plasmids derived from pM and pVP16 (Figure 2C).

pM3-VP16: positive control vector that expresses a fusion of the GAL4 DNA-BD to the VP16 AD.

pM-53: positive control vector that expresses a fusion of the GAL4 DNA-BD to the mouse p53 protein.

pVP16-T: positive control vector that expresses a fusion of the VP16 AD to the SV40 large T-antigen, which is known to interact with p53.

pVP16-CP: negative control vector that expresses a fusion of the VP16 AD to a viral coat protein, which does not interact with p53.

2.2 Protocol for Mammalian Cell Culture, Harvesting, and Transfection

Materials and Reagents: mammalian cell culture and harvesting

- Cell culture medium [e.g., Dulbecco's modified Eagle's medium (DMEM) or other appropriate growth medium].
- Fetal bovine serum (FBS), newborn calf serum, or equivalent (to supplement the growth medium)
- Phosphate-buffered saline (PBS), pH 7.4.
- 1× Trypsin/EDTA.

The reagents required for transfecting the cells of interest will depend on the method of transfection chosen. The method described subsequently is for a high-efficiency calcium-phosphate–mediated transfection (2). However, liposome-mediated transfection or electroporation may also be used. The DEAE-dextran technique has also worked well in our hands. Published protocols for these and other methods are available (2,11).

Researchers should choose a transfection method for the assay that is most appropriate to the cell line with which they are working and should be prepared to adjust conditions empirically to obtain optimum conditions. The efficiency of transfection for different cell lines may vary by several orders of magnitude. A method that works well for one host cell line may be inferior for another. Therefore, when working with a cell line for the first time, it is useful to compare the efficiencies of several transfection protocols. This can be done by assaying CAT activity following co-transfec-

Table 1. Positive Control Mammalian Two-Hybrid Experiment

Input DNA: Transfection	GAL4 DNA-BD Plasmid (10 µg[a])	VP16 AD Plasmid (10 µg[a])	Reporter Plasmid (2 µg[a])	CAT Protein or CAT activity
1A (positive control)	pM-53	pVP16-T	pG5CAT	+++
1B (negative control)	pM-53	pVP16-CP	pG5CAT	-
2 (untransfected control)	none	none	none	-[b]
3 (basal control)	pM	pVP16	pG5CAT	-[c]
4 (negative control)	pM-53	pVP16	pG5CAT	-[d]
5 (negative control)	pM	pVP16-T	pG5CAT	-[e]
6 (positive control)	pM3-VP16[f]	(pM3-VP16)	pG5CA	+++

[a]These amounts of DNA are optimized for calcium-phosphate–mediated transfection of HeLa cells. When using other transfection methods, adjust the total amount of DNA accordingly, but maintain the ratio of the different plasmids. The optimal amount of DNA may also vary with different cell types.

[b]This value is the background CAT signal in your cells. For quantitative CAT assays, subtract this value from all other experimental values.

[c]This control (or a control using just pM and pG5CAT) provides the basal expression level of CAT protein or CAT activity in your experiments. To determine the fold-activation caused by an interaction between protein X and protein Y, divide your experimental CAT values by this number. For example, if you used the Roche Molecular Product's CAT ELISA assay and obtained an absorbance of 1.5 with your experimental example and an absorbance of 0.05 with this negative control, you would conclude that the interaction between protein X and protein Y caused a 30-fold activation of CAT expression. This sort of analysis is particularly useful when comparing multiple protein interactions (e.g., the interactions between protein X and a series of protein Y mutants).

[d]Critical control to determine whether or not your protein X functions autonomously as a transcriptional activator. The level of CAT should be similar to the background value obtained in transfection 3.

[e]Critical control to determine whether or not your protein Y functions autonomously as a DNA-BD or binds directly to the DNA-BD encoded by pM. The level of CAT should be similar to the background value obtained in transfection 3.

[f]Use 20 µg when transfecting only a DNA-BD or AD vector with pG5CAT. The pM3-VP16 positive control plasmid encodes a fusion of the GAL4 DNA-BD and the VP16AD and therefore gives very strong CAT expression when cotransfected with pG5CAT.

Table 2. Recommended Set-Up For Mammalian Two-Hybrid Assays

Input DNA: Transfection	GAL4 DNA-BD Plasmid (10 μg[a])	VP16 AD Plasmid (10 μg[a])	Reporter Plasmid (2 μg[a])	CAT Protein or CAT Activity
1 (experiment)	pM-insert X	pVP16-insert Y	pG5CAT	To be determined
2 (untransfected control)	None	None	None	-[b]
3 (basal control)	pM	pVP16	pG5CAT	-[c]
4 (X control)	pM-insert X	pVP16	pG5CAT	To be determined[d]
5 (Y control)	pM	pVP16-insert Y	pG5CAT	To be determined[e]
6 (Positive control)	pM3-VP16[f]	(pM3-VP16)	pG5CAT	+++

a–f, see Table 1.

tion of the cells with the pG5CAT and pM3-VP16 plasmids. Alternatively, the host cell line can be transfected with a reporter expression vector, such as one of CLON-TECH's Great EscAPe™ vectors, which express a secreted form of alkaline phosphatase. This has the advantage that activity can be measured repeatedly over time by removing aliquots of the growth medium without harming the cells. The choice of reporter, as well as the method used to detect activity of the reporter, will also influence the method of transfection chosen because the sensitivity of the reporter assay will have an impact on the minimum percentage of transfected cells required to detect a reliable signal.

After a method of transfection is chosen, it may be necessary to optimize parameters such as cell density, the amount and purity of the DNA to be transfected, media conditions, and transfection time. Once optimized, these parameters should be kept constant to obtain reproducible results. With each method, CAT activity may be detected 48–72 h after transfection, depending on the host cell line used. A positive control experiment, such as the one detailed in Table 1, can be used at this time to confirm that the assay works in your hands as well as in your cells and also to optimize your transfection and CAT assay protocols. This experiment compares the interaction between the p53 protein and two other proteins: the SV40 large T-antigen, which is known to interact with p53, and a polyoma virus coat protein (CP), which does not interact with p53 (Reference 18) (Figure 3).

2.3 Protocol for Calcium-Phosphate–Mediated Transfection

Materials and Reagents

- 2.5 M CaCl$_2$
  Add 18.37 g of CaCl$_2$ dihydride (Sigma Chemical; tissue-culture grade) to 50 mL of H$_2$O. Filter-sterilize through a 0.45 μm nitrocellulose filter (Nalgene,

Rochester, NY, USA). Store at -20°C in 10-mL aliquots (can be frozen and thawed repeatedly).

• 2× BES-buffered saline (BBS, pH 6.95)

Component	**Final Conc.**
N, N-bis (2-hydroxyethyl)-2-aminoethane-sulfonic acid (BES; Calbiochem-Novabiochem, San Diego, CA, USA)	50 mM
NaCl	280 mM
Na_2HPO_4	1.5 mM

Note: It is critical that the pH of this solution be between pH 6.95 and 6.98. We recommend that you check each new batch of 2× BES buffer against a reference stock prepared (and tested) earlier. Filter-sterilize through a 0.45 μm nitrocellulose filter (Nalgene). Store in aliquots at -20°C (can be frozen and thawed repeatedly).

All plasmids should be CsCl-banded or of equivalent purity, diluted to a concentration of 1.0 mg/mL, and stored at 4°C. For initial experiments, we recommend that each transfection be performed with a total of 22 μg of DNA (as shown in Tables 1 and 2). However, the optimal concentration of DNA may vary with different cell types (2). The transfection procedure is as follows:

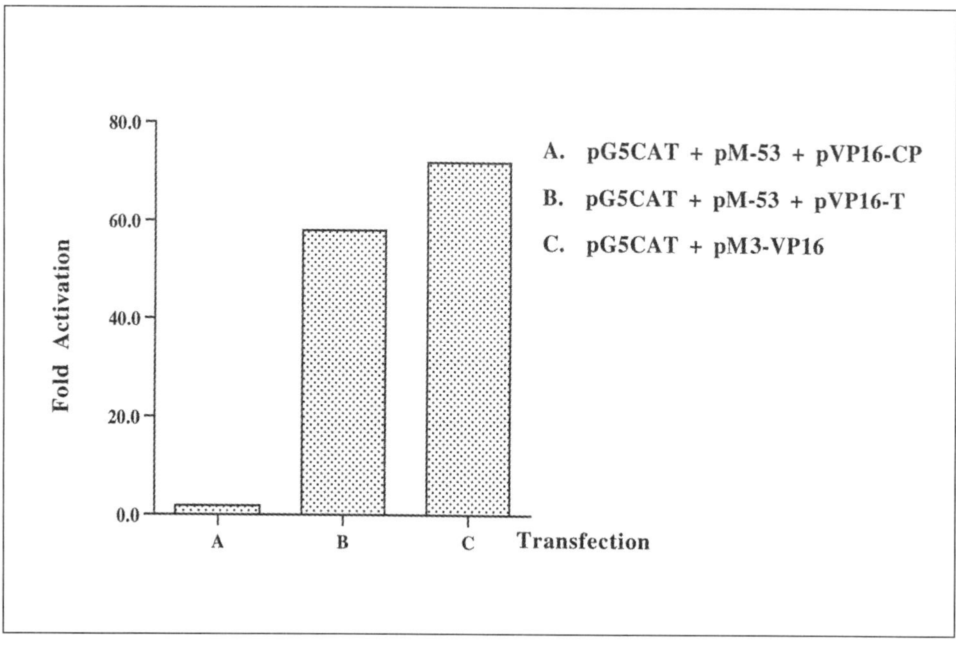

Figure 3. Demonstration of the interaction between p53 and the SV40T antigen in HeLa cells. HeLa cells, grown to 30% confluency, were transfected by the calcium phosphate method with the indicated plasmids. Each transfection used 22 μg of DNA, as detailed in Table 1. CAT assays were done using a CAT ELISA kit. The numbers shown are fold activation of *cat* gene expression.

Procedure

1. For each transfection, seed 5×10^5 exponentially growing cells in a 10-cm tissue-culture plate in 10 mL of complete medium (with serum). This should be done the day prior to the transfection so that there will be $<10^6$ cells/plate (approximately 30% confluency) just prior to transfection. This provides enough surface area on the plate for at least two more doublings.

2. Three hours prior to transfection, re-feed the cultures with fresh, warm medium. Be sure that the pH of the medium is correct to ensure high-efficiency transformation.

3. For each transfection, prepare: 22 µg DNA total per 10-cm diameter dish (see Tables 1 and 2 for details); sterile deionized water to a total volume of 450 µL; 50 µL 2.5 M $CaCl_2$; total volume: 500 µL. Vortex-mix thoroughly.

4. Add 500 µL of 2× BBS, mix well, and incubate at room temperature for 10–20 min to allow a precipitate to form.

5. Add the calcium-phosphate/DNA solution dropwise to the medium while swirling the plate.

6. Incubate overnight in a 37°C, 3% CO_2 incubator.

   Note: The level of carbon dioxide is critical. We recommend using a Fyrite gas analyzer to measure percent CO_2 prior to incubation.

7. Carefully wash the cells twice with 5 mL of PBS, then add 10 mL of complete medium (with serum).

8. Incubate the cells for an additional 24–48 h.

2.4 Protocol for CAT Assay

There are many alternatives for performing CAT assays, and the reagents required will depend on which method is chosen. At CLONTECH, we generally measure CAT protein directly by using the CAT ELISA Kit from Roche Molecular Products (Indianapolis, IN, USA). CAT activity can also be measured nonisotopically with the FAST CAT® kit from Molecular Probes (Eugene, OR, USA). The method below is for the standard, ^{14}C-chloramphenicol based assay, similar to that described by Dang et al. (7) in their original description of the mammalian two-hybrid assay. A tritium-based assay has also been described (2).

Materials and Reagents

The basic ^{14}C-chloramphenicol based CAT assay

- Tris-EDTA, sodium chloride buffer (TEN) (40 mM Tris-HCl pH 7.5, 1 mM EDTA pH 8.0, 150 mM NaCl)
- 1.0 M Tris-HCl, pH 7.8
    Dissolve 12.1 g Trizma base in 80 mL deionized water. Titrate to pH 7.8 with 12 N and then 1 N HCl. Adjust volume to 100 mL and filter-sterilize.
- 0.25 M Tris-HCl, pH 7.8

Dilute 25 mL of 1.0 M Tris-HCl pH 7.8 to 0.25 M with 75 mL H_2O. Filter-sterilize and store at 4°C.

- 200 uCi/mL [^{14}C]chloramphenicol (35–55 mCi/mmol)
- 4 mM acetyl CoA

  Dissolve 25 mg of acetyl CoA (Lithium Salt, Sigma Chemical's Catalog No. C3019) in 7.7 mL H_2O. Store at -70°C in small aliquots, which are good for six months. Thaw and use aliquots once only.

Procedure

1. Wash cells twice with ice-cold PBS, then aspirate all the PBS and add 1 mL of ice-cold TEN solution to each dish. Place dish on ice.

2. Scrape the cells off of the dish with a cell scraper and transfer to a microcentrifuge tube.

3. Centrifuge at maximum speed in a benchtop microcentrifuge for 15 s.

4. Resuspend the cell pellet in 100 µL ice-cold 0.25 M Tris-HCl, pH 7.8.

5. Extract CAT by freezing in liquid nitrogen and then thawing in a 37°C water bath three times.

6. Microcentrifuge the lysate for 5 min at 4°C. Remove and save the cell extract.

7. To assay 20 µL of the cell extract, make the following cocktail per reaction: 2 µL 200 uCi/mL [^{14}C]chloramphenicol (35–55 mCi/mmol); 20 µL 4.0 mM acetyl CoA; 32.5 µL 1.0 M Tris-HCl pH 7.8; 75.5 µL H_2O.

8. Mix 130 µL of cocktail with 20 µL extract in a microcentrifuge tube and incubate for 1 h at 37°C. (1–50 µL of extract can be assayed. However, the volume of 1.0 M Tris-HCl, pH 7.8 and H_2O should be adjusted accordingly so that the final concentration of Tris-HCl remains 0.25 M.)

9. Add 1 mL ice-cold ethyl acetate to the reaction and vortex-mix well for 30 s.

10. Microcentrifuge for 1 min and transfer the top (ethyl acetate) layer to a new tube. It is best to do this in a fume hood, since ethyl acetate vapor is toxic.

11. Dry the ethyl acetate solution to complete dryness in a SpeedVac evaporator (Savant Instruments, Holbrook, NY, USA).

12. Resuspend each sample in 30 µL ice-cold ethyl acetate.

13. Spot half of each sample 5 µL at a time to keep the area of the spot to a minimum, onto a TLC plate (Kodak). Spots should be placed 2 cm from the base of the sheet and 1.5 cm apart. This can be achieved by first marking the locations for the spots using a ruler and pencil.

14. Develop the thin-layer chromatograph (TLC) by placing it in a glass chromatography tank that has been equilibrated for 2 h with 200 mL of a 19:1 chloroform–methanol mixture. The sheet should be placed such that the liquid can move up through the sample spots.

15. Allow the chromatography to run until the liquid phase has moved 8.5 cm up the plate.

16. Expose the TLC plate to X-ray film or, preferably, to a phosphorimager screen

and obtain densitometric scans of the signal, so as to calculate percentage conversion of chloramphenicol to acetyl chloramphenicol.

Reproducible, qualitative results can be rapidly obtained using single transfections. For quantitative data, duplicate or triplicate transfections should be done. For greatest accuracy, dish to dish variations in transfection efficiency can be normalized by co-transfecting experimental plasmids with a constant amount of a second reporter under the control of a constitutive promoter. The values obtained in each sample for the primary reporter (i.e., CAT) are then normalized to the values obtained for the second reporter in the same sample. Useful vectors for this purpose are CLONTECH's pβgal-Control (Catalog No. 6047-1), or pSEAP-Control (Catalog No. 6043-1).

2.5 Additional Experiments

2.5.1 Verification of Hybrid Protein Expression (optional)

The expression of the GAL4 DNA-BD/Protein X and GAL4-AD/Protein Y fusion proteins can be verified by Western blotting analysis using soluble protein extracts from the cells. Probe the blots with CLONTECH's GAL4 DNA-BD specific or AD-specific monoclonal antibodies (Catalog No. 5399-1 and 5398-1, respectively) using standard Western blotting procedures (12).

2.5.2 Mapping Protein Structure

Once an interaction between two proteins has been detected, the Mammalian Two-Hybrid Assay provides a powerful tool to functionally dissect the specificity of the interaction using deletional or site-directed mutagenesis. CLONTECH's Transformer™ Site-Directed Mutagenesis Kit (Catalog No. K600-1) is useful for generating mutant cDNAs that can then be tested in the Mammalian Two-Hybrid Assay.

2.5.3 Confirmation in Yeast MATCHMAKER Two-Hybrid System

Interactions that are detected in the Mammalian Two-Hybrid Assay may also be detectable in a yeast-based two-hybrid assay such as CLONTECH's MATCHMAKER Two-Hybrid System 3 (Catalog No. K1612-1). Much less DNA is required for each two-hybrid experiment in yeast. Furthermore, yeast two-hybrid assays are much less expensive because they do not require tissue-culture facilities and supplies for the growth and transfection of mammalian cells. For more information on yeast two-hybrid technology, see Chapters 3, 4, and 5 elsewhere in this volume.

3. DISCUSSION

The mammalian two-hybrid assay as described has some important advantages over the yeast-based assay. Most significant being that it allows for post-translational modifications that are not replicable in yeast, thus enabling the detection of interactions dependent on such modifications. For this reason, it can detect interactions that only occur in response to a given stimulus, such as the presence of specific external growth factors or interactions that occur only in specific cell types.

The assay also provides a more rapid readout of quantitative results than the yeast assay, taking only 2–3 days as compared to over a week for the yeast assay. The most significant disadvantage is the inability to screen libraries. Thus, the basic mammalian two-hybrid assay can only be used for testing the interaction between two defined peptides. The interaction of these peptides may have been suggested by other experimental interactions, e.g., yeast two-hybrid assays or immunoprecipitation. Alternatively, the interaction may be suspected based on genetic or theoretical considerations, such as homology to other proteins with known binding specificities.

The inability to screen libraries is a significant disadvantage, due to the fact that the power of the yeast assay lies in its ability to identify novel interactions. As with the yeast assay, the use of transcriptional activation as the reporter can present a problem when intrinsic transactivation activity is present in the bait. Additionally, interactions between proteins that ordinarily reside outside the nucleus, in particular membrane proteins, may go undetected because the assay requires the hybrid proteins to translocate to the nucleus. The issue of false positives has not been as carefully studied with this assay as in the yeast system. Some authors claim that there are less false-positives in the mammalian assay (33). However, since the assay has mostly been used to test interactions between two specific proteins, rather than to identify novel interactions out of a library, this issue has not been fully explored and any conclusion is premature. Based on experience with the yeast assay, one might even expect that there is more potential for false-positives with the mammalian system because only one reporter is assayed. In contrast, the more advanced yeast assays (e.g., CLONTECH's MATCHMAKER System 3) utilize multiple reporters, each having a different minimal promoter linked to the GAL4 binding sites. This reduces the number of false-positives because oftentimes these are due to interactions of the library proteins with DNA sequences other than the GAL4 sites present in the promoter for the reporter gene.

3.1 Novel Mammalian Two-Hybrid Methods

The problems described have stimulated efforts to improve the mammalian assay. In particular, methods to enable library screening and detection of protein interactions without the use of transcription-based reporters have been sought. These efforts are discussed below, together with perspectives for future developments.

3.1.1 Karyoplasmic Interaction Selection Strategy (KISS)

The KISS two-hybrid method is a direct extension of the standard mammalian two-hybrid assay described and was developed by Dang, Fearon, and colleagues, who originally developed the basic mammalian two-hybrid assay described in this chapter (7,8). In the KISS method, the CAT reporter is replaced by a selectable marker. This marker can be a surface antigen (e.g., CD4) that permits enrichment of positive clones by flow cytometry (FACS analysis). Alternatively, an antibiotic resistance gene (e.g., hygromycin resistance) can be used to enable growth selection for interacting clones. Using the interaction between the leucine zipper domains of c-Fos and c-Jun as an example, these researchers were able to show that interactions could be detected using CD4 as a reporter (8). However, when the amount of the c-Jun-VP16 fusion was titrated—to simulate detection of a rare clone from a library—the assay was incapable of detecting a signal when the c-Jun plasmid represented less than a

100th of the mixed population. It is unlikely that such an assay would be able to pick up rare cDNAs from a traditional library. Moreover, the sensitivity of the assay is low. Transfected clones showed a broad range of reporter signal intensity from just above background to 100-fold above background. Significantly, the majority of clones showed only a tenfold increase in signal over background using a GAL4/VP16 fusion protein as a positive control. When the c-Fos/c-Jun interaction was tested, the average fold increase was only about fivefold. Weak interactions may thus be hard to detect, and this would be compounded in the case of library screening. In addition to the CD4 reporter, the authors used hygromycin resistance as a reporter. Thus, it permits selective growth, in media containing the antibiotic hygromycin, of clones transfected with plasmids that encode interacting proteins. Using this reporter, the authors were able to demonstrate the interaction of several known protein partners (Fos/Jun, CAPZ α and β, and Ankyrin and Spectrin). They estimated that the frequency of spontaneous hygromycin resistance was <1 in 5×10^6. The authors argued that this should permit library screening without generating a significant background problem from spontaneously surviving clones (8). However, the efficacy of this method for library screening was not demonstrated. Indeed, in an attempt to identify novel c-Myc interacting proteins using this reporter, Fearon et al. (8) transfected each of 30 75-cm^2 dishes of cells with 1 μg of the c-Myc bait and 1 μg of a library. Only one hygromycin-resistant clone was obtained, and this appeared to contain a hybrid protein that was interacting with the GAL4 DNA-BD rather than with c-Myc (8). As mentioned earlier, this failure may not be surprising when it is considered that for growth assays to be successful, rare cDNAs encoding interacting proteins not only need to be transfected into the cell but also need to stably integrate and be properly expressed in the cell. The probability for this in mamalian cells using standard transfection techniques is exceedingly low.

3.1.2 The Contingent Replication Assay

In the contingent replication assay, interaction between the prey and the bait results in replication of the prey construct in the mammalian cell host (32). This renders the DNA that encodes the prey insensitive to digestion with a methylation-specific restriction enzyme (*Dpn*I), which follows isolation of plasmid DNA from the transfected cells. Any such undigested plasmids can then be rescued by transformation into *Escherichia coli* (32). Replication of the prey construct is dependent on expression of the reporter, in this case SV40 T-antigen. T-antigen is a sequence-specific helicase that binds to sequences in the SV40 origin, thus causing unwinding of the two DNA strands (4). This unwinding enables host polymerases to initiate replication of the DNA at the site of unwinding (4). As with the assays described previously, expression of the T-antigen reporter is controlled by a minimal promoter linked to multiple-binding sites for GAL4. High-level expression of T-antigen only occurs when bait and prey interact to bring the AD into proximity of the promoter. Only the library plasmids can be replicated in the presence of T antigen because only they encode the SV40 origin of replication. The reporter plasmid, which expresses T-antigen, and the bait plasmid are not replicated because they do not contain the SV40 origin sequence. The use of SV40-based replication as a reporter necessitates the use of mammalian cells that are permissive to SV40 replication, such as HeLa and CV1, which may be a disadvantage for some systems where cell-type specific interactions are sought. Selection for the replicated library plasmids is possible by virtue of the fact that mammalian cells, unlike DAM$^+$ bacteria, do not methylate adenine residues in DNA. Thus, when the target plasmid is replicated, the newly synthesized copies will lack methylation and will be insensitive to digestion with the restriction enzyme *Dpn*I, which only recognizes the sequence G^{Me}ATC. Any undigested plasmids are then rescued by transformation back into *E. coli*. Because the reporter and bait constructs are not replicated, they will be digested and thus incapable of transforming an *E. coli* host. Likewise, any plasmid encoding a prey that does not interact with the bait will also be digested since it will fail to activate T-antigen expression and will also remain unreplicated. The contingent replication assay thus provides a way to enrich for interacting clones. In test assays using known interacting proteins, it was found that the assay could significantly enrich clones present at frequencies between 1/1000–1/10^6. When the interacting partner was present at a frequency greater than 1/1000, however, there seemed to be no enrichment (23). It was suggested that this is a consequence of each cell taking up many individual plasmids during the transfection. When the gene of interest is present at high frequency, most cells are likely to contain it as well as containing plasmids that encode non-interacting proteins; all of which will be replicated by virtue of being in the same cell as the plasmid that encodes the interacting partner. It may therefore be useful to do this assay at two dilutions so that low-abundance and high-abundance messages can be detected. Analysis of the *E. coli* transformants suggested that 1/10–1/40 of the colonies obtained contained clones of interest (23). Again, the reason enrichment is not 100% is likely due to the fact that more than one prey construct will enter any given cell during the transfection procedure. These will have the same access to T-antigen as the construct whose encoded protein interacts with the bait and will be replicated along with it.

The authors described two approaches that could be used to identify the true positives (23). First, clones can be pooled or assayed individually for CAT activity using the standard mammalian two-hybrid assay. This is likely to be laborious and is compounded by problems of low sensitivity if pools of clones are used, because this will reduce the effective concentration of the interacting clone. Alternatively, a colony hybridization can be done using total DNA from the *Dpn*I digestion reaction as a probe, and excess unlabeled input library DNA can be used to quench signal from unenriched clones. The basis for this approach is that positive clones are likely to have a greater representation in the *Dpn*I-digested DNA than in the original input library and will generate a probe of sufficient activity to be detected. In contrast, signal from unenriched clones will be quenched by the unlabeled input DNA. A problem with this approach is that it is also likely to detect clones of non-interacting proteins that are rescued by virtue of the fact that they are abundant in the library. For the reasons stated previously, such clones are likely to be transfected into a cell containing an interacting clone and will be replicated along with it. Alternatively, some clones may go undetected if their enrichment is slight.

Using the contingent replication assay (CRA), Nallur et al. (23) were able to identify RAP74 [a known partner of RAP30, the small subunit of the general transcription factor TFIIF (23)] from a normalized library using RAP30 as a bait. Three additional clones, one being RAP30 itself, were also identified. However, it is not yet known if these are true interacting proteins for RAP30. Thus the CRA assay can be used to identify interactions from a library. More examples of uses of this assay, which look at a range of different baits that interact with known proteins with varying affinities, will better show the general applicability of this method to the identification of novel interactions in mammalian cells.

3.1.3 Protein Fragment Complementation Assays

In protein fragment complementation assays, bait and prey proteins are expressed as hybrids fused to different fragments of a reporter gene. When bait and prey interact, they bring the reporter fragments into close proximity so that they can refold into a functional enzyme and their activity can then be detected. As such, these assays bear some similarity with the split ubiquitin system described in yeast, whereby interaction of bait and prey results in the formation of a functional ubiquitin moiety from two inactive halves (29; also see Chapter 7) and also bear similarity to a recent bacterial two-hybrid system based on trans-complementation by two fragments of *Bordetella pertussis* adenylate cyclase (15).

The protein fragment complementation approach has advantages over transcription-based assays in that protein interactions can occur in any cellular compartment—not only in the nucleus—as required by the traditional transcriptional reporter assays. Two forms of this assay have been developed, one by Michnick and colleagues in Montreal (24) and one by Blau's group at Stanford (22,25). Blau's group adapted the well-known phenomenon from bacterial genetics of α and ω complementation of β-galactosidase (LacZ), whereby N-terminal α fragments of LacZ can complement C-terminal ω fragments in *trans* to generate functional LacZ. (It is this complementation that lies at the heart of blue/white colony selection in vectors such as pBluescript from Stratagene). Blau and colleagues found that α and ω complementation could also occur in mammalian cells. Furthermore, extended α and ω fragments could be generated that had such weak affinity for each other that they did not spontaneously reassociate to give detectable levels of active enzyme (22). This raised the possibility that interactions between proteins of interest could be tested by fusing them to the non-self-associating forms of LacZ. If the proteins of interest interacted, they would cause the LacZ fragments to come into close proximity, increasing their effective local concentration and driving them to associate to form active enzyme. The method developed by Michnick and colleagues is similar, but uses fragments of either dihydrofolate reductase (DHFR) or glycinamide ribo nucleotide formyl transferase (GAR) as the reporter (24). LacZ can be detected colorimetrically, using either standard X-Gal or a fluorescent substrate (fluor-X-Gal), developed by Blau and co-workers (22). The fluorescent substrate is of particular utility because it permits detection of the interaction in live cells, which can then be isolated by FACS. DHFR can be detected using either a fluorescent substrate or survival in the presence of drugs that block endogenous DHFR activity but not that of the reporter. Using the selective growth assay for the DHFR reporter, Michnick was able to detect ras/raf interactions. Tests of the ability of this system to identify interacting clones from a library are currently ongoing (24).

There are several advantages to these methods. First, they allow detection of protein–protein interactions in their native cellular location. Moreover, by using fluorescent substrates, the location of the interaction can be visualized in live cells and the kinetics of the interaction, especially when the interaction is induced in response to an external stimulus, can be followed. This may be of considerable use in following signaling pathway activities. Neither the standard mammalian assay nor the yeast two-hybrid screen can address this important question. However, as with the other methods, if library screening is to become more readily feasible, improved methods for transferring the library plasmids into the cells are required to improve the efficiency of these systems.

3.1.4 Fluorescence Resonance Energy Transfer

Another extension of proximity-based assays to detect protein–protein interactions makes use of the phenomenon of fluorescence resonance energy transfer (FRET). In FRET, light energy emitted from a high-energy fluorophore causes excitation of a second, lower energy fluorophore when the two are close together in space (<40 Å). This results in a shift in the emission spectrum from that expected of the high-energy fluorophore to that of the lower one. The variant forms of green fluorescent protein (e.g., GFP, BFP, EGFP, YFP, CFP) provide a perfect system for such FRET-based assays (20). For example, when BFP (excitation 389 nm, emission 445 nm) is brought into proximity of EGFP (excitation 489, emission 511 nm), excitation of BFP results in a shift in the emission spectrum expected to that for EGFP (19). Using FRET, Mahajan et al. were able to show the interaction of *Bcl-2* and Bax in mitochondria using appropriate fusions of these proteins to BFP and EGFP (19). As with the protein fragment complementation assays, FRET-based assays have the advantage of being able to follow interactions taking place anywhere in the cell as well as being able to follow the kinetics of an interaction. The main question, however, is the level of sensitivity of the system. In the Bcl-2/Bax example cited, there was only a five to tenfold increase in signal associated with the interaction (19). There is currently insufficient data to determine the limit of sensitivity of this kind of assay with respect to the affinity of the interacting proteins. Use of other GFP variant pairs, such as YFP and CFP, as used by Tsien and colleagues in their Chameleons Ca^{2+} indicator assay, may give better signal to noise profiles (21). Additionally, the conformation of the interacting proteins may, in some cases, preclude productive FRET, resulting in false-negative responses. This is likely to occur if the conformations are such that the GFPs are too far apart in space.

3.1.5 Chemical Inducers of Dimerization

The creation of small molecule dimerizers such as FK1012, pioneered by Crabtree and Schreiber, is also having an impact on the design of two-hybrid assays (6,10). FK1012 is a synthetic dimer of FK506, an organic compound synthesized in certain types of soil microorganisms. FK506 binds to the immunophilin FKBP, which creates an inhibitory complex that blocks signal transduction by inhibiting calcineurin (6,26). In this way, FK506 and related compounds, such as cyclosporin A and rapamycin, act as potent immunosuppressants that have revolutionized transplant surgery (3). Since FK1012 is a dimer of FK506, it is able to bind two FKBP moieties. Crabtree and Schreiber (6) have utilized this system to cause association of specific proteins in response to treatment of cells with FK1012 by creating translational fusions of the proteins of interest and FKBP. For example, they were able to show that FK1012 could cause dimerization of cytoplasmic fragments of cell surface receptors fused to FKBP. In agreement with earlier studies that used antibodies to cell surface receptors to promote dimerization (31), FK1012-induced dimerization resulted in receptor activation and signal transduction that was independent of the receptor's natural ligand (28). By analogy, two-hybrid systems can be envisioned in which bait and prey hybrid proteins are fused to a given receptor's cytoplasmic domain. In this case, interaction of the bait and prey would result in receptor fragment dimerization, inducing signaling that could be detected by activation of a downstream reporter. A related technology is the SRS (SOS recruitment system) two-hybrid screen developed in

Karin's lab (1; also see Chapter 8). In this yeast-based two-hybrid assay, the interaction of a membrane-targeted bait protein with a hybrid protein encoding the prey fused to the signaling molecule, SOS, activates the ras signaling pathway by bringing SOS to the plasma membrane where it can stimulate ras activation. Crabtree and Schreiber have also induced transcriptional activation using small molecule dimerizers to cause the association of the GAL4-DNA-BD and the VP16 AD (13). Induced nuclear protein export has also been demonstrated using this system (17).

One additional potential use of the small molecule dimerization technology is in screening prospective compounds rapidly for functional binding to a protein of interest in a two-hybrid type scenario. In this scheme, the compound of interest would be covalently linked to a known compound such as FK506. The bait would be a Gal4-DNA-BD/FKBP hybrid and the prey could either be a specific protein or library of test proteins fused to the VP16 transactivation domain. Interaction of the compound being tested with the VP16 fusion protein would thus result in transcriptional activation. The identification of small molecules that can be used either to block protein function or to promote it via dimerization provides considerable potential for the development of new drugs and gene therapies. The use of two-hybrid approaches to identify such compounds as described is going to be of considerable importance.

4. SUMMARY AND PERSPECTIVE

The previous discussion has considered several systems that can enhance the functionality of the basic mammalian two-hybrid assay. The ability to follow the kinetics of an interaction and to determine where in the cell that interaction occurs is a clear advantage of some of these methods over the basic mammalian-based assay, as well as over yeast-based approaches.

However, convenient library screening remains an elusive goal. The most successful strategy so far is the contingent replication assay, which was able to identify a known interacting partner for RAP30 as well as several novel interacting clones from a real cDNA library. However, this technique is still rather cumbersome and, like the other methods described, has yet to come into widespread use. The main problem with library screening is the poor efficiency by which DNA can be introduced into mammalian cells by standard transfection methods. Recently, a number of researchers have been turning to retroviral-based vectors to introduce expression cassettes into mammalian cells. Retroviral-mediated gene delivery offers considerably higher efficiency transfer of genes of interest into mammalian cells than do traditional transfection methodologies.

This is clearly the next step that mammalian two-hybrid assays should take if they are to deliver libraries to a high percentage of cells in single copy. The use of regulated expression systems, such as the tetracycline-regulated expression system, to control the expression of the bait and/or the prey is also a key area for future improvement. Such regulation affords two advantages. First, toxicity due to expression of either bait or prey can be eliminated. Second, regulated expression of the bait would allow simplified detection of false-positives that provide positive signal in the absence of the bait. Improvements in the selection and screening of clones will also be required if positive clones are to be identified rapidly and with a high degree of sensitivity. In particular, brighter and more red-shifted fluorescent proteins are likely to be sought after because these have better signal to noise characteristics than GFP.

Pressure from both the pharmaceutical market for inexpensive and rapid assays to identify potential leads as well as the completion of the human genome project will undoubtedly continue to drive further improvements such as these in the mammalian two-hybrid assay. It can be expected that this will continue to be an exciting and dynamic area of research in the future.

REFERENCES

1.**Aronheim, A., E. Zandi, H. Hennemann, S.J. Elledge, and M. Karin.** 1997. Isolation of an AP-1 repressor by a novel method for detecting protein-protein interactions. Mol. Cell. Biol. *17*:3094-3102.

2.**Ausubel, F.M., R. Brent, R.E. Kingston, D.M. Moore, J.G. Seidman, J.A. Smith, and K. Struhl (Eds.).** 1994. Current Protocols in Molecular Biology. John Wiley & Sons, Inc., New York, NY.

3.**Bierer, B.E., G. Hollander, D. Fruman, and S.J. Burakoff.** 1993. Cyclosporin A and FK506: molecular mechanisms of immunosuppression and probes for transplantation biology. Curr. Opin. Immunol. *5*:763-773.

4.**Borowiec, J.A.** 1996. DNA helicases, p. 545-574. *In* M.L. DePamphilis (Ed.), DNA Replication in Eukaryotic Cells. CSHL Press, Cold Spring Harbor, NY.

5.**Chien, C.T., P.L. Bartel, R. Sternglanz, and S. Fields.** 1991. The two-hybrid system: a method to identify and clone genes for proteins that interact with a protein of interest. Proc. Natl. Acad. Sci. USA *88*:9578-9582.

6.**Crabtree, G.R. and S.L. Schreiber.** 1996. Three-part inventions: intracellular signaling and induced proximity. Trends Biochem. Sci. *21*:418-422.

7.**Dang, C.V., J. Barrett, M. Villa-Garcia, L.M.S. Resar, G.J. Kato, and E.R. Fearon.** 1991. Intracellular leucine zipper interactions suggest c-myc hetero-oligomerization. Mol. Cell. Biol. *11*:954-962.

8.**Fearon, E.R., T. Finkel, M.L. Gillison, S.P. Kennedy, J.F. Casella, G.F. Tomaselli, J.S. Morrow, and C. V. Dang.** 1992. Karyoplasmic interaction selection strategy: a general strategy to detect protein–protein interaction in mammalian cells. Proc. Natl. Acad. Sci. USA *89*:7958-7962.

9.**Fields, S. and O.-K. Song.** 1989. A novel genetic system to detect protein-protein interactions. Nature *340*:245-246.

10.**Fields, S. and R. Sternglanz.** 1994. The two-hybrid system: an assay for protein-protein interactions. Trends Genet. *10*:286-292.

11.**Freshney, R.I.** 1993. Culture of Animal Cells, 3rd Ed. Wiley-Liss, New York, NY.

12.**Harlow, E. and E. Lane.** 1988. Antibodies: A Laboratory Manual. CSHL Press, Cold Spring Harbor, NY.

13.**Ho, S.N., S.R. Biggar, D.M. Spencer, S.L. Schreiber, and G.R. Crabtree.** 1996. Dimeric ligands define a role for transcriptional activation domains in reinitiation. Nature *382*:822-826.

14.**Hope, I.A. and K. Struhl.** 1986. Functional dissection of a eukaryotic transcriptional activator protein, GCN4 of yeast. Cell *46*:885-894.

15.**Karimova, G., J. Pidoux, A. Ullmann, and D. Ladant.** 1998. A bacterial two-hybrid system based on a reconstituted signal transduction pathway. Proc. Natl. Acad. Sci. USA *95*:5752-5756.

16.**Keegan, L., G. Gill, and M. Ptashne.** 1986. Separation of DNA binding from the transcription-activating function of a eukaryotic regulatory protein. Science *231*:699-704.

17.**Klemm, J.D., C.R. Beals, and G.R. Crabtree.** 1997. Rapid targeting of nuclear proteins to the cytoplasm. Curr. Biol. *7*:638-644.

18.**Luo, Y., A. Batalao, H. Zhou, and L. Zhu.** 1997. Mammalian two-hybrid system: a complimentary approach to the yeast two-hybrid system. BioTechniques *22*:350-352.

19.**Mahajan, N.P., K. Linder, G. Berry, G.W. Gordon, R. Heim, and B. Herman.** 1988. Bcl-2 and Bax interactions in mitochondria probed with green fluorescent protein and fluorescence resonance energy transfer. Nature Biotechnol. *16*:547-552.

20.**Mitra, R.D., C.M. Silva, and D.C. Youvan.** 1996. Fluorescence resonance energy transfer between blue-emitting and red-shifted excitation derivatives of the green fluorescent protein. Gene *173*(1):13-17.

21.**Miyawaki, A., J. Llopis, R. Heim, J.M. McCaffery, J.A. Adams, M. Ikura, and R.Y. Tsien.** 1997. Fluorescent indicators for Ca^{2+} based on green fluorescent proteins and calmodulin. Nature *388*:882-887.

22.**Mohler, W.A. and H.M. Blau.** 1996. Gene expression and cell fusion analysis by lacZ complementation in mammalian cells. Proc. Natl. Acad. Sci. USA *93*:12423-12427.

23.**Nallur, G.N., H.A. Vasavada, P.R. Sankhavaram, W.J. Xu, and S.M. Weissman.** 1993. Evaluation of the contingent replication assay (CRA) and its application to the study of the general transcription initiation factor, TFIIF. Nucleic Acids Res. *21*:3867-3873.

24.**Pelletier, J.N., F.-X. Campbell-Valois, and S.W. Michnick.** 1998. Oligomerization domain-directed reassembly of active dihydrofolate reductase from rationally designed fragments. Proc. Natl. Acad. Sci. USA *95*:12141-12146.

25. **Rossi, F., C.A. Charlton, and H.M. Blau.** 1997. Monitoring protein–protein interactions in intact eukaryotic cells by beta-galactosidase complementation. Proc. Natl. Acad. Sci. USA *94*:8405-8410.
26. **Schreiber, S.L. and G.R. Crabtree.** 1992. The mechanism of action of cyclosporin A and FK506. Immunol. Today *13*:136-142.
27. **Silver, P.A., L.P. Keegan, and M. Ptashne.** 1984. Amino terminus of the yeast GAL4 gene product is sufficient for nuclear localization. Proc. Natl. Acad. Sci. USA *81*:5951-5955.
28. **Spencer, D.M., T.J. Wandless, S.L. Schreiber, and G.R. Crabtree.** 1993. Controlling signal transduction with synthetic ligands. Science *262*:1019-1024.
29. **Stagljar, I., C. Korostensky, N. Johnsson, and S. Te Heesen.** 1998. A genetic system based on split-ubiquitin for the analysis of interactions between membrane proteins *in vivo*. Proc. Natl. Acad. Sci. USA *95*:5187-5192.
30. **Takacs, A.M., T. Das, and A.K. Banerjee.** 1993. Mapping of the interacting domains between the nucleocapsid protein and the phosphoprotein of vesicular stomatitis virus by using a two-hybrid system. Proc. Natl. Acad. Sci. USA *90*:10375-10379.
31. **Ullrich, A. and J. Schlessinger.** 1990. Signal transduction by receptors with tyrosine kinase activity. Cell *61*:203-212.
32. **Vasavada, H.A., S. Ganguly, F.J. Germino, Z.X. Wang, and S.M. Weissman.** 1991. A contingent replication assay for the detection of protein-protein interactions in animal cells. Proc. Natl. Acad. Sci. USA *88*:10686-10690.
33. **Wu, L.C., Z.W. Wang, J.T. Tsan, M.A. Spillman, A. Phung, X.L. Xu, M.C. Yang, L.Y. Hwang, A.M. Bowcock, and R. Baer.** 1996. Identification of a RING protein that can interact *in vivo* with the BRCA1 gene product. Nat. Genet. *14*:430-440.

16

A Bacterial Two-Hybrid System Based on a cAMP Signaling Cascade in *Escherichia coli*

Gouzel Karimova and Daniel Ladant
Unité de Biochimie Cellulaire, Institut Pasteur, Paris, France

1. INTRODUCTION

Most biological processes involve specific protein–protein interactions. The progress in genomic sequencing has lead to the identification of a great number of putative proteins whose functions are not known (7). There is an increasing need to develop general methodologies that enable large-scale in vivo characterization of interacting proteins.

At present, the yeast two-hybrid system (8) represents the most powerful in vivo approach to screen for polypeptides that could bind to a given target protein. Bacterial equivalents to the yeast two-hybrid system have not been developed until very recently. Indeed, this technology might be one of the rare examples in which a genetic assay was developed earlier in eukaryotic cells than in *Escherichia coli*. Instead, powerful in vitro methodologies using prokaryotes have been engineered to study molecular interactions. Phage display (32), bacterial display (9), or more recently ribosome display assays (11) are currently largely utilized to screen protein–ligand interactions that are detectable under in vitro conditions.

In this chapter, we describe a novel bacterial two-hybrid system that allows an easy in vivo screening and selection of functional interactions between two proteins (17). This system, due to its sensitivity and simplicity, could have a broad application in the studies of structure/function relationships in biological macromolecules, in the functional analysis of genomes, and in high throughput screening of interacting ligands and new therapeutic agents. This system employs an *E. coli* adenylate cyclase-deficient strain (*cya*). It is based on the reconstitution of a signal transduction pathway coupled to the production of a regulatory molecule, cAMP, that triggers the transcriptional activation of catabolic operons (or specific reporter genes), thus giving rise to a selectable phenotype.

Before going into more detail about this system, we summarize two other genetic assays that can be used for the in vivo analysis of protein–protein interactions in *E. coli*.

Yeast Hybrid Technologies
Edited by L. Zhu and G.J. Hannon
© 2000 Eaton Publishing, Natick, MA

1.1 Hybrid Bacterial Repressors as Reporters of Interaction

These methods exploit the specific structural and functional features of some prokaryotic transcriptional repressors such as λcI, the repressor of bacteriophage λ, or the *E. coli* LexA repressor of the SOS system (Figure 1A):
- these proteins act as transcriptional repressors only in a dimeric form;
- they exhibit a modular structure consisting of an N-terminal DNA-binding domain (B) and a C-terminal dimerization domain (D).

Hence, the DNA-binding domain alone, which lacks dimerization capability, is unable to repress transcription. However, fusion of this domain to a polypeptide with dimerizing ability generates a functional chimeric repressor (Figure 1A). The repressor activity can be assayed by the expression of synthetic reporter genes (usually *lacZ* -β-galactosidase, under the control of the appropriate promoters) or, in the case of λcI, by determining the sensitivity to λ*vir*.

Hu and collaborators (13) first demonstrated the feasibility of this approach using λcI fusions in a study of leucine zippers. They were able to identify residues important for dimerization within the heptad repeat of a leucine zipper (13). Other groups then used a similar approach to investigate dimerization of various proteins — cyclic AMP receptor protein (16), the Rop protein (3) — or interactions between two different proteins or polypeptides such as c-Myc with Max (22) and cAMP-dependent protein kinase catalytic subunit with a peptide inhibitor (14). Using a similar approach, Bunker and Kingston performed a library screen to identify a protein that can disrupt the dimerization of a λcI -myc fusion (2).

This chimeric λ repressor assay, however, exhibits two potential limitations: *(i)* The N-terminal DNA-binding domain has an intrinsically low dimerization ability and, as a consequence, a basal repression activity exists even in the absence of interaction between the fused polypeptides; *(ii)* This approach cannot be applied to any proteins that have an intrinsic homodimerization capability. In this case, the homodimerization would obscure the heterodimerization.

These limitations have been overcome through the use of chimeric LexA repressors pioneered by Granger-Schnarr and collaborators (31). First, the LexA repressor has the advantage that its DNA-binding domain has no intrinsic dimerization ability. Second, Dmitrova et al. (4) recently developed a hybrid *lexA* operator that is repressed only upon the hetero-association of two interacting proteins fused, respectively, to a wild-type and a mutant LexA DNA-binding domain.

1.2 Hybrid Activators: Transcriptional Activation by Recruitment of RNA Polymerase

This approach, recently developed by Hochschild and collaborators, is very similar to the yeast two-hybrid system (Figure 1B). It is based on the observation that in *E. coli*, the transcription could be markedly increased through the recruitment of RNA polymerase (RNAP) at specific promoters by means of arbitrary protein–protein contacts (6).

For some *E. coli* promoters, the activation of transcription by given activators involves specific interactions between the DNA-bound activator and the C-terminal domain of the α subunit of RNA polymerase. Dove et al. (6) have shown that these contacts can be replaced by an arbitrary protein–protein interaction. They showed that when the C-terminal domain of the α subunit of RNAP is replaced by the dimer-

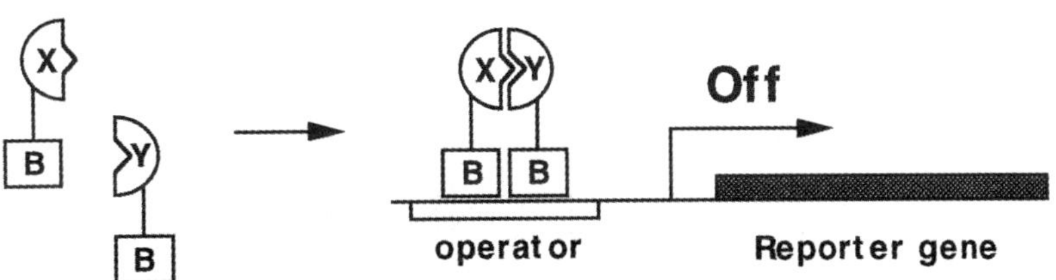

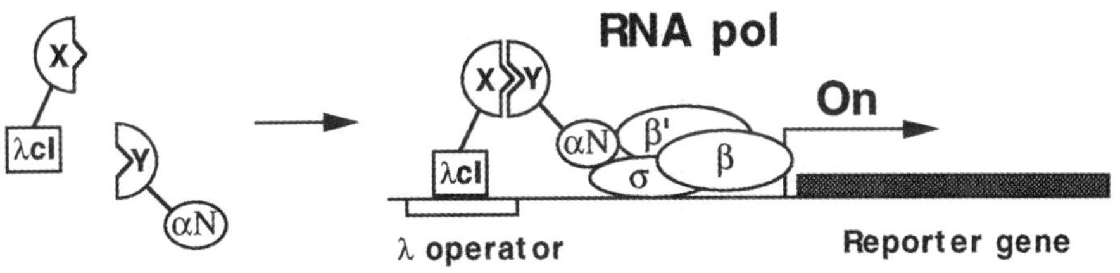

Figure 1. Bacterial genetic assays for in vivo analysis of protein–protein interactions. (A) Repressor dimerization. The B box represents the DNA-binding domain of a bacterial repressor (λcI or LexA). When B is genetically fused to interacting polypeptides X and Y, heterodimerization (or homodimerization, if X = Y) of the hybrid repressor allows binding to the λ or lexA operator site and repression of the reporter gene transcription. **(B)** Recruitment of RNA polymerase. Two chimeric proteins, consisting in λcI fused to polypeptide X and the N-terminal domain of the RNA polymerase α subunit (αN) fused to polypeptide Y, are co-expressed in *E. coli*. Interaction between X and Y will recruit the RNA polymerase at a promoter bearing a λ operator and stimulate transcription of the downstream reporter gene.

ization domain of λcI (D domain), the transcription of a reporter gene under the control of a promoter bearing a λ operator can be specifically activated by λcI. In a more recent work (5), they showed that a DNA-binding protein (λcI) fused to Gal4 could activate transcription at a promoter bearing a λ operator. This will happen in the presence of a chimeric RNAP α subunit fused to Gal11^P, a polypeptide that specifically interacts with Gal4 (Figure 1B). In the same report (5), they demonstrated that the linkage, either covalently or through non-covalent interactions (Gal4/Gal11^P), of the ω subunit of RNAP to a DNA-binding protein, λcI, can stimulate the transcription at a promoter bearing a λ operator.

These results indicate that the *E. coli* transcription machinery could be used, with modification, to study protein–protein interactions, as is the case in yeast. Although promising, the flexibility and the extent of application of this system remains to be studied.

1.3 Detection of Protein Interactions by Means of the Activation of a cAMP Signaling Cascade

We have recently described a novel two-hybrid system (17) that involves a cAMP signaling cascade. It offers some interesting possibilities compared to the classical yeast two-hybrid system as well as to the *E. coli* systems described previously. In this bacterial two-hybrid system, the proteins of interest are genetically fused to two complementary fragments of the catalytic domain of *Bordetella pertussis* adenylate cyclase (Figure 2). Interaction between the two proteins results in functional complementation between the two adenylate cyclase fragments, leading to cAMP synthesis, which in turn triggers the expression of several *E. coli* resident genes.

This genetic system is based on the following observations:

1. In *E. coli*, cAMP is a pleiotropic regulator of the expression of various genes, including genes involved in the catabolism of carbohydrates such as lactose or maltose (28,33). Hence, *E. coli* strains deficient in their endogenous adenylate cyclase (*cya*) are unable to ferment lactose or maltose.

2. *B. pertussis* (the causative agent of whooping cough) synthesizes a calmodulin-dependent adenylate cyclase toxin encoded by the *cyaA* gene (24). Its catalytic domain, located within the first 400 amino acids of this 1706 residue-long protein (10), exhibits a high catalytic activity (k_{cat} = 2000 s^{-1}) in the presence of calmodulin (CaM) and a low but detectable activity (k_{cat} = 2 s^{-1}) in the absence of this activator (18,35). When this catalytic domain is expressed in *E. coli cya* strain, which does not synthesize CaM, its residual activity (i.e., CaM-independent) is sufficient to complement the adenylate cyclase-deficiency of such strain and to restore its ability to ferment lactose or maltose (19). This can be scored either on indicator plates or on selective media.

3. The catalytic domain of CyaA has a modular structure: it consists of two complementary fragments, T25 and T18 (18,20). These two fragments, when expressed in *E. coli cya* as separate entities, are unable to recognize each other and cannot reconstitute a functional enzyme. However, when the T25 and T18 fragments are fused to peptides or proteins that interact, heterodimerization of these chimeric polypeptides results in a functional complementation, i.e., cAMP synthesis (17).

We have demonstrated that this bacterial two-hybrid system is able to reveal interactions between small peptides (GCN4 leucine zipper), bacterial proteins (tyrosyl

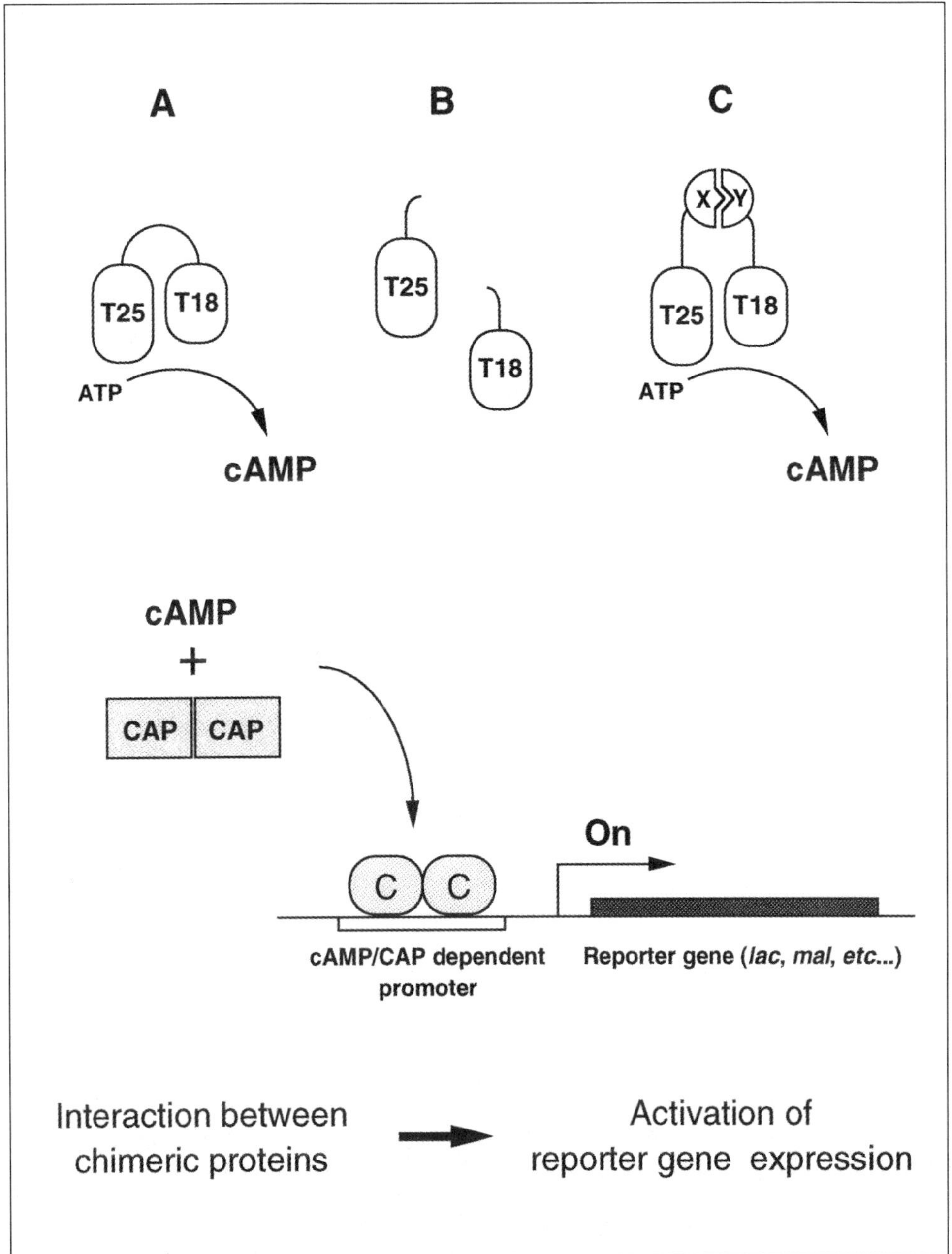

Figure 2. Principle of an *E. coli* two-hybrid system based on a cAMP signaling cascade. Upper part: in vivo complementation between the two fragments of the catalytic domain of *B. pertussis* adenylate cyclase. The two boxes represent the T25 and T18 fragments (amino acids 1–224, and 225–399 of CyaA, respectively). **(A)** The full-length catalytic domain (residues 1–399), when expressed in *E. coli*, exhibits a basal calmodulin-independent activity that results in cAMP synthesis. **(B)** T25 and T18 fragments, co-expressed as independent polypeptides, are unable to interact and remain inactive. **(C)** T25 and T18 fragments, when fused to interacting proteins X and Y, are brought into close proximity, thus restoring their adenylate cyclase activity. Lower part: readout of the complementation. cAMP, synthesized in an *E. coli cya* strain, binds to the catabolite activator protein, CAP. The cAMP/CAP complex **(C)** can then recognize specific promoters and stimulate the transcription of the corresponding genes. The reporter genes can be either endogenous *E. coli* genes under the control of cAMP/CAP, such as *lacZ* or *mal* genes, or synthetic ones, such as antibiotic resistance genes fused to a cAMP/CAP-dependent promoter.

tRNA synthetase), or eukaryotic proteins (yeast Prp11/Prp21 complex) (17).

The second part of this chapter describes the procedures and analytical methods of this bacterial two-hybrid system and is illustrated by the results of some applications. In the third part, we will discuss advantages and limitations of this approach.

2. PROTOCOLS: ANALYSIS OF PROTEIN–PROTEIN INTERACTIONS WITH A BACTERIAL TWO-HYBRID SYSTEM

2.1 Media, Strains, and Vectors Required

Media

Interaction between the two hybrid proteins results in functional complementation between the T25 and T18 fragments, which leads to cAMP synthesis that confers a cya^+ phenotype to the recipient cya bacteria. This can be easily scored either on indicator plates [i.e., Luria-Bertani (LB)–X-gal or MacConkey media supplemented with maltose or lactose] or on selective media (minimal media supplemented with lactose or maltose as unique carbon sources). The time of growth in these two types of media will vary significantly.

MacConkey medium

Adenylate cyclase-deficient bacteria are unable to ferment lactose or maltose (33); they form white colonies on MacConkey indicator media containing lactose or maltose, while cya^+ bacteria form red colonies on the same media (fermentation of the added sugar results in the acidification of the medium, which is revealed by a color change of the dye phenol red) (23). MacConkey base medium can be purchased from Difco Laboratories. Stock solutions of lactose or maltose (20% in water) are sterilized by filtration. Lactose or maltose (1% final concentration) as well as antibiotics (ampicillin and chloramphenicol at 100 µg/mL and 30 µg/mL, respectively) are added to the autoclaved MacConkey medium just before pouring plates (23).

LB–X-gal medium

In *E. coli,* the expression of the *lacZ* gene encoding β-galactosidase is positively controlled by cAMP/CAP. Hence, bacteria expressing interacting hybrid proteins will form blue colonies on rich LB medium (23) in the presence of the chromogenic substrate X-gal (5-Bromo-4-chloro-3-indolyl-β-D-galactopyranoside, 40 µg/mL), while cells expressing non-interacting proteins will remain white. IPTG (isopropyl-β-D-thiogalactopyranoside, 0.5 mM) can be included in the medium to increase β-galactosidase expression.

Complementation can be detected on these rich indicator media within 24–48 h at 30°C. It is important to avoid overcrowding the indicator plates (maximum 300–50 colonies per plate); otherwise, detection of positive clones could be difficult (23). It should be noted that after prolonged incubation (4–5 days), negative colonies (i.e., cya^-) will show a weak red (on MacConkey-maltose) or blue spot (on LB–X-gal) in the center of the colony, but they remain colorless at the periphery.

Synthetic medium

As cya^+ cells are Lac$^+$/Mal$^+$, they are able to grow on a minimal medium supplemented with lactose/maltose as unique carbon sources (23). This is a powerful selection procedure to isolate bacteria that express interacting hybrid proteins by using a standard synthetic medium, M63 (23), supplemented with 1% lactose. A stock 5× concentrated M63 medium is prepared by adding 10 g $(NH_4)_2SO_4$, 68 g KH_2PO_4, and 2.5 mg $FeSO_4.7H_2O$ to 1 L of water (adjust pH to 7.0 with KOH). Two hundred milliliters of sterile 5× M63 medium containing 1% lactose, vitamin B1 (1 µg/mL), and antibiotics (15 µg/mL chloramphenicol, 50 µg/mL ampicillin), are added to auto-claved agar (15 g in 800 mL water) before pouring plates.

Up to 10^7 cells can be plated on this selective minimal medium/lactose. Growth of Lac^+ colonies will be detected after 4–5 days of incubation at 30°C. These growing Lac^+ colonies are of two different types: truly positive cells expressing interacting hybrid proteins and false-positive cells expressing non-interacting proteins but having acquired a Lac^+ phenotype due to endogenous mutations (see discussion in part B). However, for unknown reasons, most of these false-positive clones, which appear at a frequency of about 10^{-6}, exhibit a mucoid phenotype and therefore can be easily differentiated from truly positive ones.

Bacterial Strains

Adenylate cyclase-deficient (*cya*) derivatives of various *E. coli* strains harboring either point mutations, deletions, or insertions within the structural *E. coli cya* gene have been described (28,33). In our study we used two *cya* strains. The DHP1 strain is a derivative of DH1 (F-, *glnV44(AS)*, *recA1*, *endA1*, *gyrA96 (Nal^r^)*, *thi1*, *hsdR17*, *spoT1*, *rfbD1*) that was isolated by a modified phosphomycin selection procedure (17). The TP610 strain (*cya*, *recBC*, *lacY*, *thr*, *leu*, *pro*) was isolated and characterized by Hedegaard and Danchin (12). We have observed that functional complementation between hybrid proteins was more efficient in DHP1 than in TP610, most likely because of higher stability of chimeric proteins in the former than in the latter. Indeed, when the same combination of plasmids expressing interacting hybrid proteins (e.g., GCN4 leucine zipper, see below) are cotransformed into these two strains, transformed DHP1 becomes red on MacConkey/maltose within 24 h at 30°C, while transformed TP610 requires about 40 h. Interestingly, in some *cya* strains (e.g., a *cya* derivative of HB101, obtained by the same phosphomycin selection procedure; G. Karimova, unpublished results) no complementation could be detected. The influence of a genetic background on the complementation efficiency should be kept in mind when choosing an *E. coli cya* strain. Transformations of *cya* bacteria are performed by standard techniques ($CaCl_2$ or electroporation) as described in Sambrook et al. (29).

Vectors

The basic two-hybrid technology requires co-expression of the two hybrid proteins within the same recipient *cya* bacteria. Two compatible vectors were constructed for this purpose (17) (Figure 3):

(i) pT25 encodes the T25 fragment of *B. pertussis* adenylate cyclase, which corresponds to the first 224 amino acids of CyaA (20,25). This vector is a derivative of low copy number plasmid pACYC184 (23). It carries the chloramphenicol acetyl transferase gene that confers resistance to chloramphenicol. A multicloning sequence was inserted at the 3′ end of T25 to allow construction of fusions in-frame at the C-terminal end of the T25 polypeptide.

(ii) pT18 encodes the T18 fragment (amino acid 225–399 of CyaA) (20,25) and is a derivative of pBluescript II KS (Stratagene) that expresses the β-lactamase selectable marker (ampicillin resistance). Its ColEI origin of replication is compatible with that of pT25, which carries the p15A replicon. Hence, both plasmids, pT18 and pT25, can be maintained simultaneously in the same cell. The T18 coding region is inserted downstream from the multicloning sequence of pBluescript II KS. In both vectors the expression of fusion proteins is under the control of the strong *lac* promoter.

2.2 Analytical Procedures

Cyclic AMP can be measured by an ELISA assay. Briefly, a cAMP-biotinylated-BSA conjugate—prepared as described (15) using biotinylated-BSA (Sigma Chemical, St. Louis, MO, USA) and 2′-O-Monosuccinyladenosine 3′:5′-Monophosphate (Sigma Chemical)—is coated on ELISA plates, and nonspecific protein binding sites are blocked with bovine serum albumin (BSA). Boiled bacterial cultures are then added, followed by diluted rabbit anti-cAMP antiserum (15) in 50 mM HEPES, pH 7.5, 150 mM NaCl, and 0.1% Tween 20 (HBST buffer) containing 10 mg/mL BSA. After overnight incubation at 4°C, the plates are washed extensively with HBST, then goat anti-rabbit IgG coupled to alkaline phosphatase (AP) is added and incubated for 1 h at 30°C. After washing, the AP activity is revealed by 5′-para-nitrophenyl phosphate. cAMP concentrations are calculated from a standard curve established with known concentrations of cAMP diluted in LB medium. Commercial kits for cAMP determination (ELISA or radioimmunoassays) are also available from various companies.

β-galactosidase assays are carried out on toluenized bacterial suspensions, according to Pardee et al. (27). One unit of activity corresponds to 1 nmol of ONPG hydrolyzed per min at 28°C. Other assay methods are described in Miller (23) or Sambrook et al. (29).

2.3 General Methodology to Analyze Interactions with the Bacterial Two-Hybrid System

The general methodology for the analysis of the in vivo interactions between two proteins (for instance, X and Y) is summarized in Figure 4. First, the genes encoding the two proteins of interest are cloned in-frame into pT25 and pT18. Second, the resulting hybrid protein constructs (T25-X and Y-T18) are cotransformed in competent *cya* cells, and the cotransformants are plated either on indicator plates or on selective media. The screening procedure (i.e., on indicator plates) is particularly appropriate if only a small number of cotransformants (less than 500 colonies per plate) are tested. On rich media, the cells grow faster and results can be obtained in 1–2 days of incubation. This should be the method of choice to determine in vivo interaction of two known proteins (X and Y). If X and Y interact, all colonies should give the same phenotype on the indicator plates: blue on LB–X-gal or red on MacConkey-maltose. If no interaction occurs, all colonies should be colorless. If a large number of colonies need to be tested by a library screen, it will be more advantageous to plate the transformants on minimal medium plus lactose. In this case, 10^5 to 10^6 transformants per dish can be plated. However, Lac$^+$ colonies will require at least 4–5 days to grow on this minimal medium.

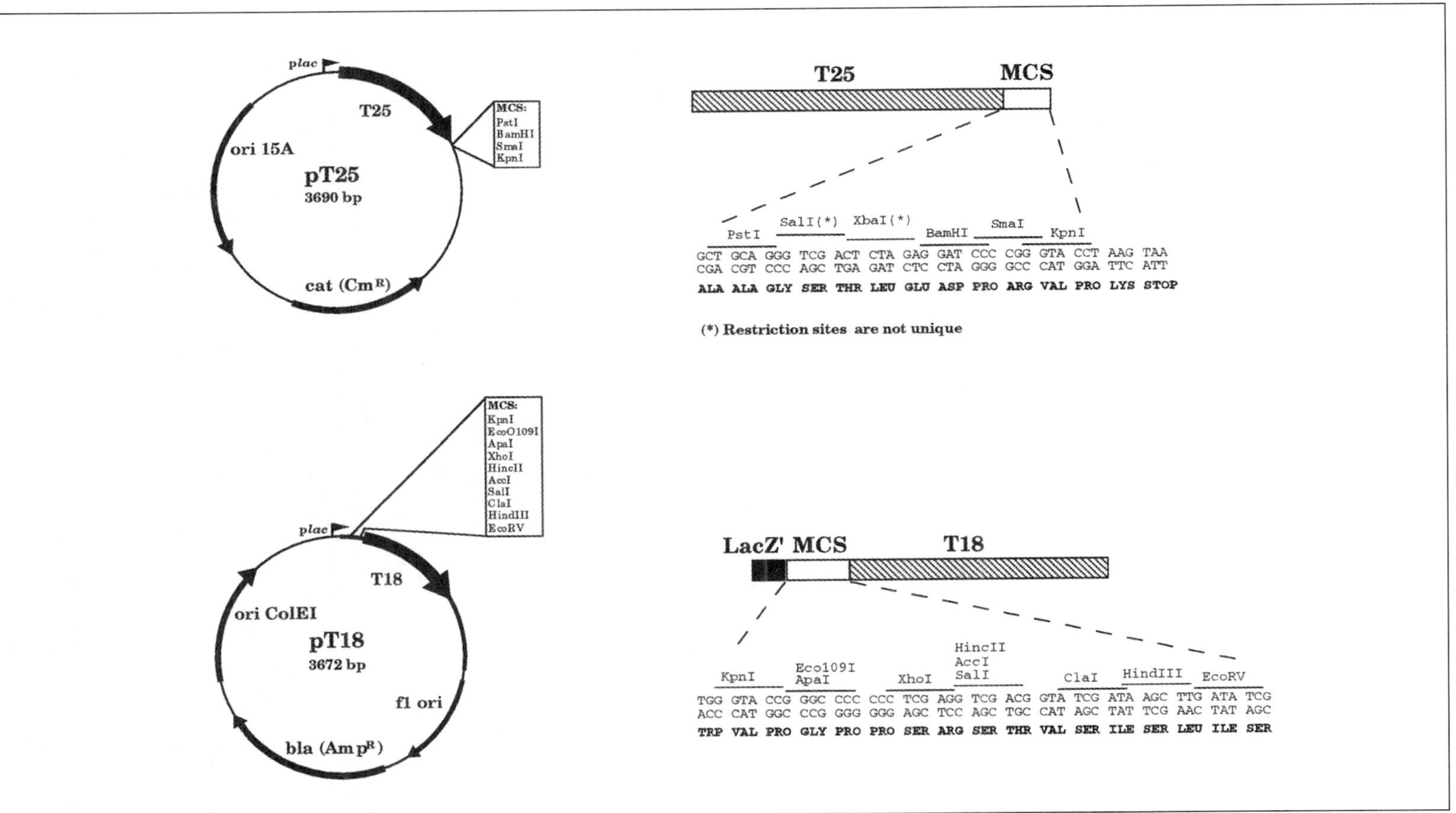

Figure 3. Cloning vectors used in the bacterial two-hybrid system. Physical maps of pT25 and pT18. The thick arrows on both plasmids represent the open reading frames of T25 and T18 fragments. The multicloning sequences (MCS) that allow insertion of foreign genes, either at the C-terminus of T25 or the N-terminus of T18, are shown in the boxes. The open reading frames overlapping the multicloning sequence of each vector are shown on the right.

3. APPLICATIONS

3.1 Monitoring Interactions Between Model Proteins

We have shown that this bacterial two-hybrid system is able to reveal interactions between small peptides (GCN4 leucine zipper), bacterial proteins (tyrosyl tRNA synthetase), or eukaryotic proteins (yeast Prp11/Prp21 complex).

It was first applied to analyze the dimerization ability of the leucine zipper motif (35 amino acids long) of GCN4 — a yeast transcriptional activator (26). Its DNA sequence was amplified by polymerase chain reaction (PCR) and subcloned separately into pT25 and pT18 (17). When the resulting plasmids — pT25-zip and pT18-zip—were cotransformed into an *E. coli* DHP1 strain and plated on MacConkey/maltose media, the resulting colonies became red after 24–30 h of growth at 30°C (Table 1). In control experiments, pT25-zip was cotransformed with pT18, and pT18-zip was cotransformed with pT25. None of the transformants gave red colonies, demonstrating that the functional complementation of T25-zip and T18-zip was mediated by the interaction of their leucine zipper motif. The efficiency of complementation was further quantified by measuring cAMP levels or β-galactosidase activities in liquid culture (Table 1).

The dimerization capability of the tyrosyl tRNA synthetase from *Bacillus stearothermophilus* (34) was also revealed with this bacterial two-hybrid system. A DNA fragment that encodes the N-terminal part (residues 1–302) of the protein was subcloned into the multicloning site of pT25 and of pT18. The resulting plasmids, pT25-Tyr and pT18-Tyr, yielded red transformants on MacConkey/maltose when cotransformed in DHP1. The transformants synthesized cAMP and expressed β-galactosidasc (Tablc 1). Control transformations confirmed that the TyrRS moiety was responsible for the functional complementation between T25-TyrRS and T18-TyrRS (Table 1). Furthermore, no complementation was observed when pT25-Tyr was cotransformed with pT18-zip or pT25-zip with pT18-Tyr. This demonstrates that the complementation was dictated by the specificity of recognition of the polypeptides fused to the two fragments, T25 and T18. We have recently extended this analysis and revealed various interactions between sub-domains of the TyrRS protein (submitted).

The interaction between the yeast-splicing factors Prp11 and Prp21 (fused to T25 and T18, respectively), previously characterized in the yeast two-hybrid assay (21), was also detected by this bacterial two-hybrid system, thus demonstrating its ability to reveal association between eukaryotic proteins (17).

More recently, we have applied this technique to analyze the dimerization capabilities of a DNA-binding protein, BvgA. BvgA is the main transcriptional activator of *B. pertussis* virulence factors (30) that belong to a large family of the bacterial two-component signal transduction regulators. Our results showed that BvgA has the capacity to dimerize (unpublished results). Experiments are in progress to delineate the amino acids that are directly involved in the dimerization of BvgA.

It is worth noting that, in most cases, the functional complementation was found to be more efficient at 30°C than at 37°C, although the precise reason behind this is still unclear.

We have not yet performed a real library screen with this bacterial two-hybrid system. However, we performed a "model" library screen to determine whether this system could be used to identify rare interacting proteins among an excess of non-interacting ones (17). DHP1 bacteria expressing interacting proteins (T25-zip and

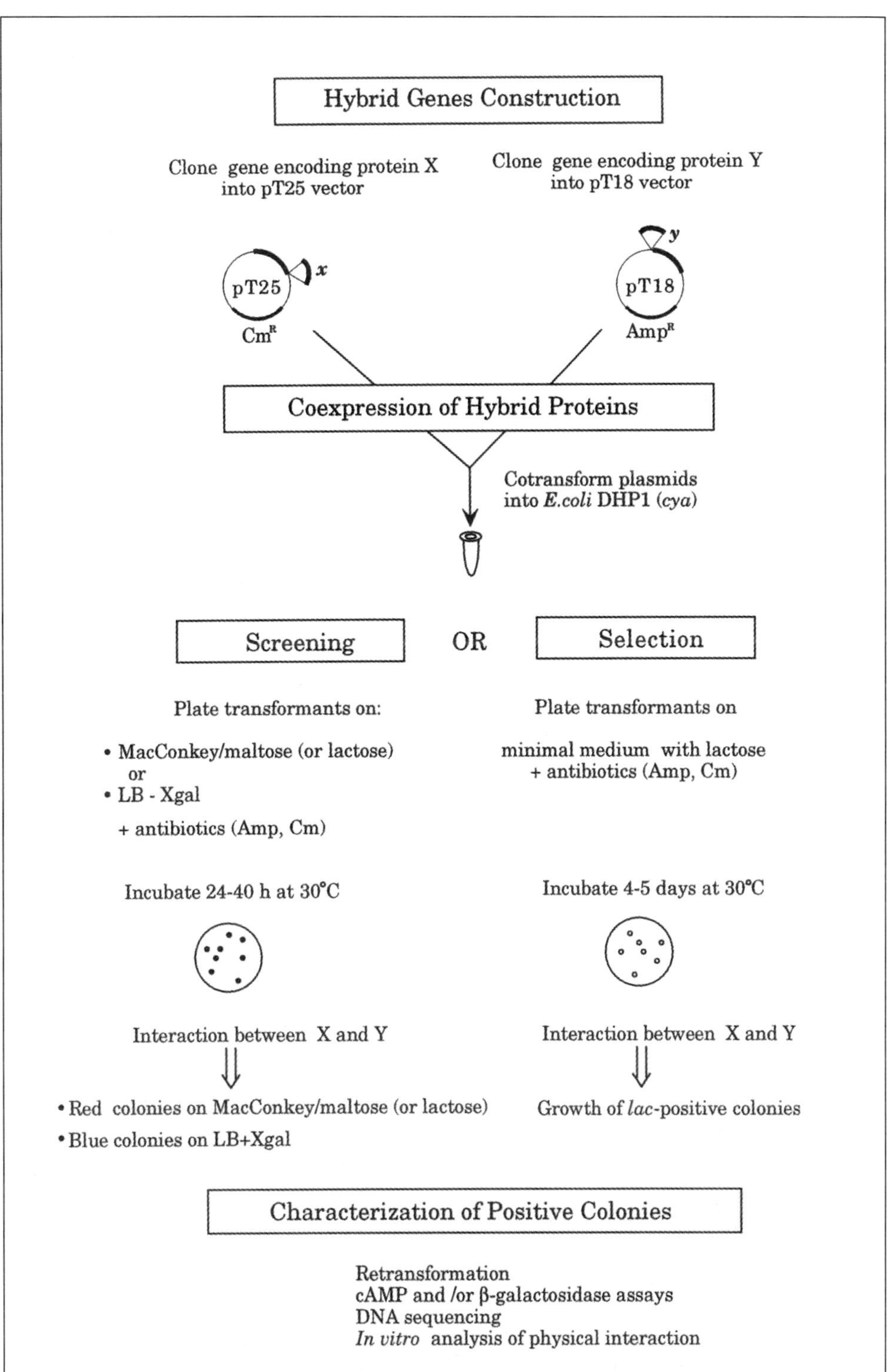

Figure 4. Flow chart: general procedure to analyze protein–protein interactions with the bacterial two-hybrid system.

Table 1. Analysis of Complementation in DHP1 Strain

Plasmids	Phenotype on MacConkey/maltose	β-Galactosidase Activity* U/mg dry weight
none	White	179
pT25 + pT18	White/72 h	130
pT25 + pT18-zip	White/72 h	183
pT25-zip + pT18	White/72 h	178
pT25-zip + pT18-zip	Red/30 h	4750
pT25-Tyr + pT18-Tyr	Red/40 h	2800
pT25-Tyr + pT18	White/96 h	193
pT25 + pT18-Tyr	White/96 h	183
pT25-Tyr + pT18-zip	White/96 h	134
pT25-zip + pT18-Tyr	White/96 h	126
pT25-prp11 + pT18-prp21	Red/40 h	850

*Bacteria were grown in LB at 30°C in the presence of 0.5 mM IPTG plus appropriate antibiotics. The results represent the average values obtained for at least five independant cultures, which differed by less than 10%.

T18-zip) were mixed with a 10^5 excess of DHP1 bacteria expressing non-interacting protein (T25 and T18); 10^7 cells from this mixture were plated on minimal media supplemented with lactose plus antibiotics. After 4–5 days at 30°C, about 100–200 Lac$^+$ colonies appeared. Plasmid DNA analysis indicated that 18 out of 20 of these colonies tested harbored pT25-zip and pT18-zip. The other two false-positive colonies (i.e., expressing non-interacting T25 and T18 fragments) appeared to represent spontaneous revertants of DHP1 to a Lac$^+$ phenotype (See DISCUSSION). This "model screening" demonstrated that bacteria expressing specific interacting proteins fused to the adenylate cyclase fragments could be selected among a large number of irrelevant clones.

4. DISCUSSION

Although this bacterial two-hybrid system has been developed only recently, we anticipate that it could provide an attractive approach to search for and analyze interacting proteins. We will now briefly examine the main advantages of this technique and its potential biases as well as the possibilities of further improvements.

4.1 Advantages:

1. The originality of this system compared to the yeast or other bacterial two-hybrid systems stems from the involvement of a signaling cascade, which utilizes the diffusable regulatory molecule, cAMP. As a consequence, the physical association of the two putative interacting proteins can be spatially separated from the transcription activation events (i.e., the readout) that are dependent on cAMP synthesis. This means that the protein–protein interaction under study

does not need to take place in the vicinity of the transcription machinery as is the case for the yeast two-hybrid system (or other bacterial two-hybrid systems mentioned in the INTRODUCTION).

Hence, it will be possible to analyze protein–protein interactions that occur either in the cytosol, at the DNA level, or at the inner membrane level. This genetic test should be of particular interest for the functional analysis of DNA binding proteins that have been found to be difficult to study with the classical yeast two-hybrid system.

Moreover, this system will offer a possibility to study bacterial proteins in their native context and to analyze in vivo both the functional activity of any given proteins and their association state (provided that the hybrid proteins maintain their functional activity). This should be particularly appropriate to examine the dynamics of the association of membrane proteins. This approach may shed new light on the role of specific protein–protein associations in the functional aspect.

Because this genetic screen is in principle an assay for proximity of the fused T25 and T18 fragments, it could be particularly suitable for analyzing co-localization of proteins of multimolecular assemblies. Such data could help to address the possible functions of unknown ORFs that are massively identified in full-genome sequencing efforts.

2. Because this genetic test is carried out in *E. coli*, it greatly facilitates the screening as well as the characterization of the interacting proteins. First, the high efficiency of transformation that can be achieved in *E. coli* allows the analysis of libraries of high complexity. This is particularly useful for *(i)* the screening of peptides from a library made from random DNA sequences that present an affinity for a given bait protein, and *(ii)* the exhaustive analysis of the network of interactions between the proteins of a given organism. Second, it will be possible to screen a library to identify a putative binding partner to a given "bait" and then to express directly the chimeric proteins to characterize their interaction by in vitro binding assays (see FUTURE DEVELOPMENTS). Third, as *E. coli* is the basic "tool" of molecular biologists, this technique could be handled easily by many researchers without a learning course necessary to master the "tricks" of yeast manipulations.

3. This two-hybrid system offers the possibility of performing both positive and negative selections (Figure 5). This chapter so far has described the positive selection that is to select cells or colonies that express **interacting** proteins. As described above, these cells synthesize cAMP and are positive in *lac* or *mal* phenotypes; they can then use these sugars as unique carbon sources.

Negative selection is to select colonies that express **non-interacting** proteins among a background of cells expressing interacting hybrids. This type of selection can be performed easily as *cya* cells are naturally resistant to bacteriophage lambda infection (33). The bacterial receptor for phage lambda, LamB, is a maltoporin that belongs to the *mal* regulon and is positively regulated by cAMP, hence, *cya* strains express a very low level of LamB. In addition, cAMP might exert some positive effects on the bacteriophage λ life cycle. Practically, upon infection by λ*vir* (at a multiplicity of 3–10 λ*vir*/cell), all *cya+* cells are lysed very efficiently and only *cya* cells are able to grow (A. Ullmann, personal communication).

This negative selection procedure will be of great interest for the fine characterization of interacting proteins. By searching for point mutations in one of the hybrid genes that abolish an established interaction between the two proteins in our assay, one can nail down the critical residues quickly. Then, compensatory mutations that restore the interaction could be selected for by using the positive selection procedure. Lastly, functional analysis of the given mutations reintroduced within the native context in the original host would help to pinpoint the physiological role of the interaction between the two components under study.

This negative selection procedure should also prove to be useful in high throughput screening for molecules that could disrupt a given interaction between two proteins of interest.

4.2 Potential Limitations

False-negative results: interacting hybrid proteins that do not yield functional complementation.

Because this assay requires the spatial proximity between the T25 and T18 fragments, there could be some internal steric constraints that decrease or abolish the adenylate cyclase enzymatic activity of the reconstituted complex. This could be the case if T25 and T18 are fused to long and asymmetric proteins that interact in such a way that the fused T25 and T18 fragments are held too far away in space to interact properly.

In our experience, functional complementation was observed with the T25-TyrRS/T18-TyrRS heterodimer (17). From the published crystal structure of the TyrRS dimer, the amino-acid residues of TyrRS connected to the T25 and T18 fragments were 85 Å apart (1).

We expect that the specific activity of any given reconstituted complex may vary considerably. One might see no functional complementation between two chimeric proteins even though they physically interact. In other words, the absence of a positive signal for complementation will not necessarily mean the absence of interactions. In reality, this rule applies similarly to all two-hybrid methodologies, although it seems the yeast system is highly tolerant to large hybrid proteins. Future work with this bacterial system will provide additional insights into its flexibility and tolerance. However, one should keep in mind that in many cases functional domains that are able to interact specifically with each other are in the range of 100–200 amino-acid residues. It is likely that most of the interactions mediated by domains of this size can be detected with the bacterial two-hybrid technique.

False-positive results: non-interacting hybrid proteins that yield complementation-positive phenotype.

False-positive colonies in this two-hybrid technique can arise from two different origins:

(*i*) The recipient cells acquire a Lac+ or Mal+ phenotype independently of the expressed hybrid proteins. This can be due to:
- Spontaneous reversion of the *cya⁻* to *cya⁺* phenotype; this reversion occurs in DHP1 at a frequency of 10^{-6}. This problem can be solved by using *E. coli* strains carrying a deletion (or an inactivating insertion) of the endogenous adenylate cyclase gene (work in progress).
- Spontaneous mutations of the CAP gene toward a cAMP-independent phenotype [so-called CRP*: the classical name of a CAP variant that is functionally active

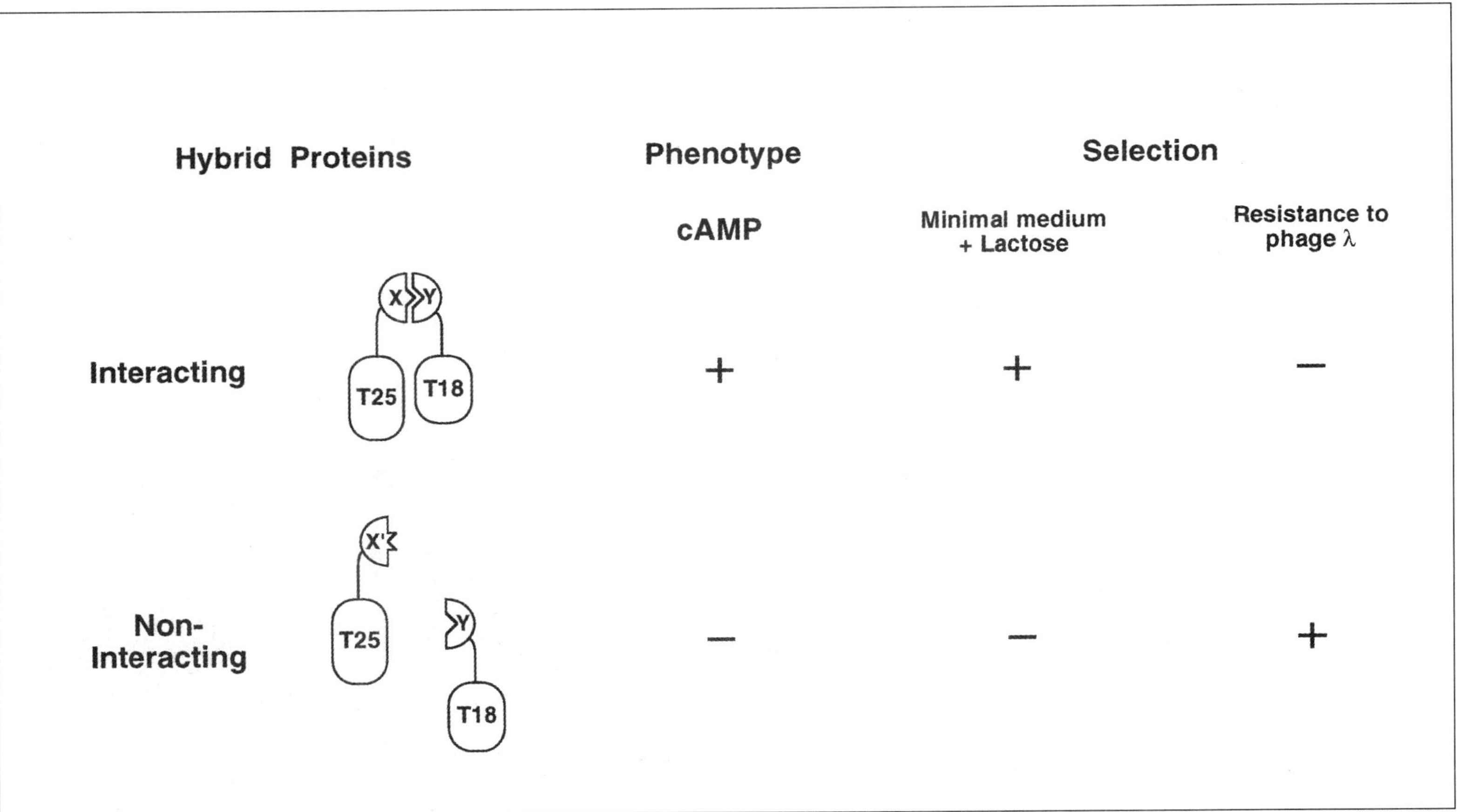

Figure 5. Versatility of the bacterial two-hybrid system. Positive and negative selections. T25-X and T18-Y represent the hybrid proteins co-expressed in DHP1. If X and Y interact, the bacteria synthesize cAMP and express catabolic operons. Expression of the *lacZ* gene allows growth on minimal medium plus lactose, while expression of the *mal* regulon results in sensitivity to phage λ. In contrast, if X and Y do not interact, the bacteria cannot grow on minimal medium plus lactose but are resistant to phage λ.

in the absense of cAMP (see Reference 33)]. This can occur at frequencies around $10^{-8}/10^{-9}$.

- Spontaneous cAMP-independent *lac* promoter mutations; this can occur at frequencies around $10^{-7}/10^{-8}$. The probability of obtaining cAMP-independent Mal+ cells should be around 10^{-21}.

In all cases, these false-positive colonies can be identified upon transformation of their plasmids into new, competent *cya* cells.

(ii) The cotransformed plasmids confer a *cya*+ phenotype, although the encoded hybrid proteins do not interact through the polypeptides fused to T25 and T18. This can be due to:

- In a library screening, the insertion in one of the cotransformed plasmids of a fragment of an adenylate cyclase that is catalytically active into *E. coli*; this may occur at a frequency of 10^{-5} to 10^{-10} or less, depending upon the origin of the screened genome (bacterial or eukaryotic).
- The insertion, in one of the cotransformed plasmids, of a gene encoding a protein that is able to bind directly to the T25 and T18 fragments and stimulate the adenylate cyclase activity; this obviously could be the case for the eukaryotic calmodulin or possibly for other calmodulin-like proteins.

In both cases, these false-positive plasmids might be easily detected upon cotransformation with the complementary plasmid expressing only the adenylate cyclase fragment (T25 or T18, not fused to the "bait" protein).

5. FUTURE DEVELOPMENTS

The basic version of this bacterial two-hybrid system might be easily improved to render it more sensitive, efficient, and versatile. Some of the possibilities that we are currently pursuing are listed below:

Strains

The readout of the complementation between T25 and T18 fragments is the transcriptional activation of cAMP/CAP-dependent genes. Until now, we have taken advantage of naturally occurring cAMP/CAP-dependent catabolic genes in *E. coli*. However, it is easy to design specific reporter cassettes in which a gene of interest is fused to a cAMP/CAP-dependent promoter. We are currently constructing such a system using an antibiotic resistance gene. This will facilitate the screening of complex libraries by a simple selection for antibiotic resistance. Alternatively, the reporter gene driven by cAMP/CAP could encode a toxic product. This could be an alternative to the use of bacteriophage λ resistance to search for chemical compounds or mutations that abolish an established interaction.

Vectors

New vectors are currently being developed to facilitate construction of gene fusions, as well as in vivo analysis and in vitro characterization of protein–protein interactions. We have recently constructed new plasmids that permit fusions at the C-terminus of T18. We have also constructed a plasmid that expresses a T25 fragment with a 6xHis tag at its N-terminus: this new construct will allow an easy in vitro characterization of the interactions between the two hybrid proteins. Hence, during the course of a library screen, one could check directly if positive clones encode truly interacting polypeptides (and eventually estimate the corresponding equilibrium/dissociation constants). Upon chromatography of the bacterial cell extracts on a Ni-column, the 6xHis-tagged T25 hybrid proteins will be specifically retained on the resin. Co-retention of the T18 chimeric proteins will undoubtedly demonstrate a direct interaction between the hybrid molecules. Alternatively, different immunological tags could be introduced in both fragments and the physical interactions could be established by co-immunoprecipitation approaches.

New Design for Increased Sensitivity

The present system relies on the basal enzymatic activity of *B. pertussis* adenylate cyclase (no CaM). We estimate, from adenylate cyclase activity measured in bacterial extracts, that about 1000–5000 T25 and T18 molecules per cell are expressed in the present design. This system could be rendered exquisitely sensitive by using the full catalytic potency of *B. pertussis* adenylate cyclase, i.e., in the presence of its natural activator, CaM. In this case, we anticipate that reconstitution of only a few hybrid molecules per cell will be sufficient to elicit a detectable signal.

Applications to Other Organisms

The basic idea of using a cAMP signaling cascade to report interaction between hybrid proteins might be extended to other organisms, like gram positive bacteria (e.g., *Bacillus subtilis*) or yeast (e.g., *Saccharomyces cerevisae*). This would allow workers to study the association of proteins of these organisms in a more native environment. In any case, these applications will require specific adaptations to a chosen organism.

6. CONCLUDING REMARKS

This bacterial two-hybrid technique should find wide applications for in vivo analysis of protein interactions. This method, as well as other similar genetic systems (bacterial, yeast, and mammalian), are likely to become essential tools to accompany the recent explosion in genomic sequencing data. It should also be applicable in drug discovery and in high throughput screening of combinatorial libraries to identify compounds that either abolish or reinforce a given interaction. Also, because this technique allows a positive selection in *E. coli*, it could be used to design a self-evolving system in which the fitter bacterial population could be selected based on a higher affinity and/or specificity of interactions between hybrid proteins.

ACKNOWLEDGMENTS

We are indebted to A. Ullmann for sharing her unpublished data, and for her constant interest and her critical reading of the manuscript. Financial support to the authors' laboratory came from the Institut Pasteur and the Centre National de la Recherche Scientifique (URA 1129).

REFERENCES

1. **Brick, P. and D.M. Blow.** 1987. Crystal structure of a deletion mutant of a tyrosyl-tRNA synthetase complexed with tyrosine. J. Mol. Biol. *194*:287–297.

2. **Bunker, C.A. and R.E. Kingston.** 1995. Identification of a cDNA for SSRP1, an HMG-box protein, by interaction with the c-Myc oncoprotein in a novel bacterial expression screen. Nucleic Acids Res. *23*:269–276.

3. **Castagnoli, L., C. Vetriani, and G. Cesareni.** 1994. Linking an easily detectable phenotype to the folding of a common structural motif. Selection of rare turn mutations that prevent the folding of Rop. J. Mol. Biol. *237*:378–387.

4. **Dmitrova, M., C.G. Younes, B.P. Oertel, D. Porte, M. Schnarr, and S.M. Granger.** 1998. A new LexA-based genetic system for monitoring and analyzing protein heterodimerization in *Escherichia coli*. Mol. Gen. Genet. *257*:205–212.

5. **Dove, S.L. and A. Hochschild.** 1998. Conversion of the omega subunit of *Escherichia coli* RNA polymerase into a transcriptional activator or an activation target. Genes Dev. *12*:745–754.

6. **Dove, S.L., J.K. Joung, and A. Hochschild.** 1997. Activation of prokaryotic transcription through arbitrary protein–protein contacts. Nature. *386*:627–630.

7. **Fields, S.** 1997. The future is function. Nat. Genet. *15*:325–327.

8. **Fields, S. and O.K. Song.** 1989. A novel genetic system to detect protein–protein interactions. Nature *340*:245–246.

9. **Georgiou, G., C. Stathopoulos, P.S. Daugherty, A.R. Nayak, B.L. Iverson, and R. Curtiss III.** 1997. Display of heterologous proteins on the surface of microorganisms: from the screening of combinatorial libraries to live recombinant vaccines. Nat. Biotech. *15*:29–34.

10. **Glaser, P., D. Ladant, O. Sezer, F. Pichot, A. Ullmann, and A. Danchin.** 1988. The calmodulin-sensitive adenylate cyclase of *Bordetella pertussis*: cloning and expression in *Escherichia coli*. Mol. Microbiol. *2*:19–30.

11. **Hanes, J. and A. Pluckthun.** 1997. *In vitro* selection and evolution of functional proteins by using ribosome display. Proc. Natl. Acad. Sci. USA *94*:4937–4942.

12. **Hedegaard, L. and A. Danchin.** 1985. The *cya* gene region of *Erwinia chrysanthemi* B374: organization and gene products. Mol. Gen. Genet. *201*:38–42.

13. **Hu, J.C., E.K. O'Shea, P.S. Kim, and R.T. Sauer.** 1990. Sequence requirements for coiled-coils: analysis with lambda repressor-GCN4 leucine zipper fusions. Science *250*:1400–1403.

14. **Jappelli, R. and S. Brenner.** 1996. Interaction between cAMP-dependent protein kinase catalytic subunit and peptide inhibitors analyzed with lambda repressor fusions. J. Mol. Biol. *259*:575–578.

15. **Joseph, E. and J.L. Guesdon.** 1982. Beta-galactosidase immunoassay for the measurement of cyclic AMP. Anal. Biochem. *119*:335–340.

16. **Joung, J.K., E.H. Chung, G. King, C. Yu, A.S. Hirsh, and A. Hochschild.** 1995. Genetic strategy for analyzing specificity of dimer formation: *Escherichia coli* cyclic AMP receptor protein mutant altered in its dimerization specificity. Genes Dev. *9*:2986–2996.

17. **Karimova, G., J. Pidoux, A. Ullmann, and D. Ladant.** 1998. A bacterial two-hybrid system based on a reconstituted signal transduction pathway. Proc. Natl. Acad. Sci. USA *95*:5752–5756.

18. **Ladant, D.** 1988. Interaction of *Bordetella pertussis* adenylate cyclase with calmodulin. Identification of two separated calmodulin-binding domains. J. Biol. Chem. *263*:2612–2618.

19. **Ladant, D., P. Glaser, and A. Ullmann.** 1992. Insertional mutagenesis of *Bordetella pertussis* adenylate cyclase. J. Biol. Chem. *267*:2244–2250.

20. **Ladant, D., S. Michelson, R. Sarfati, A.M. Gilles, R. Predeleanu, and O. Barzu.** 1989. Characterization of the calmodulin-binding and of the catalytic domains of *Bordetella pertussis* adenylate cyclase. J. Biol. Chem. *264*:4015–4020.

21. **Legrain, P. and C. Chapon.** 1993. Interaction between PRP11 and SPP91 yeast splicing factors and characterization of a PRP9-PRP11-SPP91 complex. Science *262*:108–110.

22. **Marchetti, A., M.M. Abril, B. Illi, G. Cesareni, and S. Nasi.** 1995. Analysis of the Myc and Max interaction specificity with lambda repressor-HLH domain fusions. J. Mol. Biol. *248*:541–550.

23. **Miller, J. H.** 1992. A short course in bacterial genetics. CSHL Press, Cold Spring Harbor, New York.

24. **Mock, M. and A. Ullmann.** 1993. Calmodulin-activated bacterial adenylate cyclases as virulence factors. Trends Microbiol. *1*:187–192.

25. **Munier, H., A.M. Gilles, P. Glaser, E. Krin, A. Danchin, R. Sarfati, and O. Barzu.** 1991. Isolation and characterization of catalytic and calmodulin-binding domains of *Bordetella pertussis* adenylate cyclase. Eur. J. Biochem. *196*:469–474.

26. **O'Shea, E.K., R. Rutkowski, and P.S. Kim.** 1989. Evidence that the leucine zipper is a coiled coil. Science *243*:538–542.

27. **Pardee, A.B., F. Jacob, and J. Monod.** 1959. The genetic control and cytoplasmic expression of inducibility in the synthesis of β-galactosidase of *Escherichia coli*. J. Mol. Biol. *1*:165–168.

28. **Saier, M.H.J., T.M. Ramseier, and J. Reizer.** 1996. Regulation of carbon utilization. p. 1325–1343. *In* F. C. Neidhart (Ed.), *Escherichia coli* and *Salmonella*. Cellular and Molecular Biology. ASM Press, Washington, D.C.

29. **Sambrook, J., E.F. Fritsch, and T. Maniatis.** 1989. Molecular Cloning: A Laboratory Manual. CSHL Press, Cold Spring Harbor, New York.

30. **Scarlato, V., B. Arico, M. Domenighini, and R. Rappuoli.** 1993. Environmental regulation of virulence factors in *Bordetella* species. Bioessays *15*:99–104.

31. **Schmidt-Dorr, T., P. Oertel-Buchheit, C. Pernelle, L. Bracco, M. Schnarr, and M. Granger-Schnarr.** 1991. Construction, purification, and characterization of a hybrid protein comprising the DNA binding domain of the LexA repressor and the Jun leucine zipper: a circular dichroism and mutagenesis study. Biochemistry *30*:9657–9664.

32. **Smith, G.P.** 1985. Filamentous fusion phage: novel expression vectors that display cloned antigens on the virion surface. Science *228*:1315–1317.

33. **Ullmann, A. and A. Danchin.** 1983. Role of cyclic AMP in Bacteria. p. 1–44. *In* P. Greengard and G.A. Robinson (Ed.), Advances in Cyclic Nucleotide Research. Raven Press, New York, NY.

34. **Ward, W.H., D.H. Jones, and A.R. Fersht.** 1987. Effects of engineering complementary charged residues into the hydrophobic subunit interface of tyrosyl-tRNA synthetase. Appendix: Kinetic analysis of dimeric enzymes that reversibly dissociate into inactive subunits. Biochemistry *26*:4131–4138.

35. **Wolff, J., G.H. Cook, A.R. Goldhammer, and S.A. Berkowitz.** 1980. Calmodulin activates prokaryotic adenylate cyclase. Proc. Natl. Acad. Sci. USA *77*:3841–3844.

17 Confirmation of Yeast Two-Hybrid Results: In Vitro Protein Binding Analysis

Abdallah Fanidi[1]*, Shao-bing Hua[1,2]*, Nancianne Knipfer[1], Jianing Huang[1], Jian Liao[1], and Li Zhu[1,2]
[1]*CLONTECH Laboratories, Palo Alto, and* [2]*Genetastix, San Jose, CA, USA*

**These authors contributed equally to this manuscript*

1. INTRODUCTION

Protein interactions play pivotal roles in virtually all the cellular processes. Analysis of such interactions has long been of interest to researchers in the biological sciences. Many in vitro biochemical methods have been successfully developed to study protein interactions. For example, proteins that interact with a known protein can be detected and isolated by protein affinity chromatography (8,15,22). A known protein is first covalently coupled to a solid support, such as an agarose bead column. Labeled or unlabeled protein extract is then loaded onto such a column. Unbound proteins are washed off, and retained proteins can then be eluted according to their properties (high salt buffer, cofactor). Protein interactions can also be detected by cross-linking (19), which is a powerful technique for deducing the architecture of enzyme and/or ribosomal complexes (1,4) or to detect ligand-receptor interactions (13). Affinity blotting or Far-Westerns have also been successfully used to identify calmodulin-binding proteins (9,10). A more classical and widely used method to detect protein–protein interaction is co-immunoprecipitation (21). Antibody against a specific antigen (purified protein or synthetic peptide) is first generated. Proteins interacting with such an antigen can then be co-immunoprecipitated and analyzed. Furthermore, one or more of these methods, such as cross-linking and co-immunoprecipitation, can be combined for protein interaction studies (5). A more detailed review of these in vitro techniques can be found elsewhere (14). Unfortunately, these in vitro methods all suffer from the same disadvantage, i.e., the genes that encode the interacting proteins are not readily available, making it very difficult to study the newly identified interacting proteins further. Yeast two-hybrid systems (6), on the other hand, can largely overcome such a constraint.

Yeast Hybrid Technologies
Edited by L. Zhu and G.J. Hannon
©2000 Eaton Publishing, Natick, MA

283

In the previous chapters, authors have described in great detail yeast two-hybrid systems, their related technologies, and their applications. Once clones that interact with a "bait" are isolated from a library screening, the next immediate challenge is to confirm such interactions. It is always important to have one or more independent tests for the interaction of the two proteins once genetically interacting clones are obtained. Indeed, false positive interactions do occur even when clones meet all the criteria of two-hybrid systems (see Chapter 6).

Several methods have been successfully used to confirm protein interactions identified in yeast two-hybrid systems. One method is to switch the bait clone and the library clone in their respective plasmids. The new constructs are transformed into yeast cells again to confirm the interaction. The major drawbacks of this approach, however, are that confirmation is still carried out in yeast cells, and the fusion constructs are similar to the original ones (such as GAL4 activation and binding domain fusion proteins, respectively). If the bait and its interacting partners are derived from mammalian sources, a mammalian two-hybrid system can also be used for such confirmation (see Chapter 15). A mammalian two-hybrid system is especially useful to confirm transient protein interactions.

In this chapter, we will discuss methods, which are related to the yeast two-hybrid systems, that are available to analyze in vitro protein binding. The first method is a commonly used glutathione-S-transferase (GST)-assisted in vitro affinity precipitation (3,7,18). The second is an epitope-tag–assisted co-immunoprecipitation approach. At last, we will describe a protocol to confirm interaction of proteins expressed in mammalian cells (12,17).

2. PROTOCOLS

2.1 Protocol for GST-Assisted Affinity Precipitation Method

In this method, the bait clone of interest is first fused with GST, and the fusion protein is expressed in *Escherichia coli* or in insect cells (baculovirus expression system). The fusion protein can then be purified using glutathione-agarose beads. Proteins are made from the interacting clones by in vitro transcription and translation in the presence of labeling materials such as [^{35}S]-methionine. Labeled proteins are then incubated with the purified bait protein bound to glutathione-agarose beads. After washing, bound materials are separated by sodium dodecyl sulfate-polyacrylamide gel electrophoresis (SDS-PAGE) and visualized by autoradiography (Figure 1).

2.1.1 Materials

- GST expression systems:
  - *E. coli* GST expression systems (such as the pGEX series) are commercially available from Amersham Pharmacia Biotech (Piscataway, NJ, USA).
  - Baculovirus GST expression systems are available from PharMingen (San Diego, CA, USA).
  - Glutathione-agarose beads and purification systems are available from a number of commercial sources, such as Amersham Pharmacia Biotech, PharMingen, and Sigma Chemical (St. Louis, MO, USA).

- Vectors for in vitro transcription and translation, such as the pCITE series with T7 promoters, are available from commercial sources such as Novagen (Madison, WI, USA).

  Note: If clones of interest are in CLONTECH's MATCHMAKER™ Two-Hybrid System 3 vectors, which contain T7 promoters, in vitro transcription and translation of the inserts can be directly carried out without any modifications.

- In vitro transcription and translation kits are available from Promega (Madison, WI, USA), Roche Molecular Biochemicals (Indianapolis, IN, USA) and Novagen. L-[35S]-methionine is available from commercial sources such as Amersham Pharmacia Biotech (Catalog No. AG1094, 1000 Ci/mmol).

- SDS-PAGE electrophoresis apparatus and reagents can be purchased from many providers.

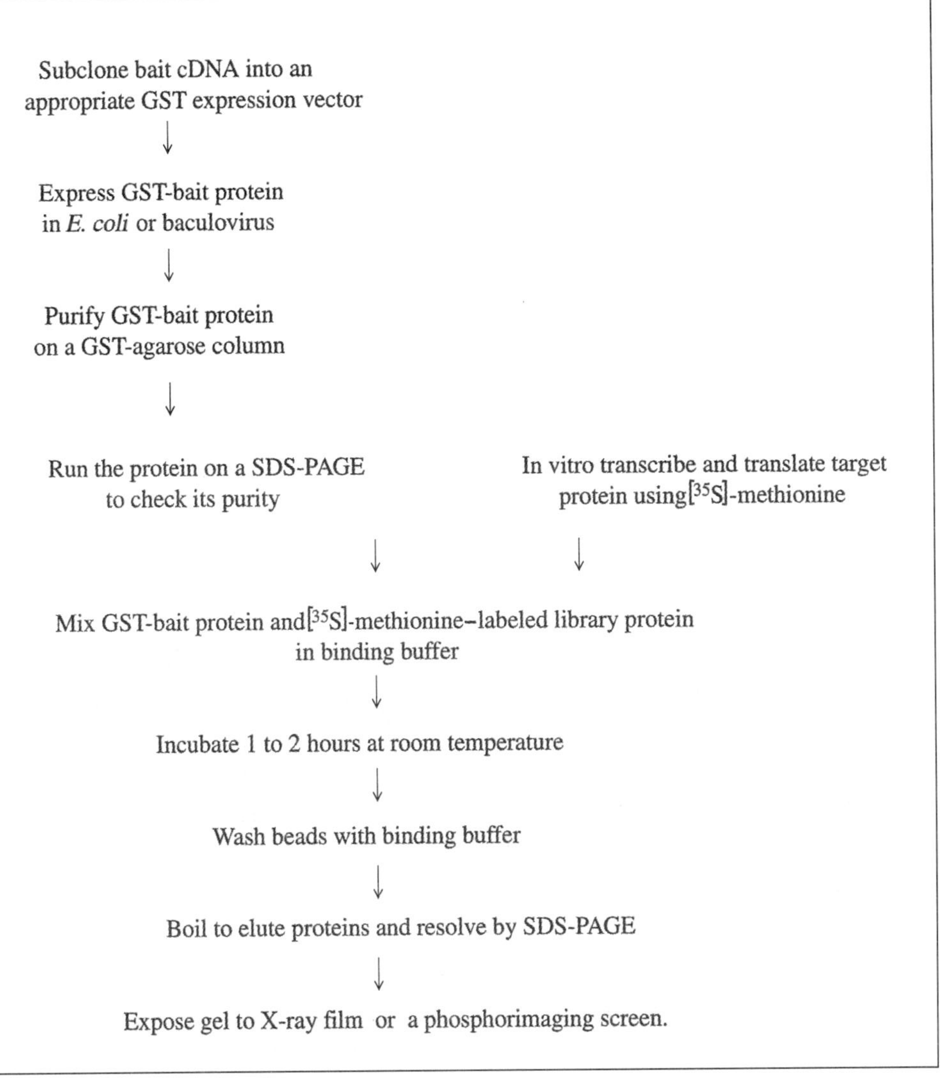

Figure 1. In vitro co-immunoprecipitation of proteins by GST-affinity precipitation assay.

- Other: gel dryer, X-ray film (such as Bio-Max MR films by Kodak), X-ray film cassettes, and film-development facilities.

2.1.2 Procedure

a. Fusion of the bait cDNA with GST

Clone the bait cDNA into an appropriate GST expression system vector using standard cloning procedures (2). Clone target cDNA into appropriate vectors, such as the pCITE series, for in vitro transcription and translation (see step c below).

b. Expression of the GST-bait recombinant protein

Express the GST-bait fusion protein with the expression system of your choice (*E. coli* or baculovirus expression system) according to the manufacturer's instructions. Purify the expressed fusion protein by glutathione-agarose beads, following the manufacturer's recommendations. Check the purity of the GST-bait fusion protein by SDS-PAGE followed by Coomassie Blue staining. For a detailed SDS-PAGE protocol, please see Ausubel et al. (2) or manuals from manufacturers of electrophoresis equipment.

c. In vitro transcription and translation of target genes cloned into appropriate vectors (see step a)

Several manufacturers provide in vitro transcription and translation systems using T7 RNA polymerase, such as the TNT Coupled Reticulocyte Lysate System from Promega, the Linked in vitro T7 Transcription and Translation Kit from Roche Molecular Biochemicals, and the Single Tube Protein System 3 for T7 from Novagen. In vitro transcription and translation should be carried out under RNase-free conditions. Plasmid DNA prepared by nucleic acid purification columns is generally pure enough for in vitro transcription and translation. However, if you find any RNase contamination in your plasmid samples, extract the samples once with 1 volume of phenol-chloroform and once with 1 volume of chloroform. Precipitate the DNA with sodium acetate and ethanol (2). Allow the sample to completely dry before adding RNase-free TE (10 mM Tris-HCl, pH 8.0; 1 mM EDTA, pH 8.0) or water.

The following protocol is based on Promega's TNT T7 Coupled Reticulocyte Lysate System. For more detailed information, please consult the manufacturer's instructions.

1. Remove the reagents from storage at -70°C. Immediately place the T7 RNA polymerase on ice. Thaw the rabbit reticulocyte lysate in hand or in a 30°C water bath and place on ice immediately.

2. Set up the following reaction in a 1.5-mL microcentrifuge tube on ice:

   25 μL TNT rabbit reticulocyte lysate

   2 μL TNT reaction buffer

   1 μL T7 RNA polymerase

   1 μL Amino acid mixture, minus methionine (1 mM)

   1 μL Ribonuclease inhibitor (40 U/μL)

   5 μL DNA template (0.1 μg/μL)

   2 μL L-[^{35}S]-methionine (1000 Ci/mmol)

3. Add RNase-free water to a final volume of 50 μL.

4. Mix gently. Do not vortex-mix. If necessary, centrifuge briefly to return the solution to the bottom of the tube. Incubate the reaction at 30°C for 60 to 120 min. Stop the reaction by placing on ice. In vitro translated products can be directly used in affinity precipitation analysis.

d. In vitro binding of proteins

1. Agarose-bound GST or GST-bait fusion protein (5–10 μg) is suspended in 50 μL of binding buffer [20 mM Tris, pH 7.5, 0.2% Nonidet P-40, 200 mM NaCl, 50 mM NaF, 0.2 mM sodium orthovanadate, 1 mM phenylmethylsulfonyl fluoride (PMSF), 2 μg/mL aprotinin, and 2 μg/mL leupeptin] and mixed with 50 μL of reticulocyte lysate containing the appropriate ^{35}S-labeled target protein.

2. After 1–2 h incubation at room temperature, the agarose beads are washed four times with binding buffer. Remove as much buffer as possible without disturbing the beads.

e. SDS-PAGE analysis

1. Add SDS-PAGE sample buffer to the beads. Load the sample onto an SDS-PAGE gel and perform electrophoresis. (We recommend using mini-gel systems with gels of approximately 0.75 mm in thickness. Use 10%–12% acrylamide gels for proteins bigger than 25 kDa and 15% acrylamide gels or higher for smaller proteins.)

2. After electrophoresis, fix the gel in fixation solution [30% ethanol (vol/vol), 10% acetic acid (vol/vol)] for 10 min at room temperature. Change fixation solution once and soak for an additional 30 min. If the gel is thicker than 0.75 mm, extend fixation time accordingly. Place the gel on a pre-wetted Whatman™ 3MM paper. Cover the gel with Saran™ wrap, and dry the gel with a gel dryer at 80°C.

3. Remove Saran wrap and expose the dried gel to an X-ray film overnight at room temperature.

Note: Certain X-ray film for ^{35}S label (such as Bio-Max MR by Kodak) are coated on a single side. Make certain the gel directly contacts the emulsion side of the film.

4. Develop film using standard techniques.

2.2 Protocol for Epitope-Tag–Assisted Co-immunoprecipitation Method

In yeast two-hybrid systems, as well as in the GST-assisted affinity precipitation procedures, genes of interest are fused to relatively large domains and/or proteins. Such domains may cause conformational changes in the protein of interest. To independently confirm protein–protein interactions identified by two-hybrid systems and to exclude the possibility that such interactions are due to conformational changes within the fusion protein, it is advised to use the small epitope-tag–assisted co-immunoprecipitation method.

In this method, small epitope tags such as c-Myc and HA (hemagglutinin) are fused to the bait and the target proteins, respectively. One of the tagged proteins, such as bait, is then expressed in the presence of labeling agents, such as [^{35}S]-methionine. The labeled bait and the target protein are incubated together. Interacting proteins can then be reciprocally precipitated and blotted using antibodies against c-Myc and HA tags. The immunoprecipitated complexes can then be separated by SDS-PAGE and visualized by autoradiography (Figure 2).

2.2.1 Materials

- Rabbit anti-HA polyclonal antibody
- Mouse anti-c-Myc monoclonal antibody
- Protein-G agarose beads
- Polymerase chain reaction (PCR) reagents
- Thermal cycler
- Appropriate vectors containing T7 promoters and Myc or HA epitope tags, such as pGBKT7 and pGADT7 from CLONTECH. A complete Co-immunoprecipitation kit is also available from CLONTECH for confirmation of protein interactions detected by a GAL4-based two-hybrid system.

See section 2.1.1 for more information on in vitro transcription and translation kits, SDS-PAGE electrophoresis apparatus, and additional materials.

2.2.2 Procedure

Once protein interactions are identified by a two-hybrid system, there are two approaches to add T7 promoters and epitope tags to the genes of interest. One can incorporate the T7 promoters and appropriate epitope tags upstream of the genes of interest by PCR. Alternatively, one can subclone the genes of interest into pGBKT7 and pGADT7 vectors, which contain c-Myc and HA tags, respectively, and then proceed to Section 2.2.2.b. However, if one identified protein interactions using MATCHMAKER Two-Hybrid System 3 vectors (i.e. pGBKT7 and pGADT7) from CLONTECH, please proceed directly to Section 2.2.2.b.

a. Incorporation of T7 Promoter and Epitope Tags to the bait and target proteins

1. Design appropriate primers for PCR. The following two forward primers contain T7 promoter and epitope tag sequences. Please notice that you will need to add at least 21 bases specific to the N-termini of the genes of interest to these primers at their 3′ ends. Please refer to your specific vectors for the reverse primers.

   - T7-Myc forward primer:

     5′-AAA ATT GTA ATA CGA CTC ACT ATA GGG CGA GCC GCC ACC ATG GAG GAG CAG AAG CTG ATC TCA GAG GAG GAC CTG + 21 bases
   - T7-HA forward primer:

     5′-AAA ATT GTA ATA CGA CTC ACT ATA GGG CGA GCC GCC ACC ATG TAC CCA TAC GAC GTT CCA GAT TAC GCT + 21 bases

     Due to the long sequences, these oligonucleotides should be purified by polyacrylamide gel electrophoresis after synthesis.

2. Set up PCRs using appropriate primers and templates. Follow the conditions recommended by the manufacturer of the PCR reagents and thermal cycler.

3. Commence thermal cycling.

4. Analyze the PCR products by electrophoresis on a 1.2% agarose gel.

5. Extract the PCR mix once with 1 volume of phenol-chloroform and once with 1 volume of chloroform. Precipitate the PCR product with sodium acetate and ethanol.

b. In Vitro Transcription and Translation

The advanced MATCHMAKER Two-Hybrid System 3 vectors, pGBKT7 and pGADT7, contain T7 promoter and epitope tags (c-Myc and HA, respectively)—sequences upstream of the multiple cloning site. Inserts cloned into these vectors can be transcribed and translated without any modifications, whereas inserts in other two-hybrid system vectors should be subcloned into pGBKT7 or pGADT7, or modified with incorporation of T7 promoters and epitope tags by PCR (see Section 2.2.2.a for details).

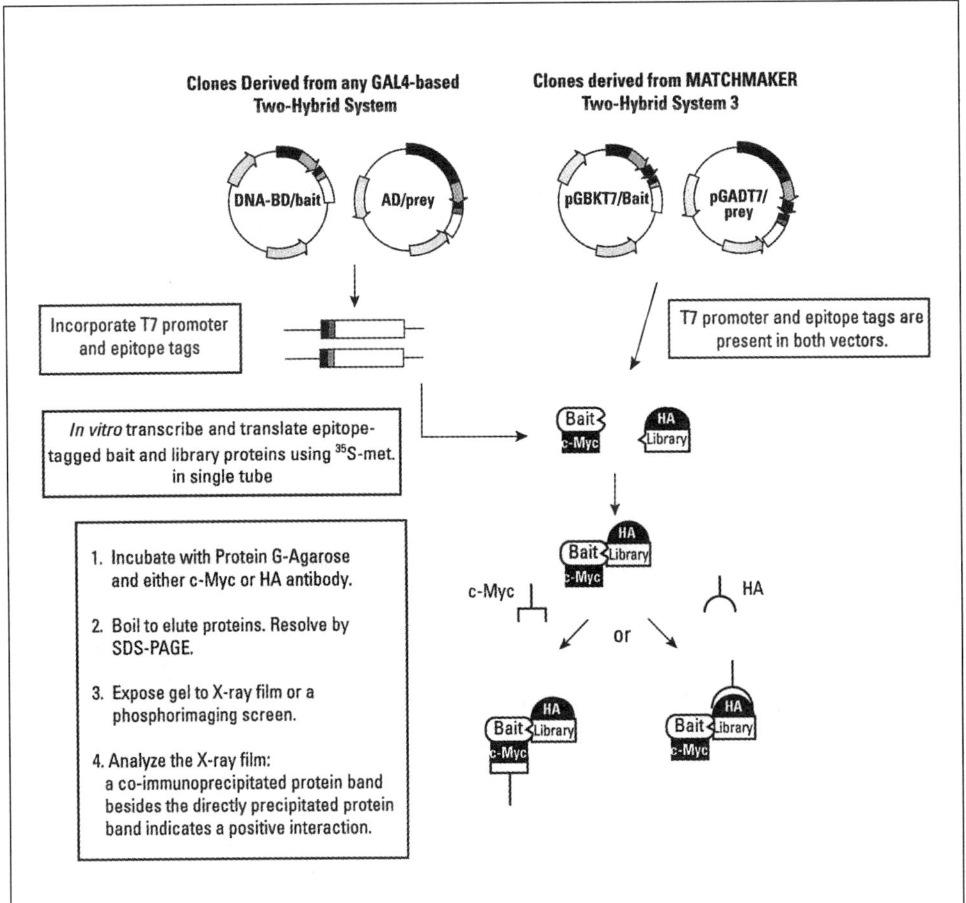

Figure 2. In vitro tag-assisted co-immunoprecipitation assay.

For in vitro transcription and translation procedures, please see Section 2.1.2.c for details.

c. Co-immunoprecipitation and SDS-PAGE Analysis

1. Set up the following reaction in a 1.5-mL microcentrifuge tube on ice:
   - 5 µL in vitro translated bait protein
   - 5 µL in vitro translated library clone (prey) protein

   Controls contain only the bait protein or only the prey protein.

2. Mix gently up and down with a pipette and incubate at 30°C for 1 h. Mix occasionally by tapping.

3. Add to the reaction mix:
   - 470 µL TBS (20 mM Tris-HCl, pH 7.5; 150 mM NaCl) containing 1 mM dithiothreitol (DTT), 5 µg/mL aprotinin, 0.5 mM PMSF and 0.1% Tween-20 (TBS-T).
   - 10 µL of a 50% (wt/vol) slurry Protein-G agarose beads pre-equilibrated in TBST.
   - 1 µg Anti-c-Myc monoclonal or HA-Tag polyclonal antibody.

     Note: During protein interactions, the c-Myc and HA tags may be buried in the complex and therefore inaccessible to antibodies. We recommend doing co-immunoprecipitations using either c-Myc or HA antibody in two separate and parallel experiments, in case one of the antibodies does not work.

4. Incubate for 90 min. Agitate continuously by inversion.

5. Centrifuge the tubes at 14 000× *g* for 2 min. Remove the supernatant and discard. Caution: Do not disturb the agarose beads.

6. Add 0.5 mL TBS-T buffer to the tube and mix well by inversion or tapping. Do not vortex-mix.

7. Repeat steps 5 and 6 three times. Steps 5 to 7 must be performed at 4°C. At the end of the washing step, remove as much washing buffer as possible, but do not disturb the agarose beads. Centrifuge the tube again if necessary.

8. Add 15 µL of SDS loading buffer to the beads. Denature samples, heat at 80°–85°C for 5 min.

9. Load 10 µL onto a slab SDS-PAGE minigel.

10. After electrophoresis, place gel in fixation solution [30% methanol (vol/vol), 10% acetic acid (vol/vol)] for 10 min at room temperature. Change the fixation solution once and soak an additional 30 min with continuous agitation. If your gel is thicker than 0.75 mm, extend the fixation time accordingly.

11. Soak the gel in a fluorographic reagent for 15–30 min to amplify the ^{35}S signal.

12. Lay gel onto a pre-wetted Whatman 3MM paper. Cover with Saran wrap and dry the gel in a vacuum dryer at 80°C.

13. Remove Saran wrap and expose gel to a phosphorimaging screen or X-ray film overnight at room temperature.

14. Develop the film using standard techniques.

Figure 3 provides an example of epitope-tag–assisted co-immunoprecipitation Method. c-Myc (Myc-tagged) and Max (HA-tagged) proteins, which are known to form a heterodimer in vitro and in vivo, were obtained by in vitro translation and labeled with ^{35}S-methionine using a Promega TNT kit. Lanes 1 and 2 represent ^{35}S-labeled c-Myc and Max proteins immunoprecipitated with a monoclonal antibody to the Myc tag and an anti-HA polyclonal antibody, respectively. Lane 3 shows the Myc-tagged Lamin C that does not interact with either c-Myc or Max. In Lane 4, c-Myc and Max proteins were incubated together at 30°C for 60 min. Immunoprecipitation was carried out using the anti-Myc epitope tag antibody that brought down the c-Myc protein as expected. Besides, Max can easily be detected in a complex with c-Myc. In lane 5, Lamin C and Max were in vitro co-immunoprecipitated in the same way as Myc and Max. However, it is evident that Lamin C and Max are unable to interact in this assay, as expected.

2.3 Protocol for Co-immunoprecipitation of Proteins Expressed in Mammalian Cells

If the bait and its interacting partners are derived from mammalian sources, it is important to confirm such interactions in mammalian cells. In addition to a mammalian two-hybrid system (see Chapter 15), confirmation by co-immunoprecipitation of proteins expressed in mammalian cells is also a common alternative (12,17,20). Using this method, both bait and interacting genes are cloned into mammalian expression vectors with different epitope tags, such as pCMV Myc and pCMV-HA from CLONTECH (Figure 4). Both constructs are then co-transfected into mammalian cells. Interaction between the bait and the target proteins is evaluated by co-immunoprecipitation. In addition, epitope-tagged proteins expressed in mammalian cells can also be used for subcellular localization studies with appropri-

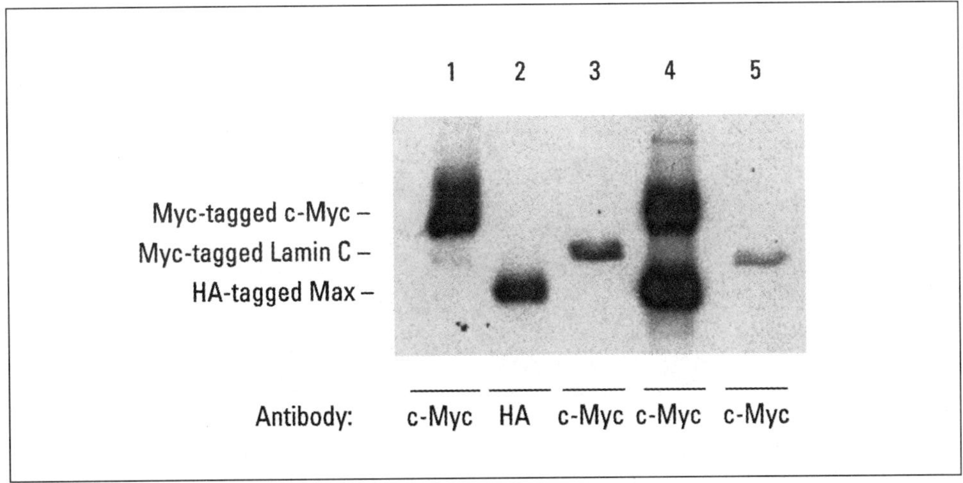

Figure 3. Demonstration of the interaction between c-Myc and Max proteins by in vitro co-immunoprecipitation. Myc-tagged c-Myc, HA-Max, and Myc-Laminc C proteins were in vitro translated from pGBKT7-cMyc, pGBKT7-Max, and pGADT7-LaminC, respectively, and labeled with ^{35}S-methionine using a TNT-T7 Coupled Reticulocyte Lysate System (Promega). An equal amount of either Myc and Max or Max and Lamin C proteins were mixed and incubated at 30°C for 60 min. Co-immunoprecipitations were carried out as described in 2.2.2.d. Lane 1: c-Myc-tagged cMyc protein; Lane 2: HA-tagged Max; Lane 3: cMyc-tagged Lamin C; Lane 4: cMyc-tagged cMyc protein + HA-tagged Max; Lane 5: cMyc-tagged Lamin C + HA-tagged Max.

ate antibodies and fluorescent dyes (20). This in vivo assay is considered a high standard for protein–protein interaction.

2.3.1 Materials

- Mammalian expression vectors containing epitope tags, such as pCMV-Myc and pCMV-HA from CLONTECH.
- Appropriate mammalian cell lines, cell culture facilities, and reagents.

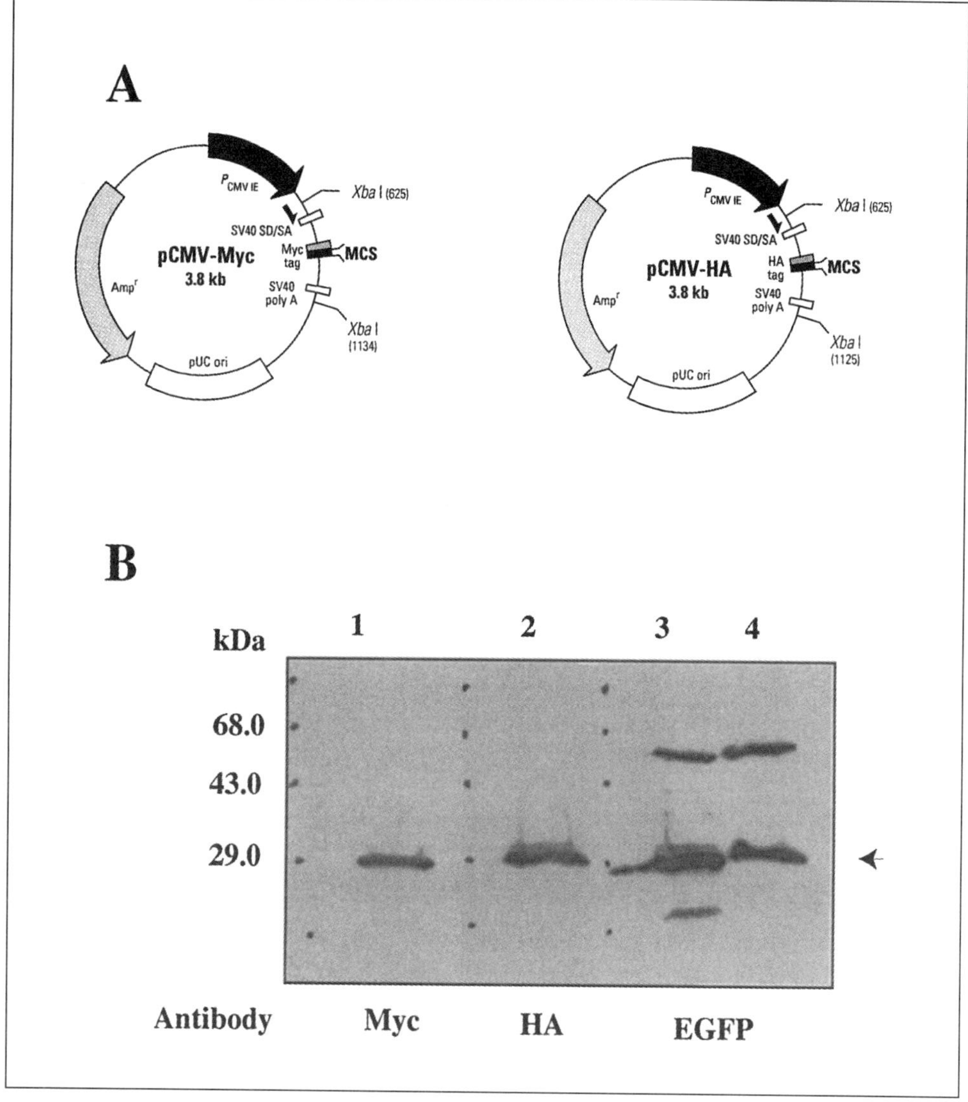

Figure 4. Western blotting analysis of tagged EGFP using pCMV-Myc and pCMV-HA mammalian expression vectors. pCMV-Myc and pCMV-HA expression vectors express proteins with N-terminal c-Myc and hemagglutinin (HA) epitope tags. (**A**) pCMV-Myc and pCMV-HA vector maps. (**B**) Western blotting analysis of tagged EGFP: pCMV-Myc-EGFP (lanes 1 and 4) and pCMV-HA-EGFP (lanes 2 and 3) were transiently transfected in HEK 293 cells, and cell lysates were analyzed by Western blotting analysis with the indicated antibodies. The arrow indicates the position of tagged proteins.

- Cell transfection reagents: Many different types of cell lines can be efficiently transfected using liposome-based reagents (16) or calcium phosphate-based reagents (2,11).
- Rabbit anti-HA polyclonal antibodies, mouse anti-c-Myc monoclonal antibody, and appropriate secondary antibodies.
- Protein-G agarose beads.
- SDS-PAGE reagents and apparatus.
- Western blotting apparatus and reagents.

2.3.2 Method

a. Cloning and Transfection

1. Transfer bait and target cDNA inserts into pCMV-Myc and pCMV-HA using standard cloning procedures (2). Purify enough plasmid DNA for transfection.

2. Plate the cells the day before the transfection experiment. The cells should be 60%–80% confluent the day of transfection.

3. Transfect the mammalian cell line of your choice in a 10-cm dish using standard transfection reagents and protocols (2), or following the instructions for the commercially available transfection reagents. 1 to 5 µg of each vector DNA is needed for a single transfection.

b. Cell Lysis

1. Harvest cells 24–72 hours after transfection.

2. Wash cells once with ice-cold phosphate-buffered saline (PBS), and pellet by centrifugation at $1000\times g$ for 5 min.

3. Resuspend and lyse the cell pellet in 1.0 mL of ice-cold lysis buffer (20 mM Tris-HCl, pH 7.5, 150 mM NaCl, 0.2% NP-40, 1 mM EDTA, 30 mM NaF) containing protease inhibitors (1 mM PMSF, 0.5 µg/mL aprotinin, 0.5 µg/mL pepstatin, and 10 µg/mL leupeptin).

4. Incubate on ice for 30 min.

5. Centrifuge the lysate at $15\,000\times g$ for 15 min at 4°C. The supernatant is ready for co-immunoprecipitation analysis.

c. Co-immunoprecipitation and Western Blotting Analysis

1. Set up the following reaction mixture in a 1.5-mL microcentrifuge tube for co-immunoprecipitation.
    - 20 µL of a 50% (wt/vol) slurry Protein-G agarose beads pre-equilibrated in the lysis buffer
    - 0.5 µg Anti-Myc monoclonal antibody (or anti-HA polyclonal antibody)
    - 0.5 mL cell lysate

2. Gently rock the tube for 2 h at 4°C.

3. Collect Protein-G agarose beads by centrifugation. Aspirate the supernatant, and wash 4 times with ice-cold lysis buffer.

4. Denature the proteins by adding 15 µL of SDS-PAGE loading buffer to the beads.

5. Load the sample onto an SDS-PAGE gel, and perform electrophoresis.

6. Transfer the separated proteins to a PVDF membrane and carry out the immunoblot analysis according to standard procedures (2).

Note: Rabbit anti-HA polyclonal antibody should be used in immunoblot analysis for samples co-immunoprecipitated with mouse anti-Myc monoclonal antibody. On the other hand, mouse anti-Myc monoclonal antibody should be used in the immunoblot analysis for samples co-immunoprecipitated with rabbit anti-HA antibodies.

ACKNOWLEDGMENT

We thank Alicia Hunner for critically reading the manuscript.

REFERENCES

1.**Aman, R.A., G.L. Kenyon, and C.C. Wang.** 1985. Cross-linking of the enzymes in the glycosome of *Trypanosoma brucei*. J. Biol. Chem. *260*:6966-6973.
2.**Ausubel, F.M., R. Brent, R.E. Kingston, D.D. Moore, J.G. Seidman, J.A. Smith, and K. Struhl.** 1994. Current Protocols in Molecular Biology. Greene Publishing Associates and John Wiley & Sons, Inc., NY.
3.**Chen, H., G. Srinivasan, and E.B. Thompson.** 1997. Protein-protein interactions are implied in glucocorticoid receptor mutant 465*-mediated cell death. J. Biol. Chem. *272*:25873-25880.
4.**Cover, J.A., J.M. Lambert, C.M. Norman, and R.R. Traut.** 1981. Identification of proteins at the subunit interface of the *Escherichia coli* ribosome by cross-linking with dimethyl 3,3′-dithiobis(propionimidate). Biochemistry *20*:2843-2852.
5.**de Gunzburg, J., R. Riehl, and R.A. Weinberg.** 1989. Identification of a protein associated with p21ras by chemical crosslinking. Proc. Natl. Acad. Sci. USA *86*:4007-4011.
6.**Fields, S. and O. Song.** 1989. A novel genetic system to detect protein-protein interactions. Nature *340*:245-247.
7.**Fiore, F., N. Zambrano, G. Minopoli, V. Donini, A. Duilio, and T. Russo.** 1995. The regions of the Fe65 protein homologous to the phosphotyrosine interaction/phosphotyrosine binding domain of Shc bind the intracellular domain of the Alzheimer's amyloid precursor protein. J. Biol. Chem. *270*:30853-30856.
8.**Formosa, T., J. Barry, B.M. Alberts, and J. Greenblatt.** 1991. Using protein affinity chromatography to probe structure of protein machines. Methods Enzymol. *208*:24-45.
9.**Geiser, J.R., H.A. Sundberg, B.H. Chang, E.G. Muller, and T.N. Davis.** 1993. The essential mitotic target of calmodulin in the 110-kilodalton component of the spindle pole body in *Saccharomyces cerevisiae*. Mol. Cell Biol. *13*:7913-7924.
10.**Glenney, J.R. Jr. and K. Weber.** 1983. Detection of calmodulin-binding polypeptides separated in SDS-polyacrylamide gels by a sensitive [125I]calmodulin gel overlay assay. Methods Enzymol. *102*:204-210.
11.**Graham, F.L. and A.J. van der Eb.** 1973. A new technique for the assay of infectivity of human adenovirus 5 DNA. Virology *52*:456.
12.**Hsu, H., H.B. Shu, M.G. Pan, and D.V. Goeddel.** 1996. TRADD-TRAF2 and TRADD-FADD interactions define two distinct TNF receptor 1 signal transduction pathways. Cell *84*:299-308.
13.**Mita, S., A. Tominaga, Y. Hitoshi, K. Sakamoto, T. Honjo, M. Akagi, Y. Kikuchi, N. Yamaguchi, and K. Takatsu.** 1989. Characterization of high-affinity receptors for interleukin 5 on interleukin 5-dependent cell lines. Proc. Natl. Acad. Sci. USA *86*:2311-2315.
14.**Phizicky, E.M. and S. Fields.** 1995. Protein-protein interactions: methods for detection and analysis. Microbiol. Rev. *59*:94-123.
15.**Ratner, D.** 1974. The interaction bacterial and phage proteins with immobilized *Escherichia coli* RNA polymerase. J. Mol. Biol. *88*:373-383.
16.**Ruysschaert, J.-M., A. El Ouahabi, V. Willeaume, G. Huez, R. Fuks, M. Vandenbranden, and P. Di Stefano.** 1994. A novel cationic amphiphile for transfection of mammalian cells. Biochem. Biophys. Res. Commun. *203*:1622-1628.

17.Schreiber-Agus, N., L. Chin, K. Chen, R. Torres, G. Rao, P. Guida, A.I. Skoultchi, and R.A. DePinho. 1995. An amino-terminal domain of Mxi1 mediates anti-Myc oncogenic activity and interacts with a homolog of the yeast transcriptional repressor SIN3. Cell *80*:777-786.

18.Song, H.Y., J.D. Dunbar, Y.X. Zhang, D. Guo, and D.B. Donner. 1995. Identification of a protein with homology to hsp90 that binds the type 1 tumor necrosis factor receptor. J. Biol. Chem. *270*:3574-3581.

19.Traut, R.R., C. Casiano, and G.N. Zecherle. 1989. Crosslinking of protein subunits and ligands by the introduction of disulphide bonds, p. 101-133. *In* T.E. Creighton (Ed.), Protein function: a practical approach. IRL Press, Oxford.

20.Wong, C. and L. Naumovski. 1997. Method to screen for relevant yeast two-hybrid-derived clones by coimmunoprecipitation and colocalization of epitope-tagged fragments—application to Bcl-xL. Anal. Biochem. *252*:33-39.

21.Yee, S.P. and P.E. Branton. 1985. Detection of cellular proteins associated with human adenovirus type 5 early region 1A polypeptides. Virology *147*:142-153.

22.Zhang, X.F., J. Settleman, J.M. Kyriakis, E. Takeuchi-Suzuki, S.J. Elledge, M.S. Marshall, J.T. Bruder, U.R. Rapp, and J. Avruch. 1993. Normal and oncogenic p21ras proteins bind to the amino-terminal regulatory domain of c-Raf-1. Nature *364*:308-313.

Novel Applications of Yeast Hybrid Methods

18

Application of Two-Hybrid System in Human Genome Protein Linkage Mapping

Shao-bing Hua[1,2], Ying Luo[1,3], Meng Sheng Qiu[1,4], Eva Chan[1,3], Helen Zhou[1,5], and Li Zhu[1,2]

[1]*CLONTECH Laboratories, Palo Alto, CA;* [2]*Genetastix, San Jose, CA;* [3]*Rigel, San Francisco, CA;* [4]*University of Louisville, Louisville, KY;* [5]*Aviron, Mountain View, CA, USA*

1. INTRODUCTION

Large-scale sequencing of the human genome through the Human Genome Project has produced a large amount of DNA sequence information. To date, nearly 50 000 unigenes (4,31) have been identified, representing approximately 50% of the estimated 100 000 human genes (36). Analysis of these sequences indicates that only a minority of these unigenes have known functions. The next great challenge will be to identify the functions of all the gene products as well as their regulatory pathways. Establishing a human genome protein–protein interaction map (protein linkage map) would greatly help to fulfill these goals. Such a linkage map will provide clues to the function of novel genes and will also reveal new functions for known genes. In addition, the linkage map will provide potential therapeutic targets for disease treatments.

Yeast two-hybrid technology, developed by Fields and coworkers (1,6,12) and described in detail in previous chapters of this book, has proven to be a successful molecular genetic approach to detect in vivo protein–protein interactions (9,18,24, 34,39). The scale-up capability of yeast two-hybrid technology over other biochemical methods, such as co-affinity purification and co-immunoprecipitation, make the technology a top choice for large-scale identification of human protein–protein interactions (10,11). In fact, a smaller-scale bacteriophage T7 protein linkage map has already been completed based on the yeast two-hybrid technology (2).

To build a complete human protein linkage map, it is critical to construct a yeast two-hybrid cDNA library that includes cDNA of all human genes. Currently available yeast two-hybrid cDNA libraries are made from single tissues or organs and, therefore, are not suitable for application in the human protein linkage map project. Instead, a comprehensive two-hybrid library derived from human EST cDNA libraries, such as those generated by Merck & Co–Washington University–National Cancer Institute (NCI) EST Project (15), would be an ideal candidate (26).

Human EST cDNA libraries from the Merck–Washington University–NCI EST

Yeast Hybrid Technologies
Edited by L. Zhu and G.J. Hannon
© 2000 Eaton Publishing, Natick, MA

Project offer several advantages: *(i)* The sequences of both ends of each EST clone are known, i.e., the identity of an interacting protein detected by the two-hybrid system will be recognized immediately. *(ii)* All EST cDNA clones are individually arrayed in *Escherichia coli* cultures, which, in theory, can be amplified infinitely; whereas amplification of traditional cDNA libraries is detrimental to the representations of all the existing cDNA inserts. *(iii)* The majority of the human EST cDNA libraries are normalized; therefore, the libraries have good representation of low abundance cDNAs (5,41). The human EST cDNA libraries generated by the Merck–Washington University–NCI EST Project consist mainly of normalized cDNA libraries derived from various tissues and organs at multiple developmental stages, thus possessing great complexity (23). These clones are from more than 210 libraries as of January, 1999. The project is targeted to cover the entire human genome. Information on the EST libraries is updated along with the progress of the genome project (see the web site **http://www.ncbi.nlm.nih.gov/dbEST/index.html** for the update information). In addition, clones of human EST libraries are commercially available.

This chapter will discuss strategies and methodologies of construction of a yeast two-hybrid cDNA library from human EST clones—as well as the library screening—for the human protein linkage map. For basic yeast maintenance and more information on the yeast two-hybrid principle, readers are encouraged to consult the book by Guthrie and Fink (19) and other chapters of this book. A yeast GAL4-based two-hybrid system is discussed in this chapter.

2. PROTOCOLS

2.1 Materials

2.1.1 Yeast strains and media

Saccharomyces cerevisiae strains Y187 (MATα, *ura3-52, his3, ade2-101, trp1-901, leu2-3,112, met⁻, gal4Δ, gal80Δ, URA3:: GAL1$_{UAS}$-GAL1$_{TATA}$ -lacZ*), Y190 (MAT**a**, *ura3-52, his3-200, ade2-101, trp1-901, leu2-3,112, gal4Δ, gal80Δ, URA3:: GAL1$_{UAS}$-GAL1$_{TATA}$ -lacZ, cyh^{r}2, LYS2:: GAL1$_{UAS}$-GAL1$_{TATA}$ -HIS3*) (20), and AH109 (MAT**a**, *ura3-52, his3-200, ade2-101, trp1-901, leu2-3,112, gal4Δ, gal80Δ, MEL1, URA3::GAL-lacZ, LYS2::GAL1$_{UAS}$-GAL1$_{TATA}$-HIS3, GAL2$_{UAS}$-GAL2$_{TATA}$-ADE2, URA3:: MEL1$_{UAS}$-MEL1$_{TATA}$-lacZ*) are commercially available from CLONTECH Laboratories (Palo Alto, CA, USA). Yeast media, such as YPD medium, minimal SD medium, and various dropout supplements are also available from commercial sources such as CLONTECH Laboratories.

2.1.2 Plasmids

The yeast expression plasmid vector that contains the GAL4 binding domain (BD) is pGBKT7, whereas the vector containing the GAL4 activation domain (AD) is pGADT7 (see chapter 4 for detailed information of both vectors). pGBKT7 and pGADT7 contain the *TRP1* and *LEU* genes, respectively, as growth selection markers so that yeast host cells (such as AH109) containing these plasmids can propagate in media lacking tryptophan (SD-W) and leucine (SD-L). All of these plasmid vectors are available from CLONTECH Laboratories. pGBKT7 contains the kanamycin resistance gene, whereas pGADT7 contains the ampicillin resistance gene, both of which allow for bacterial growth selection. pGADT7 will be used for the human uni-gene EST two-hybrid library construction (AD-library), and pGBKT7 is used for the bait library construction (BD-library). Both pGBKT7 and pGADT7 contain T7 pro-moter and epitope tags at the carboxy-termini of the GAL4 domains in their respec-tive vectors. Such elements allow expression of the cloned inserts in these vectors and further provide independent means to confirm the interactions detected by the yeast two-hybrid system.

2.1.3 Human EST libraries

The I.M.A.G.E. (Integrated Molecular Analysis of Genomes and their Expression) Consortium (29) human EST clones generated by the Merck–Washington Universi-ty–NCI EST Project are derived from more than 210 individual cDNA libraries that are made from many tissues and organs at different developmental and/or disease stages. As of May, 2000, there are more than 1 500 000 arrayed human EST clones available, which represent nearly 90 000 unigenes (see web site **http://www.ncbi. nlm.nih.gov/UniGene/index.html** for updated information). The number of unigene clones will increase along with the progress of the human EST project. The I.M.A.G.E. Consortium human EST clones are arrayed in 384-well plates. The EST cDNAs are directionally cloned into several *E. coli* vectors, including pT7T3D-Pac, Lafmid BA, pBluescript SK(-), pCMV-SPORT, and pAMP10. More than 2 000 000 5′ and/or 3′ sequence entries of human EST clones have been deposited into the EST database (dbEST).

Arrayed I.M.A.G.E. Consortium human EST cDNA may be obtained from any of the five distributors of the Consortium. Within the United States, the distributors are American Type Culture Collection, Genome Systems, and Research Genetics. In Europe, I.M.A.G.E. Consortium distributors are UK Human Genome Mapping Pro-ject Center (HGMP), Hinxton, England; and Resource Center of the German Human Genome Project (RZPD), Berlin, Germany.

2.1.4 Other required materials

- An automated, 96-channel pipetting system such as the Multimek96 system by Beckman Coulter (Fullerton, CA, USA).
- 96-well plates, petri-dishes, and other plasticware.
- Disc-tumblers from V & P Scientific (San Diego, CA, USA).
- Multi-well plate readers.
- 96-pin replicators.
- Reagents and solutions, see Chapters 3–5 for details.

2.2 Methods

2.2.1 Yeast two-hybrid library constructions

2.2.1.1 Strategies of library construction

The conventional process of transferring EST inserts from an *E. coli*-based plas-mid to the yeast expression vectors of the two-hybrid system involves multiple steps of in vitro DNA manipulation (37). Typically, plasmid DNA is first isolated from *E.*

coli culture. The EST insert is then excised from the plasmid by appropriate restriction enzyme digestion. After being retrieved, the insert DNA is ligated to an *E. coli* yeast shuttle vector and transformed into *E. coli* for confirmation and propagation. The constructed yeast two-hybrid plasmid is finally transformed into yeast cells. This cloning strategy requires compatible restriction sites between the EST fragments and the shuttle vector and also requires that the ligated gene be in-frame with upstream sequence if a fusion protein is desired. The entire process is rather time-consuming and expensive. It is not practical to use such an approach for large-scale construction of a complete yeast two-hybrid cDNA library from human EST clones. Therefore, it is necessary to use a method that is capable of rapid and efficient transferring of the EST inserts to the yeast two-hybrid vector.

It is known that the yeast *Saccharomyces cerevisiae* has a high frequency of in vivo homologous recombination (33). In fact, Ma et al. (30) and Degryse et al. (7) have used homologous recombination as a means for plasmid constructions in yeast. They have shown that yeast is capable of repairing a linearized plasmid when a restriction fragment containing appropriate homologous regions at both ends is provided. More recently, we (Hua et al., 25) have demonstrated that 30 bp of a homologous sequence at each end of a DNA fragment is sufficient to integrate the fragment into a linearized plasmid in yeast, while 60 bp of homologous sequence will deliver a high yield of recombinant transformants. Furthermore, the fidelity of rejoining, i.e., restoration of the function of the recombination region, is extremely high. We (Hua et al., 25) and others (17,28,40) noted that there is no specific sequence preference at the homologous recombination sites. Based on these observations of homologous recombination, we (Hua et al., 25) have generated a strategy to transfer EST inserts to yeast vector by integrating short regions at both ends of the EST inserts by polymerase chain reaction (PCR). These regions are homologous to the ends of the recipient vectors. The mechanism is schematically illustrated in Figure 1. Such a strategy has been successfully implemented in the construction of a modular two-hybrid library (26). The method for in vivo cloning by homologous recombination to construct a human EST-derived two-hybrid library is described in detail in this chapter.

Several issues need to be considered in the process of construction of a human EST-derived two-hybrid library. *(i)* Due to the large number of EST clones involved, it would be ideal to generate a unigene EST library from available EST clones. Establishment of such a unigene library would reduce the workload and increase the efficiency of subsequent yeast two-hybrid library construction and screening. However, a human unigene EST library is not presently available. *(ii)* Because the open reading frame information of EST clones is not available, it would be necessary to transfer EST inserts to the yeast two-hybrid expression vectors in all three possible open reading frames. *(iii)* Transfer of EST inserts to the yeast two-hybrid expression vectors should be carried out in a high throughput format. *(iv)* All unigene clones in the yeast two-hybrid library should be arrayed and amplified to maintain their maximum representations.

2.2.1.2 Human EST-derived two-hybrid library construction

A. Protocol for the Generation of Human Unigene EST Library

Procedure

1. Purchase all the human EST clones available from I.M.A.G.E. Consortium through one of its distributors.

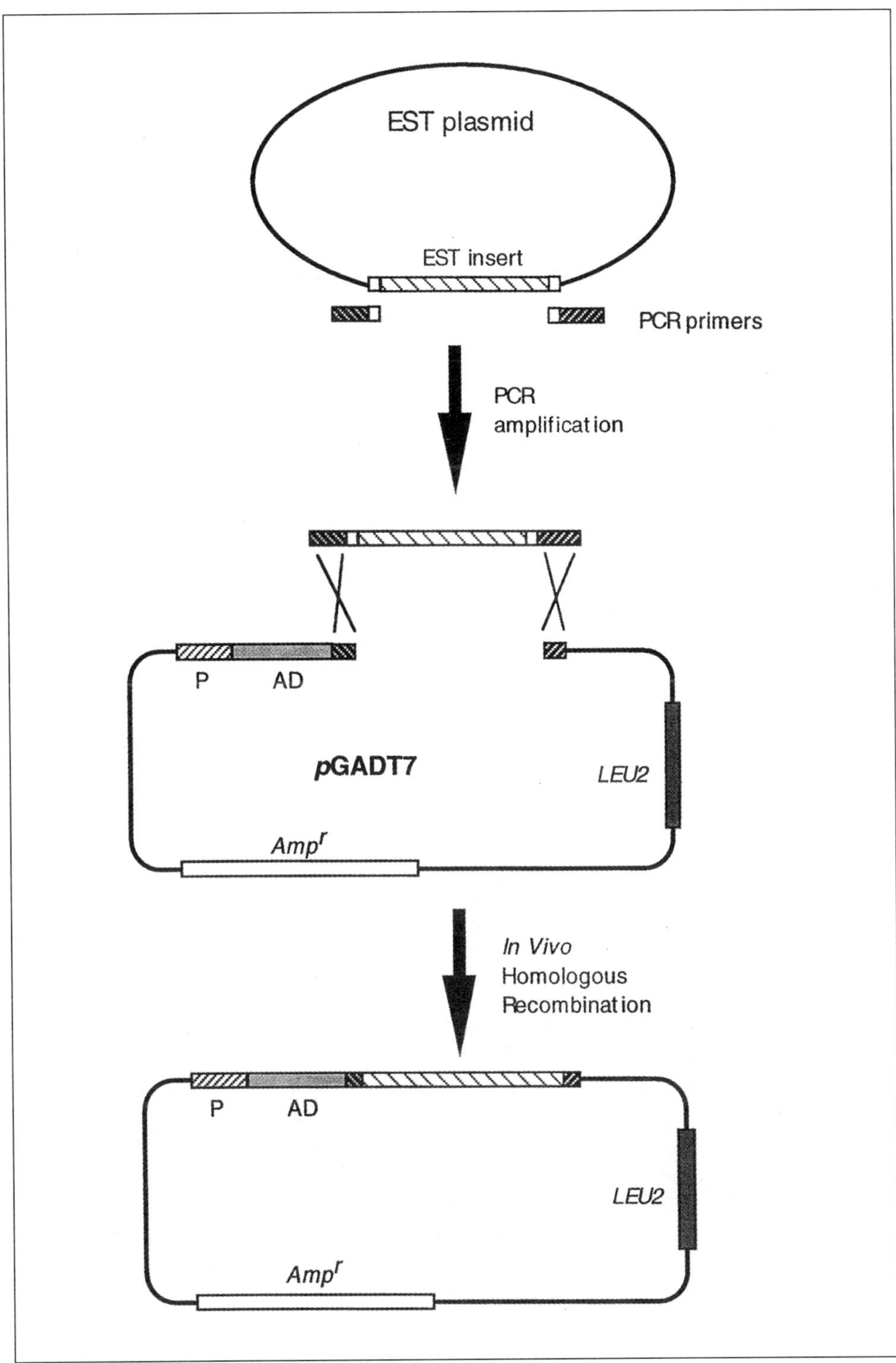

Figure 1. Schemes of transferring EST inserts from *E. coli*-based plasmid to yeast expression vectors via in vivo homologous recombination. Human EST inserts are amplified by PCR using primers that contain sequences homologous to the yeast vector (pGADT7 is shown here) at both ends. The PCR products are then mixed with linearized pGADT7 and cotransformed into yeast cells, where homologous recombination occurs.

2. Download the sequence and other related information of the EST unigene clones from the web site **http://www.ncbi.nlm.nih.gov/UniGene/**, as well as the information concerning all the available human EST clones from dbEST **http://www.ncbi.nlm.nih.gov/dbEST/index.html**.

3. After comparison of the unigene sequences with those of the dbEST by computer analysis, select two complete sets of unigene clones from the I.M.A.G.E. Consortium human EST clones. The first set of clones will contain the longest inserts derived from each unigene. A second set of unigene clones will contain inserts slightly shorter than those in the first set. It should be noted that limited duplication is not only necessary but also desirable for cross-referencing in the genome project.

4. The unigene clones can be picked by programmed robots and then arrayed in 384-well plates containing LB medium in the presence of ampicillin.

5. Incubate the unigene clones in *E. coli* cultures at 37°C until the cultures become confluent.

6. Index the human unigene EST library. Store the *E. coli* cultures at -70°C in the presence of glycerol.

B. Protocol for PCR Amplification of EST Inserts to Construct a Two-Hybrid AD-Library

Procedure

1. Oligonucleotide primers. Several plasmid vectors have been used in human EST library constructions including pT7T3D-Pac, Lafmid BA, pBluescript SK(-), pCMV-SPORT, and pAMP10. For each specific vector, four distinct oligonucleotide primers will be used for the human EST two-hybrid library construction. The first three primers are designed to amplify EST inserts in three open reading frames at their 5′ ends, whereas the fourth one corresponds to the 3′ end vector sequence. Each primer consists of two portions. The sequence at its 5′ end (64 bases) corresponds to the sequence flanking the multiple cloning sites of the yeast two-hybrid vector pGADT7, so that the PCR products can be fused into the carboxy-terminus of the GAL4 activation domain in pGADT7 by in vivo homologous recombination. The 3′ sequence (21 bases), on the other hand, corresponds to sequence flanking the EST inserts in the *E. coli* plasmid vectors. The oligonucleotide primers used for PCR amplification of EST inserts in pT7T3D-Pac vector are listed below. Bases added to allow for coding of all three open reading frames are underlined below. Primers for other vectors can be designed in a similar manner. It should be noted that no stop codons should be introduced in any of the three open reading frames. Due to their lengths, these oligonucleotides must be purified by polyacrylamide gel electrophoresis after synthesis.

1-1: 5′-GAGCGCCGCCATGGAGTACCCATACGACGTACCAGATTACGC-TCATATGGCCATGGAGGCCAGT<u>GGA</u>GGCCCTCGAGGCCAAGAATTC-3′

1-2: 5′-GAGCGCCGCCATGGAGTACCCATACGACGTACCAGATTACGC-TCATATGGCCATGGAGGCCAGT<u>GG</u>GGGCCCTCGAGGCCAAGAATTC-3′

1-3: 5′-GAGCGCCGCCATGGAGTACCCATACGACGTACCAGATTACGC-
TCATATGGCCATGGAGGCCAGT<u>G</u>GGCCCTCGAGGCCAAGAATTC-3′

1-4: 5′-GAGATGGTGCACGATGCACAGTTGAAGTGAACTTGCGGGGT-
TTTTCAGTATCTACGATTCACCCTCACTAAAGGGAATAAGCTT-3′

2. Transfer *E. coli* cultures containing unigene EST clones from each 384-well plate into four 96-well plates by a 96-pin replicator. Approximately 1 µL of *E. coli* culture will be transferred into 100 µL LB medium containing ampicillin.

3. Incubate the 96-well *E. coli* plates on a microtiter-plate shaker at 37°C overnight.

4. Set up PCR mix according to instructions of the manufacturer of the PCR reagents. Aliquot 50 µL of the PCR mix to each tube of a 96-microtube PCR plate. The following are the ingredients of PCR mix for the Advantage™ PCR kit from CLONTECH. The mix (a total of 50 µL final volume) contains 40 µL of water, 5 µL of 10× reaction buffer, 1 µL of 10 µM 5′ end primers (0.33 µL of each ORF primer, such as primers 1-1, 1-2, and 1-3 listed above), and 1 µL of 10 µM 3′ end primer, 1 µL of 10 mM dNTPs, and 1 µL of KlenTaq™ polymerase mix.

5. Using a 96-pin replicator, transfer approximately 1 µL of confluent *E. coli* culture into each tube of a 96-microtube PCR plate for direct PCR amplification.

6. Load the 96-microtube PCR plate onto a thermal cycler equipped with a 96-tube block and a heated cover (to prevent evaporation) such as the Peltier Thermal Cycler by MJ Research (Waltham, MA, USA).

7. The optimal number of amplification cycles will depend on the polymerase. For KlenTaq™ polymerase mix by CLONTECH, the two-step PCR amplification starts with denaturing at 94°C for 3 min, followed by 30 cycles of 20 s of denaturation at 94°C, and 3 min of extension at 68°C.

8. Check the amplified human EST inserts by analyzing 5 µL of each PCR product on an agarose gel.

C. Protocol for Cloning of EST Insert Into Two-Hybrid AD-Library by
Homologous Recombination

Procedure

1. Digest pGADT7 DNA with restriction enzymes *Bam*HI and *Eco*RI. After complete digestion, extract the mixture with phenol-chloroform and precipitate the DNA with ethanol. Suspend the DNA in TE buffer.

2. Mix 1 µg of the linearized pGADT7 DNA with 25 µL of each PCR product in regular 96-well plates.

3. Prepare yeast competent cells (strain Y187). Transform the competent cells with the DNA mixtures in 96-tube plates (National Scientific Supply Co., San Rafael, CA, USA) using the modified lithium acetate protocol (16,22,27). Consult chapters 3–5 for detailed protocols of yeast competent cell preparation and transformation.

4. Spread the transformants onto agar plates of SD medium lacking leucine

(SD/-Leu) using sterile glass beads (5 mm in diameter). Use one agar plate for each tube of transformants. Invert the plates, leaving the glass beads inside for harvesting in the next step. Incubate the plates at 30°C for 4–5 days. The PCR-amplified EST inserts will be integrated into the linearized pGADT7 by homologous recombination in yeast cells.

5. Scrape the yeast cells off the agar plate by shaking the plate in the presence of glass beads and 5 mL of liquid SD/-Leu medium containing 20% glycerol. Array the yeast transformants containing human EST inserts in 96-deep-well boxes (National Scientific Supply Co.) (one well/agar plate) in multiple copies.

6. The arrayed yeast clones can be individually amplified in SD/-Leu medium in 96-well plates. Due to precipitation of yeast cells at the bottom of the wells, culture medium in the 96-well plates should be shaken by micro-discs on a disc-tumbler (V & P Scientific).

7. Index the arrayed human unigene EST two-hybrid AD-library and store the library at -70°C.

After the human unigene EST two-hybrid AD-library is constructed and arrayed, two alternative methods can be used to screen the library. The first method is based on yeast transformation. In this approach, all the arrayed AD-library clones are first pooled together after amplification. The pooled yeast clones are then transformed with bait cDNA fused to the GAL4 binding domain in pGBKT7 (see Chapters 3–5 for references). The second method is based on yeast mating (3,13). Yeast cells of one mating type strain are first transformed with bait cDNA in pGBKT7. The yeast cells containing the bait cDNA are then fused to cells of the opposite mating type from the arrayed human unigene EST AD-library either by individual clone mating or by bulk mating after pooling the whole AD-library. Thus both pGADT7- and pGBKT7-derived plasmids are introduced into the same diploid yeast cells (see Section 2.2.2 for detailed procedures). The mating approach requires a two-hybrid "bait" library constructed from the human unigene EST clones.

<u>D. Protocol for Construction of Two-Hybrid Bait Library (the BD-Library)</u>

Procedure

To create the bait library, the human unigene EST inserts are first amplified by PCR and then fused to the GAL4 binding domain in pGBKT7 by homologous recombination. The procedures involved in the bait library construction are essentially the same as those described in Sections 2.2.1.2.B and C. However, the primers used for PCR amplification should contain the sequences flanking the multiple cloning site of pGBKT7 at the 5′ side of the primers. The sequences of four primers used to amplify unigene EST inserts in Lafmid vector are listed below. The first three primers are designed to generate three different open reading frames, whereas the fourth one is the reverse primer. Readers can easily design primers for other vectors used in the human EST library construction.

2-1: 5′-AGCCGCCATCATGGAGGAGCAGAAGCTGATCTCAGAGGAGGAC-CTGCATATGGCCATGGAGGCC<u>GGT</u>GACCAGGATTACGCCAAGCTT

2-2: 5′-AGCCGCCATCATGGAGGAGCAGAAGCTGATCTCAGAGGAGGAC-CTGCATATGGCCATGGAGGCC<u>GG</u>GACCAGGATTACGCCAAGCTT

2-3: 5′-AGCCGCCATCATGGAGGAGCAGAAGCTGATCTCAGAGGAGGAC-
CTGCATATGGCCATGGAGGCC<u>G</u>GACCAGGATTACGCCAAGCTT

2-4: 5′-GGCTGCAAGCGCGCAAAAAACCCCTCAAGACCCGTTTAGAGGC-
CCCAAGGGGTTATGCTAGTTATACGTTGTAAAACGACGGCCAGT

In the process of in vivo cloning by homologous recombination, *Eco*RI and *Bam*HI-digested pGBKT7 should be used. Yeast strain AH109 is recommended. In addition, the selection medium is SD lacking tryptophan (SD/-Trp). Other procedures are the same as those in Section 2.2.1.2.C. The library thus made is designated human unigene EST BD-library.

E. Protocol for Human EST Two-Hybrid Library Quality Controls

Procedure

The quality of human unigene EST two-hybrid library constructions should be monitored at several critical steps.

1. Before picking unigenes from the I.M.A.G.E. Consortium human EST clones, randomly select a certain number of unigene clones. Prepare plasmid DNA from these clones and sequence the EST inserts from both the 5′ and 3′ ends for confirmation of their locations and sequences.

2. During the process of unigene library construction, randomly select clones of arrayed unigenes and confirm their locations and sequences.

3. After PCR amplification of unigene EST clones, analyze all the PCR products on agarose gels if possible. Alternatively, analyze PCR products from 8 or 12 samples from each 96-microtube PCR plate.

4. During the process of human EST two-hybrid library construction, randomly select a number of arrayed yeast clones. Culture the yeast cells in 3 mL of SD/-Leu (for AD-library) or SD/-Trp (for BD-library) liquid media. Retrieve the plasmid DNA from the yeast cultures. Transform the retrieved plasmid DNA into *E. coli*. Pick 10–15 colonies and prepare plasmid DNA from these colonies. Sequence the EST inserts from both the 5′ and 3′ ends. The information will confirm the locations and sequences of these clones in the arrayed two-hybrid library. It will also reveal the ratio of three open reading frames in the clones.

F. Protocol for the Exclusion of Self-Activation Clones from the BD-Library

Procedure

In the two-hybrid library screening, a majority of false positives are due to transcriptional activation of reporter genes in the absence of protein–protein interactions (2). Therefore, it is necessary to exclude these self-activating clones from the human EST two-hybrid libraries.

Yeast strain AH109 used in this system contains the reporter genes *ADE2*, *HIS3*, *MEL1*, and *lacZ*. The expressions of these genes are controlled by distinct GAL4-responsive elements (see chapter 5). Self-activating clones from the previously-described BD-library are able to drive the expression of these reporter genes. These yeast cells can grow in the SD medium that lacks adenine and histidine. AH101 strain contains a wild-type *MEL1* gene that encodes secreted α-galactosidase. The α-galactosidase can be used for blue color assay in the presence of α–X-gal (5-bromo-4-chloro-3-indolyl-α-D-galactopyranoside), thus the self-activating clones can be easily identified (see detail in Chapter 4).

1. Grow the arrayed BD-library in SD/-Ade/-His/-Trp liquid medium in 96-well plates. The medium also contains 30 μg/mL α–X-gal at 30°C.

2. Exclude from the BD-library the clones that grow in the medium and develop blue color.

2.2.2 Human EST Two-Hybrid Library Screening

To identify the interacting partners of a bait protein, both the AD-library and BD-bait plasmids first need to be introduced into the same yeast cells either by transformation (46) or by yeast mating (3,13). If a bait binding domain fusion protein can interact with an activation domain fusion protein from the AD-library, the interacting fusion proteins will be able to drive the expression of reporter genes in the yeast cell. In AH109 cells, expression of *ADE2* and *HIS3* genes will allow the yeast cells to grow in SD medium lacking adenine and histidine (SD/-Ade/-His). The *MEL1* and *lacZ* gene products (α-galactosidase and β-galactosidase, respectively) can be used for blue color assay, thus further reducing the background of false positives that arise from growth selection. Because α-galactosidase is secreted from yeast cells, it can be assayed more conveniently than the β-galactosidase. Therefore, assaying β-galactosidase is not discussed in the library screening. Three approaches of library screening are described here.

2.2.2.1 Screening by individual clone mating

In this approach, individual clones of the AD-library are mated with each clone from the BD-library. The mated cells are assayed for their growth, then further assayed for the identification of the positive interactions.

1. Grow the arrayed human unigene EST AD-library in strain Y187 (mating type α) in SD/-Leu selection medium in 96-well plates overnight at 30°C to saturation.

2. Grow strain AH109 (mating type a) that contains a clone of bait cDNA from the BD-library in a large volume to saturation.

3. Transfer 2 μL each of AD-library cells and the BD-library yeast cells into fresh 96-well plates that contain 100 μL of YPD-rich medium in each well. Incubate the plates overnight at 30°C. Yeast mating occurs in the rich medium, resulting in diploid cells containing both AD- and BD-plasmids (see Chapter 5).

4. Remove the YPD medium using an automated 96-channel pipetting system.

5. Wash each well three times with 200 μL of sterile water and suspend the cells in 200 μL SD/-Ade/-Leu/-Trp/-His.

6. Transfer 2 μL of the suspended cells from each well to fresh 96-well plates containing 200 μL of SD/-Ade/-Leu/-Trp/-His in the presence of 30 μg/mL α–X-gal. Incubate the plates at 30°C for 6–9 days. Positive clones of mated yeast cells will express the *ADE2* and *HIS3* genes and therefore will grow in the

SD/-Ade/-Leu/-Trp/-His selection medium. They will also become blue due to α-galactosidase activity.

7. Examine the density of the cultures by scanning the 96-well plates with a microplate reader at λ = 405 nm. Process the scanning results by computer. Locate the wells of the positive clones.

2.2.2.2 Screening pooled AD-library by mating with bait clones

In this approach, a bait clone is mated with a pooled AD-library. The mated mixture is then plated on the selection medium and assayed for color development.

1. After being amplified individually, all the clones of human unigene EST AD-library are pooled together. Alternatively, the clones of the AD-library can be pooled into several sub-fractions.

2. Grow the pooled AD-library yeast cells (Y187) in 100 mL SD/-Leu medium to the mid-log phase (OD_{600} = 0.8).

3. Grow the individual bait clones from the BD-library (in AH109) in 100 mL SD/-Trp medium to the mid-log phase.

4. Combine an equal volume of a bait clone and the pooled AD-library in a large flask. Add 2 volumes of YPD medium to the mixture.

5. Gently shake the mixture (50–80 rpm) at 30°C overnight to allow efficient mating.

6. Centrifuge the yeast culture at 1000× g for 10 min. Wash the yeast cells with SD/-Ade/-Leu/-Trp/-His three times.

7. Resuspend the yeast cells in SD/-Ade/-Leu/-Trp/-His liquid medium containing 30 mg/mL of α–X-gal. Spread the yeast cells on agar plates of SD/-Ade/-Leu/-Trp/-His (150 mm in diameter) using sterile glass beads. Incubate the plates at 30°C for 5–9 days.

8. Pick the colonies grown on the SD/-Ade/-Leu/-Trp/-His plates that developed blue color. Grow the yeast cells in SD/-Leu liquid medium.

9. Retrieve plasmid DNA from the yeast cultures by using a yeast plasmid isolation kit (CLONTECH).

10. Transform *E. coli* cells with the retrieved plasmid DNA and spread the transformants on LB/amp plates.

11. Prepare plasmid DNA from *E. coli* cells.

12. Sequence the EST inserts.

13. Search the human EST unigene database to identify the locations and accession numbers of the positive clones.

2.2.2.3 Screening human unigene EST AD-library by transformation

This is a traditional transformation approach. Bait plasmid DNA is transformed into the pooled AD-library. Positive interactions are then identified from the library.

A. Prepare bait plasmid DNA from BD-library.

1. Culture an individual clone of the human unigene EST BD-library (see sec-

tion 2.2.1.2.D) in 5 mL of SD/-Trp liquid medium.

2. Retrieve plasmid DNA from the yeast cultures using a yeast plasmid isolation kit (CLONTECH).

3. Transform *E. coli* competent cells with the retrieved plasmid DNA.

4. Culture the transformed *E. coli* in 2 L of LB/kanamycin medium. Isolate the plasmid DNA from the *E. coli* cultures.

B. Screen the two-hybrid AD-library.

1. Amplify the EST AD-library clones in 96-well plates overnight at 30°C. Pool the AD-library.

2. Grow the pooled library cultures to mid-log phase (OD_{600}=0.6) in large flasks.

3. Prepare yeast competent cells using the lithium acetate protocol (see Chapter 3).

4. Transform the yeast competent cells with the BD-plasmid that contains a bait cDNA prepared in the previous section.

5. Spread the transformants onto agar plates of SD/-Ade/-Leu/-Trp/-His using sterile glass beads. The plates have been soaked with α-galactosidase. Incubate the plates at 30°C for 5–9 days.

6. Pick the colonies grown on the SD/-Ade/-Leu/-Trp/-His plates that developed blue color. Grow the yeast cells in SD/-Leu liquid medium.

7. Retrieve plasmid DNA from the yeast cultures by using a yeast plasmid isolation kit (CLONTECH).

8. Transform *E. coli* cells with the retrieved plasmid DNA and spread the transformants on LB/amp plates.

9. Prepare plasmid DNA from *E. coli* cells.

10. Sequence the EST inserts.

11. Based on the sequence results, identify the locations and accession numbers of the positive clones.

12. Establish the human protein linkage map database.

3. DISCUSSION

Genomics is moving from a focus on structure (mapping and sequencing) to a focus on function (21,32). Several key technologies have been implemented in functional genomics studies. For example, three recently devised methods, i.e., oligonucleotide chips (14,35), SAGE (42), and DNA microarrays (38) can be very powerful tools to obtain genome-wide mRNA expression data. These technologies can provide a view of changes in gene expression patterns in response to different physiological states (8,45). In addition, partial protein sequences from high-resolution, two-dimensional gels and electrospray mass spectrometry of protein complexes can be used to assign peptides to specific gene sequences (43,44). These technologies are mainly focused on individual genes and proteins. No information on regulatory pathways and protein–protein interactions, which play pivotal roles in physiological status, can be easily obtained from these technologies.

In this chapter, we have described a protocol for building a complete human protein linkage map based on the yeast two-hybrid technology. Yeast two-hybrid technology has been extensively used in the detection of protein–protein interactions since its establishment (12). As opposed to the conventional approach, which searches for interacting partners of a defined protein in a two-hybrid library made from a single tissue/organ, we proposed here to use human unigene EST-derived two-hybrid libraries. Such libraries are constructed from available human EST clones by employing a unique approach based on in vivo homologous recombination in yeast cells (25). Two-hybrid libraries thus constructed can be used to establish a human protein linkage map by library screening using either transformation or yeast mating. A complete human protein linkage map will provide invaluable information in the functional genomics era. For example, it will provide functional hints for novel genes and reveal new functions for known genes. Many regulatory and metabolic pathways can be established from the linkage map. Furthermore, the linkage map will also provide potential targets for therapeutic intervention.

Detection of positive interactions in the GAL4 yeast two-hybrid system is based on transcriptional activation of reporter genes upon formation of a protein complex between the interacting partners. This activation is a readout of such variables as the binding affinities of the hybrids, the protein stability of the hybrids, nuclear entry, and accessibility of the interacting surfaces to each other. It should be noted that detection of several interacting partners for a single protein does not provide evidence for exclusive or simultaneous binding. Limitations to the two-hybrid approach include positive interactions that are detected by this method but do not occur in human cells. For example, proteins derived from large gene families might fall into this category. Another potential problem is false negative results. It might be impossible to detect certain interactions that are mediated by post-translational modifications such as glycosylation and myristylation, or that occur between proteins unable to fold appropriately in an intracellular environment, or that cannot be imported into the nucleus. In addition, a certain percentage of human EST clones do not contain the complete 5′ end sequences; therefore, the amino termini of the encoded proteins are not present. Protein interactions mediated by those missing amino termini will not be detected using the current approach. Nevertheless, establishment of an EST-based human protein linkage map will make very significant contributions to the understanding of protein and cell functions in the functional genomics era.

ACKNOWLEDGMENTS

We are grateful to Ann Holtz for her enthusiastic discussions and technical expertise. We also thank Dr. John Ambroziak for critical reading of the manuscript.

REFERENCES

1. **Bartel, P., C.T. Chien, R. Sternglanz, and S. Fields.** 1993. Elimination of false positives that arise in using the two-hybrid system. BioTechniques *14*:920-924.
2. **Bartel, P.L., J.A. Roecklein, D. SenGupta, and S. Fields.** 1996. A protein linkage map of *Escherichia coli* bacteriophage T7. Nature Genet. *12*:72-77.
3. **Bendixen, C., S. Gangloff, and R. Rothstein.** 1994. A yeast mating-selection scheme for detection of protein-protein interactions. Nucleic Acids Res. *22*:1778-1779.
4. **Boguski, M.S. and G.D. Schuler.** 1995. ESTablishing a human transcript map. Nature Genet. *10*:369-371.

5. **Bonaldo, M.F., G. Lennon, and M.B. Soares.** 1996. Normalization and subtraction: two approaches to facilitate gene discovery. Genome Res. *6*:791-806.
6. **Chien, C.T., P.L. Bartel, R. Sternglanz, and S. Fields.** 1991. The two-hybrid system: a method to identify and clone genes for proteins that interact with a protein of interest. Proc. Natl. Acad. Sci. USA *88*:9578-9582.
7. **Degryse, E., B. Dumas, M. Dietrich, L. Laruelle, and T. Achstetter.** 1995. In vivo cloning by homologous recombination in yeast using a two-plasmid-based system. Yeast *11*:629-640.
8. **DeRisi, J.L., V.R. Iyer, and P.O. Brown.** 1997. Exploring the metabolic and genetic control of gene expression on a genomic scale. Science *278*:680-686.
9. **Durfee, T., K. Becherer, P.L. Chen, S.H. Yeh, Y. Yang, A.E. Kilburn, W.H. Lee, and S.J. Elledge.** 1993. The retinoblastoma protein associates with the protein phosphatase type 1 catalytic subunit. Genes Dev. *7*:555-569.
10. **Evangelista, C., D. Lockshon, and S. Fields.** 1996. The yeast two-hybrid system: prospects for protein linkage maps. Trends Cell Biol. *6*:196-199.
11. **Fields, S.** 1997. The future is function. Nature Genet. *15*:325-327.
12. **Fields, S. and O. Song.** 1989. A novel genetic system to detect protein-protein interactions. Nature *340*:245-247.
13. **Finley, R. and R. Brent.** 1994. Interaction mating reveals binary and ternary connections between Drosophila cell cycle regulators. Proc. Natl. Acad. Sci. USA *91*:12980-12984.
14. **Fodor, S.P., J.L. Read, M.C. Pirrung, L. Stryer, A.T. Lu, and D. Solas.** 1991. Light-directed, spatially addressable parallel chemical synthesis. Science *251*:767-773.
15. **Gerhold, D. and C.T. Caskey.** 1996. It's the genes! EST access to human genome content. BioEssays *18*:973-981.
16. **Gietz, D., A. St. Jean, R.A. Woods, and R.H. Schiestl.** 1992. Improved method for high efficiency transformation of intact yeast cells. Nucleic Acids Res. *20*:1425.
17. **Glasunov, A.V., M. Frankenberg-Schwager, and D. Frankenberg.** 1995. Different repair kinetics for short and long DNA double-strand gaps in *Saccharomyces cerevisiae*. Int. J. Radiat. Biol. *68*:412-428.
18. **Guan, K.L., C.W. Jenkins, Y. Li, M.A. Nichols, X. Wu, C.L. O'Keefe, A.G. Matera, and Y. Xiong.** 1994. Growth suppression by p18, a p16$^{INK4/MTS1}$- and p14$^{INK4B/MTS2}$-related CDK6 inhibitor, correlates with wild-type pRb function. Genes Dev. *8*:2939-2952.
19. **Guthrie, C. and G.R. Fink.** 1991. Methods in enzymology. vol. 194, Guide to yeast genetics and molecular biology. Academic Press. San Diego, CA.
20. **Harper, J.W., G.R. Adami, N. Wei, K. Keyomarsi, and S.J. Elledge.** 1993. The p21 *Cdk*-interacting protein Cip1 is a potent inhibitor of G1 cyclin-dependent kinases. Cell *75*:805-816.
21. **Hieter, P. and M. Boguski.** 1997. Functional genomics: it's all how you read it. Science *278*:601-602.
22. **Hill, J., K.A. Donald, and D.E. Griffiths.** 1991. DMSO-enhanced whole cell yeast transformation. Nucleic Acids Res. *19*:5791.
23. **Hillier, L., G. Lennon, M. Becker, M.F. Bonaldo, B. Chiapelli, S. Chissoe, N. Dietrich, T. DuBuque, A. Favello, W. Gish, M. Hawkins, M. Hultman, et al.** 1996. Generation and analysis of 280,000 human expressed sequence tags. Genome Res. *6*:807-828.
24. **Hsu, H., H.B. Shu, M.G. Pan, and D.V. Goeddel.** 1996. TRADD-TRAF2 and TRADD-FADD interactions define two distinct TNF receptor 1 signal transduction pathways. Cell *84*:299-308.
25. **Hua, S., M. Qiu, E. Chan, L. Zhu, and Y. Luo.** 1997. Minimum length of sequence homology required for in vivo cloning by homologous recombination in yeast. Plasmid *38*:91-96.
26. **Hua, S., Y. Luo, M. Qiu, E. Chan, H. Zhou, and L. Zhu.** 1998. Construction of a modular yeast two-hybrid cDNA library from human EST clones for the human genome protein linkage map. Gene *215*:143-152.
27. **Ito, H., Y. Fukuda, K. Murata, and A. Kimura.** 1983. Transformation of intact yeast cells treated with alkali cations. J. Bacteriol. *153*:163-168.
28. **Jha, B., F. Ahne, and F. Eckardt-Schupp.** 1993. The use of a double-marker shuttle vector to study DNA double-strand break repair in wild type and radiation-sensitive mutations of the yeast *Saccharomyces cerevisiae*. Current Genetics *23*:402-407.
29. **Lennon, G., C. Auffray, M. Polymeropoulos, and M.B. Soares.** 1996. The I.M.A.G.E. Consortium: an integrated molecular analysis of genomes and their expression. Genomics *33*:151-152.
30. **Ma, H., S. Kunes, P.J. Schatz, and D. Botstein.** 1987. Plasmid construction by homologous recombination in yeast. Gene *58*:201-216.
31. **Miller, G., R. Fuchs, and E. Lai.** 1997. IMAGE cDNA clones, Unigene clustering, and ACeDB: an integrated resource for expressed sequence information. Genome Res. *7*:1027-1032.
32. **Oliver, S.G.** 1996. From DNA sequence to biological function. Nature *379*:597-600.
33. **Orr-Weaver, T.L. and J.W. Szostak.** 1983. Yeast recombination: the association between double-strand gap repair and crossing-over. Proc. Natl. Acad. Sci. USA *80*:4417-4421.
34. **Ozenberger, B.A. and K.H. Young.** 1995. Functional interaction of ligands and receptors of the hematopoietic superfamily in yeast. Mol. Endocrinol. *9*:1321-1329.

35. **Pease, A.C., D. Solas, E.J. Sullivan, M.T. Cronin, C.P. Holmes, and S.P. Fodor.** 1994. Light-generated oligonucleotide arrays for rapid DNA sequence analysis. Proc. Natl. Acad. Sci. USA *91*:5022-5026.
36. **Rowen, L., G. Mahairas, and L. Hood.** 1997. Sequencing the human genome. Science *278*:605-606.
37. **Sambrook, J., E.F. Fritsch, and T. Maniatis.** 1989. Molecular cloning: A laboratory manual. 2nd ed. Cold Spring Harbor Press, Cold Spring Harbor, NY.
38. **Schena, M., D. Shalon, R.W. Davis, and P.O. Brown.** 1995. Quantitative monitoring of gene expression patterns with a complementary DNA microarray. Science *270*:467-470.
39. **Schreiber-Agus, N., L. Chin, K. Chen, R. Torres, G. Rao, P. Guida, A.I. Skoultchi, and R.A. DePinho.** 1995. An amino-terminal domain of Mxi1 mediates anti-Myc oncogenic activity and interacts with a homolog of the yeast transcriptional repressor SIN3. Cell *80*:777-786.
40. **Shinohara, A. and T. Ogawa.** 1995. Homologous recombination and the roles of double-strand breaks. TIBS *20*:387-391.
41. **Soares, M.B., M.F. Bonaldo, P. Jelene, L. Su, L. Lawton, and A. Efstratiadis.** 1994. Construction and characterization of a normalized cDNA library. Proc. Natl. Acad. Sci. USA *91*:9228-9232.
42. **Velculescu, V.E., L. Zhang, B. Vogelstein, and K.W. Kinzler.** 1995. Serial analysis of gene expression. Science *270*:484-487.
43. **Wilm, M., A. Shevchenko, T. Houthaeve, S. Breit, L. Schweigerer, T. Fotsis, and M. Mann.** 1996. Femtomole sequencing of proteins from polyacrylamide gels by nano-electrospray mass spectrometry. Nature *379*:466-469.
44. **Yates J.R., J.K. Eng, and A.L. McCormack.** 1995. Mining genomes: correlating tandem mass spectra of modified and unmodified peptides to sequences in nucleotide databases. Anal. Chem. *67*:3202-3210.
45. **Zhang, L., W. Zhou, V.E. Velculescu, S.E. Kern, R.H. Hruban, S.R. Hamilton, B. Vogelstein, and K.W. Kinzler.** 1997. Gene expression profiles in normal and cancer cells. Science *276*:1268-1272.
46. **Zhu, L.** 1996. Yeast GAL4 two-hybrid system, p. 173-196. *In* R. Tuan (Ed.), Methods in Molecular Biology, vol 63: Recombinant Protein Protocols: Detection and Isolation. Humana Press, Totowa, NJ.

19 A Double Interaction Screen to Isolate DNA Binding and Protein-Tethered Transcription Factors

Leslie Pick[1], Ulrike Löhr[1], and Yan Yu[2]
[1]The Brookdale Center, Department of Biochemistry and Molecular Biology, Mount Sinai School of Medicine, New York, NY; [2]OriGene Technologies, Rockville, MD, USA

1. INTRODUCTION

The expression of most eukaryotic genes is controlled by large and complex cis-acting regulatory regions. Cis-regulatory elements control transcription by interactions with sequence-specific DNA-binding proteins that positively or negatively regulate transcription by RNA polymerase II. These DNA binding transcription factors also engage in protein–protein interactions with each other, with co-activators, co-repressors, or with the basal transcription machinery to influence transcriptional activity. We have developed a modification of yeast one- and two-hybrid (4,10,11,18, 20) screens—the Double Interaction Screen (DIS)—designed to identify partners of DNA binding transcription factors (32–34). The screen relies upon the identification of a cis-acting regulatory element that is a direct target of the transcription factor in question (Protein X). This regulatory element is used in the screen to "anchor" a native full-length protein (Protein X) to DNA upstream of standard reporter genes. The regulatory element will contain binding sites for Protein X and for partners of Protein X that interact directly with DNA. In addition, the regulatory element is likely to contain binding sites for other transcription factors whose activities are independent of Protein X. Thus two baits are available in the screen: *(i)* the cis-regulatory element itself and *(ii)* Protein X, which is anchored to the regulatory element via native binding sites. This screen in theory allows for the direct isolation of all of the DNA binding proteins that interact with a cis-regulatory element of interest, even if Protein X has not yet been identified.

The scheme of the DIS is outlined in Figure 1. The cis-acting regulatory element of interest is fused upstream of reporter genes *HIS3* and/or *lacZ* to generate the first Bait, the DNA itself (DNA Bait). Protein X, the second bait (Protein Bait), is then expressed in these cells as a full-length, non-fusion protein that can bind to its native

Yeast Hybrid Technologies
Edited by L. Zhu and G.J. Hannon
© 2000 Eaton Publishing, Natick, MA

binding sites in the regulatory element. Screening of a cDNA library allows identification of three types of proteins: DNA-binding proteins that interact with the DNA but not with the Protein Bait (Figure 1C), sequence-specific DNA-binding proteins that also interact with the Protein Bait (Figure 1D), and proteins that interact with the Protein Bait only by specific protein–protein interactions (Figure 1E). We used this method previously in a screen for transcription regulators and protein partners of the *Drosophila* homeobox gene *fushi tarazu* (*ftz*). The cis-regulatory element used in the screen was a *ftz* regulatory element that had been shown to mediate autoregulation by direct binding of Ftz protein (17,24,27). We isolated activators and repressors of *ftz* transcription in the screen as well as Ftz protein partners that participate in autoregulation (33,34).

Our experience with Ftz suggests several advantages of the DIS. First, because native cis-regulatory target elements are employed, it is possible to use full-length protein baits to screen for cofactors. This should allow the protein of interest to fold in a native conformation when bound to DNA. Further, it will allow identification of protein partners that interact with any region of Protein X as well as those requiring contacts with more than one surface of the protein for stable interactions. It may also be advantageous to avoid the use of DNA-binding domain fusion proteins because fusion to heterologous DNA-binding domains can alter protein activity in some cases (12). Second, the use of a native target cis-regulatory element produces a natural environment for factor binding in that the Protein Bait binding sites are correctly spaced and interspersed with cofactor binding sites. This natural spacing results in lower basal levels of transcription activation by full-length transcription activators than is seen when multiple copies of binding sites are clustered upstream of a promoter, as in standard one- and two-hybrid screens. Third, the use of a native cis-regulatory element facilitates identification of DNA-binding protein partners that require specific binding sites for high-affinity protein–protein interaction (Figure 1D). This

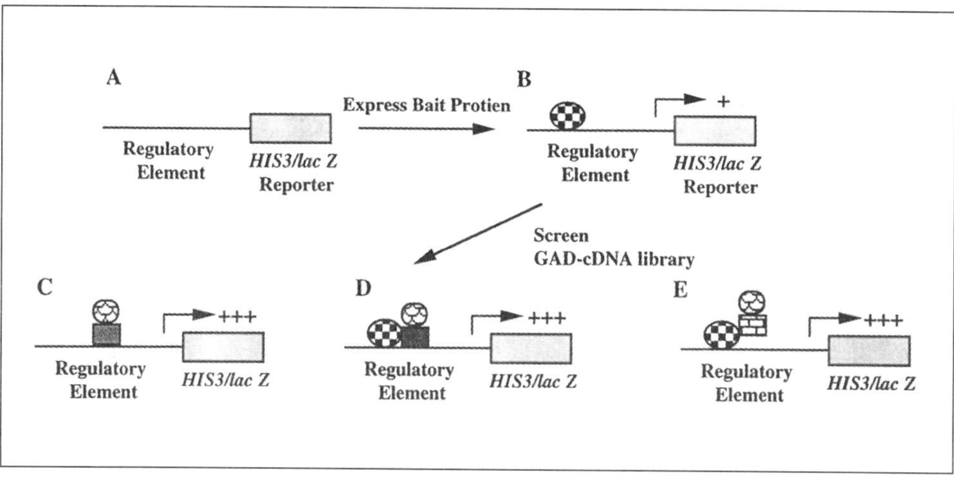

Figure 1. Scheme of the double-interaction screen. The general outline of the DIS is shown. (**A**) A cis-regulatory element is placed upstream of reporter genes and integrated into the yeast genome to generate the DNA Bait. (**B**) A protein known to bind to the regulatory element is expressed in these yeast. It binds to the regulatory element to create the Protein Bait. Expression levels are still low (+). A Gal4 Activation Domain (GAD)-tagged cDNA library is screened for activation of the *HIS3* reporter gene. Three types of factors can be isolated in the screen: (**C**) DNA-binding proteins that interact directly with the regulatory element. (**D**) Protein Bait partners that also bind to specific DNA sequences. (**E**) Protein Bait partners that interact only at the protein level.

type of protein partner may bind to DNA cooperatively with Protein X, such that binding site selection for each protein is influenced by the presence of neighboring binding sites for the specific partner. The two *Drosophila* homeodomain protein cofactors identified to date—Extradenticle (reviewed in Reference 23) and Ftz-F1 (identified in our screen for Ftz cofactors; see Reference 33), as well as the yeast homeodomain proteins MAT a1/a2 (22)—function in this way. Finally, the DIS will also allow for identification of protein partners that could be identified in standard two-hybrid screens (Figure 1E).

2. PROTOCOLS

2.1 Protocol for the Double Interaction Screen

The protocol used for the DIS is described below. Many of the methods are the same as those used for yeast one- and two-hybrid screens (2,3, and Chapters 2–6, 11) and are not described in detail here.

Method

1. *Definition of the Baits.* As summarized above and in Figure 1, two baits are used for the DIS: *(i)* a cis-acting regulatory element, and *(ii)* a DNA-binding protein that interacts directly with this element. These baits can be identified using a variety of techniques. Cis-acting regulatory elements that are important for gene expression can be identified in animals or in cell culture systems. These sequences are often identified by fusing candidate regulatory elements upstream of heterologous reporter genes. This allows for the identification of regulatory regions of several hundred base pairs or less that direct reporter gene expression in a pattern of interest. Genetic and molecular analysis may suggest candidate regulatory protein(s) that interact with these minimal cis-regulatory elements. DNase I footprinting or similar DNA-binding assays can be used to define the regions of the regulatory element that are bound by Protein X. Note that Protein X can be identified with the DIS system by screening with the DNA bait alone. When possible, the interaction of Protein X with the regulatory element and with the sequences defined in DNase I footprinting assays should be confirmed in vivo.

2. *Vectors for the DIS*

   a. DNA Bait - *HIS3* reporter. The cis-acting regulatory region of interest is inserted upstream of a *HIS3* reporter gene. We used a pUC19 backbone to create a yeast-integrating vector that contains the *HIS3* gene as reporter and the *TRP1* gene for integration into the yeast genome. A *DdeI/Xho* fragment of the *HIS3* gene, including its minimal promoter and the entire coding region (position -83 to the 3′UTR; 30), was inserted into the *Sal*I site of pUC19 to generate pUC19/*HIS3*. The regulatory element is then inserted upstream of the *HIS3* gene in the *Xba*I site of the polylinker. Regulatory elements used for the screen can be up to several hundred base pairs long. For insertion into the *Xba*I site, the regulatory element can be amplified with PCR using primers that have added *Xba*I restriction sites. Finally, a *Sac*I/*Eco*RI fragment of the

TRP1 gene from plasmid D759 is inserted into the *Sac*I/*Eco*RI sites of the pUC19 polylinker to create an integrating vector. Although we have not tested this, it should also be possible to use the *HIS3* gene for integration and selection, as suggested in the MATCHMAKER One-Hybrid Protocol.

b. DNA Bait - *lacZ* reporter. The *lacZ* reporter was derived from pLG669ZΔXho, which contains a basal *CYC1* promoter (13). The vector was modified as follows: *(i)* The 2μ region was deleted by digestion with *Eco*RI followed by re-ligation to create an integrating vector. *(ii)* The *lacZ* reporter plasmid was modified by replacing the *URA3* gene with a *TRP1* gene that was isolated from D759 as a *Hind*III fragment by partial digestion. *(iii)* The regulatory element (DNA Bait) was placed upstream of the *CYC1* promoter by insertion into the *Bam*HI/*Sal*I sites of the vector.

c. Protein Bait. The protein of interest is expressed with an *ADH* promoter using a modified YcP50 vector that carries a *URA3* gene (25). This vector was generated by inserting a *Bam*HI fragment carrying the *ADH* promoter and *ADH* terminator from the yeast expression vector pADNS (5) into YcP50. This fragment contains a polylinker between the promoter and terminator for insertion of cDNA sequences. The fragment was inserted into the *Bam*HI site of YcP50 to generate YcP50/*ADH*. To increase the versatility of this vector, the YcP50 vector had been previously modified to remove its *Eco*RI and *Hind*III sites. The cDNA encoding the Protein Bait is then inserted into the polylinker derived from pADNS. The *ADH* promoter region used here from the vector pADNS does not contain an ATG; therefore, expression is dependent on the presence of an ATG in the cDNA sequence. However, the *ADH* promoter from vector pADΔNS (6), which does include an ATG, can also be used in a similar fashion.

d. cDNA library. cDNA libraries are prepared in the λACT series (reviewed in 2; available from CLONTECH). The key feature of these libraries is the use of the Cre/Lox recombination system to generate a plasmid library that is transformed into yeast (8,9). pACT, the plasmid excised by this procedure, contains an *ADH* promoter and terminator flanking a polylinker and sequences encoding the SV40 large T-antigen nuclear localization signal (NLS) and the GAL4 Activation Domain (GAD). The pACT series of vectors also carry a *LEU2* marker for selection and a 2μ origin for autonomous replication in yeast. A number of libraries are available in this vector, including a 0–6 h *Drosophila* cDNA library generated in our laboratory. The inclusion of a GAD in the library plasmids allows for identification of cDNAs that encode weak transcription activators, transcription repressors, and partial cDNAs that include DNA-binding domains but lack native activation domains. However, transcription activators and protein partners can be isolated in the DIS without reliance on fusion to a heterologous activation domain (see DISCUSSION).

3. *Yeast Strains*

a. Integration of DNA Baits. The yeast strain used for these experiments is w3031A, which contains five auxotrophic markers: *LEU2, URA3, ADE1, TRP1,* and *HIS3*. Reporter strains are made by integration of plasmids into the *TRP1* locus after digestion with *Stu*I. *HIS3* and *lacZ* reporter are independently integrated into different yeast cells to generate two reporter strains: w3031A/DNA Bait-*HIS3* and w3031A/DNA Bait-*lacZ*. Yeast transformation, selection on SD/-Trp plates, and analysis to confirm integration of the reporter plasmids are done using standard methods (26).

b. Expression of Protein Bait. The YcP50/*ADH* vector expressing the Protein Bait is transformed into yeast cells carrying the DNA Bait-*HIS3* reporter to generate w3031A/DNA Bait-*HIS3*/YcP50/*ADH*—Protein X. Single colonies are selected on SD/-Ura,Trp plates. After restreaking, single colonies are isolated for growth tests and screening.

4. *Growth Tests*

Before transformation of the library, it is necessary to determine the background levels of transcription of the *HIS3* reporter gene. These tests should be carried out using yeast that carry the reporter alone (w3031A/DNA Bait-*HIS3*); yeast also expressing the Protein Bait (w3031A/DNA Bait-*HIS3*/YcP50/*ADH*-Protein X); and/or yeast expressing the reporter, the Protein Bait, and the empty pACT vector that will be used for the screen. Colonies are streaked onto SD/-Trp, His; SD/-Trp, His, Ura; or SD/-Trp, His, Ura, Leu plates containing varying concentrations of 3-Amino-1,2,4-triazole (3-AT). It is expected that the cis-regulatory element used will contain binding sites for yeast transcription factors, which may lead to expression of reporter genes. In our experience, concentrations of 3-AT as low as 5 mM have been sufficient to suppress growth of reporter strains carrying regulatory elements of two to three hundred base pairs in length. However, if concentrations higher than approximately 15 to 25 mM 3-AT are necessary to suppress growth, we advise trimming the cis-regulatory element to remove putative binding sites for strong yeast activators. Similarly, the bait protein may activate transcription to some extent in the yeast cells. While this could in theory be a problem that precludes the use of some protein baits in the DIS, we have found activation of reporter gene expression using native regulatory elements to be quite weak (see DISCUSSION). Therefore, the use of full-length transcription activators for the DIS has not posed any problems thus far.

5. *The Double-Interaction Screen*

a. Library Transformation. The pACT library is transformed into yeast cells (w3031A/DNA Bait-*HIS3*/YcP50/*ADH*-Protein X) as described (2,8). Yeast are grown on SD/-Trp, Ura, Leu, His plates containing 3-AT at concentrations above those necessary to suppress growth of reporter strains. We have used 25 mM 3-AT for our screens, but concentrations up to 50 mM can be used if necessary. Plates are incubated at 30°C for 7–14 days. Library screens of 10^6–10^8 cDNAs using *Drosophila* cDNA libraries have yielded no more than several hundred surviving colonies.

b. Restreaking of Positive Clones. Colonies are restreaked onto SD/-Trp, Ura, Leu, His plates containing the same concentration of 3-AT that was used for the initial selection. Most colonies from step a are expected to re-grow under these conditions.

c. Isolation of library plasmids. The overwhelming majority of false positives isolated in the DIS result from mutations in the yeast cells that allow for growth under selective conditions. Although several methods are available to identify such false positives, we have found that the most efficient screening step is to isolate plasmids from surviving colonies and re-transform them into the original reporter strain. Plasmids are isolated by standard methods from yeast cells and transformed into bacteria (XL-1 Blue) by electroporation. Using standard protocols, several hundred plasmids can be isolated in a short time and plasmid-dependence can be definitively determined in fresh reporter yeast cells.

d. Retransformation to establish plasmid dependency and Protein Bait dependency. Each isolated plasmid is transformed into (*i*) w3031A/DNA Bait-*HIS3* cells and (*ii*) w3031A/DNA Bait-*HIS3*/YcP50/*ADH*-Protein X cells and grown on SD/-Trp, Leu or SD/-Trp, Ura, Leu plates, respectively. Three colonies from each transformation are then restreaked onto plates lacking His

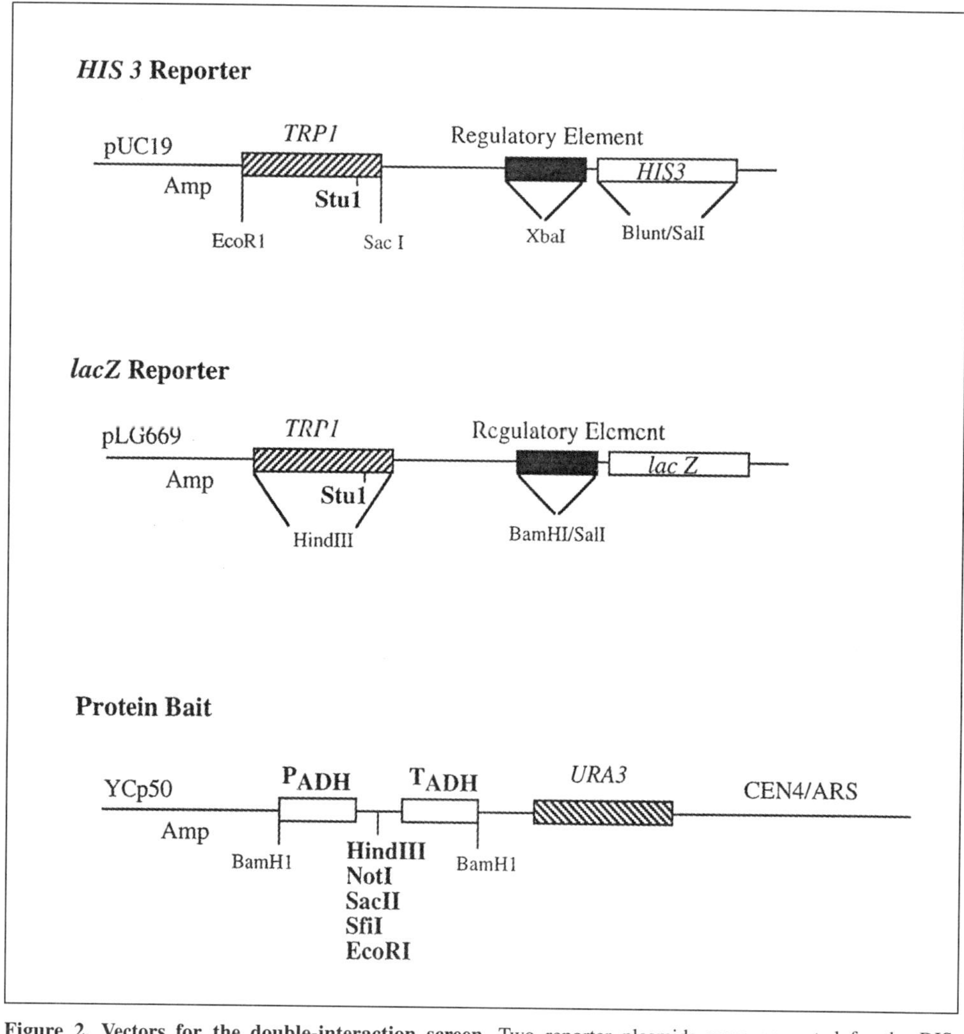

Figure 2. Vectors for the double-interaction screen. Two reporter plasmids were generated for the DIS. pUC19/*HIS3* contains a *TRP1* gene for integration and selection and a *HIS3* gene with its basal promoter as a reporter. The *lacZ* reporter also carries a *TRP1* gene for integration and selection and a *lacZ* gene with a *CYC1* basal promoter. The Protein Bait is expressed with the vector YcP50/ADH, a replicating plasmid that carries a *URA3* gene for selection. cDNAs are inserted into a polylinker between the *ADH* promoter (P_{ADH}) and terminator (T_{ADH}) for expression in yeast. (See text for details.) Restriction enzyme sites of interest are indicated. Drawings are not to scale. Note: Similar *HIS3* and *lacZ* reporter constructs available from CLONTECH contain more versatile polylinkers.

and containing 3-AT at the concentration used for the initial selection. Growth at 30°C is monitored. The following outcomes are expected:

    i. No growth on either type of plate. In this case, initial survival was plasmid-independent and probably resulted from mutations in the yeast cells. These false-positives are discarded.

    ii. Equivalent growth on both types of plates. These plasmids support growth in the absence or presence of the Protein Bait. These plasmids most likely encode DNA-binding proteins that interact with the DNA Bait used for the screen (Figure 1C). These plasmids are retained for further analysis.

    iii. Colonies grow in the absence of the Protein Bait, but growth is more robust in its presence. These plasmids are likely to encode DNA binding proteins that interact with protein X (Figure 1D). These colonies should be restreaked onto plates that contain higher concentrations of 3-AT to verify dependence on the Protein Bait. It is likely that growth at higher 3-AT concentrations will be strictly dependent on expression of both the Protein Bait and the library-encoded cDNA. These plasmids are retained for further analysis.

    An example of this category of isolate is shown in Figure 3. A cDNA isolated from the Ftz DIS encodes Ftz-F1, a protein of the nuclear hormone receptor superfamily that is a Ftz cofactor in vivo (14,19,31,33). Expression of Ftz-F1 alone activated expression of the DNA Bait-*HIS3* reporter gene, thus supporting growth in 25 mM 3-AT. However, growth was more robust when both Ftz and Ftz-F1 were expressed. This interaction was clearer when growth was tested in the presence of 50 mM 3-AT.

    iv. Colonies grow only in the presence of Protein X. These plasmids are likely to encode cofactors for the Protein Bait that do not themselves bind to DNA (Figure 1E). This can occur if the full-length protein isolate does not contain a DNA-binding domain or if the cDNA encodes a truncated version of the protein that retains its protein–protein interaction domain but lacks the DNA-binding domain. These plasmids are retained for further analysis.

e. Tentative classification of positive clones. At this stage, most of the plasmids retained encode either DNA-binding proteins or partners for Protein X, although a small number of false-positives may remain. Plasmids can be divided into two groups before further analysis:

    Class 1: Protein Bait-independent: candidate DNA-binding proteins (Group *ii.*)

    Class 2: Protein Bait-dependent: candidate protein partners that may or may not also bind DNA (Groups *iii* and *iv*)

6. *Other tests to verify isolation of real DNA-binding proteins and Protein Bait partners:*

a. DNA sequence analysis. It is expected that a relatively small number of plasmids (<50) will survive the tests described above. DNA sequence analysis is therefore usually feasible at this step and can provide a tremendous amount of information about the nature of the isolated plasmids. Although we have found that very few false-positives survive through this step of screening, we consider it reasonable to eliminate any isolates at this stage that encode proteins known to be exclusively cytoplasmically localized. In addition, a number of proteins that are isolated as frequent false-positives in yeast two-hybrid screens, such as ribosomal proteins, may survive through to this step.

It is very likely that at least one of the isolated plasmids will encode a previously identified DNA-binding protein or will share homology with a previously characterized DNA-binding domain. This will verify that the basic steps of the DIS have worked properly.

   b. Tests with the *lacZ* reporter gene. Isolated plasmids are transformed into w3031A/DNA Bait-*lacZ* cells with and without Protein Bait, as in step 5d. Liquid assays for β-galactosidase activity are carried out using standard methods (1). Use of a simple, spectrophotometric assay allows for quantitation of β-galactosidase activity units. As discussed above, even for strong transcription activators, activity on native regulatory elements that contain dispersed binding sites for the Bait Protein appears to be weaker than when multimerized copies of individual binding sites are used. To obtain reliable units, it is therefore usually necessary to incubate with the substrate o-Nitrophenyl β-D Galactopyranoside (OPNG) overnight at 30°C for β-galactosidase assays with a native regulatory element. In this case, the factor "t" (time of incubation, see MATCHMAKER Protocol) should be dropped from the equation to calculate units, since the measurements are essentially done at saturation.

   Results from these assays should be in general agreement with the growth assay using the *HIS3* reporter described previously. However, as discussed subsequently, we were surprised to find that this *lacZ* reporter assay is not always a good predictor of real- versus false-positives (see DISCUSSION). We have therefore hesitated to eliminate any isolates from the screen solely on the basis of the test with the *lacZ* reporter.

   c. Independent Methods to verify positive interactions. As in standard one- and two-hybrid screens, further tests using independent methods are essential at this step to rule out isolation of cDNAs encoding non-specific binding proteins.

(Class I) Protein Bait-independent: Candidate DNA binding proteins. That these isolates encode DNA binding proteins will in many cases be supported by sequence analysis that reveals consensus DNA-binding domains. To demonstrate that these proteins bind to the regulatory element of interest, two experiments are recommended. First, the transcription activation seen in yeast should be dependent on inclusion of the regulatory element DNA in the reporter plasmid. Second, protein expressed in vitro should bind to the regulatory element directly. Any number of standard expression systems can be used to produce protein for DNase I footprinting or other types of DNA-binding assays. These assays will verify direct binding to DNA and will also define the binding site(s) for the protein isolated from the DIS. Transcription activation by the protein, in yeast cells, in other cell systems, or in whole animals, should be dependent on the presence of these binding sites.

(Class II) Protein Bait-dependent: Candidate protein partners. These may or may not also bind directly to DNA. Protein–protein interactions should be verified using

at least one independent method such as co-immunoprecipitation or "GST-pull down" assays (1; Chapter 17). Standard yeast two-hybrid tests to rule out non-specific protein–protein interactions with an unrelated protein bait (see Chapter 6) cannot be used in the DIS because DNA-binding domain fusion proteins were not used as initial baits. However, if other proteins have been identified that bind to the DNA Bait used in the DIS, these can be used to verify the specificity of interaction with Protein X. Finally, for those proteins that appear to bind both DNA and Protein X, in vitro DNA-binding assays should be carried out to define these interactions and assess cooperative DNA binding.

3. DISCUSSION

We have described a Double Interaction Screen designed to identify both DNA binding and protein-tethered transcription factors. The DIS combines many of the features of standard yeast one- and two-hybrid screens with two major modifications. First, the DIS makes use of a native cis-regulatory element that may contain binding sites for a number of proteins. Second, the DIS utilizes a full-length protein bait that is anchored to DNA via native binding sites in the regulatory element. As such, this screen is specific for transcription factors, but allows for the simultaneous isolation of a variety of types of factors in one screen (Figure 1). In our first use of the DIS (34), a screen of fewer than 10^6 transformants resulted in the isolation of at least four DNA-binding proteins that interact with the cis-regulatory element utilized (Figure 1C). These include both transcription activators and at least one transcription repressor that was isolated because of fusion to the GAD in pACT. Further, this screen identified two partners for the homeodomain protein Ftz—the protein bait used. One of these partners, Ftz-F1, has been shown to bind to the regulatory element bait used in the screen and to interact with Ftz by cooperative DNA binding (33; Figure 1D). The second protein partner currently being characterized also appears to be a DNA-binding Ftz cofactor (J. Hama, Y. Yu, and L. Pick, unpublished).

3.1 Advantages and Disadvantages of Native DNA and Protein Baits

The design of the DIS allows for the use of native cis-acting regulatory elements as baits to isolate DNA binding transcription factors. To be most effective in isolating molecules that are important for gene regulation in vivo, the screen should be carried out with a regulatory element that has been shown to function in vivo to direct reporter gene expression in a pattern reminiscent of the gene from which it is derived. For the DIS to isolate *ftz* regulators and cofactors, we used a large regulatory element (>300 bp) to direct reporter gene expression in yeast. This fragment was chosen because it directs reporter gene expression in a *ftz*-like and Ftz protein-dependent pattern in vivo in transgenic *Drosophila* embryos (15,16,27).

Overall, we have found it to be advantageous to use a large regulatory element for isolating DNA-binding proteins. However, there are several potential drawbacks to this design. Large regulatory elements may contain binding sites for yeast proteins. If such endogenous proteins are strong transcription activators, this would preclude carrying out a DIS. In such a case, we would recommend trimming the regulatory element to try to remove the binding sites for these yeast activators. However, there may be cases where the essential regulatory sequences necessary for the screen will overlap with these binding sites; in such cases, alternative strategies would have to be used. Another possible difficulty with yeast proteins binding to the regulatory element is that one could isolate proteins in a DIS that interact with these yeast proteins. These would not be ruled out by standard re-screening, because they would continue to behave as plasmid-dependent, protein bait-independent isolates (*Class I*). Thus far, we have not encountered any false-positives in our screens that are obviously explained by this type of interaction. In fact, in contrast to expectations, our level of isolation of false positives with the DIS has been very low.

Similarly, the DIS uses a native full-length protein bait. While having the obvious advantage of avoiding fusion to a heterologous DNA-binding domain, there are also some potential problems with this approach. First, expression of full-length transcription factors in yeast may lead to toxicity if these proteins fortuitously influence the expression of endogenous yeast genes. We have in fact found that expression of some *Drosophila* transcription factors in yeast does slow growth both on plates and in liquid culture significantly. This can pose obvious problems for a screen that involves growth selection and will likely preclude the use and isolation of some full-length transcription factors. Second, the expression of full-length activators that bind to the DNA Bait might have been expected to induce very high levels of reporter-gene activation, making a further screen of a cDNA library unrealistic. However, we have found that when native regulatory elements are used, single proteins—even strong transcription activators—activate reporter gene expression relatively weakly. This is illustrated for Ftz homeodomain protein in Figures 3 and 4, but was also observed for proteins with different types of DNA-binding domains such as Ftz-F1 (zinc finger DNA-binding motif) and Adf-1 (novel Myb-like DNA-binding motif; 16).

The DNA bait in the DIS with Ftz was a >300 bp fragment that contained five endogenous Ftz protein binding sites (24). Much to our surprise, however, Ftz activation of the *HIS3* and *lacZ* reporter genes fused to this regulatory element was extremely weak; in fact, levels of reporter gene expression were barely distinguishable in the absence and presence of Ftz (Figure 3). In contrast to this, Ftz strongly activated expression of a reporter gene that contained six concatamerized copies of a home-

| | Growth in 3-AT | | | |
	1 mM	25 mM	50 mM	50 mM
Regulatory Element/HIS3	+	−	−	
FTZ	+	−	−	
FTZ-F1	+++	++	+	
FTZ + FTZ-F1	+++	+++	++	

Figure 3. Ftz-F1 was isolated as a Ftz partner in a DIS. A 323 bp fragment from the *ftz* proximal enhancer was inserted upstream of a *HIS3* reporter gene in pUC19/HIS3. This fragment contains five binding sites for Ftz protein. After integration of the reporter gene into yeast cells, growth was tested in the absence of histidine and presence of increasing concentrations of 3-AT. As shown, cells grew in 1 mM 3-AT. Expression of Ftz protein only weakly activated reporter gene expression. One isolate from the DIS was Ftz-F1, an orphan member of the nuclear hormone receptor family. Expression of Ftz-F1 alone activated reporter gene expression, allowing for good growth in the presence of 25 mM 3-AT. However, interaction with Ftz was suggested by better growth when Ftz was also expressed. When cells were grown in the presence of 50 mM 3-AT (right panel), growth was clearly more robust when Ftz and Ftz-F1 were both expressed than when either protein was expressed alone. This figure was modified with permission from Reference 33.

odomain binding site upstream of the *HIS3* reporter (34). In addition, all other proteins with related homeodomains that were tested strongly activated expression of this reporter (Figure 4). This observation suggests that the use of multiple copies of single binding sites fused directly upstream of a basal promoter may lead to artificially high levels of transcription activation in cells. In contrast to this, while native regulatory element targets of transcription activators may also contain multiple binding sites, these sites are likely to be dispersed within native regulatory elements that also contain binding sites for other proteins. The multiple native binding sites may also have varying affinities for the regulatory proteins, which may decrease the efficiency of occupancy of the regulatory element. Thus, overall, our observations suggest that the combination of a native regulatory element with full-length native proteins may in fact lead to a lower level of promoter loading than in standard one- and two-hybrid screens, which use multiple copies of single binding sites upstream of reporter genes. This in turn may explain why very few false-positives were isolated in the DIS.

3.2 Isolation of Reverse Orientation Inserts in the DIS

One of the surprising findings in the DIS carried out with Ftz was that five isolates contained cDNAs that were inserted into the pACT vector in reverse orientation (32). Two of these were isolated as Ftz-dependent transcriptional activators: Ftz-F1 and a newly identified protein, FIP2. Three reverse orientation isolates activated the DNA Bait-*HIS3* reporter in a Ftz-independent fashion. The possibility of functional reverse orientation inserts being expressed was noted in the original two-hybrid literature (4) and has also recently been reported by another group (21). In these cases, it is likely that transcription is initiating from within the *ADH* terminator that is present in the pACT vectors. In addition, it has been shown that yeast transcription can initiate from non-conventional promoters such as poly dA·dT (29,30). Whereas the precise transcription initiation site in these cases has not been mapped, protein products of such cDNAs would not carry a GAD and therefore could only be isolated if they encode transcriptional activators. In keeping with this, two of the five reverse orientation isolates from the DIS with Ftz encode Ftz-F1, a strong transcriptional activator in yeast (16; Figure 3). We note that the only other report, to our knowledge, of the isolation of a reverse orientation insert from a yeast screen was a human homolog of Ftz-F1 (21). However, we have verified the activity of at least one other reverse orientation isolate

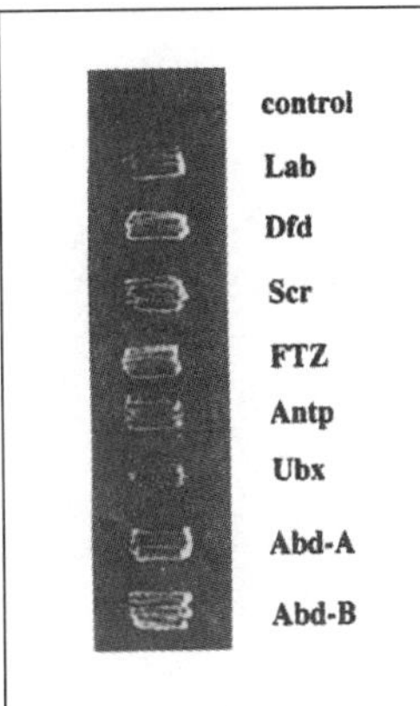

Figure 4. Homeodomain proteins strongly activate reporter genes with multiple copies of a canonical DNA binding site. Six copies of a typical homeodomain binding site (7) were placed upstream of the *HIS3* gene in pUC19/HIS3. Growth of yeast cells was tested in the absence of histidine and presence of 25 mM 3-AT. No growth was detected with cells carrying the reporter gene and empty YcP50/ADH expression vector. Various homeodomain proteins of the Antennapedia class (28) were expressed in these yeast using YcP50/ADH: Labial, Deformed, Sex combs reduced, Ftz, Antennapediea, Ultrabithorax, Abdominal-A, or Abdominal-B. Each of these proteins appeared to activate transcription of the reporter gene, supporting growth in 25 mM 3-AT.

that does not encode Ftz-F1 from our screen, so this phenomenon cannot be specific for Ftz-F1 family proteins.

None of the reverse orientation isolates activated expression of the *lacZ* reporter gene in the DIS above background levels, whereas all of the forward orientation isolates from the DIS did activate this reporter significantly (32). Because expression levels from a non-specific promoter in the reverse orientation isolates are expected to be much lower than from the strong *ADH* promoter used for transcription in the forward orientation, this suggests a differential sensitivity of the two basal promoters used for the *HIS3* and *lacZ* reporters (*HIS3* and *CYC1*, respectively) to levels of activator protein bound to upstream sequences. This suggestion is supported by the finding that Ftz-F1 strongly activates transcription of the *lacZ* reporter when it is expressed in yeast with the *ADH* promoter with or without the GAD (16), although the reverse orientation isolates of Ftz-F1 did not activate this reporter.

In light of these findings, we have not used the *lacZ* reporter assay as a criterion to eliminate false-positives from the DIS. Rather, we suspect that reverse orientation isolates may encode real transcription factors that carry both specific DNA binding domains and endogenous activation domains. Low levels of *lacZ* gene expression in such cases may reflect only low levels of expression of these cDNAs, rather than low or non-specific activities. Therefore, we recommend analyzing any such reverse orientation isolates by expressing them in the "forward orientation", in expression vectors such as pADNS, before eliminating them from the screen (also see Chapter 6).

4. CONCLUDING REMARKS

The Double Interaction Screen is a general method for isolating DNA binding transcription factors and protein partners of known DNA binding proteins. In theory, this method should allow for isolation of all the transcription factors that interact with a regulatory element of interest. Multiple activators and repressors of transcription that interact with the regulatory element via direct contacts with the DNA or via interactions with a previously identified DNA-bound protein factor can be isolated in a single screen. Repeated rounds of screening with newly identified DNA binding proteins can then identify new protein partners that participate in transcription regulation. Since several selectable markers are available, it should also be possible to express more than one regulatory protein in yeast cells to carry out screens for multimeric transcription complexes.

ACKNOWLEDGMENTS

We thank Ron Kohanski and Rob Krauss for comments on the manuscript. Studies reported here were supported by a grant to L.P. from the National Institutes of Health (HD 27937).

REFERENCES

1. **Ausubel, F.M., R. Brent, R.E. Kingston, D.P. Moore, J.G. Seidman, J.A. Smith, and K. Struhl.** 1992. Current protocols in molecular biology. John Wiley & Sons, New York.
2. **Bai, C. and S.J. Elledge.** 1996. Gene identification using the yeast two-hybrid system. Methods Enzymol.

273:331-347.

3. **Brent, R. and R.L. Finley.** 1997. Understanding gene and allele function with two-hybrid methods. Ann. Rev. Genet. *31*:663-704.

4. **Chien, C.T., P.L. Bartel, R. Sternglanz, and S. Fields.** 1991. The two-hybrid system: A method to identify and clone genes for proteins that interact with a protein of interest. Proc. Natl. Acad. Sci. USA *88*:9578-9582.

5. **Colicelli, J., C. Birchmeier, T. Michaeli, K. O'Neill, M. Riggs, and M. Wigler.** 1989. Isolation and characterization of a mammalian gene encoding a high-affinity cAMP phosphodiesterase. Proc. Natl. Acad. Sci. USA *86*:3599-3603.

6. **Colicelli, J., C. Nicolette, C. Birchmeier, L. Rodgers, M. Riggs, and M. Wigler.** 1991. Expression of three mammalian cDNAs that interfere with RAS function in *Saccharomyces cerevisiae*. Proc. Natl. Acad. Sci. USA *88*:2913-2917.

7. **Desplan, C., J. Theis, and P.H. O'Farrell.** 1985. The *Drosophila* developmental gene, *engrailed*, encodes sequence-specific DNA binding activity. Nature *318*:630-635.

8. **Durfee, T., K. Becherer, P.L. Chen, S.H. Yeh, Y. Yang, A.E. Kilburn, W.H. Lee, and S.J. Elledge.** 1993. The retinoblastoma protein associates with the protein phosphatase type 1 catalytic subunit. Genes Dev. *7*:555-569.

9. **Elledge, S.J., J.T. Mulligan, S.W. Ramer, M. Spottswood, and R.W. Davis.** 1991. λYES: A multifunctional cDNA expression vector for the isolation of genes by complementation of yeast and *Escherichia coli* mutations. Proc. Natl. Acad. Sci. USA *99*:1731-1735.

10. **Fields, S. and O.K. Song.** 1989. A novel genetic system to detect protein-protein interaction. Nature *340*:245-246.

11. **Fields, S. and R. Sternglanz.** 1994. The two-hybrid system: an assay for protein-protein interactions. Trends Genet. *10*:286-292.

12. **Golemis, E.A. and R. Brent.** 1992. Fused protein domains inhibit DNA binding by LexA. Mol. Cell. Biol. *12*:3006-3014.

13. **Guarente, L. and M. Ptashne.** 1981. Fusion of *Escherichia coli lacZ* to the cytochrome c gene of *Saccharomyces cerevisiae*. Proc. Natl. Acad. Sci. USA *75*:2199-2203.

14. **Guichet, A., J.W.R. Copeland, M. Erdelyi, D. Hlousek, P. Zavorszky, J. Ho, S. Brown, A. Percival-Smith, H.M. Krause, and A. Ephrussi.** 1997. The nuclear receptor homologue Ftz-F1 and the homeodomain protein Ftz are mutually dependent cofactors. Nature *385*:548-552.

15. **Han, W., Y. Yu, N. Altan, and L. Pick.** 1993. Multiple proteins interact with the *fushi tarazu* proximal enhancer. Mol. Cell. Biol. *13*:5549-5559.

16. **Han, W., Y. Yu, K. Su, R.A. Kohanski, and L. Pick.** 1998. A binding site for multiple transcriptional activators in the *fushi tarazu* proximal enhancer is essential for gene expression in vivo. Mol. Cell. Biol. *18*:3384-3394.

17. **Hiromi, Y. and W.J. Gehring.** 1987. Regulation and function of the *Drosophila* segmentation gene *fushi tarazu*. Cell *50*:963-974.

18. **Inouye, C., P. Remondelli, M. Karin, and S. Elledge.** 1994. Isolation of a cDNA encoding a metal response element binding protein using a novel expression cloning procedure: the one hybrid system. DNA Cell Biol. *13*:731-742.

19. **Lavorgna, G., F.D. Karim, C.S. Thummel, and C. Wu.** 1993. Potential role for a FTZ-F1 steroid receptor superfamily member in the control of *Drosophila* metamorphosis. Proc. Natl. Acad. Sci. USA *90*:3004-3008.

20. **Li, J.J. and I. Herskowitz.** 1993. Isolation of ORC6, a component of the yeast origin recognition complex by a one-hybrid system. Science *262*:1870-1874.

21. **Li, M., Y-H Xie, Y.-Y. Kong, X. Wu, L. Zhu, and Y. Wang.** 1998. Cloning and characterization of a novel human hepatocyte transcription factor, hB1F, which binds and activates enhancer II of Hepatitis B virus. J. Biol. Chem. *273*:29022-29031.

22. **Li, T., M.R. Stark, A.D. Johnson, and C. Wolberger.** 1995. Crystal structure of the MATa1/MAT alpha 2 homeodomain heterodimer bound to DNA. Science *270*:262-269.

23. **Mann, R.S. and S.-K. Chan.** 1996. Extra specificity from *extradenticle*: the partnership between HOX and PBX/EXD homeodomain proteins. TIGS *12*:258-262.

24. **Pick, L., A. Schier, M. Affolter, T. Schmidt-Glenewinkel, and W.J. Gehring.** 1990. Analysis of the *ftz* upstream element: germ layer-specific enhancers are independently autoregulated. Genes Dev. *4*:1224-1239.

25. **Rose, M.D., P. Novick, J.H. Thomas, D. Botstein, and G.R. Fink.** 1987. A *Saccharomyces cerevisiae* genomic plasmid bank based on a centromere-containing shuttle vector. Gene *60*:237-243.

26. **Rothstein, R.** 1991. Targeting, disruption, replacement, and allele rescue: integrative DNA transformation in yeast. Methods Enzymol. *194*:281-301.

27. **Schier, A.F. and W.J. Gehring.** 1992. Direct homeodomain-DNA interaction in the autoregulation of the *fushi tarazu* gene. Nature *356*:804-807.

28. **Scott, M.P., J.W. Tamkun, and I.G.W. Hartzell.** 1989. The structure and function of the homeodomain.

Biochim. Biophys. Acta. *989*:25-48.

29.**Singer, V.L., C.R. Wobbe, and K. Struhl.** 1990. A wide variety of DNA sequences can functionally replace a yeast TATA element for transcriptional activation. Genes Dev. *4*:636-645.

30.**Struhl, K.** 1985. Nucleotide sequence and transcriptional mapping of the yeast *pet56-his3-ded1* gene region. Nucleic Acids Res. *13*:8587-8601.

31.**Ueda, H., S. Sonoda, J.L. Brown, M.P. Scott, and C. Wu.** 1990. A sequence-specific DNA-binding protein that activates *fushi tarazu* segmentation gene expression. Genes Dev. *4*:624-635.

32.**Yu, Y.** 1996. Ph.D. Thesis: Upstream and parallel regulators of *Drosophila fushi tarazu* are identified by a double interaction screen. Mount Sinai School of Medicine of the City University of New York.

33.**Yu, Y., W. Li, K. Su, W. Han, M. Yussa, N. Perrimon, and L. Pick.** 1997. The nuclear hormone receptor FTZ-F1 is a cofactor for the *Drosophila* homeodomain protein Ftz. Nature *385*:552-555.

34.**Yu, Y., M. Yussa, J. Song, J. Hirsch, and L. Pick.** 1999. A Double Interaction Screen identifies positive and negative *ftz* regulators and Ftz-interacting proteins. Mech. Dev. *83*:95-105.

20 | Yeast Two-Hybrid Screening of Associated Proteins and Characterization of Mutant Androgen Receptors

Kazuo Nishimura, Erik R. Sampson, Shian-Jang Yan, Eungseok Kim, Xin Wang, Hiroshi Uemura, Chihuei Wang, Naohiro Fujimoto, Hong-Yo Kang, Shuyuan Yeh, and Chawnshang Chang
George Whipple Laboratory for Cancer Research, Departments of Pathology, Urology, and Biochemistry, University of Rochester, Rochester, NY, USA

1. BACKGROUND AND INTRODUCTION

The yeast two-hybrid system is an in vivo genetic assay to detect protein–protein interactions. Unlike traditional biochemical methods, it enables sensitive detection of weak and transient protein interactions (9). It has been used to test interactions among known proteins, to define critical regions or residues involved in interactions, and most importantly, to screen libraries for novel interactions.

Based on the yeast two-hybrid system, several modified methods have been developed, such as the three-hybrid system to detect RNA–protein interactions (22) and the one-hybrid system to detect DNA–protein interactions (24). Furthermore, as yeast have also proved to be a convenient system to study the mechanism of steroid hormones, the combination of the yeast two-hybrid screening and steroid hormone regulation has allowed us to isolate several hormone-dependent receptor cofactors that play important roles in the regulation of steroid hormone functions.

The androgen receptor (AR) belongs to the nuclear receptor superfamily that regulates hormone-responsive genes in a ligand-inducible manner (4). The AR is composed of discrete domains including ligand binding, dimerization, DNA binding, and transactivation domains. Androgen binding to AR induces a conformational change

Yeast Hybrid Technologies
Edited by L. Zhu and G.J. Hannon
© 2000 Eaton Publishing, Natick, MA

329

in AR that allows the androgen-AR complex to directly interact with an androgen response element (ARE) and to regulate target gene transcription (1). The detailed mechanism by which DNA-bound nuclear receptors activate their target genes, however, remains unclear.

Androgen is a steroid hormone whose receptor is known to be involved in two important diseases: Testicular Feminization Syndrome (20) and Kennedy's Disease, also known as Spinal/Bulbar Muscular Atrophy (16). The cloning of the full-length of the AR cDNA (4) eventually led to the determination of the genetic basis of these two diseases. The development of an immunoassay for the AR and the discovery of AR mutations in prostate tumors also provided great insight into the progression of prostate cancer and its change from an androgen-dependent to an androgen-independent stage. While prostate cancer has become the most common male malignancy in the United States, the effectiveness of "maximal androgen ablation therapy" combined with surgical or medical castration and antiandrogens, such as casodex, hydroxyflutamide (HF), or flutamide, is still very limited. Most of the patients will no longer respond to this therapy within 18–24 months of treatment.

Labrie et al. demonstrated that maximal androgen ablation therapy caused more than 90% reduction in serum testosterone (T) level, but only a reduction of 26%–49% in serum adrenal androgen levels. Blocking the remaining serum adrenal androgens in the patient's serum may therefore be worth consideration. The prostate cancer patients' serum Δ5-androstenediol also remains high during androgen ablation therapy, and flutamide cannot completely block Δ5-androstenediol–mediated AR transcriptional activity. Therefore, the development of new drugs to block Δ5-androstenediol's androgenic activity in the presence of androgen receptor associated proteins (ARA) may become a very important step in the treatment of prostate cancer (5).

During the cloning of the AR, the first nuclear orphan receptor (no known ligand has been found) and a new member of the steroid receptor superfamily was discovered and named Testicular Receptor 2 (TR2). Recently, it was discovered that TR2 and Testicular Receptor 4 (TR4) could function as repressors, through heterodimerization with the AR or estrogen receptor (ER), to inhibit androgen and estrogen signal pathways. Conversely, the AR and ER can also function as repressors to inhibit the TR2/TR4 signal pathways (18). These interesting findings have expanded the classical view that the AR and ER act only as receptor/transcription factors to regulate their own target genes.

Some of the basal transcriptional factors may be targets for regulation by DNA-binding transcriptional factors such as steroid receptors (2), which often require coactivators to mediate their effect on the basal transcription machinery (19,26,27). Although the molecular mechanism of the coactivator is still unknown, it is possible that it may form a bridge between the steroid receptor and the basal transcription complex or may stabilize a direct interaction (12,14). So far, many putative coactivators have been identified using the yeast two-hybrid system (2,14). These proteins interact with the hormone-binding domain of nuclear receptors in the presence of cognate ligands.

The isolation of a group of androgen receptor associated proteins (ARAs), including ARA70 (25), ARA24, ARA54, ARA55, ARA160, and the tumor suppressors, Rb and BRCA1 (29), represent other key discoveries in the andrology field. The impact of these discoveries is enormous, since they not only prove that the maximal or proper androgen activity requires more than just the AR, but also demonstrate that the specificity of sex hormones or antiandrogens can be modulated by some selective androgen receptor coactivators. The coactivator's functions help explain why flutamide, an antiandrogen used widely in prostate cancer therapy during androgen ablation therapy, may become an androgenic compound to stimulate androgen target genes (27).

The discovery of ARAs has also successfully contributed to an understanding of how other signaling pathways, such as Her2/Neu, IL-6, MAP Kinase, PI3-Kinase, and Akt Kinase, can stimulate or repress prostate cancer cell growth. As demonstrated in our recent findings, the reason these growth factors can modulate AR transcriptional activity is that growth factors may allow the AR to differentially recruit coactivators by phosphorylation of the AR at some specific amino acids (26). This important finding may eventually lead to the development of drugs to block the growth-factor–mediated prostate cancer growth by preventing the interaction between the AR and ARAs.

We report here how to isolate androgen-dependent AR cofactors using the yeast two-hybrid method (Figure 1) and how to apply the yeast ARE-based one-hybrid assay for functional studies. We also use the yeast system to identify AR mutants that may respond differently to hormone treatment (Figure 2). In all these cases, yeast has been proven to be a very powerful system to study the mechanism of androgen-AR (A-AR) function and regulation. As A-AR action plays a central role in the development of prostate cancer, further understanding of the A-AR functional mechanism may allow us to develop new therapies to treat prostate cancer.

We were also able to demonstrate that Δ5-androstenediol, a precursor of T, is a natural hormone with androgenic activity that can be enhanced by selective ARAs (19).

2. PROTOCOLS

2A. Protocol for the Isolation of Androgen-Dependent AR cofactors

Procedure

2A. 1. Construction of DNA-BD/target plasmid.

We recommend a two-hybrid system kit developed by Dr. Stephan J. Elledge (Baylor College of Medicine). The hormone-binding domain (HBD) of AR amino acids (aa) 614–919 (25), wild type, or LNCaP mutant should be cloned into the GAL4 DNA-binding domain (GAL4DBD) fusion vector pAS2 (8). Each androgen receptor hormone-binding domain (ARHBD) should be fused with GAL4DBD in the proper reading frame.

2A. 2. Estimation of autonomous activation of bait plasmid by small-scale transformation.

1. Transform the yeast strain Y190 with bait plasmid only and plate on SD/-Trp plates.

2. Wait for several days until yeast colonies grow to 2–3 mm in diameter.

3. Perform colony-lift filter β-galactosidase (β-gal) assays to estimate the autonomous activation of ARHBD. If the autonomous activation level is significantly positive, the bait cannot be used directly in subsequent screenings. Such baits need to be eliminated or modified (Figure 1).

2A. 3. Screening a cDNA library by sequential transformation.

The cDNA library (prostate and brain cDNA libraries were gifts from Dr. Stephen J. Ellege) was made in the λ vector ACT (activation domain). The λ vector ACT library was converted into a plasmid ACT (pACT2) library by Cre-Lox recombination. The converted library was then transformed into Y190 pre-transformed with pAS2-ARHBD.

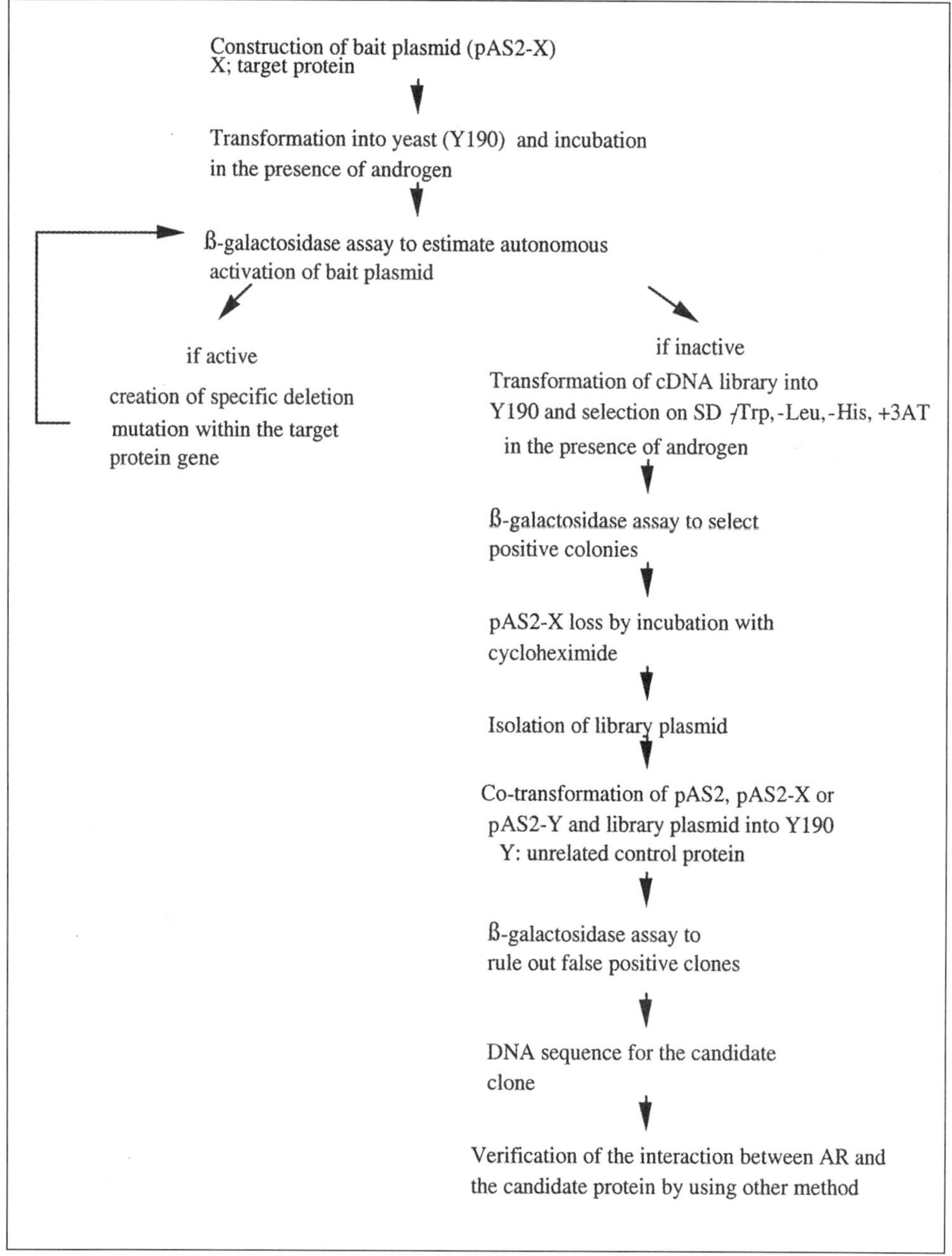

Figure 1. Procedure for the isolation of androgen-dependent AR cofactors.

2A. 4. Estimation of specific interaction in the presence of different hormones.

Small-scale transformation and β-gal assays can be used to check the specific interaction between the HBD of other steroid receptors and the isolated clone. To quantify each result, β-gal liquid assays can be applied as previously described (8).

1. Co-transform Y190 with pAS2-HBD of AR, ER, GR, PR, or TR4, plus the candidate plasmid (see Table 1) at small scale.

2. Spread the transformed cells onto SD/-Trp/-Leu/-His, 25 mM 3-AT plates containing the appropriate ligand, steroid hormones, or analogs for each nuclear receptor (1 nM–1 μM).

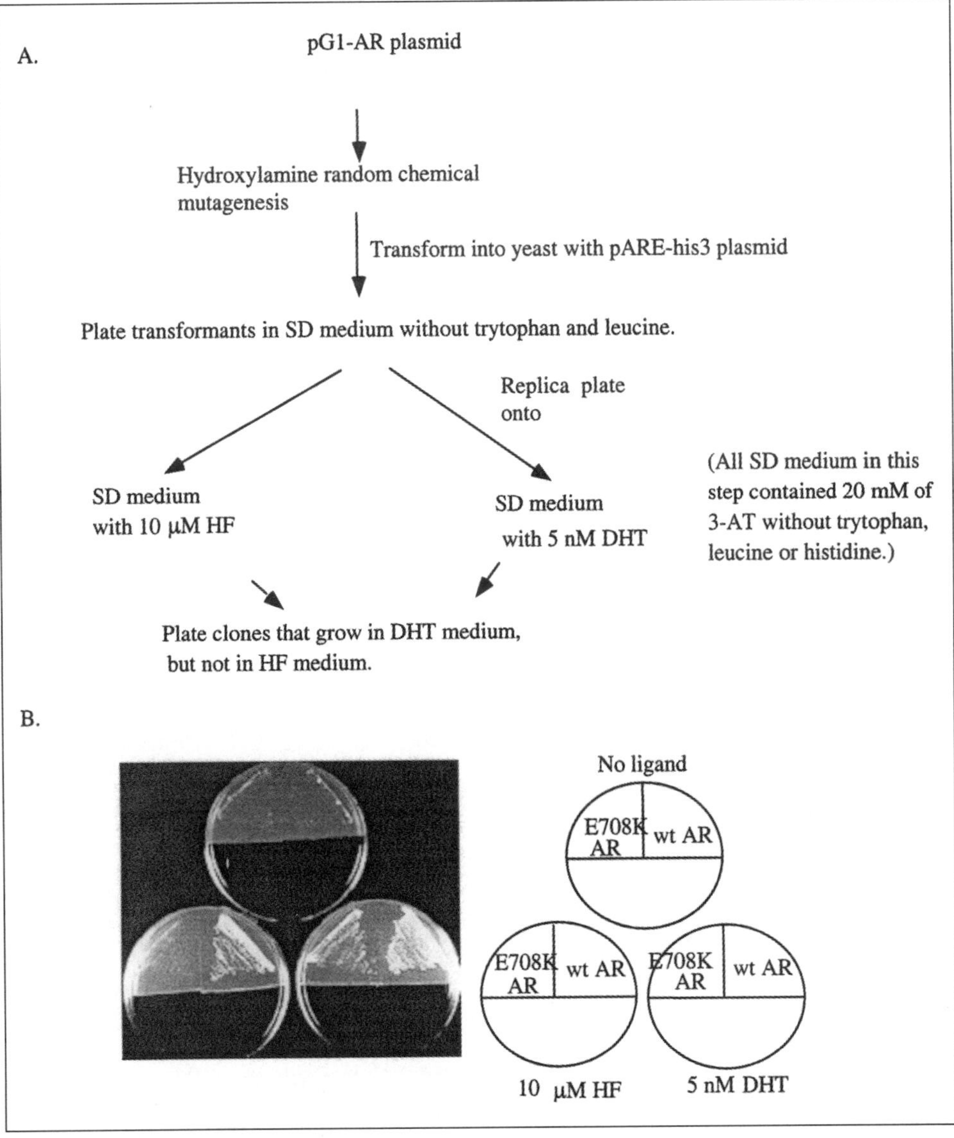

Figure 2. (A) Procedure for the isolation of mutated ARs that respond to DHT, but not to HF. (B) Results of growth of the yeast transformed with wt AR/E708K AR in the absence or presence of HF or DHT.

Table 1. DNA Mixture for Yeast Transformation

1. Estimation of autonomous activation of bait plasmid

    1 pAS2, 0.2 µg; carrier DNA, 0.1 mg

    2 pAS2-ARHBD, 0.2 µg; carrier DNA, 0.1 mg

2. Library screening

    1 Library plasmid, 500 µg; carrier DNA, 20 mg

3. Elimination of false positive clone

    1 pAS2, 0.2 µg; pACT-C, 0.2 µg; carrier DNA, 0.1 mg

    2 pAS2-ARHBD, 0.2 µg; pACT-C, 0.2 µg; carrier DNA, 0.1 mg

    3 pAS2-Y, 0.2 µg; pACT-C, 0.2 µg; carrier DNA, 0.1 mg

    C: candidate protein. Y: unrelated protein.

4. Estimation of the specific interaction

    1 pAS2-ARHBD, 0.2 µg; pACT-C, 0.2 µg; carrier DNA, 0.1 mg

    2 pAS2-ERHBD, 0.2 µg; pACT-C, 0.2 µg; carrier DNA, 0.1 mg

    3 pAS2-GRHBD, 0.2 µg; pACT-C, 0.2 µg; carrier DNA, 0.1 mg

    4 pAS2-PRHBD, 0.2 µg; pACT-C, 0.2 µg; carrier DNA, 0.1 mg

3. Incubate for 3–7 days until colonies grow >2 mm in diameter.

4. Perform the β-gal assay.

2A. 5. ARE–β-gal assays (see Table 1).

We recommend the yeast strain Acy40 provided by Dr. Avrom J. Caplan (3) for this assay. Acy40 is derived from the W3031b strain with a copy of pPGK-ARE-LacZI integrated into the URA3 locus of the yeast chromosome. This integrated reporter gene mimics the endogenous androgen target genes in mammalian cells. The expression of AR in this strain activates reporter β-gal gene expression in the presence of androgen. We used pG1-hARa yeast expression vector to express human AR, which is regulated by the yeast G3PDH promoter. The AR cofactor is expressed in another yeast expression plasmid, pYX243, which is regulated by the yeast pG1 promoter. The pG1 promoter is tightly controlled by galactose uptake and therefore we can induce the expression of the AR cofactor immediately before androgen treatment. The pG1-AR plasmid is selected by the *trp1* gene, whereas the pYX-243-AR cofactor plasmid is selected by the *leu2* gene. Both genes are replicated in yeast via the 2 µ origin of replication.

1. These two plasmids can be transformed into yeast Acy40 sequentially or simultaneously.

2. It is advised to plate several colonies of each transformant, as well as to make frozen stocks.

3. To ensure a stable and consistent expression level, all transformants must be restreaked from the frozen stocks and grown fresh on SD/-Trp/-Leu plates

before the experiment.

4. Inoculate a fresh colony from a restreaked SD/-Trp/-Leu plate and grow in 5 mL SD/-Trp/-Leu/2% glucose liquid media at 30°C overnight.

5. Centrifuge the yeast cells at $1000\times g$ at room temperature for 5 min.

6. Wash the pellet with 25 mL of sterile, distilled water and pellet the cells again.

7. Wash the cells once more with sterile, distilled water and pellet the cells.

8. Resuspend the cells with SD/-Trp/-Leu with 2% galactose and adjust the OD_{660} to approximately 0.7.

9. Grow the cultures at 30°C for 3 h to induce the expression of the AR cofactor protein.

10. Split each transformant into two tubes and treat one of them with androgen, for example, 50 nM dihydrotestosterone (DHT).

11. Incubate all cultures at 30°C for 6 more h.

12. Measure the OD_{660} of each tube. It should be between 0.7–1.2.

13. Transfer 1 mL of each sample into a 1.5-mL microcentrifuge tube.

14. Centrifuge the cells at $5000\times g$ for 1 min.

15. Wash the pellet once with distilled water.

16. Resuspend the cells in 100 µL Z buffer.

17. Freeze-thaw the cells 5 times. Freeze with liquid nitrogen and thaw in a 37°C H_2O bath.

18. Mix well and transfer 4 µL of androgen-treated sample or 20 µL of untreated sample into a new tube.

19. Keep all lysates on ice, and fill each sample to 50 µL with Z buffer.

20. Add 350 µL of Z buffer containing β-mercaptoethanol (2.7 µL/1 mL Z buffer).

21. Add 160 µL o-Nitrophenyl β-D-Galactopyranoside (ONPG) solution to each tube in an ice bath.

22. Incubate all samples at 30°C for 10–20 min until the samples develop a yellowish color.

23. Chill all samples on ice and add 400 µL of sodium carbonate to stop the reaction.

24. Vortex-mix each sample briefly. Centrifuge at $20\,800\times g$ for 5 min, and measure the OD_{420} value.

25. The cell number ($Y \times 10^7$ cells/mL) can be normalized from the OD_{660} value (X) using the formula: $Y = 0.276 \times 10^{0.826X}$

26. The β-gal activity must be normalized with cell numbers used in each β-gal reaction to obtain crude data.

27. The yeast that were transformed with the pG1AR and pYX243 parental plasmids usually have fivefold induction upon addition of 50 nM DHT after a 6-h treatment. The degree of DHT induction reflects the effect of particular AR cofactors on AR transactivational activity.

2B. Protocol for the Mutagenesis and Isolation of Mutated ARs

Procedure

2B. 1. Expression and Reporter Plasmids in Yeast Cells.

pG1-hAR was a gift from Dr. A. J. Caplan (3). The pLeu2 plasmid was constructed by inserting a *Bgl*II fragment containing polycloning sites from pACTII (CLONTECH) into a pGAD424 (CLONTECH), previously cut by *Sph*I to remove the *Sph*I-*Sph*I small fragment. Both the *Bgl*II fragment and the cut pGAD424 vector were treated with Klenow and then joined by blunt end ligation. pHis3 was constructed by inserting a *Bam*HI and *Sal*I cut fragment from pRS315 (24) into pLeu2 that was previously cut by *Bam*HI and *Sal*I. The *Bam*HI-*Sal*I fragment from pRS315 contains the *Gal*1 minimal promoter with the *His3* gene. The pHis3 vector was digested by *Bam*HI and ligated with two copies of C3 ARE oligonucleotide (Biotechnology Center, University of Wisconsin, Madison, WI, USA). The resulting plasmid, pARE-his3, was used as a reporter plasmid in AR mutation screening experiments.

2B. 2. Hydroxylamine Mutagenesis of Plasmid DNA (see Table 2).

1. Prepare the hydroxylamine solution just before use.

2. Add 100 µg of $CsCl_2$-purified pG1-AR plasmid DNA to 5 mL hydroxylamine solution.

3. Incubate at 70°C for 1 h.

4. Precipitate the DNA by adding 5 mL of isopropanol and 0.5 mL of 3.0 M sodium acetate.

5. Centrifuge the precipitated DNA at 20 800× *g* for 15 min. Remove the supernatant.

6. Allow the pellet to air-dry and resuspend in 100 µL 1× TE.

2B. 3. Yeast Strains and Yeast Transformation Method.

The yeast strain, W3031b (*ade2-1 leu2-3.112 his3-11.15 trp1-1 ura3-1 can1-100*) was also a gift from Dr. A.J. Caplan (3). Reporter plasmid pARE-His3 should be first transformed into W3031b. Selected transformants are grown and re-transformed with 20 µg of mutagenized pG1-AR plasmid library.

2B. 4. Screening for AR Mutations.

Plate W3031b transformants onto SD medium minus leucine and tryptophan and grow for 2 days at 30°C. These colonies are then replicated on SD/-Trp, -Leu, -His in the presence of 20 mM of 3-aminotriazole (3-AT; Sigma Chemical, St. Louis, MO, USA) with ligands as indicated in Figure 2 and allowed to grow for 2 days at 30°C. To ensure that the phenotypes of AR mutations are due to DNA mutations, plasmids should be recovered by the boiling method (see 2B. 5) and transformed into W3031b to confirm their phenotypes. Finally, AR coding regions should be sequenced to confirm the presence of mutations.

2B. 5. DNA Preparation from Yeast (Table 3).

1. Grow yeast overnight at 30°C in the medium that maintains selection for the

Table 2. Method for Hydroxylamine Mutagenesis

1. Hydroxylamine solution:

 Hydroxylamine HCl 3.5 g

 NaOH 1.8 g

 Distilled water 50 mL

 Dissolve the solids in the water. Adjust to pH 7.0. Prepare just before use.

2. $CsCl_2$-purified pG1-AR plasmid DNA 100 µg

3. Isopropanol

4. 1× TE:

 10 mM Tris-HCl (pH 8.0)

 1 mM EDTA

5. 3.0 M Sodium acetate

    plasmid DNA.

2. Fill a 1.5-mL microcentrifuge tube with the culture and pellet the cells by centrifugation.

3. Decant the supernatant.

4. Add 0.2 mL of yeast lysis buffer (Table 3), 0.2 mL phenol:chloroform:isoamyl alcohol (25:24:1), and acid-washed glass beads.

5. Vortex-mix for 2 min.

6. Place tube in boiling water for 30 s.

7. Centrifuge at $20\,800\times g$ for 15 min.

8. Collect the supernatant and add 16 µL of 3.0 M sodium acetate and 160 µL of isopropanol.

9. Centrifuge at $20\,800\times g$ for 15 min.

10. Decant the supernatant and wash the pellet with 70% alcohol.

11. Air-dry the pellet and dissolve it in 50 µL 1× TE.

3. RESULTS AND DISCUSSION

3A. Androgen-Dependent AR cofactors

The hypothesis that steroid receptors may require coactivators was derived from the observations that excess receptor could inhibit its own transactivation effect as well as transactivation by other transactivators (11). So far, many cofactors (2,14) including SRC-1 (21), GRIP-1 (12), and TIF-1 (7) have been cloned by yeast two-hybrid screening using NRHBD as bait in the presence of its ligand. Some co-repressors (14), such as N-CoR (13) and SMRT (6), have also been cloned. These co-repressors were isolated by yeast two-hybrid screening using NR as bait without ligand.

We used the yeast strain Y190, which contains *lacZ* and *HIS3* as reporter genes

Table 3. Materials for DNA Preparation from Yeast

1. Yeast lysis buffer:

   2% Triton-X-100

   1% SDS

   100 mM NaCl

   1× TE

2. Phenol:chloroform:isoamyl alcohol (25:24:1).

3. Acid-washed glass beads

with pAS2 as the bait plasmid (8). After Y190 is transformed with the pAS2 target gene, this strain should be checked for self-activation and also checked for its growth on SD/-His with different concentrations of 3-AT. According to Chang Bai and Stephen J. Elledge (1), a 3-AT concentration of 25 to 50 mM is typically sufficient. If the bait gene activates *lacZ*, deletions for a specific domain responsible for the autonomous activation must be made. If it does not activate *lacZ*, the plasmid can be used as bait for the subsequent screening of a cDNA library.

pAS2 carries the *CYH2* gene, which confers cyclohexamide sensitivity on suitable resistant strains such as Y190. Therefore, only Y190 yeast cells that have spontaneously lost the pAS2-target plasmid while retaining the library plasmid can grow on SD/-Leu plates containing cyclohexamide. This feature is very useful to selectively eliminate bait plasmid from yeast and for subsequent testing of the specificity of interaction (8).

The quality of the cDNA library is crucial. Durfee et al. (8) have constructed several excellent libraries in λ ACT in which cDNA clones are fused with the GAL4 activation domain. In vivo Lox-mediated excision allows the direct generation of the activation domain library. The cDNA library is amplified in *E. coli* and purified by the CsCl method to obtain a library plasmid of good quality, ready for use in yeast transformation.

We used sequential-transformation library screening because this method gave us a high transformation efficiency while using a small amount of cDNA library plasmid compared to simultaneous transformation. However, if the target protein can inhibit yeast growth, sequential transformation cannot be applied, but simultaneous transformation can still be tested.

After primary cDNA library screening, false positive clones must be eliminated. Although recovery efficiency may be low, cDNA library plasmid can be purified from yeast and transformed into *E. coli*. The library plasmid should be re-transformed into yeast with pAS2-ARHBD, pAS2 (no fusion protein), or pAS2-Y (non-related protein) as controls. β-gal assays can then be performed. False positive clones can be eliminated according to the results of β-galactosidase assays. pAS2 itself has weak transactivation activity assayed by *lacZ* activity (1), and care must be taken when pAS2 is used as a control to eliminate false positives. After a specific positive clone is obtained, the interaction of the two proteins should be checked by other independent methods such as co-immunoprecipitation, Far Western assay, or mammalian two-hybrid assay.

To date, our laboratory has identified several ARAs using the yeast two-hybrid screening strategy (10,15,17,20,28,29). ARA70, the first ARA identified, was screened out of a human brain cDNA library using a GAL4DBD-AR fusion protein as bait. The interaction of ARA70 with AR was then compared with that of other nuclear receptors using the yeast two-hybrid system (Figure 3). Based on the results of this assay, ARA70 was shown to be a ligand-dependent ARA.

ARA55 was also identified by the yeast two-hybrid screening using a human prostate cDNA library. The ligand-dependent interaction between ARA55 and AR was demonstrated both in the yeast two-hybrid assay and in the mammalian two-hybrid assay using a chloramphenicol acetylase (CAT) reporter (Figure 4). Other ARAs identified using the yeast two-hybrid system include ARA54, ARA24,

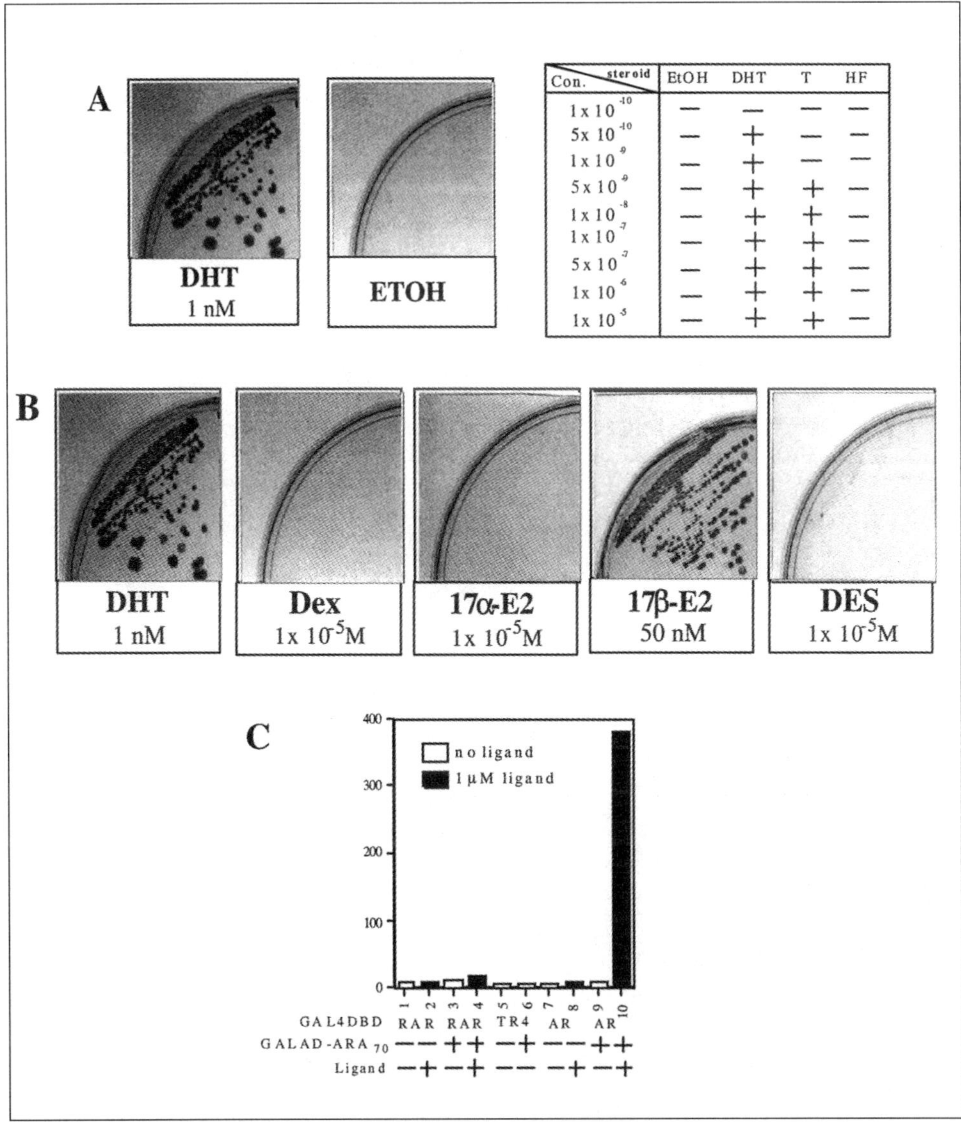

Figure 3. (A) ARA70 interacts with AR in the presence of 1.0 nM DHT. (B) ARA70 can interact with AR in the presence of DHT or 17β-E2, but not in the presence of 17α-E2, Diethylstilbesterol (DES), or Dexamethasone (DEX). (C) ARA70 cannot interact with TR4 or the LBD of RAR, with or without DHT. ARA70 can interact with the LBD of AR in the presence of 1 μM DHT.

ARA160, and SRC-1 (15,20,21). Moreover, proteins with significantly different functions, such as Rb, a regulator of the cell cycle, and CBP/p300, a histone acetly-transferase, have been identified as AR cofactors (28,29). The transcriptional activation effects of several of these coactivators on AR are shown in Figure 5 (29).

Although yeast two-hybrid screening has provided a sensitive method to clone ARAs, so far only a fraction of these putative coactivators have been shown to enhance the ligand-dependent AR transactivation in mammalian cells (23). Unless the yeast two-hybrid results can be confirmed by other more direct functional assays, one should be cautious in interpreting these primary screening results.

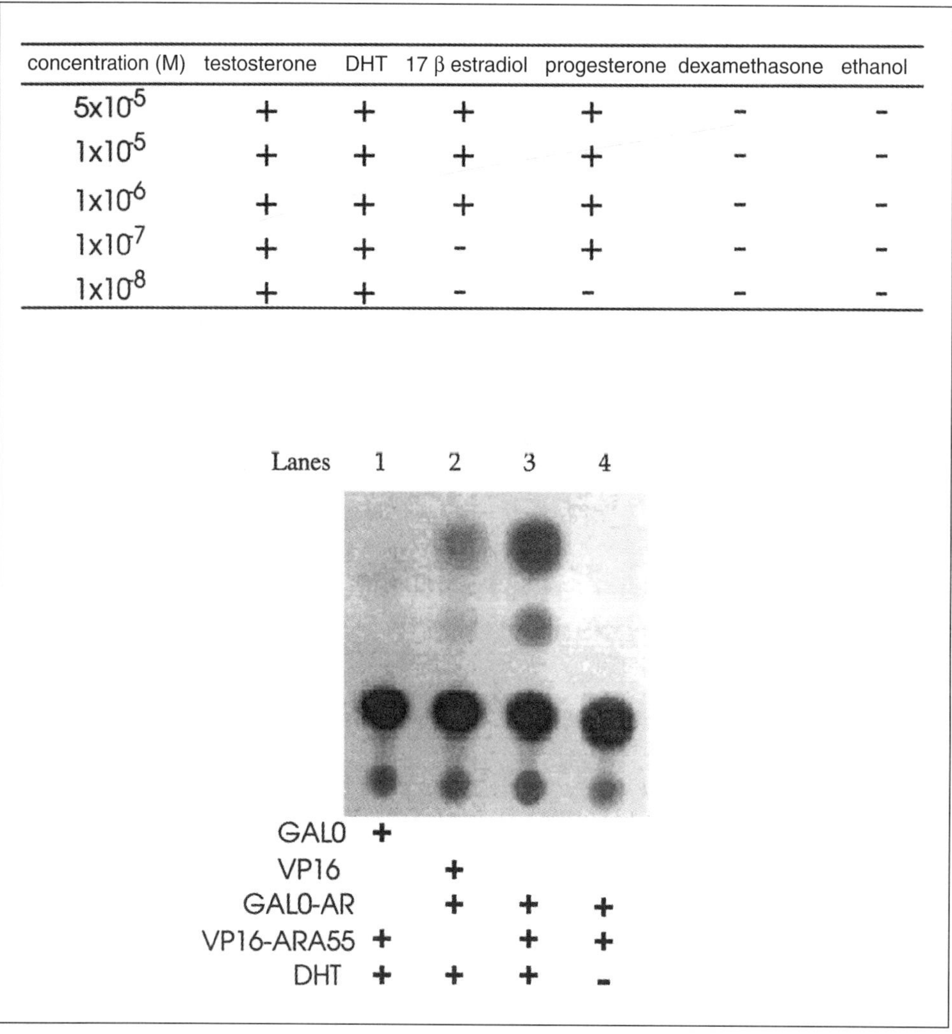

concentration (M)	testosterone	DHT	17 β estradiol	progesterone	dexamethasone	ethanol
$5{\times}10^{5}$	+	+	+	+	−	−
$1{\times}10^{5}$	+	+	+	+	−	−
$1{\times}10^{6}$	+	+	+	+	−	−
$1{\times}10^{7}$	+	+	−	+	−	−
$1{\times}10^{8}$	+	+	−	−	−	−

Figure 4. (Upper panel) Summary of ARA55 and AR yeast two-hybrid interaction data. ARA55 can interact with AR in the presence of androgen (T or DHT) as well as in the presence of 17β-E2 or progesterone. ARA55 cannot interact with AR in the presence of dexamethasone or ethanol. **(Lower panel)** Mammalian two-hybrid assay data. The ARA55-VP16 fusion protein does not interact with the GAL4 BD (GAL0) protein alone (Lane 1). The GAL0-AR fusion protein slightly interacts with VP16 (Lane 2) and strongly interacts with the VP16-ARA55 fusion protein in the presence of DHT (Lane 3). VP16-ARA55 does not interact with GAL0-AR in the absence of DHT (Lane 4). The assay utilized a CAT reporter plasmid.

3B. Modified Yeast One-Hybrid Assay System for Mutant AR Screening

The yeast ARE–β-gal assay provides a simple system to assay the effects of individual AR cofactors on androgen-AR transactivation. So far it has given us consistent results. The reporter gene ARE-His3 responds to androgen, and AR in yeast cells faithfully mimics androgen target genes in mammalian cells. In this system, both AR and AR cofactor plasmids are expressed by high copy number plasmids and are maintained in yeast cells by different nutritional markers. This system provides a higher sensitivity than assay systems that employ a chromosome-integrated reporter gene. We have also generated a high copy-number AR-His3 reporter plasmid and used it for a one-hybrid screening against a mutagenized co-activator library.

Using this modified one-hybrid yeast system, we have succeeded in isolating one AR mutant, E708K AR. At very high concentrations of HF, HF acts as an agonist for wild-type AR (wt AR). However, this effect is not observed for the E708K AR mutant. For this E708K AR mutant, HF acts as an antagonist at all concentrations. However, wt AR and E708K AR have similar binding affinity toward DHT and HF, and Western blotting suggests that DHT and HF have similar effects on the expression level of wt AR and E708K AR. Thus, E708K mutation of AR may cause a subtle change of AR conformation. It may be very useful to design a more effective androgen antagonist (Wang, C. and Chang, C., unpublished data).

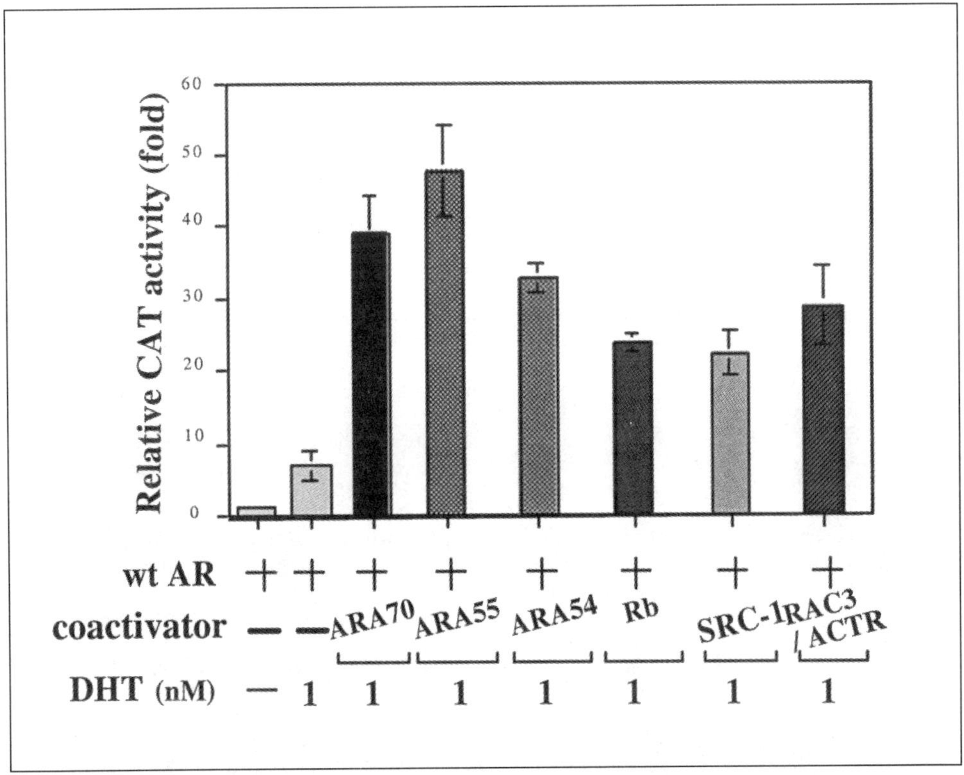

Figure 5. The effects of various coactivators on the transactivational activity of AR in the presence of DHT. AR alone (Lane 1). AR activity in the presence of 1 nM DHT (Lane 2). AR activity in the presence of 1 nM DHT and ARA70, ARA55, ARA54, pRb, SRC-1, or RAC3 (Lanes 3–8). AR transactivational activity was assayed using a CAT reporter plasmid. The relative CAT activity was calculated by phosphorimager quantitation.

Our system has two basic components: *(i)* a mutational AR library created by hydroxylamine and *(ii)* a reporter gene containing ARE. The simplicity of this method is based on the ease of yeast manipulation. Nonetheless, there is a limited range of possible mutations produced in AR using hydroxylamine mutagenesis. Other mutagenesis protocols, such as PCR-based mutagenesis (19) or growth of the pG1-AR plasmid DNA in a mutator strain of bacteria (14), might yield a wider range of AR mutations. It is also possible to produce AR mutants that may not occur physiologically. The functional relevance of these mutants should be confirmed by biological or biochemical experiments. Finally, we hope that the isolation of these mutants will be useful for the study of AR structure by X-ray crystallography or NMR.

Yeast do not have endogenous steroid hormone systems. The results observed in the yeast two-hybrid assay may therefore only represent the effect of a particular AR cofactor on AR activity. Yeast cells contain a basal transcriptional machinery very similar to the mammalian counterpart, yet some of the transcription factors are only remotely homologous to their human counterpart. For example, yeast TBP has a conserved core domain homologous to human TBP, but its N-terminal domains are distinct. On one side, yeast obviously does not have all essential players for complete steroid function. On the other side, yeast provides a simple and clean assay background for determining certain, but not all, effects of AR cofactors compared with mammalian cells. The latter may be completely faithful, but may not be as sensitive due to the presence of a full array of many functionally related factors/cofactors.

REFERENCES

1. **Bai, C. and S.J. Elledge.** 1996. Gene identification using the yeast two-hybrid system. Methods in Enzymology *273*:331-347.
2. **Beato, M. and A. Sanchez-Pancheco.** 1996. Interaction of steroid hormone receptors with the transcription complex. Endocr. Rev. *17*:587-609.
3. **Caplan, A.J., E. Langley, E.M. Wilson, and J. Vidal.** 1995. Hormone-dependent transactivation by human androgen receptor is regulated by a dnaJ protein. J. Biol. Chem. *270*:5251-5257.
4. **Chang, C., J. Kokontis, and S. Liao.** 1988. Molecular cloning of human and rat complementary DNA encoding androgen receptors. Science *240*:324-326.
5. **Chang, H.C., H. Miyamoto, S. Yeh, H. Lardy, and C. Chang.** Suppression of Delta5-Androstenediol-induced AR transactivation by selective steroids in human prostate cancer cells. Proc. Natl. Acad. Sci. USA *96*:11173-11177.
6. **Chen, J.D. and R.M. Evans.** 1995. A transcriptional co-repressor that interacts with nuclear hormone receptors. Nature *377*:454-457.
7. **Douarin, B.L., C. Zechel, J.-M. Garnie, Y. Lutz, L. Tosa, B. Pierrat, D. Heery, H. Gronemeyer, P. Chambon, and R. Losson.** 1995. The N-terminal part of TIF1, a putative mediator of ligand-dependent activation function (AF-2) of nuclear receptors, is fused to B-raf in the oncogenic protein T18. EMBO J. *14*:2020-2033.
8. **Durfee, T., K. Becherer, P.L. Chen, S.H. Yeh, Y. Yang, A.E. Kilburn, W.H. Lee, and S.J. Elledge.** 1993. The retinoblastoma protein associates with the protein phosphatase type 1 catalytic subunit. Genes Dev. *7*:555-569.
9. **Fields, S. and O. Song.** 1989. A novel genetic system to detect protein-protein interactions. Nature *340*:245-246.
10. **Fujimoto, N., S. Yeh, H.-Y. Kang, S. Inui, H.-C. Chang, A. Mizokami, and C. Chang.** 1999. Cloning and characterization of androgen receptor coactivator, ARA55, in human prostate. J. Biol. Chem. *274*:8316-8321.
11. **Gill, G. and M. Ptashne.** 1988. Negative effect of transcriptional activator GAL4. Nature *334*:721-724.
12. **Hong, H., K. Kohli, A. Trivedi, D.L. Johnson, and M.R. Stallcup.** 1996. GRIP, a novel mouse protein that serves as a transcriptional coactivator in yeast for the hormone-binding domain of steroid receptors. Proc. Natl. Acad. Sci. USA. *93*:4948-4952.
13. **Horlen, A.J., A.M. Naar, T. Heinzel, J. Torchia, B. Gloss, R. Kurokawa, A. Ryan, Y. Kamei, M. Soderstrom, C.K. Glass, and M.G. Rosenfeld.** 1995. Ligand-independent repression by the thyroid hor-

mone receptor mediated by a nuclear receptor co-repressor. Nature *377*:397-403.

14. **Horwitz, K.B., T.A. Jackson, D.L. Bain, J.K. Richer, G.S. Takimoto, and L. Tung.** 1996. Nuclear coactivators and corepressors. Mol. Endocrinol. *10*:1167-1177.

15. **Hsiao, P.-W. and C. Chang.** 1999. Isolation and characterization of ARA160 as the first androgen receptor N-terminal-associated coactivator in human prostate cell. J. Biol. Chem. *274*:22373-22379.

16. **Hsiao, P.-W., D.-L. Lin, R.Nakao, and C. Chang.** 1999. The linkage of Kennedy's neuron disease to ARA24, the first identified androgen receptor polyglutamine region-associated coactivator. J. Biol. Chem. *274*:20229-20234.

17. **Kang, H.-Y., S. Yeh, F. Naohiro, and C. Chang.** 1999. Cloning and characterization of human prostate coactivator ARA54, a novel protein that associates with the androgen receptor. J. Biol. Chem. *274*:8570-8576.

18. **Lee Y, C.-R. Shyr, T. Thin, and C. Chang.** 1999. Convergence of two corepressors through heterodimer formation of androgen receptor and TR4 orphan nuclear receptor: a new signaling pathway in the steroid receptor superfamily. Proc. Natl. Acad. Sci. USA *96*:14724-14729.

19. **Miyamoto H, S. Yeh, H. Lardy, E. Messing, and C. Chang.** 1998. Delta5-Androstenediol is a natural hormone with androgenic activity in human prostate DU145 cells. Proc. Natl. Acad. Sci. USA *19*:11083-11088.

20. **Mowszowicz, I., H. Lee, M.C. Portois, F. Kuttenn, and C. Chang.** 1993. Complete androgen insensitivity due to a single base substitution in exon 8 of the steroid-binding domain of the androgen receptor. Endocrine *1*:203-209.

21. **Onate, S.A., S.Y. Tsai, M-J. Tsai, and B.W. O'Malley.** 1995. Sequence and characterization of a coactivator for the steroid hormone receptor superfamily. Science *270*:1354-1357

22. **SenGupta, D.J., B. Zhang, B. Kraemer, P. Pochart, S. Fields, and M. Wickens.** 1996. A three-hybrid system to detect RNA-protein interactions in vivo. Proc. Natl. Acad. Sci. USA *93*:8496-8501.

23. **Walfish, P.G., T. Yoganathan, Y.-F. Yang, H. Hong, T.R. Butt, and M.R. Stallcup.** 1997. Yeast hormone response element assays detect and characterize GRIP1 coactivator-dependent activation of transcription by thyroid and retinoid nuclear receptors. Proc. Natl. Acad. Sci. USA *94*:3697-3702.

24. **Wang, M.M. and R.R. Reed.** 1993. Molecular cloning of the olfactory neuronal transcription factor Olf-1 by genetic selection in yeast. Nature *364*:121-126.

25. **Yeh, S. and C. Chang.** 1996. Cloning and characterization of a specific coactivator, ARA70, for the androgen receptor in human prostate cells. Proc. Natl. Acad. Sci. USA *93*:5517-5521.

26. **Yeh, S., H. Lin, H.-Y. Kang, M. Lin, and C. Chang.** 1999. From HER2/Neu signal cascade to androgen receptor and its coactivators: a new pathway by induction of androgen target genes through MAP kinase in prostate cancer cells. Proc. Natl. Acad. Sci. USA *96*:5458-5463.

27. **Yeh, S., H. Miyamoto, and C. Chang.** 1997. Hydroxyflutamide may not always be a pure antiandrogen. Lancet *349*:852-853.

28. **Yeh, S., H. Miyamoto, K. Nishimura, H.-Y. Kang, J. Ludlow, P. Hsiao, C. Wang, C.-Y. Su, and C. Chang.** 1998. Retinoblastoma, a tumor suppressor, is a coactivator for the androgen receptor in human prostate cancer DU145 cells. Biochemical & Biophysical Research Communications *248*:361-367.

29. **Yeh, S., H.-Y. Kang, H. Miyamoto, K. Nishimura, H.-C. Chang, H.-J. Ting, M. Rahman, H.-K. Lin, N. Fujimoto, Y.-C. Hu, A. Mizokami, et al.** 1999. Differential induction of androgen receptor transactivation by different androgen receptor coactivators in human prostate cancer DU145 cells. ENDO. *11*:195-202.

21 Hunting of Caspase Substrates

Shinji Kamada[1] and Yoshihide Tsujimoto[2]
[1]Department of Medical Genetics, Biomedical Research Center, Osaka University Graduate School of Medicine; [2]CREST of Japan Science and Technology Corp. (JST), Osaka, Japan

1. OVERVIEW

The yeast two-hybrid system has been widely and successfully used to identify proteins with an ability to bind to any protein target of interest under investigation. However, this approach has not yet been used to identify enzyme substrates. In this chapter, we introduce a modified yeast two-hybrid system to identify enzyme substrates, using caspases as an example. Caspases are cysteine proteases, which play a critical role in the execution of apoptosis by cleaving a set of cellular proteins and have attracted much attention. Because caspases consist of a large and small subunit, we made two major modifications: *(i)* both large and small subunits of active caspases were expressed in yeast under ADH1 promoters, and the small subunit was fused to the LexA DNA-binding domain, and *(ii)* a point mutation was introduced, which substituted serine for the active site cysteine and thereby prevented proteolytic cleavage of the substrates, possibly stabilizing the enzyme-substrate complexes in yeast. After screening a mouse embryo cDNA expression library using the bait plasmid for caspase-3, we obtained 13 clones that encoded proteins that bound to caspase-3 and showed that 10 clones were cleaved by recombinant caspase-3 in vitro. Our results indicate that this cloning method is useful to identify substrates of caspases and thus possibly of other enzymes.

2. INTRODUCTION

The yeast two-hybrid technique is the prototype of a genetic assay system to detect protein–protein interaction and has been successfully used to identify novel proteins that interact with proteins under examination. Although various versions of the system have been developed as described in other chapters of this book, the yeast system had not been successfully applied to identify enzyme substrate. We recently described a new version of the yeast two-hybrid system that can be used for this purpose. We have successfully applied the new version of the yeast two-hybrid system to

Yeast Hybrid Technologies
Edited by L. Zhu and G.J. Hannon
© 2000 Eaton Publishing, Natick, MA

isolate genes for substrate of caspases, which are crucial proteases involved in apoptosis and one of our major scientific interests.

Apoptosis is a fundamental cellular process involved in a variety of biological phenomena including morphogenesis and maintenance of tissue homeostasis. *ced-3* is a gene of the nematode *Caenorhabditis elegans* that is required for cell death; it encodes a protein resembling mammalian interleukin-1β (IL-1β)-converting enzyme (ICE), a cysteine protease that cleaves the inactive 31-kDa precursor of IL-1β to generate the active cytokine (7,12).

Thirteen mammalian caspases homologous to CED-3 have been identified and have distinct roles in apoptosis and inflammation (8). Caspases share similarities in amino acid sequence, structure, and substrate specificity (5). They are all expressed as dormant proenzymes and are proteolytically activated in response to various death signals (5,8). The active form is composed of two subunits, large and small, both of which are derived from the proenzyme by cleavage at the C-terminal side of specific Asp residues (5,7,8,12). Caspase-3, in many cell types, cleaves target proteins in a cascade that ultimately leads to cell death by apoptosis (9). There is great interest in these downstream targets for general study and as possible therapeutic targets. Here we described a system for cloning the substrates of caspase-3 (1), which can easily be adapted to other caspases.

The cloning method is summarized in Figure 1. Since caspases are composed of two subunits, both subunits must be expressed simultaneously in yeast. Therefore, both large and small subunits of caspase-3 were separately expressed in yeast under ADH1 promoters to maintain an equimolar ratio of large to small subunits. Because the N-terminal parts of the small subunits in the active forms of caspase-1 and caspase-3 seem to face the outside of the complexes—judging from their three-dimensional crystal structures (6,10,11)—the small subunit was fused to the LexA DNA-binding domain. Furthermore, possibly to stabilize the enzyme-substrate complex in yeast, a point mutation was introduced that substituted serine for the active site cysteine and thereby prevented proteolytic cleavage of the substrates.

3. PROTOCOLS

3.1 Construction of the Bait Plasmid

The strategy for construction of the bait plasmid is shown in Figure 2. Polymerase chain reaction (PCR) was used to obtain the cDNA fragments corresponding to p17 and p12 subunits of caspase-3. The PCR primers employed for human caspase-3 were p17-5′ (5′-CGGATATCACCATGTCTGGAATATCCCTG-3′) and p17-3′ (5′-CGGATATCTTAGTCTGTCTCAATGCC-3′), corresponding to the N-terminal and C-terminal regions of the p17 subunit of caspase-3 cDNA, as well as p12-5′ (5′-CGGAATTCAGTGGTGTTGATGAT-3′) and p12-3′ (5′-CGGGATCCTTAGTGA-TAAAAATAGAG-3′), corresponding to the N-terminal and C-terminal regions of the p12 subunit of caspase-3 cDNA. p17-5′ primer has an *Eco*RV site with a Kozak sequence, p17-3′ primer has an *Eco*RV site with stop codon, p12-5′ primer has an *Eco*RI site, and p12-3′ primer has a *Bam*HI site with a stop codon at the 5′ end.

A fragment encoding caspase-3-p17^m and caspase-3-p12 was generated by PCR using *caspase-3* cDNA bearing an active site mutation (Cys163 to Ser) (2,3) as the template. The *Eco*RI-*Bam*HI fragment encoding caspase-3-p12 was cloned into the

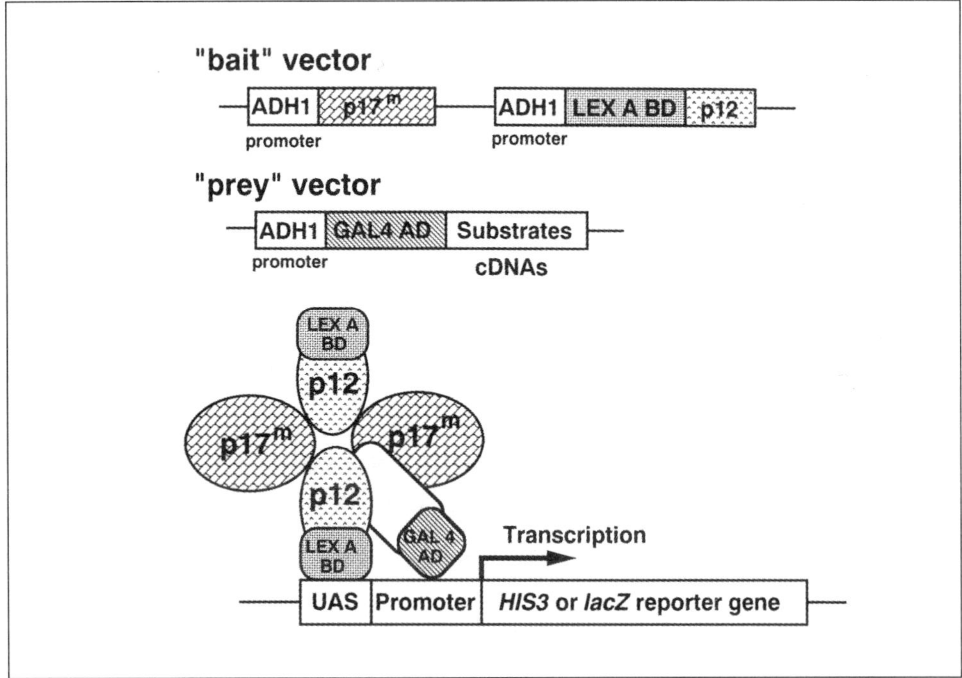

Figure 1. Diagram of the method of cloning genes of caspase targets using the modified yeast two-hybrid system. The bait vector contained sequences for both the large (p17) and small (p12) subunits of active caspase-3, which were expressed separately in yeast under ADH1 promoters, and the small subunit was fused to the LexA DNA-binding domain. A point mutation was introduced into the active site cysteine of p17 to prevent proteolytic cleavage of the substrates allowing formation of a stable enzyme-substrate complex in yeast.

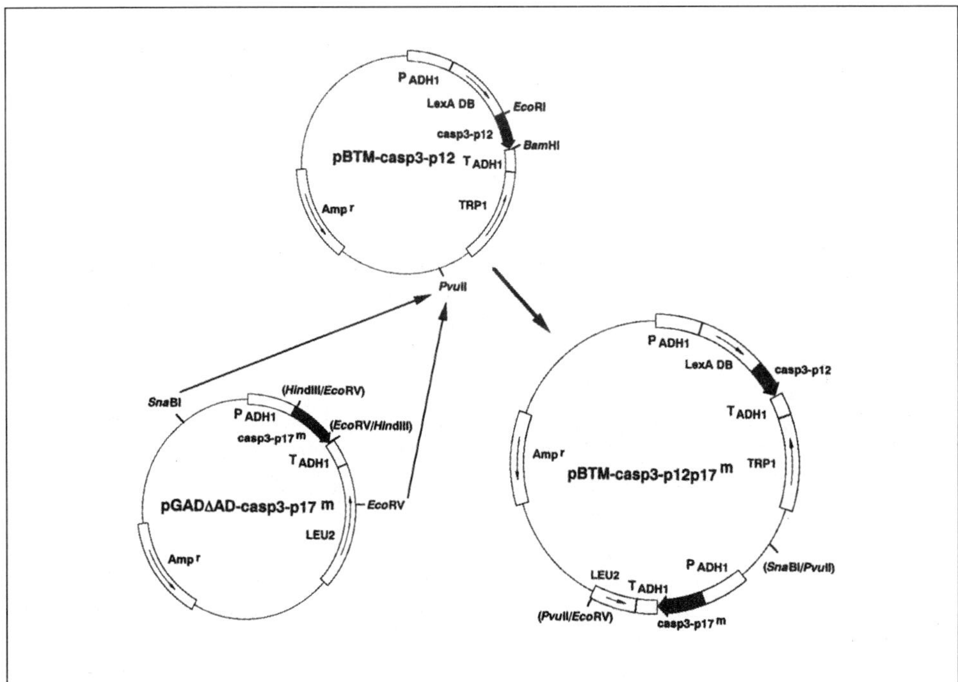

Figure 2. Construction of bait plasmid. For further description, see text.

*Eco*RI-*Bam*HI site of pBTM$_{116}$ to generate pBTM-casp3-p12. The *Eco*RV fragment encoding caspase-3-p17^m was cloned into the blunt-ended *Hind*III site of pGAD$_{10}$ lacking the Gal4 activation domain to generate pGADΔAD-casp3-p17^m. The *Sna*BI-*Eco*RV fragment bearing caspase-3-p17^m under control of the ADH1 promoter from pGADΔAD-casp3-p17^m was cloned into the *Pvu*II site of pBTM-casp3-p12 to generate pBTM-casp3-p12p17^m.

3.2 Yeast Two-Hybrid Screening

Yeast two-hybrid screening using pBTM-casp3-p12p17^m as the bait was performed with cDNA expression libraries fused to the Gal4 activation domain in the pGAD$_{10}$ plasmid following the MATCHMAKER™ Two-Hybrid System Protocol (CLONTECH Laboratories, Palo Alto, CA, USA) in L40 cells (*MATa trp1 leu2 his3 ade2 LYS2::lexA-HIS3 URA3::lexA-lacZ*).

We have screened a mouse 11-day embryo cDNA expression library with pBTM-casp3-p12p17^m. Sixty-nine positive clones were obtained by screening 4.2×10^7 transformants. From the size of the inserted fragments, the restriction enzyme digestion pattern, and partial DNA sequencing, the 69 clones were divided into 13 groups.

3.3 Purification of Recombinant His$_6$-Tagged Caspase-3

DNA sequence encoding the 29–277th amino-acid residue of caspase-3 was amplified by PCR and subcloned into the *E. coli* expression vector pRSET A (Invitrogen, Carlsbad, CA, USA), in which the protease sequence was placed under the control of the T7 promoter and joined in-frame to sequences encoding an N-terminal fusion peptide. The N-terminal fusion peptide includes an ATG translation initiation codon and the sequence for six consecutive histidine residues that function as a metal-binding domain in the translated proteins. The resultant plasmid was transformed into *E. coli* strain JM109. Induction of recombinant His-tagged protease was achieved according to the manufacturer's instructions. Cells were harvested and recombinant protease was purified essentially as described elsewhere (4). Fractions showing proteolytic activity in an in vitro cleavage assay were pooled and dialyzed against buffer A [50 mM HEPES, 0.1 M NaCl, 10% (vol/vol) glycerol, pH 7.5] to remove excess imidazole and were stored at -80°C.

3.4 In Vitro Cleavage of Candidate Substrates by Recombinant Caspase-3

The *Eco*RI fragments of positive clones after screening were inserted into the *Eco*RI site of pRSET B (Invitrogen), in which the sequences were under control of the T7 promoter and joined in-frame to an ATG translation initiation codon. [^{35}S]-methionine–labeled proteins were prepared using expression plasmids and a TNT T7-coupled in vitro transcription and translation system (Promega Corp., Madison, WI, USA) according to the manufacturer's instructions and were subjected to cleavage by recombinant caspase-3. Recombinant caspase-3 was preincubated at room temperature for 15 min with or without 1 µM Ac-DEVD-CHO (Peptide Institute, Japan) in 25 µL of a buffer [50 mM HEPES, 0.1 M NaCl, 10% (vol/vol) glycerol, pH 7.5, and 10 mM dithiothreitol (DTT)], and the reaction was initiated by addition of 5 µL of [^{35}S]-methionine–labeled protein. After incubation at 37°C for 2 h, the reaction was stopped by the addition of sodium dodecyl sulfate-polyacrylamide gel electrophoresis (SDS-PAGE) sample buffer (2% SDS, 10% glycerol, 100 mM DTT, 60 mM Tris-HCl, pH 6.8, and 0.001% bromphenol blue), and cleaved products were analyzed by 12% SDS-PAGE.

Figure 3 shows the cleavage of candidate substrates from a mouse 11-day embryo cDNA expression library. Ten of 13 candidate substrates were cleaved by recombinant caspase-3 and cleavage was inhibited in the presence of 1 μM Ac-DEVD-CHO, suggesting that these proteins were possible substrates for caspase-3. These results indicated that our method could successfully identify caspase substrates.

4. DISCUSSION

We established a novel method for cloning the genes of caspase substrates using the yeast two-hybrid system and obtained several candidate substrates for caspase-3. Although the yeast two-hybrid system has already proven very useful to identify interacting molecules, this is the first successful application of the yeast two-hybrid system to identify enzyme substrates. This approach to find enzyme substrates may have a broad applicability to other polypeptide modifying enzymes such as acetylases, isomerases, and carboxylases.

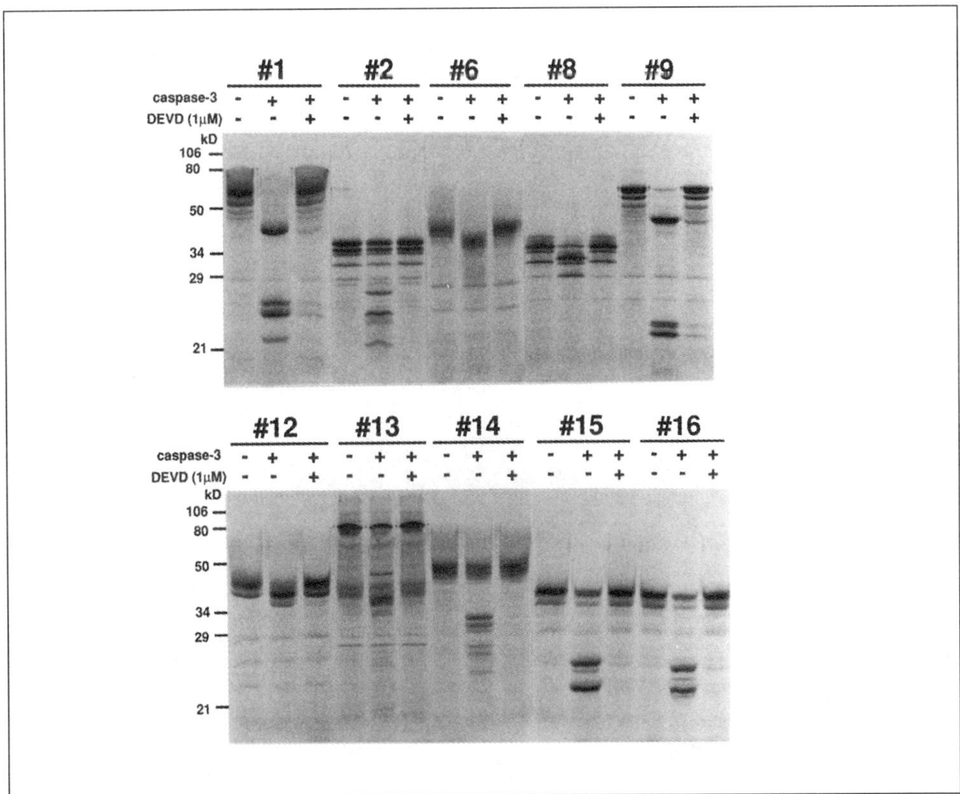

Figure 3. In vitro cleavage of candidate proteins by recombinant active caspase-3. [35S]-methionine–labeled proteins produced from positive clones by in vitro transcription/translation were incubated with 0.6 μg of purified recombinant caspase-3 in the presence or absence of 1 μM Ac-DEVD-CHO for 2 h, and the reaction products were analyzed by 10% SDS-PAGE. (Reproduced from Reference 1.)

ACKNOWLEDGMENTS

This work was supported in part by grants-in-aid for Scientific Research on Priority Areas and for COE Research from the Ministry of Education, Science, Sports, and Culture of Japan.

REFERENCES

1. **Kamada, S., H. Kusano, H. Fujita, M. Ohtsu, R.C. Koya, N. Kuzumaki, and Y. Tsujimoto.** 1998. A cloning method for caspase substrates that uses the yeast two-hybrid system: cloning of the antiapoptotic gene gelsolin. Proc. Natl. Acad. Sci. USA *95*:8532-8537.

2. **Kamada, S., M. Washida, J.-I. Hasegawa, H. Kusano, Y. Funahashi, and Y. Tsujimoto.** 1997. Involvement of caspase-4(-like) protease in Fas-mediated apoptosis. Oncogene *15*:285-290.

3. **Kamada, S., Y. Funahashi, and Y. Tsujimoto.** 1997. Caspase-4 and caspase-5, members of the ICE/Ced-3 family of cysteine proteases, are CrmA-inhibitable proteases. Cell Death Differ. *4*:473-478.

4. **Lippke, J.A., Y. Gu, C. Sarnecki, P.R, Caron, and M.S.-S.Su.** 1996. Identification and characterization of CPP32/Mch2 homolog 1, a novel cysteine protease similar to CPP32. J. Biol. Chem. *271*:1825-1828.

5. **Nicholson, D.W. and N.A. Thornberry.** 1997. Caspase: killer proteases. Trends Biochem. Sci. *22*:299-306.

6. **Rotonda, J., D.W. Nicholson, K.M. Fazil, M. Gallant, Y. Gareau, M. Labelle, E.P. Peterson, D.M. Rasper, R. Ruel. J.P. Vaillancourt, N.A. Thornberry, and J.W. Becker.** 1996. The three-dimensional structure of apopain/CPP32, a key mediator of apoptosis. Nat. Struct. Biol. *3*:619-625.

7. **Thornberry, N.A., H.G. Bull, J.R. Calaycay, K.T. Chapman, A.D. Howard, M.J. Kostura, D.K. Miller, S.M. Molineaux, J.R. Weigner, J. Aunins, et al.** 1992. A novel heterodimeric cysteine protease is required for interleukin-1β processing in monocytes. Nature *356*:768-774.

8. **Thornbery, N.A. and Y. Lazebnik.** 1998. Caspases: enemies within. Science *281*:1312-1316.

9. **Villa, P., S.H. Kaufmann, and W.C. Earnshaw.** 1997. Caspases and caspase inhibitors. Trends Biochem. Sci. *22*:388-393.

10. **Walker, N.P.C., R.V. Talanian, K.D. Brady, L.C. Dang, N.J. Bump, C.R. Ferenz, S. Franklin, T. Ghayur, M.C. Hackett, L.D. Hammill, et al.** 1994. Crystal structure of the cysteine protease interleukin-1β-converting enzyme: a $(p20/p10)_2$ homodimer. Cell *78*:343-352.

11. **Wilson, K.P., J.F. Black, J.A. Thomson, E.E. Kim, J.P. Griffith, M.A. Murcko, S.P. Chambers, R.A. Aldape, S.A. Raybuck, and D.J. Livingston.** 1994. Structure and mechanism of interleukin-1β converting enzyme. Nature *370*:270-275.

12. **Yuan, J., S. Shaham, S. Ledoux, H.M. Ellis, and H.R. Horvitz.** 1993. The C. elegans cell death gene *ced-3* encodes a protein similar to mammalian interleukin-1β-converting enzyme. Cell *75*:641-652.

Appendices

APPENDIX A. Two-Hybrid Cloning Vectors

Vector	System[a]	Description	Selection on SD Medium	Size (kb)	Diagnostic R.E. Sites (kb)	GenBank Accession #	References (Plasmid name in reference)
pACT	GAL4 2H	GAL4 AD, *LEU2*, amp[r], HA epitope tag	-Leu	7.65	*Eco*R I (3.0, 3.05, 1.6)	not available	(4)
pACT2[b]	GAL4 2H	GAL4 AD, *LEU2*, amp[r], HA epitope tag	-Leu	8.1	*Hin*d III (7.3, 0.8)	U29899	(20)
pAS2-1[b,c]	GAL4 2H	GAL4 BD, *TRP1*, amp[r], *CYH[s]2*	-Trp	8.4	*Hin*d III (4.6, 2.2, 0.9, 0.7)	U30497	(15)
pB42AD[b]	LexA 2H	acidic activator B42, *TRP1*, amp[r], HA epitope tag	-Trp	6.45	*Hin*d III (3.4, 2.1, 0.6, 0.35)	not available	(pJG4-5 in Reference 13)
pGAD10	GAL4 2H	GAL4 AD, *LEU2*, amp[r]	-Leu	6.6	*Hin*d III (5.9, 0.7)[d]	U13188	(2)
pGAD424[b]	GAL4 2H	GAL4 AD, *LEU2*, amp[r]	-Leu	6.6	*Hin*d III (5.9, 0.7)[d]	U07647	(2)
pGAD GH	GAL4 2H	GAL4 AD, *LEU2*, amp[r]	-Leu	7.9	*Hin*d III (7.1, 0.5, 0.3)	not available	(14)
pGAD GL	GAL4 2H	GAL4 AD, *LEU2*, amp[r]	-Leu	6.9	*Hin*d III (6.1, 0.5, 0.3)	not available	(14)
pGBT9[b]	GAL4 2H	GAL4 BD, *TRP1*, amp[r]	-Trp	5.5	*Hin*d III (4.6, 0.9)	U07646	(2)
pGilda	LexA 2H	LexA, *HIS3*, amp[r]	-His	6.6	*Hin*d III (0.2, 6.3)	not available	(12,10)
pLexA[b]	LexA 2H	LexA, *HIS3*, amp[r]	-His	10.2	*Hin*d III (5.2, 4.8, 0.2)	not available	(pEG202 in Reference 13)

a Key to system abbreviations: GAL4 2H = GAL4-based yeast two-hybrid systems; LexA 2H = LexA-based yeast two-hybrid systems.
b Additional information and restriction maps of these vectors are in Chapter 2.
c pAS2-1 is a derivative of the plasmid described in this reference; the plasmid was modified at CLONTECH. The *Eco*R I site is unique in pAS2-1.
d pGAD424 is linearized by digestion with *Sal*I; pGAD10 does not contain a *Sal*I site.

APPENDIX B. Yeast One-Hybrid System Cloning/Reporter and Control Plasmids

Vector	Description	Selection on SD Medium	Size (kb)	Diagnostic R.E. Sites (kb)	GenBank Accession #	Reference
pHISi[a]	*HIS3* under control of cloned target element; *URA3*, amp[r]	-Ura, -His[b]	6.7	*Eco*R I/*Xho*I (5.7, 1.0)[d]	not available	(1)
pHISi-1[a]	*HIS3* under control of cloned target element; amp[r]	-His[b]	5.4	*Eco*R I/*Xho*I (4.4, 1.0)	not available	(1)
pLacZi[a]	*lacZ* under control of cloned target element; *URA3*, amp[r]	-Ura	6.4	*Eco*R I/*Xho*I (6.4, 0.04)[c]	not available	(22)
p53HIS	*HIS3* under control of p53 binding sites in pHISi; *HIS3, URA3*, amp[r]	-Ura, -His	6.7	*Eco*R I/*Xho*I (5.7, 1.0)		(22)
p53BLUE	*lacZ* under control of p53 binding sites in pLacZi, *URA3*, amp[r]	-Ura	6.4	*Eco*R I/*Xho*I (6.3, 0.1)		(22)
pGAD53m	murine p53$_{(72-300)}$ fused to GAL4 AD; *LEU2*, amp[r]	-Leu	7.6	*Eco*R I[d] (7.6)		(22)

a In the one-hybrid system, a putative recognition sequence (the target DNA sequence) must be cloned into the MCS of one of the reporter plasmids. The construct is then used to generate the necessary yeast reporter strain to detect specific DNA–protein interactions.

b Leaky *HIS3* expression in these plasmids permits its use as a selectable marker on SD/-His (without 3-AT).

c In addition, pHISi and pLacZi have a single *Sma*I site, which makes it possible to distinguish them from p53HIS and p53Blue, which are not cut by *Sma*I.

d pGAD53m is not cut by *Xho*I.

APPENDIX C. Yeast Two-Hybrid Reporter and Control Plasmids

Vector[a]	System	Description	Selection on SD Medium	Size (kb)	Diagnostic R.E. Sites (kb)	References (Plasmid name in reference)
p8op-lacZ	LexA 2H	lacZ under control of lexA$_{op(x8)}$; URA3, amp^r	-Ura	~10.3	Hind III (6.3, 2.1, 1.9)	(pSH18-34 in Reference 5)
pB42AD-T[a]	LexA 2H	SV40 large T-antigen$_{(87-708)}$ in pB42AD, TRP1, amp^r	-Trp	8.5	Hind III (3.4, 2.1, 1.0, 0.9, 0.6, 0.5)	(19,3)
pCL1	GAL4 2H	wild-type full-length GAL4 gene in a YCp50 derivative, LEU2, amp^r	-Leu	~15.3	Hind III (~11.2, 2.8, 1.8)	(8)
pLAM5'[a]	GAL4 2H	Human lamin C$_{(66-230)}$ in pGBT9; TRP1, amp^r	-Trp	6.0	Hind III (4.7, 0.8, 0.6)	(2)
pLAM5'-1[a]	GAL4 2H	Human lamin C$_{(72-300)}$ in pAS2-1; TRP1, amp^r	-Trp	~9.0	Hind III (4.6, 2.2, 0.9, 0.85, 0.4)	(2)
pLexA-53[a]	LexA 2H	murine p53$_{(72-300)}$ in pLexA HIS3, amp^r	-His	11.1	Hind III (5.7, 5.2, 0.2)	(16)
pLexA-Lam[a]	LexA 2H	Human lamin C$_{(72-300)}$ in pLexA; HIS3, amp^r	-His	10.6	Hind III (5.2, 4.3, 0.9, 0.2)	(2)
pLexA-Pos	LexA 2H	LexA/GAL4 fusion gene; HIS3, amp^r	-His	~13.5	Hind III (6.0, 4.5, 3.0)	(pSH17-4 in Reference 11)
pTD1	GAL4 2H	SV40 large T antigen$_{(87-708)}$ in pGAD3F; LEU2, amp^r	-Leu	~15.0	Hind III (12, 1.3, 1.2, 0.5)	(19,3)
pTD1-1[a]	GAL4 2H	SV40 large T antigen$_{(87-708)}$ in pACT2; LEU2, amp^r	-Leu	~10.0	Hind III (7.3, 1.2, 1.0, 0.5)	(19)
pVA3	GAL4 2H	murine p53$_{(72-300)}$ in pGBT9; TRP1, amp^r	-Trp	6.4	Hind III (4.6, 1.8)	(16)
pVA3-1[a]	GAL4 2H	murine p53$_{(72-300)}$ in pAS2-1; TRP1, amp^r	-Trp	9.4	Hind III (4.6, 2.2, 1.7, 0.9)	(16,3)

a These plasmids are derivatives of the plasmids described in the indicated references; plasmids were modified at CLONTECH.

APPENDIX D. Yeast Reporter Strains in the One- and Two-Hybrid Systems

Strain	System	Genotype[a]	Reporter(s)[b]	Transformation Markers[c]	References
SFY526	GAL4 2H	MATa, ura3-52, his3-200, ade2-101, lys2-801, trp 1-901, leu2-3, 112, can^r, gal4-542, gal80-538, URA3::GAL1$_{UAS}$-GAL1$_{TATA}$-lacZ	lacZ	trp1, leu2	(15)
HF7c	GAL4 2H	MATa, ura3-52, his3-200, ade2-101, lys2-801, trp1-901, leu2-3, 112, gal4-542, gal80-538, LYS2::GAL1$_{UAS}$-GAL1$_{TATA}$-HIS3, URA3::GAL4$_{17-mers(x3)}$-CYC1$_{TATA}$-lacZ	HIS3, lacZ cyh^{r}2	trp1, leu2,	(6) C. Giroux, personal communication
Y187	GAL4 2H	MATα, ura3-52, his3-200, ade 2-101, trp1-901, leu2-3, 112, gal4Δ, met$^-$, gal80Δ, URA3::GAL1$_{UAS}$-GAL1$_{TATA}$-lacZ	lacZ	trp1, leu2	(15)
CG-1945[d]	GAL4 2H	MATa, ura3-52, his3-200, ade2-101, lys2-801, trp1-901, leu2-3, 112, gal4-542, gal80-538, cyh^{r}2, LYS2::GAL1$_{UAS}$-GAL1$_{TATA}$-HIS3, URA3::GAL4$_{17-mers(x3)}$-CYC1$_{TATA}$-lacZ	HIS3, lacZ cyh^{r}2	trp1, leu2,	Craig Giroux, personal communication
Y190	GAL4 2H	MATa, ura3-52, his3-200, ade2-101, lys2-801 trp1-901, leu2-3, 112, gal4Δ, gal80Δ, cyh^{r}2, LYS2::GAL1$_{UAS}$-HIS3$_{TATA}$-HIS3, URA3::GAL1$_{UAS}$-GAL1$_{TATA}$-lacZ	HIS3, lacZ cyh^{r}2	trp1, leu2,	(15,9)
EGY48	LexA 2H	MATα, ura3, his3, trp1, LexA$_{op(x6)}$-LEU2	LEU2, ura3	his3, trp1,	(5,21)
YM4271	MM 1H	MATa, ura3-52, his3-200, lys2-801, ade2-101, ade5, trp1-901, leu2-3, 112, tyr1-501, gal4Δ, gal80Δ, ade5::hisG		his3, trp1	
PJ69-2A[e]	GAL4 2H	MATa, trp1-901, leu2-3, 112, ura3-52, his3-200, gal4Δ, gal80Δ, LYS2::GAL1$_{UAS}$-GAL1$_{TATA}$-HIS3, GAL2$_{UAS}$-GAL2$_{TATA}$-ADE2	HIS3, ADE2	trp1, ura3, leu2	(17)

a The trp1, his3, gal4, and gal80 mutations are all deletions; leu2-3, 112 is a double mutation. The LYS2 gene is nonfunctional in the HF7c and CG-1945. See Chapter 2 for more information on the promoters of the reporter genes.
b See Appendix C for more information on reporter genes and their phenotypes.
c Genes that are used as selection markers in this system.
d CG-1945 is a derivative of HF7c (6).
e The ade2-101 gene of the precursor strain was replaced (by recombination) with the GAL2-ADE2 reporter construct.

APPENDIX E.1. Selected Yeast Genes and their Associated Phenotypes

Wild type	Allele Mutant	Phenotype of mutant	
TRP1	*trp1-901*	Trp⁻	Requires tryptophan (Trp) in the medium to grow, i.e., is a Trp auxotroph
LEU2	*leu2-3, 112*	Leu⁻	Requires leucine (Leu) to grow, i.e., is a Leu auxotroph
HIS3	*his3-200*	His⁻	Requires histidine (His) to grow, i.e., is a His auxotroph
URA3	*ura3-52*	Ura⁻	Requires uracil (Ura) to grow, i.e., is a Ura auxotroph
LYS2	*lys2-801*	Lys⁻	Requires lysine (Lys) to grow, i.e., is a Lys auxotroph
ADE2	*ade2-101*	Ade⁻	Requires adenine (Ade) to grow; i.e., is an Ade auxotroph; in addition, confers a pink or red colony color to colonies growing on media low in adenine. The red pigment is apparently an oxidized, polymerized derivative of 5-amino-imidazole ribotide, which accumulates in vacuoles (23,24).
GAL4	*gal4-542* (or *gal4Δ*)	Gal⁻	Deficient in regulation of galactose-metabolizing genes (9,18)
GAL80	*gal80-538* (or *gal80Δ*)	Gal⁻	Deficient in regulation of galactose-metabolizing genes (GAL genes are constitutively expressed)
CYH^{s}2	*cyh^{r}2*	Cyh^r	Resistant to cycloheximide

APPENDIX E.2. Yeast Hybrid Reporter Genes and their Phenotypes

Reporter Gene	Gene Description	Positive Phenotype[a]	Negative Phenotype[a]
lacZ	Encodes β-galactosidase	LacZ⁺ • Blue colony • β-gal activity above background	LacZ⁻ • White colony • Undetectable or background level of β-gal activity
HIS3	Confers His prototrophy	His⁺ • Grows on SD/-His	His⁻ • Does not grow on SD/-His[b]
LEU2	Confers Leu prototrophy	Leu⁺ • Grows on SD/-Leu	Leu⁻ • Does not grow on SD/-Leu
ADE2	Confers Ade prototrophy	Ade⁺ • Grows on SD/-Ade	Ade⁻ • Does not grow on SD/-Ade • Pink or red colony color when grown on medium (such as YPD) low in Ade

a Relative levels of background expression and reporter gene induction are dependent on the promoter constructs controlling them.

b 5–60 mM 3-AT may be required to suppress leaky *HIS3* expression in certain host strains and transformants and to obtain an accurate His⁻ phenotype.

APPENDIX F. Yeast Media

• YPD Medium

YPD medium is a blend of peptone, yeast extract, and dextrose in optimal proportions for growth of most strains of *Saccharomyces cerevisiae*. If you purchase YPD medium from a commercial supplier, prepare the medium according to the instructions provided. Alternatively, prepare your own YPD mixture as follows:

20	g/L	Difco peptone
10	g/L	Yeast extract
20	g/L	Agar (for plates only)

Add H_2O to 950 mL. Adjust pH to 5.8 if necessary, and autoclave. Allow medium to cool to approximately 55°C and then add dextrose (glucose) to 2% (50 mL of a sterile 40% stock solution).

Note: If you add the sugar solution before autoclaving, autoclave at 121°C for 15 min; autoclaving at a higher temperature for a longer period of time or repeatedly may cause the sugar solution to darken and will decrease the performance of the medium. Note that YPD from some commercial sources already contains glucose.

- (Optional) For adenine-supplemented YPD (YPDA), prepare YPD medium as above. After autoclaved medium has cooled to 55°C, add 15 mL of a 0.2% adenine hemisulfate solution per liter of medium. (Final concentration is 0.003%, in addition to the trace amount of Ade that is naturally present in YPD).
- (Optional) For kanamycin-containing medium, prepare YPD as above. After autoclaved medium has cooled to 55°C, add 0.2–0.3 mL of 50 mg/mL kanamycin (final concentration 10–15 mg/L).

• SD Medium

Synthetic dropout (SD) is a minimal medium used in yeast transformations to select and test for specific phenotypes. SD medium is generally prepared by combining a minimal SD base (providing a nitrogen base, a carbon source, and in some cases, ammonium sulfate) with a stock of "dropout" solution that contains a specific mixture of amino acids and nucleosides (recipe below). If you purchase minimal SD base from a commercial supplier, prepare the medium according to the instructions provided. Alternatively, prepare SD medium as follows:

6.7	g	Yeast nitrogen base without amino acids (Difco Catalog No. 0919-15-3)
20	g	Agar (for plates only)
850	mL	H_2O
100	mL	of the appropriate sterile 10× dropout solution or the required amount of commercially supplied DO supplement powder (see package instructions).

- Adjust pH to 5.8 if necessary and autoclave. Allow medium to cool to approximately 55°C before adding 3-AT, cycloheximide, additional adenine, or X-gal.
- Add the appropriate sterile carbon source, usually dextrose (glucose) to 2%, unless specified otherwise for your application. Adjust final volume to 1L if necessary.

Notes:

- If you add the sugar solution before autoclaving, autoclave at 121°C for 15 min; autoclaving at a higher temperature for a longer period of time or repeatedly may cause the sugar solution to darken and will decrease the performance of the medium. Note that SD minimal base from some commercial suppliers already contains a carbon source.
- If you purchase galactose separately, it must be highly purified and contain <0.01% glucose.
- (Optional) For 3-AT-containing medium, add the appropriate amount of 1.0 M 3-AT stock solution and swirl to mix well. The concentration of 3-AT used in the medium depends on the yeast strain and, to some extent, on the presence of transforming plasmid(s).

Notes:

- 3-AT is heat-labile and will be destroyed if added to medium warmer than 55°C.
- 3-AT, a competitive inhibitor of the yeast HIS3 protein (His3p), is used to inhibit low levels of His3p expressed in a leaky manner in some reporter strains (7,4).
- (Optional) For cycloheximide-containing medium, add the appropriate amount of 1 mg/mL cycloheximide stock solution and swirl to mix well. The concentration of cycloheximide used in the medium depends on the yeast strain.

Notes:

- Cycloheximide is heat-labile and will be destroyed if added to medium warmer than 55°C.
- Cycloheximide-containing medium is used for selection of yeast strains, such as Y190 and CG-1945, which carry the cyh^{r}2 allele.
- (Optional) If you wish to add excess adenine to SD medium, add 15 mL of 0.2% adenine hemisulfate solution per Liter of medium.
- Pour plates and allow medium to harden at room temperature. Store plates inverted, in a plastic sleeve, at 4°C.

- **SD/Gal/Raff/X-gal Plates**

Prepare SD medium as described above except use 725 mL of H$_2$O and do not adjust the pH. Autoclave and cool to approximately 55°C. Then add:

	Final concentration	To prepare 1 L of medium
Galactose	2%	50 mL of 40% stock
Raffinose	1%	25 mL of 40% stock
10× BU salts	1×	100 mL of 10× stock
X-Gal	80 mg/L	4 mL of 20 mg/mL

Pour plates and allow medium to harden at room temperature. Store plates inverted, in a plastic sleeve in the dark, at 4°C for up to two months. Adjust final volume to 1L if necessary.

Notes:

- Galactose must be highly purified and contain <0.01% glucose.

- If the medium is too hot (i.e., >55°C) when the salt solution is added, the salts will precipitate. Also, X-Gal is heat labile and will be destroyed if added to hot medium.
- BU salts must be included in the medium to adjust the pH to approximately 7.0, which is closer to the optimal pH for β-gal activity, and to provide the phosphate necessary for the β-gal assay to work.
- As the plates age, salt crystals will form in the medium. These do not affect the performance of the medium or the results of the β-gal assay.
- If you are assaying for expression of a *lacZ* reporter gene in a system that requires expression of a protein from an intact yeast GAL1 promoter (such as in Brent's original LexA Two-Hybrid System), you must use 2% galactose + 1% raffinose as the carbon sources instead of glucose. If you are not using SD/Gal/Raff Minimal Base from a commercial supplier, be sure to obtain high-quality galactose that is not contaminated by glucose.

Stock Solutions for use with SD Media

- 1.0 M 3-AT (3-amino-1,2,4-triazole; Sigma Catalog No. A-8056); prepare in deionized H_2O and filter sterilize. Store at 4°C. Store plates containing 3-AT sleeved at 4°C for up to 2 months.
- 10× BU Salts
  Dissolve the following components in 1 L (total) of H_2O:

      70 g $Na_2HPO_4 \cdot 7H_2O$
      30 g NaH_2PO_4

  Adjust to pH 7.0, then autoclave and store at room temperature.
- Carbon sources, filter sterilized or autoclaved:

  Note: Autoclave at 121°C for 15 min; autoclaving at a higher temperature for a longer period of time or repeatedly may cause the sugar solution to darken and will decrease the performance of the medium.
  - 40% Dextrose (glucose)
  - 40% Galactose (for LexA Two-Hybrid System; D(+) Galactose, e.g., Sigma Catalog No. G-0750)
  - 40% Raffinose (for LexA Two-Hybrid System)
- 1 mg/mL (1000×) **Cycloheximide** (CHX; Sigma Chemical; Catalog No. C-7698); prepare in deionized H_2O and filter sterilize. Store at 4°C for up to 2 months. Store plates containing CHX sleeved at 4°C for up to 1 month.
- 50 mg/mL **kanamycin** (kan); prepare in deionized H_2O and filter sterilize. Store at -20°C indefinitely. Store plates containing kan sleeved at 4°C for up to 1 month.

- **X-gal** (20 mg/mL in DMF)

  Dissolve 5-bromo-4-chloro-3-indolyl-β-D-galactopyranoside (X-gal) in N,N-dimethylformamide (DMF). Store in the dark at -20°C.

- **10× Dropout (DO) supplements**

  10× dropout solutions contain all but one or more of the following components. A combination of a minimal SD base and a DO supplement will produce a synthetic, defined minimal medium lacking one or more specific nutrients. The specific components omitted depends on the selection medium desired. To prepare SD/-Trp/-Leu, for example, use a 10× dropout supplement lacking Trp and Leu. If a component is not indicated as missing, then it is assumed to be present in the medium. Many of the commonly used 10× dropout supplements are commercially available. Alternatively, prepare your own DO supplements as described below. 10× dropout supplements may be autoclaved and stored at 4°C for up to 1 y.

Note: Serine, aspartic acid, and glutamic acid have been left out of this recipe because they make the media too acidic. The yeast can synthesize these amino acids endogenously.

	10× Concentration	Sigma Chemical Catalog No.
L-Isoleucine	300 mg/L	I-7383
L-Valine	1500 mg/L	V-0500
L-Adenine hemisulfate salt	200 mg/L	A-9126
L-Arginine HCl	200 mg/L	A-5131
L-Histidine HCl monohydrate	200 mg/L	H-9511
L-Leucine	1000 mg/L	L-1512
L-Lysine HCl	300 mg/L	L-1262
L-Methionine	200 mg/L	M-9625
L-Phenylalanine	500 mg/L	P-5030
L-Threonine	2000 mg/L	T-8625
L-Tryptophan	200 mg/L	T-0254
L-Tyrosine	300 mg/L	T-3754
L-Uracil	200 mg/L	U-0750

REFERENCES FOR APPENDICES

1. **Alexandre, C., D.A. Grueneberg, and M.Z. Gilman.** 1993. Studying Heterologous Transcription Factors in Yeast. METHODS: A Companion to Methods in Enzymology 5:147-155.
2. **Bartel, P.L., C.-T. Chien, R. Sternglanz, and S. Fields.** 1993. Using the Two-Hybrid System to Detect Protein-Protein Interactions, p. 153-179. *In* D.A. Hartley (Ed.), Cellular Interactions in Development: A Practical Approach. Oxford University Press, Oxford.
3. **Chien, C.T., P.L. Bartel, R. Sternglanz, and S. Fields.** 1991. The two-hybrid system: A method to identify and clone genes for proteins that interact with a protein of interest. Proc. Natl. Acad. Sci. USA 88:9578-9582.
4. **Durfee, T., K. Becherer, P.L. Chen, S.H. Yeh, Y. Yang, A.E. Kilbburn, W.H. Lee, and S.J. Elledge.** 1993. The retinoblastoma protein associates with the protein phosphatase type 1 catalytic subunit. Genes Devel. 7:555-569.
5. **Estojak, J., R. Brent, and E.A. Golemis.** 1995. Correlation of two-hybrid affinity data with in vitro measurements. Molec. Cell. Biol. 15:5820-5829.
6. **Feilotter, H.E., G.J. Hannon, C.J. Ruddel, and D. Beach.** 1994. Construction of an improved host strain for two-hybrid screening. Nucleic Acids Res. 22:1502-1503.
7. **Fields, S.** 1993. The two-hybrid system to detect protein-protein interactions. METHODS: A Companion to Meth. Enzymol. 5:116-124.
8. **Fields, S. and O. Song.** 1989. A novel genetic system to detect protein-protein interactions. Nature 340:245-247.
9. **Flick, J.S. and M. Johnston.** 1990. Two systems of glucose repression of the GAL1 promoter in *Saccharomyces cerevisiae*. Mol. Cell. Biol. 10:4757-4769.

10. **Gimeno, R.E., P. Espenshade, and C.A. Kaiser.** 1996. COPII coat subunit interactions: Sec24p and Sec23p bind to adjacent regions of Sec16p. Mol. Biol. Cell 7:1815-1823.

11. **Golemis, E.A., J. Gyuris, and R. Brent.** 1994. Interaction trap/two-hybrid systems to identify interacting proteins, Ch. 13 & 14. *In* Current Protocols in Molecular Biology. John Wiley & Sons.

12. **Golemis, E.A., J. Gyuris, and R. Brent.** 1996. Analysis of protein interactions; and Interaction trap/two-hybrid systems to identify interacting proteins, Ch. 20.0 & 20.1. *In* Current Protocols in Molecular Biology. John Wiley & Sons.

13. **Gyuris, J., E. Golemis, H. Chertkov, and R. Brent.** 1993. Cdi1, a human G1 and S phase protein phosphatase that associates with Cdk2. Cell 75:791-803.

14. **Hannon, G.J., D. Demetrick, and D. Beach.** 1993. Isolation of the Rb-related p130 through its interaction with CDK2 and cyclins. Genes Devel. 7:2378-2391.

15. **Harper, J.W., G.R. Adami, N. Wei, K. Keyomarsi, and S.J. Elledge.** 1993. The p21 Cdk-interacting protein Cip1 is a potent inhibitor of G1 cyclin-dependent kinases. Cell 75:805-816.

16. **Iwabuchi, K., B. Li, P. Bartel, and S. Fields.** 1993. Use of the two-hybrid system to identify the domain of p53 involved in oligomerization. Oncogene 8:1693-1696.

17. **James, P., J. Halladay, and E.A. Craig.** 1996. Genomic libraries and a host strain designed for highly efficient two-hybrid selection in yeast. Genetics 144:1425-1436.

18. **Johnston, M., J.S. Flick, and T. Pexton.** 1994. Multiple mechanisms provide rapid and stringent glucose repression of GAL gene expression in *Saccharomyces cerevisiae*. Mol. Cell. Biol. 14:3834-3841.

19. **Li, B. and S. Fields.** 1993. Identification of mutations in p53 that affect its binding to SV40 T antigen by using the yeast two-hybrid system. FASEB J. 7:957-963.

20. **Li, L., S.J. Elledge, C.A. Peterson, E.S. Bales, and R.J. Legerski.** 1994. Specific association between the human DNA repair proteins XPA and ERCC1. Proc. Natl. Acad. Sci. USA 91:5012-5016.

21. **Liu, J., T.E. Wilson, J. Milbrandt, and M. Johnston.** 1993. Identifying DNA-binding sites and analyzing DNA-binding domains using a yeast selection system. *In* Methods: A Companion to Methods Enzymol. 5:125-137.

22. **Luo, Y., S. Vijaychander, J. Stile, and L. Zhu.** 1996. Cloning and analysis of DNA-binding proteins by yeast one-hybrid and one-two-hybrid systems. BioTechniques 20:564-568.

23. **Smirnov, M.N., V.N. Smirnov, E.I. Budowsky, S.G. Inge-Vechtomov, and N.G. Serebrjakov.** 1967. Red pigment of adenine-deficient yeast *Saccharomyces cerevisiae*. Biochem. Biophys. Res. Commun. 27:299-304.

24. **Weisman, L.S., R. Bacallao, and W. Wickner.** 1987. Multiple methods of visualizing the yeast vacuole permit evaluation of its morphology and inheritance during the cell cycle. J. Cell Biol. 105:1539-1547.

Index